AF327230

Environment Control for Animals and Plants

Louis D. Albright
Professor,
Department of Agricultural and Biological Engineering
Cornell University
Ithaca, NY 14853-5701

An ASAE Textbook
Number 4 in a series published by

The American Society of Agricultural Engineers
2950 Niles Road
St. Joseph, Michigan 49085-9659

Pamela DeVore-Hansen, Editor
Technical Publications
August 1990

Environment Control for Animals and Plants

Louis D. Albright
Professor,
Department of Agricultural and Biological Engineering
Cornell University
Ithaca, NY 14853-5701

An ASAE Textbook
Number 4 in a series published by

The American Society of Agricultural Engineers
2950 Niles Road
St. Joseph, Michigan 49085-9659

Pamela DeVore-Hansen, Editor
Technical Publications
August 1990

Dedication
To the memory of David C. Sprague

PREFACE

The process of designing environment control for agricultural buildings has undergone a significant evolution during the past decade. Much new information has become available for analyzing heat and fluid flows in buildings and designing environment control systems. Personal computers have become everyday tools capable of computationally intensive design techniques. Rules of thumb have frequently been adopted in design to avoid the drudgery of repeated applications of extensive, repetitive calculation procedures to assess the "best" design for a given purpose. With a computer, many potential designs can be investigated to determine effects of changes of pertinent design variable values. Repeated simulations can lead to better understanding of system behavior, and result in designs having a significantly greater chance of success.

However, when the subject of environment control is taught, computer use must not be permitted to obscure underlying fundamentals. Design cannot be based on a "black box" from which answers are magically received. Thus, in this text, the emphasis is on principles. Computer applications follow and permit design options to be explored. For example, it is instructive to complete a ventilation graph by hand once or twice. Beyond that the process is unwelcome drudgery. Who would enthusiastically develop a dozen or more ventilation graphs by hand to explore how design assumptions might affect the minimum ventilation rate? Computer programs which accompany this text are intended for that second level of student work – exploring options after basic understanding is achieved.

Topics included in this text follow approximately what is covered in a senior-level engineering design course in environment control for agricultural buildings, offered by the Department of Agricultural and Biological Engineering at Cornell University. Prerequisites for the course include thermodynamics, and an introduction to psychrometrics and the problem solving process. Many topics in the course are standard in virtually all agricultural engineering environment control courses. However, some topics may not be. They have been chosen from research literature of recent years to show how engineering techniques can be used to quantify environment control and lead to better designs.

The text and accompanying exercises have been used in the classroom at Cornell in trial runs in an effort to expose contradictions, inconsistencies, and other errors. That sieve has, without doubt, been imperfect. Comments and suggestions are welcome.

An attempt has been made to include many worked examples in the text. Some are elementary, but others are extensive and require making assumptions.

Few undergraduates have a natural ability to make reasonable engineering assumptions. Experience is needed before reasonable assumptions can be made. Learning to recognize when assumptions are needed, and how to make them, are a critical part of an engineer's education. Too often in engineering courses, all information is provided and the work is reduced to mere mathematical manipulation. Thus, a need to make assumptions is built into many of the examples, and classroom discussion of them may be pedagogically useful (especially when there is disagreement about data that are assumed).

Emphasis in this text is on physical aspects of environment control with limited attention to the many important biological factors which must be understood if environment control is to be a success. Other excellent books exist which focus on biological matters. *Environment Management in Animal Agriculture*, by S.E. Curtis, and *Environmental Aspects of Housing for Animal Production*, edited by J.A. Clark, are recommended.

No attempt has been made to include exhaustive lists of references. Rather, some references applicable to the subject of each chapter have been listed as starting points for literature searches. References at the end of Chapter 1 comprise a basic library of agricultural engineering environment control materials.

The SI system is used exclusively. Most students use the SI system in science and engineering courses. The slow transition to the SI system in the United States must surely reach completion after a generation or two of persons are raised to be comfortable with metric units. However, it may be useful to assign some homework exercises in IP units so students become familiar with both. This, as much as anything, should demonstrate the greater simplicity and benefits of the SI system.

Programs accompanying this text were developed using Turbo Pascal 4.0 (© Borland International, Inc.) on an IBM PC, and are provided in executable form. Two files, PSYFUNC and PSYPROC, are included as Pascal code to be used as Include files, and as uncompiled units to be used with Turbo Pascal. As they are not provided in compiled form, the units must be compiled to disk using the version of Pascal available for program development (for example, units compiled using Turbo Pascal 4.0 are not compatible with Turbo Pascal 5.0). In addition, versions of the psychrometric functions and procedures are provided in a form suitable to be used as subroutines for Fortran on a personal computer. The subroutines were adapted from the Pascal versions by Dr. Richard Gates of the Department of Agricultural Engineering at the University of Kentucky, and his generosity in sharing them is greatly appreciated.

The programs have been tested on several IBM and IBM-compatible computers, including an IBM PC, AT, PS/2, Zenith, and AST. The programs were written for a color monitor, but that is not a requirement. A graphics card (CGA, EGA, VGA, or Hercules) is required for the programs with graphical output (VENTGRPH, SYSCHAR1, SYSCHAR2, POLYNOM, and WEATHER). Program GRAPHID is included as a utility to identify the graphics card present in a computer, if any.

Developing a text can never be accomplished as an individual effort. Many thanks are due to colleagues within Cornell University, and without, with whom discussions over the years have helped clarify the subject matter in my own mind. Special thanks is given to those many undergraduate and graduate students whose questioning has forced continued reexamination, reinterpretation, and winnowing of information.

Special thanks must be expressed to several who have contributed greatly through reviewing this text and offering many constructive comments and suggestions. At Cornell, two doctoral students, Alain Rousseau and Arend Jan Both, were especially keen-eyed and helpful. Dr. Robert Bottcher of North Carolina State University, and Dr. Donald Colliver of the University of Kentucky were two reviewers for ASAE who were extraordinarily thorough and thoughtful with their comments. To these four, especially, thank you!

Finally, on the personal side, my wife and best friend, Marilyn, has cheerfully lived through this endeavor and has contributed in so many ways to help me complete this work. Without her support, love and encouragement, progress would have been much slower, and perhaps nonexistent.

Louis D. Albright
Ithaca, NY 14853-5701
November 1, 1989

TABLE OF CONTENTS

PREFACE iii

CHAPTER 1 - APPLICATIONS OF ENVIRONMENT CONTROL IN
                        AGRICULTURE

1-1.   General 1
1-2.   Mechanically Ventilated Barns 1
1-3.   Naturally Ventilated Barns 2
1-4.   Greenhouses 3
1-5.   Horticultural Product Storage 4
1-6.   Plant and Animal Growth Chambers 4
1-7.   Food Storage 5
1-8.   Animal Laboratories 5
1-9.   Zoo Animals 5
          References 6

CHAPTER 2 - PSYCHROMETRICS

2-1.   Introduction 7
2-2.   Psychrometric Properties 8
          2-2.1.   Dry-Bulb Temperature 8
          2-2.2.   Water Vapor Saturation Partial Pressure 8
          2-2.3.   Relative Humidity 8
          2-2.4.   Humidity Ratio 9
                       Example 2-1 10
          2-2.5.   Psychrometric Chart 10
          2-2.6.   Degree of Saturation 11
                       Example 2-2 14
          2-2.7.   Specific Volume and Density 14
                       Example 2-3 15
          2-2.8.   Dew Point Temperature 15
                       Example 2-4 16
          2-2.9.   Enthalpy 16
                       Example 2-5 17
          2-2.10. Wet-Bulb Temperature 17
                       Example 2-6 19
2.3   Psychrometric Properties Examples 19
2.4   Measuring Psychrometric Properties 21
2-5.   Psychrometric Processes 22
          2-5.1.   Sensible Heating and Cooling 22

		Example 2-7	22
		Example 2-8	23
	2-5.2.	Cooling With Dehumidification	24
		Example 2-9	25
		Example 2-10	26
	2-5.3.	Adiabatic Mixing	27
		Example 2-11	29
		Example 2-12	30
	2-5.4.	Evaporative Cooling	31
		Example 2-13	32
		Example 2-14	33

2-6.   Integrated Psychrometric Processes 34

            Example 2-15 39

2-7.   Cooling and Dehumidification Revisited 41

2-8.   Computerized Psychrometric Functions 41

2-9.   Computerized Psychrometric Procedures 42

2-10.   Example using PSYPROC 42

            Example 2-16 46

2-11.   Program PLUS 46

    Symbols 46

    Exercises 48

    References 48

CHAPTER 3 - HEAT TRANSFER BASICS

3-1.   Introduction 49

    3-1.1.   Thermal Conduction 49

    3-1.2.   Thermal Convection 50

    3-1.3.   Thermal Radiation 51

3-2.   Conduction Heat Transfer 51

    3-2.1.   The Heat Conduction Equation 51

    3-2.2.   Temperature Fields 53

            Example 3-1 53

            Example 3-2 53

    3-2.3.   Conduction Heat Transfer 53

            Example 3-3 56

            Example 3-4 57

    3-2.4.   Resistances in Series 58

            Example 3-5 58

    3-2.5.   Resistances in Parallel 60

            Example 3-6 60

3-3.   Convective Heat Transfer 63

    3-3.1.   Natural Convection 63

               Example 3-7 66
               Example 3-8 67
               Example 3-9 68
      3-3.2.   Forced Convection 71
               Example 3-10 72
               Example 3-11 73
3-4.   Radiation Heat Transfer 77
      3-4.1.   General 77
      3-4.2.   Emitted Thermal Radiation 78
               Example 3-12 79
      3-4.3.   Reflected and Transmitted Thermal Radiation 79
      3-4.4.   Absorptance and Emittance 80
      3-4.5.   Angle Factors 80
               Example 3-13 82
               Example 3-14 82
      3-4.6.   Thermal Radiation Exchange 85
               Example 3-15 85
3-5.   Mixed Mode Heat Transfer 88
               Example 3-16 88
               Example 3-17 91
3-6.   Program BALANCE 95
               Example 3-18 96
3-7.   Combined Convective and Radiation Surface Coefficients 97
3-8.   Thermal Resistance of Plane Airspaces 99
               Example 3-19 100
3-9.   Thermal Radiation Exchange with Gases 100
    Symbols 101
    Exercises 102
    References 104

CHAPTER 4 - STEADY-STATE THERMAL ANALYSIS

4-1.   Introduction 107
4-2.   Heat Transfer Through Walls 107
               Example 4-1 107
               Example 4-2 109
               Example 4-3 111
               Example 4-4 113
4-3.   Heat Transfer Through Ceilings 115
4-4.   Heat Transfer Through Glazings 115
      4-4.1.   General 115
      4-4.2.   R-values 116
               Example 4-5 116

4-5.   Heat Transfer Through Doors 117
        Example 4-6 117
4-6.   Heat Transfer Through Floors on Grade 118
4-7.   Heat Transfer Through Basement Walls and Floors 120
        Example 4-7 121
4-8.   Program RVALUE 123
        Example 4-8 124
4-9.   Sol-Air Temperature 126
        Example 4-9 127
4-10.   Heat Exchangers 128
        4-10.1. Overall Heat Transfer 128
        4-10.2. Overall Heat Transfer Coefficient 129
                 Example 4-10 129
        4-10.3. Logarithmic Mean Temperature Difference 130
                 Example 4-11 131
        4-10.4. Heat Exchanger Effectiveness and the NTU Method 132
                 Example 4-12 134
                 Example 4-13 136
        4-10.5. Program XCHANGER 137
                 Example 4-14 137
        Symbols 138
        Exercises 139
        References 140

CHAPTER 5 - STEADY STATE ENERGY AND MASS BALANCES

5-1.   Introduction 143
5-2.   Components of the Sensible Energy Balance 147
        5-2.1.   Sensible Heat Produced by Animals, q_s 147
                 Example 5-1 149
                 Example 5-2 149
        5-2.2.   Mechanically Produced Heat, q_m 150
                 Example 5-3 151
        5-2.3.   Solar Heat Gain, q_{so} 151
                 Example 5-4 153
                 Example 5-5 154
                 Example 5-6 155
        5-2.4.   Heating System, q_h 157
        5-2.5.   Ventilation, q_{vi} and q_{vo} 157
        5-2.6.   Structural Heat Loss, q_w 157
        5-2.7.   Heat Exchange with the Floor, q_f 158
        5-2.8.   Evaporation, q_e 158
5-3.   Uses of the Sensible Energy Balance 158

		Example 5-7	159
		Example 5-8	161
		Example 5-9	161
5-4.	Components of the Mass Balance, Humidity		163
	5-4.1.	Moisture Production from Animals	163
		Example 5-10	163
		Example 5-11	164
	5-4.2.	Ventilation, m_{vi} and m_{vo}	164
5-5.	Uses of the Mass Balance, Moisture		165
		Example 5-12	165
5-6.	Components of the Mass Balance, Carbon Dioxide		166
	5-6.1.	Carbon Dioxide Produced by Animals	167
5-7.	Uses of the Mass Balance, Carbon Dioxide		167
		Example 5-13	167
		Example 5-14	169
	Symbols		170
	Exercises		170
	References		171

CHAPTER 6 - VENTILATION RATES

6-1.	Introduction		173
6-2.	Design Weather Data		174
6-3.	Animal Housing Ventilation		175
	6-3.1.	General	175
	6-3.2.	Maximum Ventilation Rate	176
		Example 6-1	177
		Example 6-2	180
	6-3.3.	Minimum Ventilation Rate	181
		Example 6-3	184
6-4.	Computerized Procedures to Determine Ventilation Rates		187
	6-4.1.	Polynomial Regression	187
		Example 6-4	188
	6-4.2.	Program VENTGRPH	189
		Example 6-5	190
6-5.	Staging Between Minimum and Maximum Ventilation Rates		191
		Example 6-6	195
6-6.	Supplemental Heat		196
		Example 6-7	197
6-7.	Greenhouse Ventilation		200
		Example 6-8	200
	Symbols		201
	Exercises		201
	References		202

CHAPTER 7 - THERMAL INSULATION AND MOISTURE
              BARRIERS

7-1.   Introduction 205
7-2.   Optimum Insulation Thickness 205
         Example 7-1 208
         Example 7-2 208
7-3.   Condensation on Wall Surfaces 211
         Example 7-3 212
7-4.   Condensation Within Walls 213
         7-4.1.   Water Vapor Diffusion 214
         Example 7-4 215
         7-4.2.   Condensation Rates 216
         Example 7-5 217
         7-4.3.   Vapor Retarding Barriers 220
         Example 7-6 221
       Symbols 221
       Exercises 222
       References 223

CHAPTER 8 - AIR INLETS AND OUTLETS

8.1.   Introduction 225
8-2.   Fluid Mechanics Basics 226
         Example 8-1 227
         Example 8-2 229
8-3.   Air Jets 230
         8-3.1.   Air Jet Velocity Decay and Profiles, Wall Jets 231
         Example 8-3 232
         8-3.2.   Entrainment into Wall Jets 233
         Example 8-4 234
         8-3.3.   Air Jet Velocity Decay and Profiles, Free Round Jets 235
         Example 8-5 236
         8-3.4.   Nonisothermal Wall Jets, Temperature Profiles and
                   Heat Transfer Coefficients 237
         Example 8-6 239
8-4.   Slotted Air Inlets 242
         8-4.1.   Slotted Inlet Operation 242
         8-4.2.   Airflow Through Hinged-Baffle and Center-Ceiling
                   Slotted Inlets 244
8-5.   Air Infiltration 245
8-6.   System Characteristic Technique 247
         8-6.1.   The Technique 247

Example 8-7 248
8-6.2. Program SYSCHARl 253
Example 8-8 254
8-7. Air Outlets 255
Symbols 257
Exercises 257
References 259

CHAPTER 9 - AIR DISTRIBUTION

9-1. Introduction 261
Example 9-1 264
9-2. Inlet Air Jet Stability, the Effect of Thermal Buoyancy 265
9-3. Inlet Air Jet Stability, the Effect of Obstructions 266
Example 9-2 268
9-4. Air Mixing Pattern Stability 268
Example 9-3 269
9-4.1. Air Recirculation to Promote Mixing Pattern Stability 270
9-5. Inlet Jet Momentum Numbers as a Criterion to Control Air
Inlet Baffles 271
9-5.1. Program SYSCHAR2 273
9-5.2. Interpreting Results of Program SYSCHAR2 274
Example 9-4 274
9-6. Characterizing Incomplete Mixing 276
9-6.1. Short Circuiting 276
9-6.2. Secondary Recirculation Zones 277
9-6.3. Air Mixing Zones in Series 278
9-7. Airflow Visualization 278
9-8. Air Mixing Models 279
9-8.1. Idealized Air Mixing Models 279
9-8.2. Plug Flow, Perfect Non-mixing 280
9-8.3. Perfect Mixing 280
9-8.4. Dispersed Plug Flow 280
9-8.5. Mixed Flow 281
9-8.6. Mixing Factor Concept 282
9-8.7. Other Models 282
Symbols 283
Exercises 284
References 285

CHAPTER 10 - VENTILATION CONTROL AND
                QUANTIFICATION OF PERFORMANCE

10-1.   Introduction 287
10-2.   Duty Factors, Staged Fan Systems Using Single-Speed Fans 290
10-3.   Ventilating Efficiency Ratios and the Cost of Ventilation 294
        Example 10-1 294
10-4.   The Cost of Ventilating Animal Housing, an Example 295
        Example 10-2 295
10-5.   Program DUTYFACT 298
        Example 10-3 299
        Example 10-4 302
10-6.   Quantifying Environment Control Effectiveness 303
        10-6. 1. Program WEATHER 309
        10-6. 2. Example use of WEATHER 310
        Example l0-5 310
10-7.   Ventilation Control in Greenhouses 312
        Symbols 314
        Exercises 314
        References 315

CHAPTER 11 - NATURAL VENTILATION

11-1.   Introduction 319
11-2.   Natural Ventilation Due to Thermal Buoyancy, an
        Empirical Approach 322
        Example 11-l 322
11-3.   Thermal Buoyancy, the Stack Effect 324
        Example 11-2 326
11-4.   Thermal Buoyancy, the Concept of a Neutral Pressure Plane 329
11-5.   Wind Pressure on Buildings 334
        Example 11-3 337
11-6.   Natural Ventilation Due to the Wind, an Empirical Approach 337
        Example 11-4 338
11-7.   Natural Ventilation Due to the Wind, the Wind Pressure
        Coefficient Method 338
        Example 11-5 340
11-8.   Combined Thermal Buoyancy and Wind Ventilation, Quadrature . 342
        Symbols 343
        Exercises 343
        References 345

CHAPTER 12 - AIR FLOW IN DUCTS

12-1.   General 347

12-2.   The Bernoulli Equation 347

       12-2.1.   Pressure Change Examples 349

       12-2.2.   System Characteristics 351

12-3.   Friction Losses 352

       12-3.1.   Darcy-Wiesbach and Colebrook Equations 352

              Example 12-1 353

       12-3.2.   Friction Charts 355

              Example 12-2 356

       12-3.3.   Rectangular Ducts 357

              Example 12-3 357

       12-3.4.   Computational Form for Velocity Pressure and
                  Reynolds Number 358

       12-3.5.   Program FRICTION 358

              Example 12-4 359

12-4.   Dynamic Losses 359

              Example 12-5 360

12-5.   Combined Friction and Dynamic Losses, an Example ... 363

              Example 12-6 363

12-6.   System Effect Factors 366

12-7.   Design of Air Duct Systems 368

              Example 12-7 369

12-8.   Computerized Design Procedures 377

12-9.   Fan Operation Cost 377

12-10. Fan Laws 378

              Example 12-8 378

12-11. Perforated Polyethylene Air Ducts 379

       Symbols 381

       Exercises 381

       References 383

CHAPTER 1
APPLICATIONS OF ENVIRONMENT CONTROL
IN
AGRICULTURE

1-1. General

Heating, ventilating, and air conditioning comprise a major field of engineering. Environment control also forms a major area of agricultural engineering. Animals and plants which are today commercially important originated under wild conditions, but their economic usefulness in the wild was often limited. For example, dairy cows during the era of the American Revolution were raised primarily outdoors, were approximately waist high, weighed less than 300 kg, and produced little milk by today's standards. Chickens of 200 years ago, raised outdoors, produced eggs only during a limited season of the year. The ancestors of many of today's floriculture crops were tropical, flowers two centuries ago meant seasonal blooms grown in a family garden. Environment control (or modification) has made animal agriculture and the greenhouse industry much of what they are today.

Designing environment control systems for agriculture requires understanding the complex interactions between the biological system within the space and the environment provided to that system. The biotic system will have specific environmental needs and its growth, production, and well-being will be strongly influenced by the environment. Concomitantly, the biotic system will strongly influence conditions within the airspace. The goal of environment control should be to create a balance favorable to both the biological and physical systems. To be able to create such a balance successfully requires understanding physics, thermodynamics, mathematics through calculus, fluid mechanics, heat transfer, mass transfer, psychrometrics, refrigeration, weather phenomena, control theory, and environmental biology. Other subjects could be added for specific applications.

Environment control has many uses in agriculture. Some examples follow.

1-2. Mechanically Ventilated Barns

The arrival of electricity on farms permitted a major change in the way animals could be housed. When dairy herd and laying flock sizes (for examples) were small, animals were housed in barns ventilated only by wind and some thermal buoyancy effects. During summer, cows were kept outdoors except to be milked, and chickens were brought indoors only at night, if at all. During cold weather animals were brought indoors, but space per animal was generous. Too many animals in a small space would have quickly led to unhealthy conditions in the aerial environment, and perhaps even to suffocation if conditions were extreme. The reason was inadequate air exchange, otherwise known as poor ventilation.

When electricity became available, fans were able to provide a steady and reliable rate of ventilation. More cows could be kept in a barn, and could be kept inside all summer if desired (the so-called "zero pasture" method). Laying hens could be housed at much greater animal stocking densities in tiered cages, and caged layer operations evolved. Large groups of pigs could be confined all year without inducing fatal epidemics of respiratory diseases. New developments in genetics provided animals (especially chickens) better suited for confined housing. Veterinary medicine developed means to cope with the new methods of housing and health problems specific to them.

Mechanically ventilated animal housing is the norm today. Exceptions exist, of course. In warm climates such as in California, Texas, and Florida, large dairy herds are kept outdoors, usually under sunshades. In colder climates, large dairy herds are often housed in naturally ventilated barns. However, many dairy cows, most swine, and virtually all laying hens spend their entire lives inside mechanically ventilated buildings.

With proper design, conditions inside mechanically ventilated buildings can be equivalent to or even better than outdoor conditions. Properly chosen ventilation rates and good air distribution will keep air temperatures to within a few degrees of outdoor air, and even several degrees below if cooling is used. The building provides protection from the sun and unfavorable weather.

1-3. Naturally Ventilated Barns

Natural ventilation is defined as that induced by so-called natural means – wind or thermal buoyancy effects. There has been a return to natural ventilation, especially for dairy herds, but in a way much different from the naturally ventilated, small barns of several generations ago. Modern natural ventilation systems usually are for large herds. A naturally ventilated barn is not designed today for multiple purposes, such as hay, grain, and machinery storage. It is specialized to provide a proper environment for the animals, and efficient herd management. Even milking is done in a separate section of the building – the milking parlor.

With a larger herd (80 milking cows is a commonly chosen dividing point above which naturally ventilated housing is considered) the cows can be divided into groups for feeding, and milking is simplified by using a milking parlor. Building construction costs can be less, and herd management may be simplified when naturally ventilated barns are used. However, environment design and control in a naturally ventilated barn is not necessarily simpler than in a mechanically ventilated barn. While simple field recommendations can be adopted, thorough analysis of airflow in a naturally ventilated airspace is complex and control is difficult.

Most naturally ventilated barns shade animals from the sun and wind, but air temperature is not controlled. Such barns are especially suited to providing good conditions in warm weather, but are less successful in ameliorating extreme cold. Thus, in most natural ventilation applications in cold weather, animals eat significantly more feed because of the cold, and increased feed cost tends to balance savings arising from not needing electricity for ventilation fans.

1-4. Greenhouses

Protected cultivation of crops is a technology known as long ago as the Roman Empire. Large conservatories were a fashion among the wealthy during the nineteenth century. However, it has been only during the past two or three generations that a large industry has evolved which raises flowers, foliage plants, vegetables, and herbs out-of-season in greenhouses.

Commercial greenhouses require exacting control of air temperature, light conditions, fertility, and diseases as a minimum. Suitable environmental conditions for greenhouses are usually confined to much narrower ranges than is the case for animal housing. The large and constantly changing nature of solar input adds a complicating feature to environment control. However, all these must be dealt with successfully if crops are to be grown with quantity, quality, and timing.

The need for quantity in greenhouse production is obvious. Most plants respond to the rule of chemistry that reaction rates double for every 10 C temperature rise (up to the point where high temperatures become damaging). However, quantity is not sufficient for profit if quality is missing. Quality is an ephemeral aspect of plant production which is critical to profits. Environmental conditions are critical in determining quality with many greenhouse crops. Finally, especially with ornamental crops, neither quantity nor quality assures profit if timing is incorrect. For example, consider the market for poinsettias the week after Christmas. There is none. Many factors determine crop timing; environment control and environmental uniformity are major ones.

Heating, cooling, ventilation, and air distribution are critical for environment control in greenhouses. While it is possible to maintain suitable conditions for plant growth during the summer in cool climates by natural ventilation alone, mechanical ventilation is necessary for most greenhouses. If the climate is sufficiently dry, evaporative cooling is recommended for hot weather environment modification. Winter heating can be sufficiently expensive that enhancement of the natural solar heating of greenhouses may be attractive. Many opportunities exist for innovative designs of greenhouse environment control methods.

1-5. Horticulture Product Storage

Horticultural products for fresh consumption are usually stored as living organisms. If they are not dried for storage, such as grain would be, a carefully designed environment control system is necessary if storage is to be for more than a limited time.

Product storage has evolved from simple root cellars, where the only environment control was temperature modification. Today many horticultural products are stored in controlled conditions from soon after harvest until sale to consumers. Because the products are alive, they require moisture and oxygen; produce water vapor, carbon dioxide, and other gases; and are subject to infectious agents. Aerial environment requirements for long-term storage are exacting and the relative humidity often must be high enough to prevent drying out, but low enough to prevent spoilage.

A more recent development has been controlled atmosphere for very long-term storage. Apples can be kept for several months after harvest by using only refrigeration. To have fresh apples all year requires controlled atmosphere storage. Longer storage is possible when air provided to the stored product is modified by removing most of the oxygen, adding carbon dioxide, and modifying other constituents according to the needs of the product. The result is a slowing of the stored product's metabolism so that natural sugars, etc., are consumed more slowly than would be the case in simple refrigerated storage. This maintains eating quality. The product continues to live, but in a more dormant state. Apples are the primary crop stored in a controlled atmosphere, some cabbage is stored this way, and controlled atmosphere storage may be adapted to other crops in the future. It is a high technology form of environment control.

1-6. Plant and Animal Growth Chambers

Certain applications of environment control for animals and plants require more careful control than can reasonably be obtained in barns and greenhouses. Examples are conditions for research purposes, and growth rooms to provide high quality seedlings for later transplanting to greenhouses.

Growth chambers used for research are typically small, and often are purchased as manufactured units. Air temperature, relative humidity, and light levels are typically controlled and can be programmed. Purchase cost is high and operating costs can be very high if careful control is not effected. Control of other environmental conditions such as the carbon dioxide level, air speed, and mean radiant temperature of the surroundings can also be provided in growth chambers if desired.

Germination and growth rooms for greenhouse operations are a recent innovation in total environment control for plant growth. The only light is from lamps, and conditions are carefully controlled so seedlings can be produced quickly and uniformly. Although operating costs are high, plant number density is so high the cost per unit of production may be acceptable. In such facilities, problems of environmental control often focus on the need for uniformity of aerial and light conditions as well as the removal and possible later use of heat produced by artificial lighting.

1-7. Food Storage

Much processed food must be stored in controlled environments prior to sale and consumption. Some foods require only refrigeration, others must be frozen. Storages range from large, walk-in coolers, to specifically adapted trucks for transport, to specialized storage compartments in supermarkets. To design food storages usually requires an intimate knowledge of environment control equipment as well as a firm understanding of energy balances and environment control principles.

1-8. Animal Laboratories

Many unique problems arise in designing housing for research animals in laboratories. Concerns for animal welfare have frequently dictated a limited range of permitted conditions. The wide variety of animal types which may be in one airspace may introduce seemingly conflicting requirements. Special concern for odor abatement may be required. In many cases, laboratories are designed by architects or mechanical engineers with a limited knowledge of the special problems of animal housing.

1-9. Zoo Animals

Housing for zoo animals is a problem which has received little attention from agricultural engineers. Yet, some zoo animals remain indoors all year, and most others are brought indoors for at least part of the winter in cold climates. Unless housing has been designed by a person knowledgeable in animal confinement problems, unhealthy conditions easily arise.

Zoo animals are typically worth a great deal of money as individuals and may be difficult or impossible to replace; some are members of endangered species. They are often kept indoors in small, individual pens when not exhibited to the public. Distributing fresh air uniformly is difficult. Odor control is of great concern. Disease transmission by aerosols (and other means) must be prevented. Temperature and light control may be critical for health or reproduction. These problems, and others, make environment control for zoos a unique challenge.

REFERENCES

ASAE. 1981. Farmstead engineering. Proceedings of the American Society of Agricultural Engineers Farmstead Engineering Conference. American Society of Agricultural Engineers, St. Joseph, MI.

ASHRAE. 1989. Handbook of fundamentals. American Society of Heating, Refrigerating, and Air Conditioning Engineers, Atlanta, GA.

ASHRAE. 1987. HVAC systems and applications. American Society of Heating, Refrigerating, and Air Conditioning Engineers, Atlanta, GA.

Clark, J.A. 1981. Environmental aspects of housing for animal production. Butterworths, London.

Curtis, S.E. 1983. Environmental management in animal agriculture. The Iowa State University Press, Ames, IA.

Esmay, M.L. and J.E. Dixon. 1986. Environmental control for agricultural buildings. The AVI Corporation, Westport, CT.

Hellickson, M.L. and J.N. Walker, eds. 1983. Ventilation of agricultural structures. American Society of Agricultural Engineers, St. Joseph, MI.

Langhans, R.W. 1990. Greenhouse management. Halcyon Press, Ithaca, NY.

Masterlerz, J.A. 1977. The greenhouse environment. John Wiley & Sons, New York.

Midwest Plan Service. 1983. MWPS-1, Structures and environment handbook. Midwest Plan Service, Iowa State University, Ames, IA.

Scott, N.R. 1984. Livestock buildings and equipment: A review. J. Agr. Engr. Res. 29:93-114.

CHAPTER 2
PSYCHROMETRICS

2-1. Introduction

Predominantly, environmental control means control of the aerial environment. The fluid of interest is moist air, and the study of moist air is psychrometrics. Part of psychrometrics deals with methods to determine moist air properties at known states and part deals with determining properties at an unknown state after a defined change from a known state – the study of psychrometric processes.

Standard applications of gas laws can be found in basic physics and thermodynamics texts. Boyle's law, Charles' law, Dalton's law, and the perfect gas law apply with sufficient accuracy to atmospheric air to be acceptable for most environmental control applications. Errors introduced by assuming air acts as a perfect gas are generally less than 1%.

Dry air is defined as the state of air when all moisture and contaminants (pollutants, dust, etc.) have been removed; its composition is effectively constant. Carbon dioxide shows a seasonal and geographic variation and a long-term trend upward but today averages approximately 345 ppm (527 mg/kg). (Before the Industrial Revolution, the atmospheric carbon dioxide level is estimated to have been approximately 270 ppm.) The apparent molecular weight of standard dry air is 28.9645 based on a weighted average of all components. Its gas constant is 287.055 J/kgK. Dry air is heavier than water vapor and lighter than carbon dioxide. The gas constant of water vapor is 461.52 J/kgK and water's molecular weight is 18.01534.

Standard conditions are frequently chosen for psychrometric calculations. The Standard Atmosphere is a typical choice; the Standard Atmosphere is at 15 C temperature and 101.325 kPa pressure, which is sea level. Standard conditions for other elevations are in Table 2-1.

Table 2-1. Properties of the standard atmosphere at various elevations.

Elevation above Sea Level, m	Temperature, Celsius	Pressure, Pa
-500	18.2	107,478
0	15.0	101,325
500	11.8	95,461
1000	8.5	89,874
2000	2.0	79,495
3000	-4.5	70,108
4000	-11.0	61,640
5000	-17.5	54,020

(adapted from the ASHRAE Handbook of Fundamentals, 1989)

2-2. Psychrometric Properties

Moist air is a mixture of dry air and water vapor. Numerous properties describe the state of air and its degree of saturation with water vapor. In most cases, knowledge of two of the properties permits determination of the others.

2-2.1. Dry-Bulb Temperature. The reference for specifying ordinary air temperature is dry-bulb temperature – the temperature sensed by an ordinary thermometer at thermal equilibrium with air.

2-2.2. Water Vapor Saturation Partial Pressure. The greatest amount of moisture which can be held by dry air occurs at saturation, defined as the state of equilibrium between moist air and free water on a flat surface. The air and water must be at the same temperature – the temperature at which water would boil were it at a pressure equal to the water vapor saturation partial pressure. If water vapor partial pressure is less than the water vapor saturation partial pressure at the same temperature, the vapor is superheated.

Dry-bulb temperature determines the water vapor saturation partial pressure, p_{ws}. The water vapor saturation partial pressure can be determined (in Pa) from the following for temperatures, T, in Kelvin:

$$\ln(p_{ws}) = A_1 / T + A_2 + A_3 T + A_4 T^2 + A_5 T^3 + A_6 T^4 + A_7 \ln(T). \quad (2\text{-}1)$$

In the dry-bulb temperature range from -100 to 0 C, water vapor saturation of air is over ice and the coefficients of Equation 2-1 have the following values:

$$
\begin{aligned}
A_1 &= -5.6745359 &&E + 03 \\
A_2 &= 6.3925247 &&E + 00 \\
A_3 &= -9.677843 &&E - 03 \\
A_4 &= 0.6221570 &&E - 06 \\
A_5 &= 2.0747825 &&E - 09 \\
A_6 &= -0.9484024 &&E - 12 \\
A_7 &= 4.1635019 &&E + 00
\end{aligned}
$$

Over water, in the temperature range from 0 to 200 C, coefficients to calculate water vapor saturation partial pressure are:

$$
\begin{aligned}
A_1 &= -5.8002206 &&E + 03 \\
A_2 &= 1.3914993 &&E + 00 \\
A_3 &= -48.640239 &&E - 03 \\
A_4 &= 41.764768 &&E - 06 \\
A_5 &= -14.452093 &&E - 09 \\
A_6 &= 0.0 \\
A_7 &= 6.5459673 &&E + 00
\end{aligned}
$$

A graph of water vapor saturation partial pressure based on Equation 2-1 is in Figure 2-1.

2-2.3. Relative Humidity. The actual partial pressure of water vapor in moist but unsaturated air, p_w, is related to the property of air called the relative humidity, ϕ, where

$$\phi = p_w / p_{ws} \qquad (2\text{-}2)$$

and where the partial pressure of the water vapor and its partial pressure at saturation are defined at identical temperatures and atmospheric pressures. Relative humidity is a measure of the degree to which air is saturated, for ϕ equals 0.0 for dry air and 1.0 when air is completely saturated.

Relative humidity also equals the mole fraction, x, of water vapor in a given air sample divided by the mole fraction were the air to be saturated at the same temperature and pressure,

$$\phi = x_w / x_{ws}. \qquad (2\text{-}3)$$

2-2.4. Humidity Ratio. A second measure of the water vapor content of air is the humidity ratio, W, which is the mass of water vapor evaporated into a unit mass of dry air, kg/kg. The ratio of molecular masses of water vapor and (averaged) air is $18.01534 / 28.9645 = 0.62198$. Thus,

$$W = 0.62198 x_w / x_a. \qquad (2\text{-}4)$$

If air and water vapor are assumed to act as perfect gases,

$$p_a V = n_a RT, \qquad (2\text{-}5)$$

$$p_w V = n_w RT, \text{ and} \qquad (2\text{-}6)$$

$$pV = nRT, \qquad (2\text{-}7)$$

where p is atmospheric pressure ($p = p_a + p_w$) and n is the number of moles of moist air ($n = n_a + n_w$).

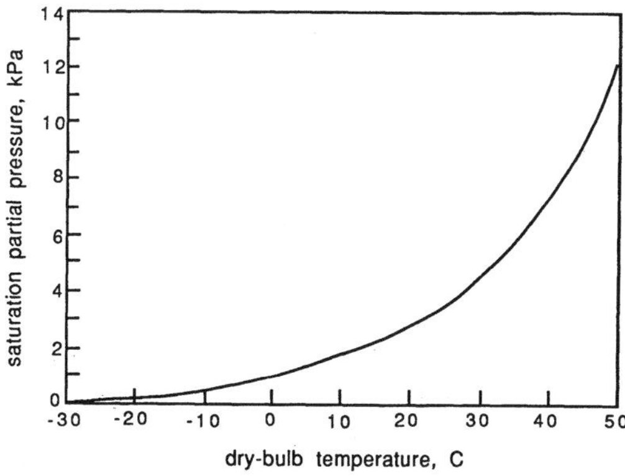

FIGURE 2-1. Water vapor saturation pressure, kPa, as a function of temperature, Celsius.

Thus,

$$x_a = p_a / p; \quad x_w = p_w / p; \qquad\qquad (2\text{-}8)$$

and

$$W = 0.62198 p_w / (p - p_w);$$
$$W_s = 0.62198 p_{ws} / (p - p_{ws}). \qquad\qquad (2\text{-}9)$$

If dry-bulb temperature, atmospheric pressure, and relative humidity are known, the humidity ratio can be calculated. The equations can also be used to calculate relative humidity if dry-bulb temperature, atmospheric pressure, and humidity ratio are known. This is frequently useful in environmental control calculations, as shall be seen.

Example 2-1

Problem: Moist air at standard atmospheric pressure is at 20 C dry-bulb temperature and 50% relative humidity. What is the humidity ratio of the air?

Solution: The humidity ratio is calculated from the partial pressure of the water vapor which can be obtained knowing the dry-bulb temperature and relative humidity.

From Equation 2-1 for a dry-bulb temperature of 293.15 K, the water vapor saturation partial pressure is

$$
\begin{aligned}
p_{ws} = \exp(&- 5.8002206E + 03/293.15 + 1.3914993 \\
&- (48.640239E - 03)(293.15) + (41.764768E - 06)(293.15)^2 \\
&- (14.452093E - 09)(293.15)^3 + 6.5459673 \ln(293.15)) \\
&= \exp(7.757) = 2339 \text{ Pa.}
\end{aligned}
$$

By the definition of relative humidity, Equation 2-2, the actual partial pressure of the water vapor is

$$p_w = \phi p_{ws} = (0.50)(2339) = 1169 \text{ Pa} = 1.169 \text{ kPa.} \qquad (2\text{-}10)$$

The humidity ratio for an atmospheric pressure of 101.325 kPa is

$$W = 0.62198(1.169) / (101.325 - 1.169) = 0.00726 \text{ kg/kg}$$

2-2.5. Psychrometric Chart. Dry-bulb temperature and humidity ratio form the x and y axes of the psychrometric chart, a graph which can be used to determine many of an air sample's psychrometric properties if two of its properties are known. An outline of the psychrometric chart is in Figure 2-2.

Two standard psychrometric charts for sea level conditions are in Figures 2-3 and 2-4. (Charts for elevations above sea level are also available as engineering resources.) Dry-bulb temperature is the ordinate; humidity ratio is the abscissa. These two scales are divided evenly which makes scales for other psychrometric properties nonlinear. Copies of the chart are available from ASHRAE, the American Society of Heating, Refrigerating, and Air Conditioning Engineers.

The line of water vapor saturation defines the upper boundary of the psychrometric chart, and lines of constant relative humidity parallel the saturation line. (Note: The degree of saturation and water vapor partial pressure are not on the psychrometric chart and must be calculated if needed.)

Although emphasis in this text will be on using equations and a computer for psychrometric calculations, it is necessary also to be able to use the psychrometric chart. Consider again Example 2-1, air at sea level contains a dry-bulb temperature of 20 C and relative humidity of 50%. These conditions are found on Figure 2-3 – normal temperatures. When the intersection of the two given conditions is located on the chart, and a horizontal line is followed to the abscissa, a humidity ratio of approximately 0.0073 kg/kg is found.

Note that although dry-bulb temperature and humidity ratio form the x and y axes of the psychrometric chart, the independent variables used to graph the chart in the Mollier form are humidity ratio and enthalpy. One result of this is that lines of constant dry-bulb temperature are neither parallel to each other nor perpendicular to the x axis. Other forms of the psychrometric chart can be found which are graphed using other independent variables, but the Mollier form is the ASHRAE standard. Fewer thermodynamic approximations are required when the chart is drawn in this form.

2-2.6. Degree of Saturation. Relative humidity was defined as the ratio of partial vapor pressures and is a measure of the extent to which air is saturated with water vapor. The degree of saturation, μ, is a separate means to describe

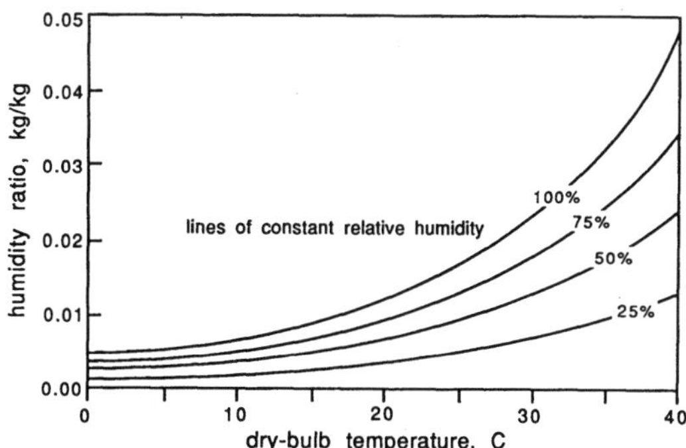

Figure 2-2. Basic psychrometric chart showing lines of constant relative humidity.

11

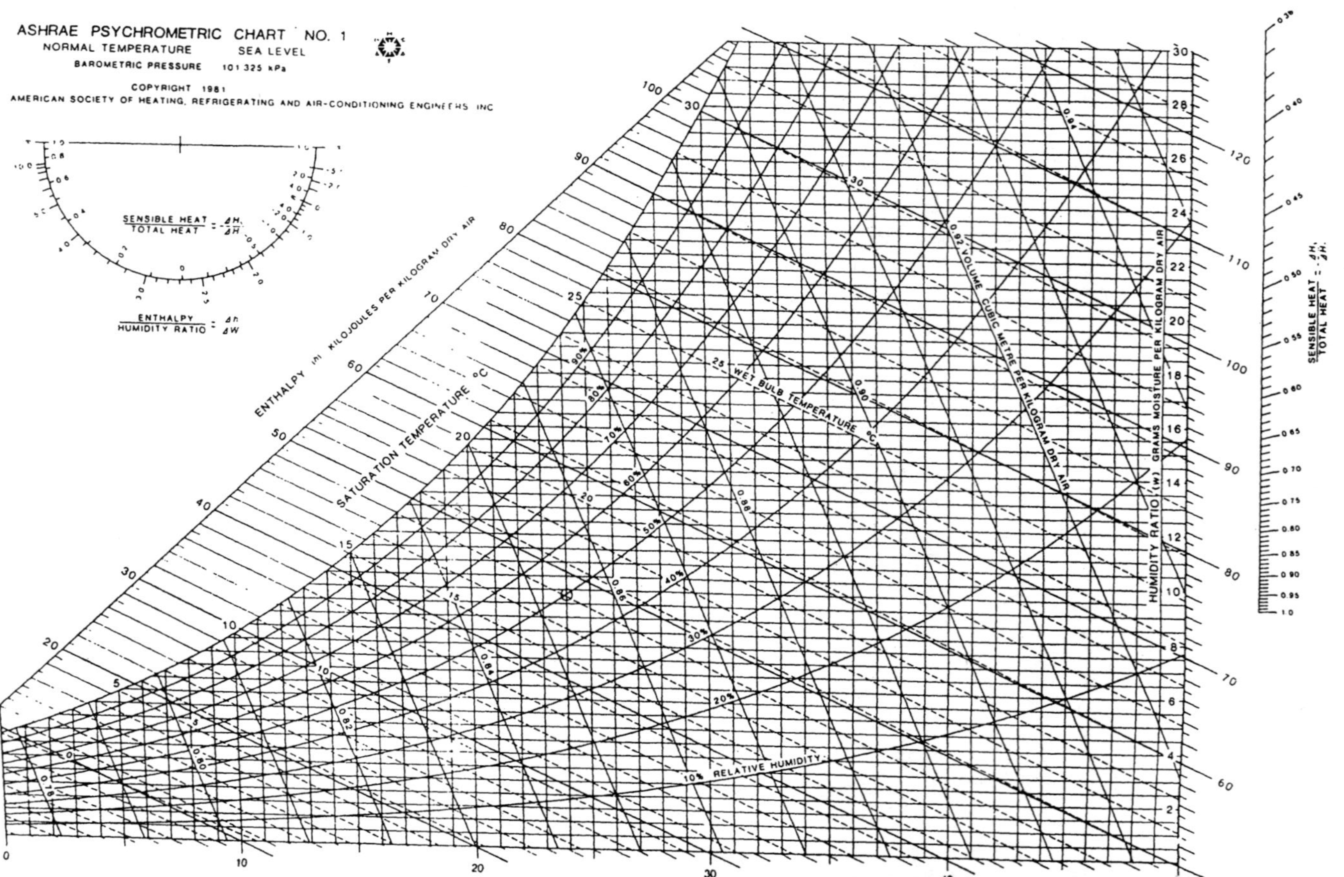

ASHRAE PSYCHROMETRIC CHART NO. 1
NORMAL TEMPERATURE   SEA LEVEL
BAROMETRIC PRESSURE   101 325 kPa
COPYRIGHT 1981
AMERICAN SOCIETY OF HEATING, REFRIGERATING AND AIR-CONDITIONING ENGINEERS INC
SENSIBLE HEAT / TOTAL HEAT
ENTHALPY / HUMIDITY RATIO
ENTHALPY IN KILOJOULES PER KILOGRAM DRY AIR
SATURATION TEMPERATURE °C
WET BULB TEMPERATURE °C
VOLUME CUBIC METRE PER KILOGRAM DRY AIR
HUMIDITY RATIO (W) GRAMS MOISTURE PER KILOGRAM DRY AIR
SENSIBLE HEAT / TOTAL HEAT
RELATIVE HUMIDITY

Figure 2-3

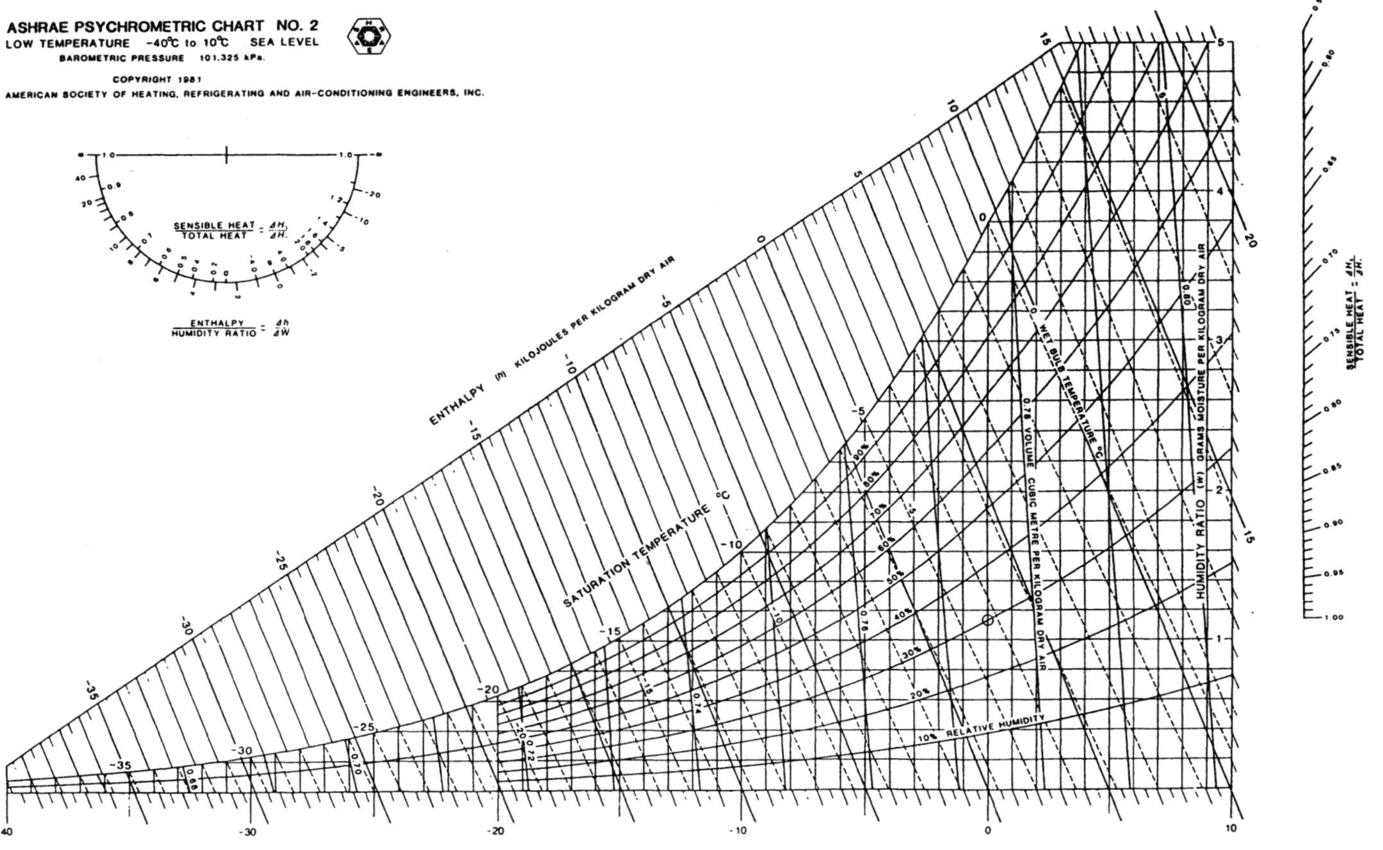

ASHRAE PSYCHROMETRIC CHART NO. 2
LOW TEMPERATURE -40°C to 10°C SEA LEVEL
BAROMETRIC PRESSURE 101.325 kPa.
COPYRIGHT 1981
AMERICAN SOCIETY OF HEATING, REFRIGERATING AND AIR-CONDITIONING ENGINEERS, INC.
SENSIBLE HEAT / TOTAL HEAT
ENTHALPY / HUMIDITY RATIO
ENTHALPY (h) KILOJOULES PER KILOGRAM DRY AIR
SATURATION TEMPERATURE °C
WET BULB TEMPERATURE °C
VOLUME CUBIC METRE PER KILOGRAM DRY AIR
RELATIVE HUMIDITY
HUMIDITY RATIO (W) GRAMS MOISTURE PER KILOGRAM DRY AIR
SENSIBLE HEAT / TOTAL HEAT

Figure 2-4

the extent of saturation. It is defined as

$$\mu = W / W_s \qquad (2\text{-}11)$$

which is not the same as relative humidity. The two are related by

$$\phi = \mu / (1 - (1 - \mu)(p_{ws} / p)), \text{ and} \qquad (2\text{-}12a)$$

$$\mu = \phi (1 - p_{ws} / p) / (1 - \phi p_{ws} / p). \qquad (2\text{-}12b)$$

The degree of saturation and relative humidity are equivalent for dry air and saturated air, but they differ at intermediate moisture contents because of the nonlinear relationship between humidity ratio and water vapor partial pressure.

Example 2-2

<u>Problem:</u> To continue the example of air at 20 C and 50% relative humidity, determine the degree of saturation of the air.

<u>Solution:</u> The water vapor saturation partial pressure was calculated as 2.339 kPa and the actual water vapor partial pressure as 1.169 kPa. For the Standard Atmosphere, W was previously calculated as 0.00726 kg/kg. Equation 2-9 can be used again to find

$$W_s = 0.62198(2.339) / (101.325 - 2.339) = 0.01470 \text{ kg/kg},$$

and by Equation 2-11, the degree of saturation, 0.494, is slightly different from the relative humidity. To complete the circle, Equation 2-12a yields a relative humidity value of 0.5.

2-2.7. Specific Volume and Density. A physical property of moist air useful for environmental calculations is density, ρ, or its inverse, specific volume, v. Specific volume is based on a unit mass of dry air, thus,

$$v = V_a / M_a = V / 28.9645 \, n_a. \qquad (2\text{-}13)$$

The perfect gas relationship, with $p = p_a + p_w$, leads to

$$v = R_a T / (p - p_w) \qquad (2\text{-}14)$$

which, when combined with the definition of the humidity ratio, provides the following computational equation for specific volume of dry air, and through its inverse, density:

$$v_{dry\ air} = (1 / p)R_a T(1 + 1.6078 \, W). \qquad (2\text{-}15)$$

Note: In psychrometric equations, specific quantities are based on a unit mass of dry air, so, although units of v are in m^3/kg dry air, this is volume of moist air per unit mass of dry air. Dry air forms the basis of specific quantities because it is unchanged as the humidity changes during a psychrometric process. The value of specific volume used for environmental control calculations is typically based on moist air. Correcting the specific volume of dry air to the specific volume of moist air leads to

$$v = ((1 / p)R_aT(1 + 1.6078 \, W)) / (1 + W). \qquad (2\text{-}16)$$

The psychrometric chart includes specific volume of moist air as a parameter. Steep, diagonal lines from upper left to lower right define specific volume. Air density, ρ, is its reciprocal. Note: In Equations 2-15 and 2-16, atmospheric pressure is in pascals not kilopascals.

Example 2-3

<u>Problem:</u> For the previous example of 20 C and 50% relative humidity, the psychrometric chart shows a specific volume of approximately 0.84 m^3/kg. Determine the specific volume of the moist air using psychrometric equations instead.

<u>Solution:</u> Applying Equation 2-16 yields the following estimate of specific volume of the moist air:

$$v = \frac{(1/101,325)(287.055)(293.15)(1 + (1.6078)(0.00726))}{1 + 0.00726}$$

$$= 0.83 \, \text{m}^3/\text{kg}.$$

2-2.8. Dew Point Temperature. When moist air is cooled sufficiently, water vapor condenses. Rain, snow, fog, and dew are formed this way. The temperature at which condensation occurs is the dew point temperature, t_d. When the state of air is below the saturation line on the psychrometric chart (as it is in normal situations), the dew point temperature is less than the dry-bulb temperature. That is, air in typical situations must be cooled at least a little before water begins to condense. The difference between dry-bulb and dew point temperatures is determined by the moisture content of the initial air sample relative to the moisture content at saturation at the same dry bulb temperature. Of course, at saturation the dew point and dry-bulb temperatures are the same.

When air is cooled, initially, there is no moisture content change. The relative humidity rises because cooler air has less moisture-holding capacity than does warmer air. However, the humidity ratio remains unchanged until air is cooled to the point of saturation, where water begins to condense out of the air if cooling is below the dew point temperature. If this reasoning is applied to the psychrometric chart, the dew point temperature is at the intersection of the saturation line and the line of constant humidity ratio (a horizontal line) which passes through the state defining the original condition of the air.

Because dew point temperature is a function of the partial pressure of water vapor in the air, it may be calculated from one of the following equations.

For conditions of frost between - 60 C and 0 C, the dew point in Celsius is

$$t_d = -60.45 + 7.0322 \ln(p_w) + 0.3700 (\ln(p_w))^2. \qquad (2\text{-}17)$$

For temperatures between 0 and 70 C

$$t_d = -35.957 - 1.8726 \ln (p_w) + 1.1689 (\ln(p_w))^2, \qquad (2\text{-}18)$$

where p_w has units of pascals.

Example 2-4

<u>Problem:</u> Continuing with the example of air at 20 C and 50% relative humidity, the psychrometric chart shows the dew point temperature to be approximately 9 C. Check this value by calculation.

<u>Solution:</u> For a partial pressure of water vapor of 1169 Pa, as calculated in Example 2-1, the dew point can be found using Equation 2-18.

$$\begin{aligned} t_d &= -35.957 - 1.8726 \ln(1169) + 1.1689 (\ln(1169))^2 \\ &= 9.14 \text{ C.} \end{aligned}$$

2-2.9. Enthalpy. The enthalpy, h, of moist air is a property useful in quantifying psychrometric processes which involve thermal energy exchanges. Enthalpy is an extensive property, thus, the enthalpy of a mixture equals the sum of enthalpies of the parts – the dry air, h_a, and water vapor, h_w:

$$h = h_a + h_w. \qquad (2\text{-}19)$$

Enthalpy in psychrometrics is referenced to 0 C (in the SI system) and water is assumed to be liquid at that temperature. The specific heat of dry air is 1.006 kJ/kg-K and the specific heat of water vapor is 1.805 kJ/kg-K. The specific heat of dry air actually varies from 1.006 at 0 C to 1.009 at 50 C, but the change is so slight it usually is ignored.

Note: Psychrometric calculations in the IP system of units reference enthalpy to 0 F. Conversion of enthalpy values from one set of units to the other is not straightforward. When converting from one system of units to the other, the enthalpy difference between 0 C and 0 F must always be included in calculations of enthalpy. Calculations of enthalpy changes do not require this correction.

The heat of vaporization of water at 0 C is 2501 kJ/kg. Heat of vaporization, kJ/kg, is a function of temperature and can be described by

$$h_{fg} = 2501 - 2.42t \qquad (2\text{-}20)$$

for dry-bulb temperatures, t, between 0 and 65 C. Because of its weak dependence on temperature, the heat of vaporization is often treated as a constant, 2501 kJ/kg.

Using the data as given, enthalpy of moist air in kJ/kg of dry air can be calculated if dry-bulb temperature and the humidity ratio are known.

$$h = 1.006t + W(2501 + 1.805t). \qquad (2\text{-}21)$$

It should be noted that in Equation 2-21, the combination $1.006 + 1.805W$ is frequently treated as the specific heat (kJ/kg-K) of moist air. However, in many practical applications, the value 1.006 kJ/kg-K is used for convenience to represent the specific heat of moist air. The error in this assumption is small. Note that in Equation 2-21, enthalpy is calculated assuming the water contained in the air evaporates at 0 C and is then heated, along with the air, to t. This is appropriate as enthalpy is a function of state, not path.

In Figures 2-3 and 2-4, lines of constant enthalpy are included and are the solid lines from the lower right to the upper left corner of the chart.

Example 2-5

<u>Problem:</u> Returning to the example of 20 C and 50% relative humidity, the chart in Figure 2-4 can be used to estimate an enthalpy content relative to 0 C of 39 kJ/kg of dry air. Check this value by calculation.

<u>Solution:</u> Equation 2-21 can be used to obtain a computed value:

$$\begin{aligned} h &= 1.006\,(20) + 0.00726\,(2501 + (\,1.805)\,(20)) \\ &= 38.5 \text{ kJ/kg.} \end{aligned}$$

2-2.10. Wet-Bulb Temperature. Another measure of the moisture content of air is the thermodynamic wet-bulb temperature. For any sample of moist air, a mental experiment could be performed which would isolate the air adiabatically from its surroundings, but permit contact with a water source on a flat surface (to eliminate surface tension effects) for water vapor exchange, with the limitation that thermal energy needed to evaporate water would come only from the air. Except for energy used for evaporation, no other thermal energy could be exchanged between the air and water.

If the air sample were not initially saturated, evaporation would occur with thermal energy from the air being the source of energy for evaporation. Water vapor would mix with the air and the humidity ratio would increase. The process would continue until the air became saturated. The air would have cooled, for sensible heat would have been transformed into latent heat. The cooling process would proceed until saturation. If the air were very dry initially, a great deal of cooling would occur. If the air were near saturation, little cooling would be possible. The limiting temperature, at which the air would become saturated, is the wet-bulb temperature.

The assumption of adiabatic conditions infers enthalpy is conserved in the evaporation process. Enthalpy (in the form of sensible heat) from the air used for evaporation returns to the air as latent heat. However, although enthalpy is conserved overall, the air gains enthalpy slightly in the process, and the water loses a like amount. The difference is the enthalpy content of the evaporated water prior to evaporation.

The wet-bulb temperature for given conditions of moist air can be determined from the psychrometric chart. Diagonal dashed lines from the lower right to upper left on the chart intersect the saturation line at the wet-bulb temperature corresponding to the initial conditions. There is a unique wet-bulb temperature for any given set of initial conditions although the reverse is not true. A given wet-bulb temperature applies to any of the combinations of conditions which lie along the wet-bulb line. Note that lines of constant wet-bulb temperature do not correspond to lines of constant enthalpy. The difference is the enthalpy content of water prior to evaporation. However, the two lines are nearly parallel, and on some psychrometric charts they are used interchangeably (sometimes with a correction factor introduced).

Wet-bulb temperature can be determined using equations, but the computational process is iterative. The process of adiabatic saturation of air is described by

$$h_s^* = h + h_w^*(W_s^* - W) \qquad (2\text{-}22)$$

which is a statement of the enthalpy change of air as it saturates adiabatically. The humidity ratio starts at W and in the constant pressure process increases from W to W_s^* at saturation (the * superscript indicates conditions at the wet-bulb temperature) while enthalpy of the air increases slightly from h to h_s^*. Saturation is at the wet-bulb temperature, t^*.

Enthalpy of water, h_w^*, is found from

$$h_w = 4.186t; \ h_w^* = 4.186t^*. \qquad (2\text{-}23)$$

A procedure to determine wet-bulb temperature is to search for a temperature, t*, such that Equation 2-22 is satisfied. An ad hoc search procedure is to begin the search slightly above the air's dry-bulb temperature, and decrement the candidate temperature until the equation is satisfied to within an acceptable level of precision. Values for atmospheric pressure, initial dry-bulb temperature, initial enthalpy, and initial humidity ratio are required to begin the process. Accelerated search techniques sweeping in alternate directions with ever-decreasing step sizes, or using the bisection method, can reduce the time required for convergence, but searching in one direction with small decrements of temperature will lead to a solution as long as the initial direction of search is in the proper direction.

Example 2-6

If we return to the example of air at 20 C dry-bulb temperature and 50% relative humidity, the psychrometric chart in Figure 2-3 shows a wet-bulb temperature of approximately 13.5 C. A solution for wet-bulb temperature using a computerized procedure to solve Equation 2-22, program PLUS, is 13.8 C. Program PLUS uses computerized psychrometric functions and processes described later in this chapter.

2-3. Psychrometric Properties Examples

Throughout Section 2-2, moist air at 20 C and 50% relative humidity was examined for its other psychrometric parameters.

To provide benchmarks for practice in determining the parameters, Table 2-2 contains data for six different conditions. The data can be used with the psychrometric charts (where possible) and psychrometric equations to be sure both are understood. However, at this point it will not be possible to determine wet-bulb temperatures using the equations unless programmed on a computer. Case 1 is the example which has been followed through the descriptions of psychrometric parameters.

Data in Table 2-2 can help develop an intuitive understanding of psychrometrics. If relative humidity rises while air temperature and atmospheric pressure are constant (case #2 compared to #1), the humidity ratio, dew point temperature, wet-bulb temperature, and enthalpy increase. If air temperature decreases while atmospheric pressure and relative humidity remain the same (case #3 compared to #1), the humidity ratio, dew point temperature, wet-bulb temperature, and enthalpy decrease. These changes are obvious from examination of the psychrometric chart.

Conditions #1, #5, and #6 have identical dry-bulb temperatures and relative humidities. However, condition #1 represents sea level and conditions #5 and #6 represent increasing elevation. In spite of the atmospheric pressure decrease, dew point temperatures are the same. Why? Recall that dew point temperature is a function only of water vapor partial pressure, which is a function only of dry-bulb temperature and relative humidity. Atmospheric pressure does not affect the vapor pressure of water.

The humidity ratio increases as atmospheric pressure falls, while dry-bulb temperature and relative humidity remain constant. We have already learned that water vapor partial pressure does not change with atmospheric pressure. What does? Specific volume (or density) change causes the difference. At high elevations, there is less dry air in a unit volume but the same amount of water vapor if temperatures are identical. The result is more water vapor per unit mass of dry air.

Can the same explanation be used for the enthalpy increase as the pressure decreases? It can, but only in part because the ratio of humidity ratios (0.0120 / 0.0073, for example) does not equal the ratio of enthalpies (50.6 / 38.5). If exactly the same explanation were true, the ratios would be identical. Think about where the rest of the difference lies. Hint: The specific heat of dry air does not change as a function of altitude, but the humidity ratio does change if temperature and relative humidity remain constant.

Finally, it may not be intuitively obvious that wet-bulb temperature should decrease as atmospheric pressure decreases for the same dry-bulb temperature and relative humidity. However, it has been seen that when atmospheric pressure is reduced, the moisture carrying capacity of a unit mass of dry air

Table 2-2. Six example psychrometric combinations.						
Parameter	#1	#2	#3	#4	#5	#6
atmospheric pressure, kPa	101.325	101.325	101.325	89.874	89.874	61.64
dry-bulb temperature, C	20	20	-10	10	20	20
relative humidity, %	50	80	50	20	50	50
humidity ratio, kg/kg	0.0073	0.0117	0.0008	0.0017	0.0082	0.0120
dew point temperature, C	9.1	16.3	-17.5	-10.9	9.1	9.1
wet-bulb temperature, C	13.8	17.7	-11.5	2.0	13.5	12.6
enthalpy, kJ/kg	38.5	49.8	-8.1	14.4	40.9	50.6
specific volume, m^3/kg	0.833	0.833	0.746	0.909	0.943	1.37

increases because of the decrease of density. Air at low pressure can gain relatively more water vapor when brought from 50% relative humidity to saturation. When more water is evaporated, more thermal energy is needed for evaporation. Thermal energy must come from the air in an adiabatic saturation process and, as a consequence, the air is cooled further. More cooling reduces the moisture carrying capacity, and a balance is reached in case #6, 1.2 K below the wet-bulb temperature of case #1.

2-4. Measuring Psychrometric Properties

Instruments have been developed to measure several of the psychrometric properties. The most obvious is dry-bulb temperature which can be measured by simple thermometers, thermocouples, thermistors, resistance thermometers, and more recently, solid state devices for computerized measurements. In practical measurements, it is important to ensure the temperature sensor is not influenced significantly by thermal radiation such as solar heating.

Dew point temperatures can be measured by dew point hygrometers. These are integrated instruments which sense the presence of dew on a chilled mirror by measuring the intensity of light reflected from the mirror. When dew forms, light is scattered more than when the mirror is clear. The mirror is alternately cooled and warmed; dew alternately forms and evaporates. The temperature of the mirror is measured, and the centerpoint of the temperature cycling provides an estimate of the temperature at which dew forms – the dew point temperature.

Relative humidity can be measured several ways. The traditional means is to move air around a wick-covered, wetted thermometer at a speed of approximately 3 m/s and simultaneously around a dry thermometer. The thermometers can be aspirated using fans or relative air movement can be effected by whirling the pair, attached to a base and handle, through the air. In either way, the dry thermometer will reach approximate equilibrium with the dry-bulb temperature of the air, and the wetted thermometer will reach approximate equilibrium with the wet-bulb temperature. Evaporation of water within the wick reduces the wick temperature, and thereby the thermometer temperature, to approximately the thermodynamic wet-bulb temperature.

Accuracy of the wet-bulb estimate is improved if the wick is clean and the water pure and distilled. When the two thermometers reach equilibrium states, the readings can be used to determine relative humidity using a psychrometric chart or a distillation of the chart in the form of a slide rule or data table.

More recent methods to measure relative humidity rely on solid state devices and electronics. Several physical principles can be used. One is to measure the change of electrical properties of a matrix as water molecules diffuse into and out of the matrix in response to changes of the air's moisture content. Another is to measure the change of electrical properties of a suitable material as water molecules adhere to its surface. Depending on the specific device, either capacitance or inductance is the electrical property which is measured. Careful calibration of each sensor is required, and frequently these sensors cannot tolerate conditions at or near saturation where they may lose calibration permanently. At present the measurement of relative humidity in the dusty, very humid conditions of animal housing and greenhouses is still a problem without a universally satisfactory solution. New solid state humidity sensors are now coming onto the market which are less likely to exhibit problems near saturation. Older humidity measuring instruments, which relied on contraction and relaxation of a fiber (such as a human hair) as humidity changed, were notoriously unreliable for use in barns and greenhouses.

2-5. Psychrometric Processes

A psychrometric process can be defined as a change in the state of moist air caused by adding or removing, either individually or in combination, thermal energy or water vapor. When a barn is ventilated, animals in the barn add sensible heat and moisture to the air. Solar input to a greenhouse adds sensible heat and causes water to evaporate to form water vapor (directly and indirectly through transpiration). If the initial state of the air is known, and rates of heat and water vapor changes can be determined, the state of air within the conditioned space can be calculated from psychrometric processes. Alternately, if the beginning and end states are known, the sensible and latent heat additions (or removals) to achieve the end state can be determined.

2-5.1. Sensible Heating and Cooling. When only sensible heat is added or removed from air, the process follows a horizontal line on the psychrometric chart. The only exception is cooling to a temperature lower than the dew point which will be discussed later. As air is heated, the humidity ratio does not change and relative humidity falls. The reverse occurs upon cooling.

For steady flow conditions, heat which is exchanged, q, causes a temperature change as follows:

$$\Delta T = q / \dot{M}_a c_{pa}, \text{ or } \Delta T = q / \Delta h \qquad (2\text{-}24)$$

where $\dot{M}_a$ is the mass flow rate of air involved in the process, and c_{pa} is the specific heat of the air at constant pressure.

Example 2-7

Problem: Determine the final state of moist air originally at 20 C and 50% relative humidity if 10 kJ are removed from 1.2 kg of the air.

Solution: The specific heat of air is approximately 1.006 kJ/kgK, thus, the temperature change using Equation 2-24 is

$$\Delta T = (-10 \text{ kJ}) / (1.2 \text{ kg})(1.006 \text{ kJ/kgK}) = -8.28 \text{ K},$$

and the final temperature is 20 C - 8.28 K = 11.72 C. If the process is tracked on a psychrometric chart, it is as shown in Figure 2-5. The humidity ratio remains constant at 0.00726 kg/kg, and the relative humidity increases to approximately 85%.

An alternate means to follow the process on a psychrometric chart begins with the enthalpy change,

$$\Delta h = q / \dot{M}_a$$
$$= (-10 \text{ kJ}) / (1.2 \text{ kg}) = -8.33 \text{ kJ/kg}. \qquad (2\text{-}25)$$

The initial enthalpy is 38.5 kJ/kg and can be found from the chart at the initial conditions. The final enthalpy is 38.5 - 8.3 = 30.2 kJ/kg. The intersection of this new enthalpy value, and a line of constant humidity ratio drawn through the initial conditions of 20 C and 50% relative humidity, provide an estimate of the end point of the process – a dry-bulb temperature of slightly less than 12 C.

Example 2-8

<u>Problem:</u> Rework Example 2-7 using the psychrometric equations rather than the psychrometric chart.

<u>Solution:</u> The end state of Example 2-7 can also be determined using psychrometric equations, as would be done in a computer implementation of psychrometric processes. The first step is to calculate the end point temperature, 11.72 C, as was done in Example 2-7.

The humidity ratio is calculated from conditions at the beginning of the process. At a dry-bulb temperature of 20 C (293.15 K), the water vapor saturation partial pressure with the coefficients for temperatures above 0 C is calculated using Equation 2-1

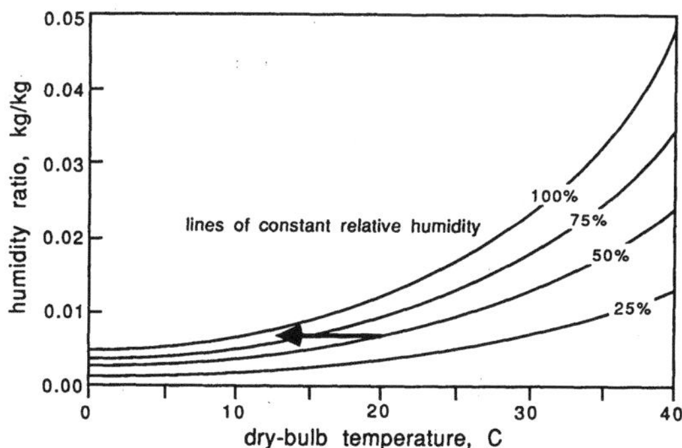

Figure 2-5. Sensible cooling, see Examples 2-7 and 2-8.

$$\ln(p_{ws}) = -\,5.8002206E+03\,/\,293.15 + 1.3914993 - 48.640239E-03\,(293.15)$$
$$+\,41.764768E06\,(293.15)^2$$
$$-\,14.452093E-09\,(293.15)^3 + 6.5459673\,\ln(293.15)$$

and $p_{ws} = 2339$ Pa.

The relative humidity is 50%, thus,

$$p_w = 0.50(2.339) = 1.169 \text{ kPa.}$$

Water vapor partial pressure remains constant during sensible cooling; at 11.72 C the partial pressure is still 1.169 kPa because the dew point temperature still has not been reached. (This can be checked in a computer program by calculating the dew point temperature once the water vapor partial pressure has been determined.)

The water vapor saturation partial pressure decreases, however, because of the lower dry-bulb temperature at the end of the process (284.87 K). Equation 2-1 can be used again with the new temperature to yield a water vapor saturation partial pressure of 1377 Pa. At this new state, the relative humidity, by its definition, is

$$\phi = 100(1.169 \text{ kPa} / 1.377 \text{ kPa}) = 84.9\%.$$

If other psychrometric parameters at the original or new state are desired, they can be calculated within the same computer program.

2-5.2. Cooling With Dehumidification. When air cools, the humidity ratio remains unchanged until the air is cooled to saturation. If air continues to cool after the dew point is reached, the process follows the saturation line and water is removed by condensation. If condensation occurs, the final state is at the intersection of the saturation line and the final dry-bulb temperature. The amount of water removed from the air equals the difference between the initial and final humidity ratios multiplied by the mass of dry air involved in the dehumidification.

Dehumidification can be traced on a psychrometric chart in two ways. If the final temperature is known – as it might be if air were passed through a refrigeration device and cooled to the temperature of the refrigeration coils – the process can be traced as shown in Figure 2-6. The initial point is known; the final point is at the intersection of the saturation line and the dry-bulb temperature to which the air is cooled.

The enthalpy change can be separated into its sensible and latent components. Psychrometric properties are functions of state not path. The dehumidification path from 1 to 2 in Figure 2-6 is equivalent to the path from 1 to 3 to 2. The path from 1 to 3 is only a process of water vapor removal (a latent heat removal which can be quantified by the enthalpy change from 1 to 3). The path from 3 to 2 is only temperature change, a sensible heat removal, quantified by the enthalpy change from 3 to 2.

Example 2-9

<u>Problem:</u> Air at 30 C and 60% relative humidity in a plant growth chamber is cycled past the cooling coils and is returned back to the chamber at a temperature of 15 C. Using the psychrometric chart, determine the psychrometric properties of the air after it is cooled, the sensible and latent heat removed, and the water vapor condensed per kg of dry air moved past the coils.

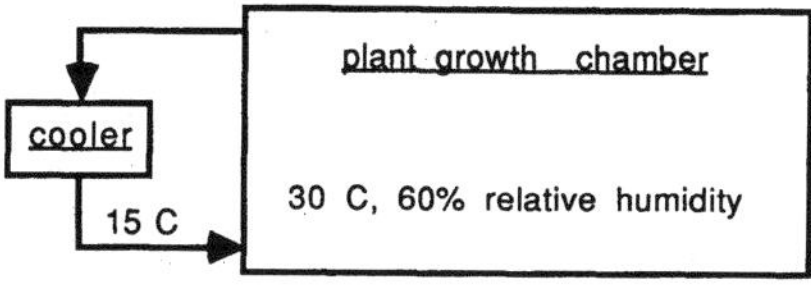

<u>Solution:</u> Point 1 on Figure 2-6 is at the intersection of the lines for 60% relative humidity and 30 C dry-bulb temperature. For this condition, dew point temperature is approximately 21.5 C; further cooling to 15 C must be accompanied by condensation. Point 2 is at 15 C on the saturation line. At point 2, air has the following properties:

dry-bulb temperature: 15 C
dew point temperature: 15 C
wet-bulb temperature: 15 C
relative humidity: 100 %
humidity ratio: 0.010648 kg/kg
specific volume: 0.83 m^3/kg
enthalpy: 42.0 kJ/kg

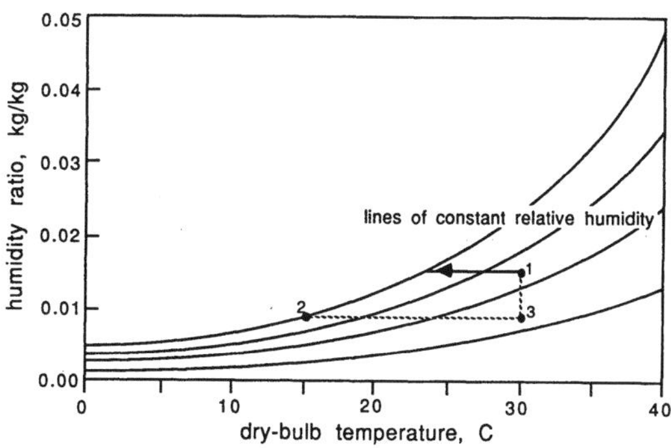

Figure 2-6. Cooling with dehumidification, see Examples 2-9 and 2-10.

25

The removal of sensible and latent heat is represented by lines 3-2 and 1-3, respectively, on Figure 2-6. The enthalpy at point 2 has already been determined to be 42.0 kJ/kg. At points 1 and 3 the enthalpies are 71.2 and 57.4 kJ/kg, respectively. For every kg of air moved past the cooling coils, the sensible heat removal is

$$h_S = (1 \text{ kg})(57.4 \text{ kJ/kg} - 42.0 \text{ kJ/kg}) = 15.4 \text{ kJ},$$

and the latent heat removed is

$$h_l = (1 \text{ kg})(71.2 \text{ kJ/kg} - 57.4 \text{ kJ/kg}) = 13.8 \text{ kJ}.$$

The quantity of water removed can be determined from the humidity ratio change. The humidity ratios at points 1 and 2 are 0.016042 and 0.010648 kg/kg, respectively. The difference is 0.005394 kg for 1 kg of dry air.

Water removal can also be determined from the enthalpy change if the heat of vaporization is known. From Equation 2-20, at 15 C the heat of vaporization is 2465 kJ/kg. When 13.8 kJ are removed, 13.8 kJ / 2465 kJ/kg = 0.0056 kg of water are condensed. The slight difference between the two estimates of the amount of condensed water arose from imprecision in reading the psychrometric chart. However, the degree of accuracy is normally adequate for environmental control applications. Note: We have assumed ideal conditions in the heat exchanger with all air cooled to 15 C and no short circuiting and mixing. Mixing will be addressed later.

Example 2-10

<u>Problem:</u> Repeat Example 2-9 using psychrometric equations in a way suitable for computer implementation.

<u>Solution:</u> A dehumidification process can be implemented readily on a computer if the final temperature is known. The final dry-bulb temperature, relative humidity, dew point, and wet-bulb temperatures are known a priori, although the psychrometric equations should also provide the correct answers. Again, we are assuming all the air is cooled to the same final temperature, and there is no short circuiting of part of the air with less condensation in that air.

A procedure suitable for computerized implementation is summarized in Table 2-3. Other sequences using the equations can also work. In some cases the sequence is unique – in others it is not. For example, specific volume can be calculated any time after the dry-bulb temperature and humidity ratio are known; further calculations do not depend on it. Dew point temperature can be calculated any time after water vapor partial pressure is known. However, water vapor saturation partial pressure must be calculated before humidity ratio can be determined, and enthalpy must be calculated before wet-bulb temperature. These are examples where proper sequences must be followed.

Table 2-3. Example computational sequence to determine psychrometric properties after a process of cooling and dehumidifying moist air.

<u>Given:</u>
1. The initial state was 30 C dry-bulb temperature and 60% relative humidity.
2. The final state is 15 C at saturation.

<u>Procedure:</u>

For the initial state of the air,
1. Determine the water vapor saturation partial pressure using Equation 2-1, the actual water vapor partial pressure using the definition of relative humidity, and the dew point of the original conditions using Equation 2-18.
2. Check to determine whether the final state is at a dry-bulb temperature below the dew point temperature of the original state. This is a check of whether there actually will be condensation. If there will be condensation, continue with the steps listed below.
3. Specify final relative humidity as 100% and dry-bulb temperature as 15 C (based on given conditions).

For the final state of the air,
4. Calculate water vapor saturation partial pressures using Equation 2-1.
5. Set the dew point temperature equal to the dry-bulb temperature.

For both the beginning and final states of the air,
6. Calculate humidity ratios using Equation 2-9.
7. Calculate specific volumes using Equation 2-16.
8. Calculate enthalpies using Equation 2-21.
9. Calculate wet-bulb temperatures using Equation 2-22 and the iterative procedure as described. The wet-bulb temperature of the final state can be set immediately equal to the final dry-bulb temperature.
10. Calculate enthalpy of the state where dry-bulb temperature is still 30 C, but the humidity ratio is at its final value (point 3 on Figure 2-6).
11. Calculate enthalpy changes due to sensible and latent heat removal.
12. Calculate the amount of water removed. Use the difference of humidity ratios multiplied by the mass of air.

2-5.3. Adiabatic Mixing. Conditions frequently arise in environmental control where two streams of air are mixed adiabatically. An example is a calf nursery where air is recirculated to maintain a well-mixed condition in the nursery, and fresh air is added to the recirculated airstream with an equal mass flow rate of recirculated air being exhausted to the outdoors to maintain steady conditions. Recirculation tempers fresh air in cold weather to avoid drafts on the calves and provides more mixing than might be possible when only a small amount of fresh air is needed. Evenly distributing a small flow rate of air in a ventilated airspace with good air mixing is difficult (shown in detail in later chapters).

The process of adiabatic mixing is sketched in Figure 2-7. In the example shown in the sketch, warmer air (state 2) is mixed with colder air (state 1), and the warmer air contains more moisture than does the colder air. Intuitively, one might expect the resulting mixture (state 3) to be at a dry-bulb temperature intermediate to dry-bulb temperatures of states 1 and 2, and the same to be true for the humidity ratio of the mixture.

In many practical processes, mixing is essentially adiabatic and the following equations, where 3 is the mixed state, apply:

$$\dot{M}_{a1}h_1 + \dot{M}_{a2}h_2 = \dot{M}_{a3}h_3 \qquad (2\text{-}26)$$

$$\dot{M}_{a1}W_1 + \dot{M}_{a2}W_2 = \dot{M}_{a3}W_3 \qquad (2\text{-}27)$$

$$\dot{M}_{a1} + \dot{M}_{a2} = \dot{M}_{a3} \qquad (2\text{-}28)$$

Eliminating $\dot{M}_{a3}$ from Equations 2-26 and 2-27 yields the following relationships to determine the humidity ratio and enthalpy of state 3:

$$(W_2 - W_3)/(W_3 - W_1) = \dot{M}_{a1} / \dot{M}_{a2} \qquad (2\text{-}29)$$

$$(h_2 - h_3)/(h_3 - h_1) = \dot{M}_{a1} / \dot{M}_{a2} \qquad (2\text{-}30)$$

Equations 2-29 and 2-30 are linear, thus, state 3 must lie along a straight line connecting states 1 and 2 on the psychrometric chart. The location of point 3 on the line is determined by the ratio of the mass flow rates of the two airstreams, $\dot{M}_{a1} / \dot{M}_{a2}$.

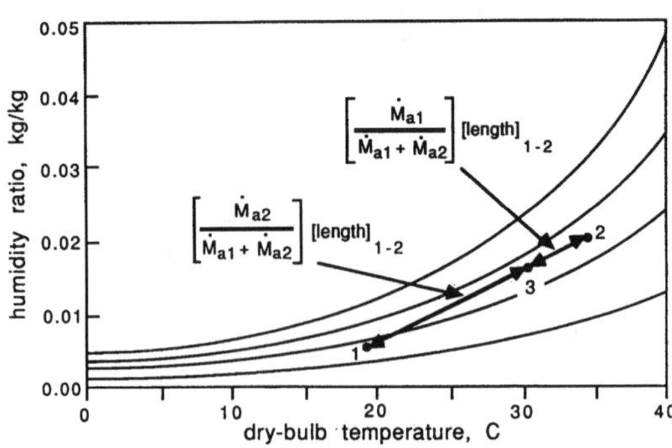

Figure 2-7. A process of adiabatic mixing.

Example 2-11

<u>Problem:</u> A ventilation system for a calf nursery draws 1 m³/s of air from the nursery, exhausts 0.2 m³/s of the air to the outside, and replaces the exhausted air with fresh air. Air in the nursery is at 20 C and 50% relative humidity, and outdoor air is at 5 C and 80% relative humidity. Determine properties of the mixed air when it is returned to the nursery.

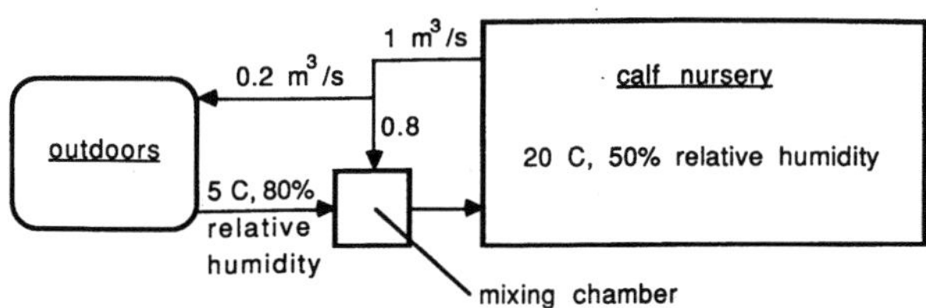

<u>Solution:</u> The first step is to determine properties of the outdoor and nursery air. Volumetric airflow rate is denoted by V.

Note: The usual convention in stating such problems is to specify the volumetric airflow rate based on conditions at the inlet of the fan. In this example, fresh air is specified (or measured) at outdoor conditions, recirculated and exhausted air are specified at <u>indoor conditions</u>.

The psychrometric chart for normal temperatures can be used to obtain appropriate data as follows:

$$\dot{V}_1 = \text{unknown}, \qquad \dot{V}_2 = 0.8\,\text{m}^3/\text{s},$$
$$W_1 = 0.00436\,\text{kg/kg}, \qquad W_2 = 0.00735\,\text{kg/kg},$$
$$h_1 = 16.0\,\text{kJ/kg}, \qquad h_2 = 38.8\,\text{kJ/kg},$$
$$\qquad\qquad\qquad\qquad V_2 = 0.8\,\text{m}^3/\text{kg}.$$

Volumetric flow rates are given, but psychrometric calculations are based on mass flow. Mass flow equals volumetric flow divided by specific volume, thus,

$$\dot{M}_{a1} = (0.2\,\text{m}^3/\text{s})/(0.83\,\text{m}^3/\text{kg}) = 0.24\,\text{kg/s, and}$$
$$\dot{M}_{a2} = (0.8\,\text{m}^3/\text{s})/(0.83\,\text{m}^3/\text{kg}) = 0.96\,\text{kg/s}.$$

Equation 2-29 can be rearranged to solve for W_3 as follows:

$$W_3 = (\dot{M}_{a1}W_1 + \dot{M}_{a2}W_2)/(\dot{M}_{a1} + \dot{M}_{a2}). \tag{2-31}$$

Equation 2-30 can be similarly rearranged:

$$h_3 = (\dot{M}_{a1}h_1 + \dot{M}_{a2}h_2)/(\dot{M}_{a1} + \dot{M}_{a2}). \tag{2-32}$$

Equations 2-31 and 2-32 can be solved to determine two of the properties of air at state 3.

$$W_3 = \frac{(0.24 \text{ kg/s})(0.00436 \text{ kg/kg}) + (0.96 \text{ kg/s})(0.00735 \text{ kg/kg})}{0.24 \text{ kg/s} + 0.96 \text{ kg/s}}$$

$$= 0.00675 \text{ kg/kg}.$$

$$h_3 = \frac{(0.24 \text{ kg/s})(16.0 \text{ kJ/kg}) + (0.96 \text{ kg/s})(38.8 \text{ kJ/kg})}{0.24 \text{ kg/s} + 0.96 \text{ kg/s}}$$

$$= 34.2 \text{ kJ/kg}.$$

With two properties of state 3 known, others can be determined from the psychrometric chart:

dry-bulb temperature = 17 C,
wet-bulb temperature = 12 C,
dew point temperature = 8 C,
relative humidity = 55%,
specific volume = 0.825 m³/kg,
density = 1.212 kg/m³,

and the volumetric airflow rate after mixing is

$$\dot{V}_3 = 0.825 \text{ m}^3/\text{kg}(0.24 \text{ kg/s} + 0.96 \text{ kg/s}) = 0.99 \text{ m}^3/\text{s}.$$

Example 2-12

<u>Problem:</u> Repeat Example 2-11 using psychrometric equations in a way suitable for computer implementation.

<u>Solution:</u> In an adiabatic mixing process, humidity ratios and enthalpies of the two separate airstreams must first be determined, followed by humidity ratio and enthalpy of the mixed state, based on Equations 2-31 and 2-32.

A sequence to accomplish the task is listed in Table 2-4. Notes related to sequencing included in Example 2-10 also apply to this example.

Table 2-4. Example computational sequence to determine psychrometric properties after a process of adiabatic mixing of two airstreams.

<u>Given:</u>
    1.   State 1 represents air at 5 C and 80% relative humidity.
    2.   State 2 represents air at 20 C and 50% relative humidity.

<u>Procedure:</u>

For states 1 and 2,

3. Determine water vapor saturation partial pressures using Equation 2-1, actual water vapor partial pressures using the definition of relative humidity, humidity ratios from Equation 2-9, and enthalpies from Equation 2-21.
4. Determine the specific volume of the moist air at condition 2 using Equation 2-16.

For state 3,
5. Calculate the humidity ratio and enthalpy of the mixed state as described in Example 2-11.
6. Rearrange Equation 2-21 to solve for the dry-bulb temperature of the mixture.

$$t = (h - 2501W)/(1.006 + 1.805W). \qquad (2\text{-}33)$$

7. The next step is a check to catch the unlikely, but possible, situation of a mixed state above the saturation line. Both states 1 and 2 are near saturation, the straight line connecting them may pass above the saturation line.
8. Calculate the water vapor saturation partial pressure corresponding to the dry-bulb temperature determined in step 6. Use Equation 2-9 to calculate the humidity ratio at saturation at the calculated dry-bulb temperature of state 3. Check to determine whether the humidity ratio calculated in step 5 is less than that calculated in this step, and if so, continue.
9. Calculate the dew point temperature of state 3 using Equation 2-17 or 2-18.
10. Calculate the relative humidity at state 3 using Equation 2-2.
11. Calculate the specific volume at state 3 using Equation 2-15, corrected for humidity ratio.
12. Calculate the wet-bulb temperature of state 3 using Equation 2-22 and the iterative procedure as described.

2-5.4. Evaporative Cooling. Numerous applications of environment control in agriculture require more cooling than can be provided by ventilation alone. Temperature requirements for best animal production typically are within the range from 10 to 30 C. Day temperature requirements for greenhouse crops are seldom higher than 25 C. Whenever outdoor temperatures are warm, maintaining conditions within these ranges may become impossible unless artificial cooling is used.

Mechanical refrigeration is normally used to cool air within buildings for human habitation. However, people have been conditioned to expect a relatively narrow and carefully controlled range of comfort conditions. Such careful control has not been found necessary for animals and plants. This is fortunate because mechanical refrigeration would be expensive to install and energy-intensive to operate to meet the large cooling loads encountered in modern barns and greenhouse ranges.

An alternative means of cooling widely used in agriculture is evaporative cooling. It is most effective in dry climates, but uses have been found even in humid climates during the hottest part of the day when relative humidity may be significantly below saturation. Devices and design methods to provide evaporative cooling will be described in more detail later. For now it is enough to know evaporative cooling is achieved by placing outdoor air in contact with a wetted medium for a specific length of time to permit a process of adiabatic saturation to proceed nearly to completion.

Practical considerations preclude devices which permit air to attain the wet-bulb temperature completely. A measure of efficiency is involved. A well-designed evaporative cooler can be expected to achieve up to 80% of the "wet-bulb depression". For a given state of unsaturated air, the wet-bulb depression is defined as the difference between the dry-bulb and wet-bulb temperatures. For example, air at 35 C and 20% relative humidity has a wet-bulb temperature of 18.8 C. The difference between 35 C and 18.8 C is the wet-bulb depression, 16.2 C. An evaporative cooler which could achieve 80% cooling efficiency would cool the air by 0.80 (16.2 C) = 13 C. Air would exit the cooler at 35 C - 13 C = 22 C. Animals in a barn ventilated with fresh air at 35 C would probably be subjected to considerable heat stress, but would not be stressed at 22 C.

Obviously evaporative cooling is most effective in dry climates, but even in humid climates, such as the southeastern United States, relative humidity during midday may be only 50%. If air were at 35 C and 50% relative humidity, the wet-bulb temperature would be 26 C, and an 80% efficient cooler could provide ventilation air cooled to 28 C which is a significant improvement. Of course, careful economic and engineering analysis would be required to determine the usefulness of evaporative cooling in humid climates.

Example 2-13

Problem: When ambient conditions are 35 C and 25% relative humidity, determine the dry-bulb temperature to which ventilation air could be cooled if drawn through an evaporative cooler with an efficiency of 75%. Calculate how much water must be added to each cubic meter of air drawn through the cooler.

Solution: On the psychrometric chart, the wet-bulb temperature corresponding to the given conditions is 20 C. The wet-bulb depression is 35 C - 20 C = 15 C. A cooler operating at 75% efficiency will cool the air 0.75(15 C) = 11 C and reduce the dry-bulb temperature to 35 C - 11 C = 24 C.

At initial conditions, the specific volume is 0.88 m^3/kg and the humidity ratio is 0.00876 kg/kg. At a wet-bulb temperature of 20 C and dry-bulb temperature of 24 C on the psychrometric chart, the humidity ratio is 0.01320 kg/kg.

A cubic meter of air will contain 1.0 m^3/(0.88 m^3/kg) = 1.14 kg. The change of humidity ratio is 0.01320 kg/kg - 0.00876 kg/kg = 0.00444 kg/kg. Thus, the total water gained by each cubic meter of outdoor air will be (1.14 kg)(0.00444 kg/kg) = 0.005 kg, or approximately 0.005 L.

Example 2-14

Problem: Repeat Example 2-13 using psychrometric equations in a method suitable for computer implementation.

Solution: A suitable sequence is listed in Table 2-5. Because the process begins with the known dry-bulb temperature and relative humidity and proceeds along the constant wet-bulb line, checks are not needed to stop the calculation procedure if impossible conditions arise. Notes in previous examples regarding the calculation sequence apply.

Table 2-5. Example computational sequence to determine the final state of air after an evaporative cooling process and the amount of water required for evaporation into each cubic meter of air at the state before cooling.

Given:
1. The initial state is air at 35 C dry-bulb temperature and 25% relative humidity.
2. The efficiency of the evaporative cooling process is 75%.

Procedure:

For the initial state of the air,

3. Calculate the water vapor saturation partial pressure using Equation 2-1, the actual water vapor partial pressure using Equation 2-2, and the humidity ratio using Equation 2-9.
4. Calculate the enthalpy using Equation 2-21.
5. Calculate the wet-bulb temperature using Equation 2-22 and the iterative procedure as described.
6. This determines the initial state of the air. The evaporative cooling process can now be quantified by the wet-bulb depression, the actual cooling based on cooler efficiency, and the final dry-bulb temperature of air after passing through the evaporative cooler. The procedure in Example 2-13 may be used.
7. At wet-bulb temperature and saturation, calculate the humidity ratio using Equation 2-9. The difference between this value of humidity ratio, and the humidity ratio of air before evaporative cooling, is the maximum possible addition of water vapor to the air. This difference, multiplied by the evaporative cooler efficiency, is the actual water vapor added to the air.
8. Calculate the specific volume of the air prior to cooling using Equation 2-16.
9. Calculate water vapor added per unit volume of air at the initial state using the procedure in Example 2-13.

2-6. Integrated Psychrometric Processes

A real ventilation situation is likely to involve several psychrometric processes occurring simultaneously or sequentially. This section provides an example of a more complex process. Each component has been seen previously in concept or example.

Example 2-15

Problem: You are designing an evaporative cooling system to ventilate a controlled environment, totally enclosed, poultry house in a hot and dry climate. Summer design weather conditions for your location are 40 C dry-bulb temperature with a relative humidity of 20%. When the evaporative cooling system operates, you expect to have 75 m^3/s of outdoor air drawn through the evaporative pads and 15 m^3/s of outdoor air drawn through the poultry house by the mechanism of infiltration. Infiltration circumvents the evaporative cooling process. Each volumetric flow rate is determined based on outdoor air conditions.

The evaporative cooling pads will provide 75% efficiency if properly maintained. More water will flow through the cooling pads than will be needed for evaporation to avoid salt build-up within the pads as water evaporates. Excess water will flush salts from the pads. A (water flow/evaporation rate) ratio of 3.0 is desired.

The only significant heat gains in the poultry house are the sensible and latent heat gains from the birds. Because ventilation is so rapid (approximately one air change per minute) as a first approximation we may neglect heat gains or losses through the building shell. The birds are expected to produce 250 kW of total heat of which 40% is latent heat and the rest sensible heat. (Sources of animal heat production data and its partitioning between sensible and latent heats will be discussed in a later chapter.)

Air inside the building can be assumed well mixed (everywhere uniform). Standard atmospheric pressure is expected; the location is near sea level. For the analysis, complete the following three calculations:
  1. Determine the psychrometric properties of air being drawn into the poultry house (infiltration plus cooled air mixed together before heat and humidity are added by the birds).
  2. Determine the psychrometric properties and volumetric flow rate of air exhausted from the building (after sensible and latent heat are added by the birds).
  3. Determine the rate (L/s) at which water must be supplied to the evaporative cooling pads by a water pump.

Solution: The ventilation process is sketched in Figure 2-8. Air at state 1 is at outdoor conditions, 40 C and 20% relative humidity. State 2 describes air which has been drawn through the evaporative cooler, and state 3 describes the mixture of cooled air and infiltration air. State 3 is the effective state of the ventilation air as it enters the poultry house. State 4 describes air expelled from the poultry house after gaining sensible and latent heat from the birds.

The change of state from 1 to 2 is evaporative cooling, from 1 and 2 to 3 is adiabatic mixing, and from 3 to 4 is a process of heating and humidifying.

Ventilation starts at state 1, and psychrometric properties at outdoor conditions may be obtained from either the psychrometric chart or equations.

At state 1:

> (dry-bulb temperature = 40 C),
> (relative humidity = 20%),
> water vapor saturation partial pressure = 7.38 kPa,
> water vapor partial pressure = 1.48 kPa,
> humidity ratio = 0.009199 kg/kg,
> specific volume = 0.893 m^3/kg,
> enthalpy = 63.9 kJ/kg,
> wet-bulb temperature = 22 C,
> dew point temperature = 12.6 C.

The mass flow rate through the evaporative cooler is

$$\dot{M}_{al} = (75 \text{ m}^3/\text{s})/(0.893 \text{ m}^3/\text{kg}) = 84 \text{ kg/s},$$

and the mass flow rate entering the poultry house by infiltration is

$$\dot{M}'_{al} = (15 \text{ m}^3/\text{s})/(0.893 \text{ m}^3/\text{kg}) = 17 \text{ kg/s}.$$

The next step is to determine the properties of air after it leaves the evaporative cooler (state 2). The wet-bulb temperature of state 1 is 22 C which is a wet-bulb depression of 40 C - 22 C = 18 K. Evaporative cooling is 75% efficient, thus, the actual cooling effect is 0.75(18 K) = 13.5 K. As a consequence, the dry-bulb

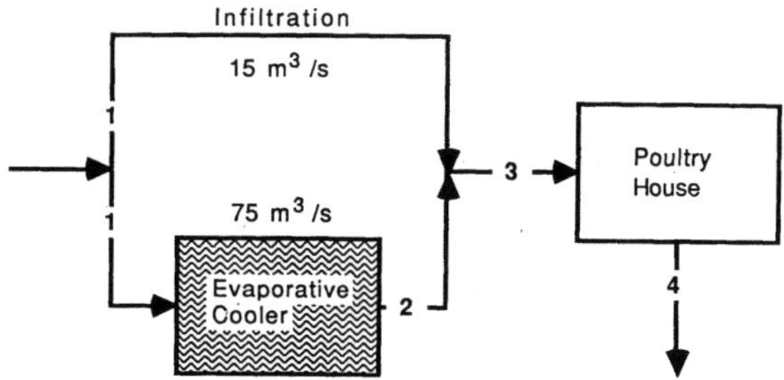

Figure 2-8. Ventilation with infiltration and evaporative cooling; see Example 2-15.

temperature of air leaving the evaporative cooler is $40 \text{ C} - 13.5 \text{ K} = 26.5 \text{ C}$, and the wet-bulb temperature is still 22 C. With two properties at state 2, the rest can be found from the psychrometric chart or psychrometric equations. The properties are:

    (dry-bulb temperature = 26.5 C),
    (wet-bulb temperature = 22 C),
    water vapor saturation partial pressure = 3.46 kPa,
    water vapor partial pressure = 2.35 kPa,
    humidity ratio = 0.014783 kg/kg,
    relative humidity = 67.9%,
    enthalpy = 64.3 kJ/kg,
    dew point temperature = 20.0 C,
    specific volume = 0.855 m³/kg.

The state of each airstream involved in the mixing process is now known. The state of the mixture can be determined. From Equation 2-31,

$$W_3 = \frac{((84 \text{ kg/s})(0.014783 \text{ kg/kg}) + (17 \text{ kg/s})(0.009199 \text{ kg/kg}))}{84 \text{ kg/s} + 17 \text{ kg/s}}$$

$$= 0.013843 \text{ kg/kg}.$$

From Equation 2-32,

$$h_3 = \frac{((84 \text{ kg/s})(64.3 \text{ kJ/kg}) + (17 \text{ kg/s})(63.9 \text{ kJ/kg}))}{84 \text{ kg/s} + 17 \text{ kg/s}}$$

$$= 64.2 \text{ kJ/kg}.$$

Other properties of state 3 may now be found and are:

    (humidity ratio = 0. 013843 kg/kg),
    (enthalpy = 64.2 kJ/kg),
    dry-bulb temperature = 28.7 C,
    wet-bulb temperature = 22 C,
    dew point temperature = 18.9 C,
    water vapor saturation partial pressure = 3.94 kPa,
    water vapor partial pressure = 2.21 kPa,
    relative humidity = 56%,
    specific volume = 0.862 m³/kg.

Note the wet-bulb temperature does not change in this mixing process. In general, a change would be expected, but in this example two airstreams each with a wet-bulb temperature of 22 C are mixed. One would not anticipate the wet-bulb temperature of the mixture to differ from that of the mixing streams. This observation provides a useful check to detect possible errors in using the psychrometric chart or psychrometric equations.

After adiabatic mixing, the next process is one of heating and humidifying. As previously stated, psychrometric properties at the end of a process are not functions of the path of the process. Heating and humidifying can be treated separately.

There are 101 kg/s of air at state 3 (and at state 4). To this air is added 150 kW (kJ/s) of sensible heat and 100 kW of latent heat. The dry-bulb temperature change of the heated air can be determined from

$$\Delta T = q_{sensible} / \dot{M}_a c_p, \tag{2-34}$$

where the specific heat, c_p, is 1.006 kJ/kgK. Thus,

$$\Delta T = (150 \text{ kJ/s})/((101 \text{ kg/s})(1.006 \text{ kJ/kgK})) = 1.5 \text{ K}.$$

The dry-bulb temperature of air within the poultry house is 28.7 C + 1.5 K or 30.2 C.

Heat of vaporization of water at 30.2 C, from Equation 2-20, is

$$h_{fg} = 2501 - (2.42)(30.2) = 2428 \text{ kJ/kg},$$

and the humidity ratio change resulting from the birds' latent heat addition is

$$\begin{aligned} \Delta W &= q_{latent} / \dot{M}_a h_{fg} \\ &= (100 \text{ kJ/s})/((101 \text{ kg/s})(2428 \text{ kJ/kg})) = 0.000408 \text{ kg/kg}. \end{aligned} \tag{2-35}$$

The humidity ratio of air within the poultry house equals the humidity ratio of air entering the house (state 3) plus the change or 0.013843 + 0.000408 = 0.014251 kg/kg. Two properties are now known, the others are:

(dry-bulb temperature = 30.2 C),
(humidity ratio = 0.014251),
water vapor saturation partial pressure = 4.30 kPa,
water vapor partial pressure = 2.27,
relative humidity = 53%,
dew point temperature = 19.4 C,
enthalpy = 66.8 kJ/kg,
wet-bulb temperature = 22.7 C,
specific volume = 0.870 m^3/kg.

The volumetric flow rate of air expelled from the poultry house is

$$\dot{V} = (101 \text{ kg/s}) (0.870 \text{ m}^3/\text{kg}) = 87.9 \text{ m}^3/\text{s}$$

which is slightly less than the total volumetric rate of outdoor air entering the building (75 m^3/s + 15 m^3/s = 90 m^3/s) . The difference is due to changes of dry-bulb temperature and moisture content.

The humidity ratio change of air passing through the evaporative cooler is 0.014783 kg/kg - 0.009199 kg/kg = 0.005584 kg/kg, based on data obtained in the analysis of the evaporative cooling process. Total airflow rate through the cooler is 84 kg/s, thus, water evaporates at a rate of

$$\dot{M}_{water} = (84 \text{ kg/s}) (0.005584 \text{ kg/kg}) = 0.47 \text{ kg/s}.$$

The pumping rate is to be triple the above rate, or approximately 1.41 kg/s (1.5 L/s). In an actual design, this pumping rate for the water pump would be specified at a pressure drop calculated separately for the plumbing system to be used.

Using the psychrometric chart. To this point, the solution has been presented with little detail of how psychrometric properties were determined after parameters of each process were calculated. The traditional means to trace processes and obtain values of properties has been the psychrometric chart. The processes of this example are traced on the psychrometric chart in Figure 2-9.

To use the chart, the first step is to locate state 1 – the outdoor air. The volumetric airflows of the two airstreams are given at state 1. The specific volume can be estimated and the respective mass flow rates determined.

State 2 is found 75% of the way (measured) from state 1 to the wet-bulb temperature of state 1.

State 3 is determined by the mixing process and lies along the line connecting states 1 and 2. The ratio $\dot{M}_{a1} / (\dot{M}_{a1} + \dot{M}_{a2})$ multiplied by the distance between states 1 and 2 is used to measure how far state 3 is from state 2. A ruler can be used, or dry-bulb temperature can be used as the scale. It is easy to measure in the wrong direction in this step. As drawn in Figure 2-9, state 3 is much nearer state 2 than state 1. This should be intuitively correct, for a greater mass of air at state 2 is involved in the mixing process. This intuitive check is useful to avoid inadvertent error.

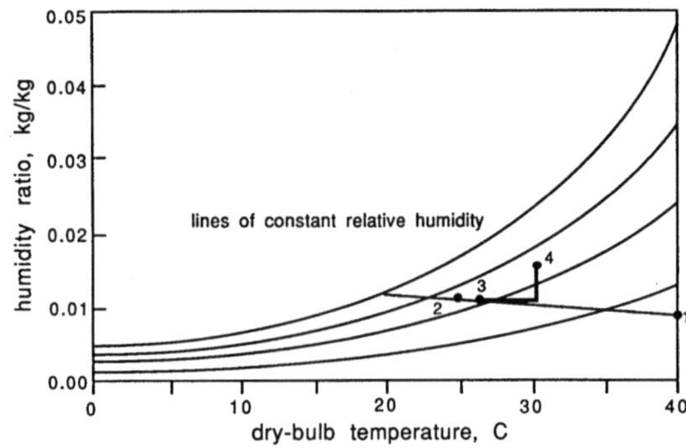

Figure 2-9. Evaporative cooling and reheating process, see Example 2-15.

Starting at state 3, the dry-bulb temperature and humidity ratio increases are calculated as done above and measured on the psychrometric chart to reach state 4.

Using the psychrometric equations. Example 2-15 could be readily implemented on a computer. Two computational sequences which would be required have been presented previously in this chapter. They may be integrated into the solution procedure described above. The sequence in Table 2-5 can be used to determine properties after the evaporative cooling process. Table 2-4 can be used to provide a suitable sequence to determine properties after mixing to reach state 3.

Dry-bulb temperature and humidity ratio for state 4 can be determined as done above. When they are known, other properties can be found using the sequence in Table 2-6.

Table 2-6. Example computational procedure to determine psychrometric properties when the dry-bulb temperature and humidity ratio are known.

<u>Given:</u> Dry-bulb temperature and humidity ratio.

<u>Procedure:</u>
1. Calculate the water vapor saturation partial pressure using Equation 2-1.
2. Calculate the water vapor partial pressure using Equation 2-9.
3. Calculate relative humidity using Equation 2-2.
4. Calculate specific volume using Equation 2-16.
5. Calculate dew point temperature using Equation 2-17 or 2-18.
6. Calculate enthalpy using Equation 2-21.
7. Calculate wet-bulb temperature using Equation 2-22 and the iterative procedure as described.

2-7. Cooling and Dehumidifying Revisited

Cooling and dehumidifying have previously been described as a process of first cooling to the dew point temperature, then following the saturation line to the

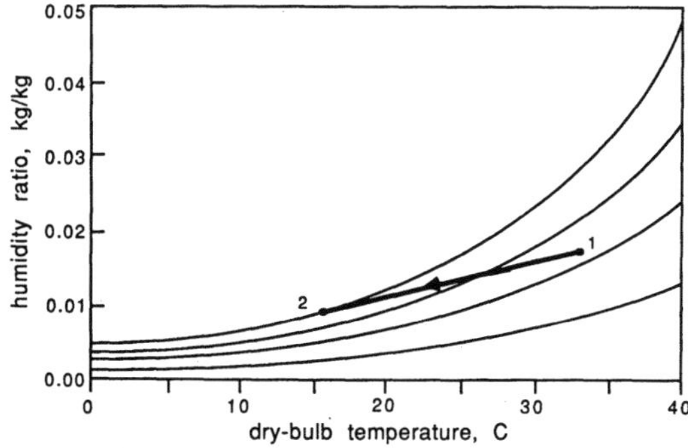

Figure 2-10. Simultaneous cooling and dehumidification, the straight-line law.

final dry-bulb temperature. While this may be correct in concept and suitable for many psychrometric calculations, it does not accurately describe the path followed by a mass of air on the average in contact with a surface below its dew point temperature.

A method to analyze combined heat and mass transfer (condensation) is the so-called "straight-line" law. This law states that when moist air loses heat and water vapor simultaneously to a surface below the air's dew point temperature, the process follows a straight line drawn between the original state of the air and saturation at the temperature of the surface. The process is sketched in

Table 2-7. Functions contained in computer file PSYFUNC.

Function	Action
PWR(x,y:real)	Raises a real number x to a real power y.
DEWPOINT(t,p:real)	Calculates dew point temperature (C) based on dry-bulb temperature, t (C), and water vapor pressure, p (kPa). Applies for dry-bulb temperatures between -60 to 70 C. See Equations 2-17 and 2-18.
PSAT(t:real)	Calculates partial pressure (kPa) of water vapor at saturation at dry-bulb temperature, t (C). Applies for dry-bulb temperatures between -40 and 120 C. See Equation 2-1.
ENTHALPY(t,w:real)	Calculates enthalpy of moist air (kg/kg) at dry-bulb temperature, t (C), and humidity ratio w, (kg/kg). See Equation 2-21.
SPEC_VOLUME (t,w,pa:real)	Calculates the specific volume (m^3/kg) of dry air at dry-bulb temperature, t (C), humidity ratio, w (kg/kg), and air pressure, pa (kPa). See Equation 2-15.
HUM_RATIO(p,pa:real)	Calculates the humidity ratio of moist air (kg/kg) at air pressure, pa (kPa), and water vapor partial pressure, p (kPa). See Equation 2-9.
WET_BULB(t,w,hh,pa:real)	Uses iterative procedure to determine wet-bulb temperature (C) when dry-bulb temperature, t (C), humidity ratio, w, (kg/kg), enthalpy, hh (kJ/kg) and air pressure, pa (kPa), are known. As written, determines wet-bulb temperature to nearest 0.01 C. See Equation 2-22.
REL_HUM(mu,pws,pa:real)	Calculates relative humidity (decimal) of moist air when degree of saturation, mu (dimensionless), water vapor saturation partial pressure, pws (kPa), and air pressure pa (kPa), are known. See Equation 2-12a.
DEG_SAT(rh,pws,pa:real)	Calculates the degree of saturation when the relative humidity, rh (decimal), water vapor saturation partial pressure, pws (kPa) and air pressure, pa (kPa), are known. See Equation 2-12b.

Figure 2-10. Such processes occur frequently in agricultural buildings, such as in air-air heat exchangers for energy conservation, and when condensation occurs on the inner surfaces of outside walls during cold weather. We will return to this subject after discussing heat transfer.

2-8. Computerized Psychrometric Functions

The file PSYFUNC contains psychrometric functions that can be used to develop psychrometric computer programs. See Table 2-7 for the functions contained in PSYFUNC.

2-9. Computerized Psychrometric Procedures

The file PSYPROC contains Pascal procedures that can be used in locally written programs to calculate psychrometric properties when two of the properties are known.

One other variable used in file PSYPROC is airp, the air pressure (kPa). The 12 variables (properties plus airp) should be declared as fields of a record in a program that uses PSYPROC and PSYFUNC. The 12 variables should be passed among the main program and its procedures as a variable of the type defined by the record. See Table 2-9 for an example where a suitable record type, and variable of the type, are defined and used to call the psychrometric procedures.

Numerous checks are incorporated into the procedures to screen out combinations of psychrometric property values which are not possible. The procedures and their uses are listed below. For each procedure, the listed psychrometric properties must be known before the procedure can be called. The procedure calculates values for the other properties. The entire file can be

Table 2-8. Properties required to use computer files PSYFUNC and PSYPROC.

Property	Name	Units
dry-bulb temperature	dbt1	C
relative humidity	rh 1	(decimal)
water vapor saturation partial pressure	ppsat1	kPa
water vapor partial pressure	pp1	kPa
humidity ratio	w1	kg/kg
dew point temperature	dpt1	C
enthalpy	h1	kJ/kg
wet-bulb temperature	wbt1	C
degree of saturation	mu 1	(decimal)

Two additional properties are calculated but cannot be included as known properties to begin the process of calculating the others.

Property	Name	Units
specific volume of dry air	spvol1	m^3/kg
density of moist air	dens1	kg/m^3

included in a locally written program by using the Include function of Turbo Pascal, or appropriate options available through other Pascal editors. Iterative search techniques are used in many of these procedures.

Procedure	Properties Which Must Be Known
DB_RH	dry-bulb temperature and relative humidity
DB_W	dry-bulb temperature and humidity ratio
DB_WB	dry-bulb and wet-bulb temperatures
DB_DP	dry-bulb and dew point temperatures
DB_H	dry-bulb temperature and enthalpy
RH_W	relative humidity and humidity ratio
RH_WB	relative humidity and wet-bulb temperature
RH_DP	relative humidity and dew point temperature
RH_H	relative humidity and enthalpy
W_WB	humidity ratio and wet-bulb temperature
W_H	humidity ratio and enthalpy
WB_DP	wet-bulb and dew point temperatures
WB_H	wet-bulb temperature and enthalpy
DP_H	dew point temperature and enthalpy

2-10. Example Using PSYPROC

Example 2-16

<u>Problem:</u> Develop a computer program to solve the ventilation problem described in Example 2-15 minus infiltration air. Write the program in Pascal. The program should be sufficiently user friendly and flexible to request the following input:

    outdoor air temperature,
    outdoor relative humidity,
    atmospheric pressure,
    airflow rate through the cooling pads (m^3/s),
    cooling system efficiency,
    latent heat production by the birds,
    sensible heat production by the birds.

Output of the program should be indoor air temperature and relative humidity.

<u>Solution:</u> In Table 2-9 there is a listing of a simple program written for Turbo 3 (© Borland International) Pascal. Files PSYFUNC.PAS and PSYPROC.PAS are in the program through the use of Include files. A copy of a screen showing input and output of the program is included in the table. Note: The calculation procedure in the program is only one of many possible paths to the solution.

```pascal
program Example2_16;

CONST      cpair = 1006.0;

TYPE       psyprops = RECORD              {Properties required in psychrometric
                                           procedures in PSYPROC.PAS.  Records
                                           are easier to pass among procedures
                                           than are large groups of variables.}

                      airp,               {atmospheric pressure, kPa}
                      dbt,                {dry-bulb temperature}
                      rh,                 {relative humidity}
                      ppsat,              {water vapor saturation pressure}
                      pp,                 {water vapor pressure}
                      w,                  {humidity ratio}
                      dpt,                {dew point temperature}
                      h,                  {enthalpy}
                      wbt,                {wet bulb temperature}
                      mu,                 {degree of saturation}
                      spvol,              {specific volume}
                      dens:               {air density}
                            real
                   END;

           probdata = RECORD              {data specific to the problem}

                      dbtout,             {outdoor dry-bulb temperature}
                      dbtin,              {indoor dry-bulb temperature}
                      rhout,              {outdoor relative humidity}
                      rhin,               {indoor relative humidity}
                      wout,               {outdoor humidity ratio}
                      win,                {indoor humidity ratio}
                      airp,               {atmospheric pressure, kPa}
                      volairflow,         {volumetric air flow rate}
                      massairflow,        {mass rate of air flow}
                      effic,              {evaporative cooling efficiency}
                      latentheat,         {addition of latent heat to the space}
                      sensibleheat:       {addition of sensible heat to the space}
                            real
                   END;


VAR        runagain:        boolean;
           airproperties:   psyprops;
           otherproperties: probdata;


{$IPSYFUNC.PAS}
{$IPSYPROC.PAS}

PROCEDURE DrawBox(x1,y1,x2,y2:integer);

    VAR   i: integer;

    BEGIN

       gotoxy(x1,y1);
       write(chr(218));
       gotoxy(x2,y1);
       write(chr(191));
       gotoxy(x1,y2);
       write(chr(192));
       gotoxy(x2,y2);
       write(chr(217));

       FOR i := (x1+1) TO (x2-1) DO
          BEGIN
             gotoxy(i,y1);
             write(chr(196));
             gotoxy(i,y2);
             write(chr(196));
          END;
```

```pascal
      FOR i := (y1+1) TO (y2-1) DO
         BEGIN
            gotoxy(x1,i);
            write(chr(179));
            gotoxy(x2,i);
            write(chr(179));
         END;

   END;   {PROCEDURE DrawBox}



FUNCTION DoAgain: Boolean;

   VAR  choice: char;

   BEGIN

      DrawBox(10,22,67,24);
      gotoxy(14,23);
      write('Enter <y> to run the program again, <n> if not: ');
      readln(choice);
      IF upcase(choice) = 'Y' THEN DoAgain := TRUE
      ELSE DoAgain := FALSE;

   END;   {FUNCTION DoAgain}

PROCEDURE Input(VAR airproperties:psyprops; VAR otherdata:probdata);

   {NOTE: no error trapping is included here, so program will bomb
          if other than numbers are entered}

   BEGIN


      WITH otherdata DO BEGIN

         gotoxy(10,1);
         textbackground(lightgray);
         textcolor(black);
         write(' SAMPLE PROGRAM TO SOLVE EXAMPLE 2-16, EVAPORATIVE COOLING ');
         textbackground(black);
         textcolor(lightgray);
         DrawBox(10,3,67,17);
         gotoxy(15,4);
         write('Please enter atmospheric pressure, kPa: ');
         readln(airproperties.airp);
         gotoxy(15,6);
         write('Please enter outdoor air temperature, C: ');
         readln(dbtout);
         gotoxy(15,8);
         write('Please enter outdoor relative humidity, %: ');
         readln(rhout);
         gotoxy(15,10);
         write('Please enter volumetric air flow rate, m3/s: ');
         readln(volairflow);
         gotoxy(15,12);
         write('Please enter cooling system efficiency, %: ');
         readln(effic);
         gotoxy(15,14);
         write('Please enter sensible heat production, W: ');
         readln(sensibleheat);
         gotoxy(15,16);
         write('Please enter latent heat production, W: ');
         readln(latentheat);

      END;   {WITH}

   END;   {PROCEDURE Input}


PROCEDURE Output(otherdata:probdata);

   BEGIN

      DrawBox(14,18,63,21);
      gotoxy(16,19);
      write('     Indoor air temperature is: ', otherdata.dbtin:7:1,' C');
      gotoxy(16,20);
```

```pascal
        write('      Indoor relative humidity is: ', otherdata.rhin:5:1,' %');

END;   {PROCEDURE Output}

PROCEDURE Calculate(props:psyprops; VAR data:probdata);

   VAR   hfg: real;    {heat of vaporization of water}

   BEGIN

      WITH data DO BEGIN

         props.dbt := dbtout;
         props.rh := rhout/100.0;
         DB_RH(props);
         dbtin := dbtout - (effic/100.0)*(dbtout - props.wbt);
         massairflow := volairflow*props.dens;
         props.dbt := dbtin;
         DB_WB(props);
         dbtin := props.dbt + sensibleheat/(massairflow*cpair);
         hfg := 2501.0 - 2.42*dbtin;
         win := props.w + latentheat/(massairflow*hfg*1000.0);
         props.dbt := dbtin;
         props.w := win;
         DB_W(props);
         rhin := props.rh*100.0;

      END;   {WITH}

   END;   {PROCEDURE Calculate}


{*************************** MAIN PROGRAM **************************************}

BEGIN

   runagain := TRUE;
   window(1,1,80,25);

   WHILE runagain DO BEGIN
      clrscr;
      input(airproperties,otherproperties);
      calculate(airproperties,otherproperties);
      output(otherproperties);
      RunAgain := DoAgain;
   END;   {WHILE}

   clrscr;

END.
```

```
       SAMPLE PROGRAM TO SOLVE EXAMPLE 2-16, EVAPORATIVE COOLING

     ┌────────────────────────────────────────────────────────────┐
     │    Please enter atmospheric pressure, kPa: 101.325           │
     │                                                              │
     │    Please enter outdoor air temperature, C: 40               │
     │                                                              │
     │    Please enter outdoor relative humidity, %: 20             │
     │                                                              │
     │    Please enter volumetric air flow rate, m3/s: 75           │
     │                                                              │
     │    Please enter cooling system efficiency, %: 75             │
     │                                                              │
     │    Please enter sensible heat production, W: 150000          │
     │                                                              │
     │    Please enter latent heat production, W: 100000            │
     └────────────────────────────────────────────────────────────┘

          ┌──────────────────────────────────────────────┐
          │    Indoor air temperature is:    28.3 C       │
          │    Indoor relative humidity is:  63.2 %       │
          └──────────────────────────────────────────────┘

      ┌──────────────────────────────────────────────────────┐
      │   Enter <y> to run the program again, <n> if not:     │
      └──────────────────────────────────────────────────────┘
```

2-11. Program PLUS

The program PLUS (Psychrometric Look-Up Substitute) is available as a substitute for the psychrometric chart and is an implementation of the psychrometric equations described in this chapter. It is provided as a tool for work related to future chapters rather than an auxiliary to this chapter. Contained within PLUS are the functions and procedures described in the previous sections of this chapter.

Program PLUS is self-contained and information windows are provided to assist users. It would be a useful exercise to go now to PLUS, become familiar with its operation, and check its results against some of the examples described above.

SYMBOLS

c_p	specific heat, J/kgK or kJ/kgK
h	enthalpy, J/kg or kJ/kg
M	mass, kg
$\dot{M}$	mass flow rate, kg/s
n	number of moles
p	pressure, Pa or kPa
q	thermal energy, joules
R	gas constant, J/kgK
t	temperature, C
T	absolute temperature, K
V	volume, m^3
v	specific volume, m^3/kg
$\dot{V}$	volumetric flow rate, m^3/s
W	humidity ratio, kg/kg
x	mole fraction
μ	degree of saturation
ρ	density, kg/m^3
ϕ	relative humidity

EXERCISES

1. Use the psychrometric charts to determine how much water vapor must be added to 5 kg of air at -15 C and 50% relative humidity, if it is to be conditioned to 25 C and 20% relative humidity.

2. Determine the final state of air resulting from an adiabatic mixing process where 20 kg of air at 10 C and 80% relative humidity is mixed with 4 kg of air at 30 C and 50% relative humidity. Sketch the process on a psychrometric chart.

3. Determine the final dry-bulb temperature of air at 35 C and 40% relative humidity which is cooled evaporatively at 80% efficiency.

4. Calculate, based on psychrometric equations, the humidity ratio, enthalpy, specific volume, and dew point temperature of air at 27.4 C dry-bulb temperature and 43% relative humidity when the atmospheric pressure is 92 kPa.

5. Develop a computer program to calculate relative humidity, humidity ratio, enthalpy, specific volume, and dew point temperature when dry-bulb and wet-bulb temperatures are known.

6. Develop a computer program to solve an adiabatic mixing problem such as described in Example 2-11. Retain sufficient flexibility in the input of the program so that dry-bulb temperatures, relative humidities, and volumetric flow rates of each airstream can be independently specified. That is, do not write the program specifically for the conditions of Example 2-11.

7. In a refrigeration unit, 2.8 m³/s of air at 14 C dry-bulb temperature, 85% relative humidity, and standard atmospheric pressure enters the unit. The air is cooled by the refrigeration process to 4 C dry-bulb temperature, and there is no reheating. Using the psychrometric chart, sketch the process and determine the refrigerating capacity of the unit in kW, the rate of water removal from the air in kg/s, and the volumetric flow rate of air leaving the refrigerator.

8. A very well-insulated building (no heat loss or moisture diffusion through the structural cover) is being ventilated. The building is located at an elevation of 1750 m above sea level. Water vapor within the building is produced at a rate of 0.018 kg/s and sensible heat at a rate of 82 kW. If outdoor relative humidity remains constant at 80% and the indoor dry bulb temperature remains constant at 20 C, determine the outdoor temperature at which indoor relative humidity is minimized. Explore the outdoor air temperature range between - 25 to 15 C. Program PLUS may be useful in determining psychrometric properties quickly. Describe (in words) why there is a relative humidity minimum instead of a monotonic change as outdoor air temperature changes from - 25 to 15 C.

9. Consider the same building as in Problem 8 (very well insulated). If outdoor air relative humidity is 80% and outdoor air temperature is -10 C, determine the ventilation rate (to the nearest kg/s) at which the indoor relative humidity will be maximized. What will be the indoor temperature and relative humidity at that ventilation rate? Program PLUS may be helpful to find psychrometric properties for many conditions quickly. Explore the ventilation rate range between 5 and 20 kg/s. Describe (in words) why there is a relative humidity maximum instead of a monotonic change as the ventilation rate changes from 5 to 20 kg/s.

REFERENCES

ASHRAE. 1977. Brochure on psychrometry. American Society of Heating, Refrigerating, and Air Conditioning Engineers, Atlanta, GA.

ASHRAE. 1989. Handbook of Fundamentals. American Society of Heating, Refrigerating, and Air Conditioning Engineers, Atlanta, GA.

Kusuda, T. 1970. Algorithms for psychrometric calculations. NBS Publication BSS-21. Superintendent of Documents, U.S. Government Printing Office, Washington, DC.

Nelson, R. M. and M. Pate. 1986. A comparison of three moist air property formulations for computer applications. ASHRAE Transactions 92 (1B):435-447.

Wexler, A., R. Hyland and R. Stewart. In Preparation. Thermodynamic properties of dry air, moist air and water and SI psychrometric charts. American Society of Heating, Refrigerating, and Air Conditioning Engineers, Atlanta, GA.

Wilhelm, L.R. 1976. Numerical calculation of psychrometric properties in SI units. Transactions of the ASAE 19(2): 318-321, 325.

CHAPTER 3
HEAT TRANSFER BASICS

3-1. Introduction

Building environment analysis and design must be based on understanding the three modes of heat transfer: conduction, convection, and radiation. Each of the three can be present independently, but usually at least two, and frequently all three, are present simultaneously and contribute significantly to determining environmental conditions inside buildings.

Heat transfer mechanisms involving phase changes will not be discussed now. They occur frequently in environmental situations, especially related to changes of thermal energy between animals and their environment, and we will return briefly to them later.

Heat transfer differs from thermodynamics in a fundamental way, even though the two topics appear similar. Classical thermodynamics is based on the concept of equilibrium, or infinitesimal deviations from equilibrium. Heat transfer arises only from non-equilibrium, specifically, from finite differences of temperature.

Heat transfer may be steady state, where temperatures and heat fluxes do not change as functions of time. Heat transfer may be steady-periodic, where conditions change with time in a regular fashion and periodically return to their starting conditions (as in response to daily cycles of outside conditions). Heat transfer may be strictly transient with neither steady-periodic nor steady-state assumptions adequate to represent the process.

Recent research has applied transient heat transfer analysis to the design of environmental control methods for large, thermally massive buildings, but for most engineering designs of agricultural buildings, steady-state analysis is adequate at least as a close approximation. This text will focus on steady-state analysis.

3-1.1. Thermal Conduction.
Thermal conduction is diffusion of thermal energy through a continuous, frequently stationary medium, and depends on properties of that medium. For example, the loss of heat from the indoor surface of a building's outer wall to the outdoor surface is by conduction if the wall is solid.

In a kinetic sense, conduction heat transfer is often viewed as a cascading transfer of energy of motion among particles on an atomic level. Atoms at higher temperatures are believed to possess more kinetic energy than those at lower temperatures, energy which is transferred to less active neighbors through elastic collision in a fluid, or oscillations of atoms and transport of free electrons in a solid matrix. Heat transfer by conduction is always from regions of higher temperatures to regions of lower temperatures; the second law of thermodynamics would be violated otherwise.

Thermal conduction is the only means of heat transfer through solid, opaque objects. Conduction can also occur in fluids, but dominates overall heat transfer in fluids only in laminar flow, and in the laminar sublayer of the turbulent boundary layer adjacent to solid objects.

3-1.2. Thermal Convection. Thermal convection is a process involving fluids. The term refers loosely either to thermal energy transfer from place to place within a fluid, or between a fluid and a solid surface. In this text, the term convective heat transfer will be used to denote transfer of thermal energy between a fluid and a solid. The term convection will be used to denote transport of thermal energy through fluids by large scale eddying motions. For example, a dairy cow loses heat to the surrounding air, and the heat may be eventually lost to the outdoors by conduction through the building's wall. Transfer of the heat from the cow's pelt to her surrounding air is by convective heat transfer, transfer through the air from the vicinity of the cow to the vicinity of the wall is by convection, and transfer from the air to the wall surface is by convective heat transfer.

The mechanism of thermal convection which is most important in determining building environment is convective heat transfer. The transfer process involves the boundary layer. In turbulent fluid flow, eddies in the main part of the flow exchange thermal energy between the boundary layer and surrounding fluid, and penetrate within the outer regions of the boundary layer. See Figure 3-1; a turbulent boundary layer is sketched. Within the laminar sublayer of the turbulent boundary layer, thermal energy exchange reduces to thermal conduction. Ultimately, all convection between fluids and solid objects is limited by the conduction process. As fluid velocity increases and the laminar sublayer thins, the region of conduction shrinks, turbulent eddies push farther into the boundary layer, and convective heat transfer is enhanced.

Thermal convection and convective heat transfer are classified in two ways. If a fan or pump causes fluid motion, the process is termed "forced convection". If

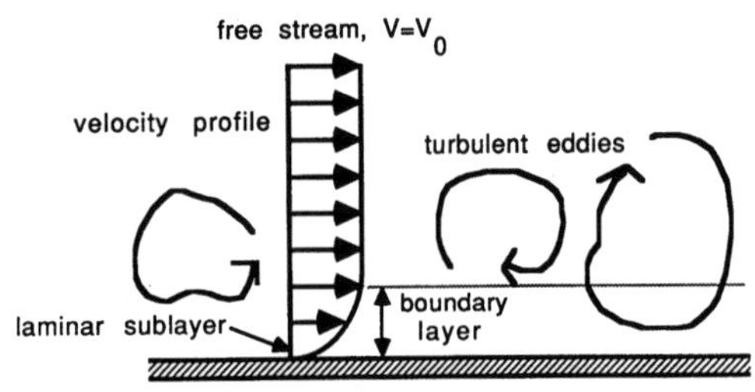

Figure 3-1. Profile of a turbulent boundary layer.

fluid motion is induced by density differences within the fluid, usually caused by temperature differences, the process is termed "natural convection", or "free convection".

3-1.3. Thermal Radiation. Thermal radiation differs markedly from both conductive and convective heat transfer processes. While conduction and convection require the presence of matter, thermal radiation heat transfer requires the absence of an intervening absorbing medium.

Radiation heat transfer occurs when electromagnetic energy leaves one object and is intercepted and absorbed by another. All objects at temperatures above absolute zero emit thermal radiation, thus radiation heat transfer is an exchange process. Objects at high temperatures emit more electromagnetic radiation than those at low temperatures; the overall process between two objects is a net transfer of thermal energy from objects at higher temperatures to those at lower temperatures. When thermal radiation leaves an object, the energy content of that object is lessened, and the object exhibits a lower temperature. The converse occurs when an object absorbs thermal radiation.

The wavelength of electromagnetic radiation traditionally considered to be involved in radiation heat transfer is the infrared band. See Figure 3-2. The lower limit of infrared radiation is at the upper limit of the visible light range, approximately 0.8 microns wavelength. The upper limit of thermal radiation is less well defined, but extends past 10 microns, a wavelength which typifies thermal radiation emitted by objects at earth temperature. However, thermal radiation exchange is not limited to these wavelength bands. Visible light and ultraviolet radiation from the sun, for examples, are electromagnetic radiation and, when absorbed, are converted to thermal energy.

3-2. Conduction Heat Transfer

3-2.1. The Heat Conduction Equation. The heat conduction equation for isotropic solids is derived in most heat transfer texts. One form is

$$\nabla^2 t + q_{gen} / k = (\alpha)^{-1} \delta t / \delta \tau, \qquad (3-1)$$

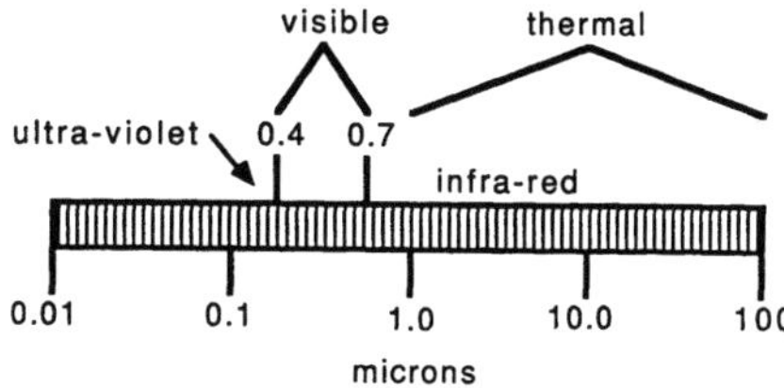

Figure 3-2. The electromagnetic radiation band.

where t is temperature, q_{gen} is the rate of internal heat generation, W/m^3; k is thermal conductivity, W/mK; α is thermal diffusivity, s/m^2; and τ is time, s.

Thermal conductivity is an intensive property of material which is a proportionality constant relating heat flux by conduction, and the temperature gradient,

$$k = - q'' / (dt/dn) \tag{3-2}$$

where n is the direction of the positive temperature gradient. Heat flux, q", is thermal energy transferred per unit area per unit time (W/m^2, for example). Thermal conductivity values for real materials are determined empirically. Thermal conductivity may be a function of time, temperature, moisture content, direction, or location. However, for most applications in environmental engineering, it is adequate to assume thermal conductivity is a relatively constant scalar. The negative sign is introduced in Equation 3-2 so a positive heat flux corresponds to a negative temperature gradient. Heat always flows "downhill".

In the SI system, units for thermal conductivity are W/mK. To follow convention, temperature units in thermal conductivity, and other parameters to follow, will be expressed in K when a temperature difference is represented and C for actual temperature. Each degree difference in K is equivalent to a degree difference in C.

Appendix Tables A3-1 and A3-2 contain thermal conductivity data for common engineering materials.

Thermal diffusivity is a combination of thermal parameters,

$$\alpha = k / \rho c_p, \tag{3-3}$$

where ρ is mass density and c_p is specific heat. Thermal diffusivity is a measure of how rapidly thermal energy can penetrate a solid material and is important in time-dependent heat conduction problems.

The Laplacian, $\nabla^2 t$, for one-dimensional heat transfer is

$$\nabla^2 t = \frac{1}{n^m} \frac{d}{dn} \left(n^m \frac{dt}{dn} \right), \begin{array}{l} m = 0 \text{ cartesian} \\ m = 1 \text{ cylindrical} \\ m = 2 \text{ spherical} \end{array} \tag{3-4}$$

Many engineering problems can be analyzed using simplified forms of the heat conduction equation. If there is no source term, the Fourier equation is obtained

$$\nabla^2 t = (\alpha)^{-1} \delta t / \delta \tau, \tag{3-5}$$

which applies to time-dependent, conduction heat transfer problems.

If conduction is steady-state, but a uniformly distributed heat source is present, the Poisson equation is obtained

$$\nabla^2 t + q_{gen} / k = 0. \qquad (3\text{-}6)$$

This situation can arise, for example, when solar energy is absorbed within the glass or plastic glazing on a greenhouse.

If conduction is steady-state and there are no internal, uniformly distributed heat sources, the Laplace equation results

$$\nabla^2 t = 0. \qquad (3\text{-}7)$$

The Laplace equation applies to most thermal energy exchanges by conduction in agricultural buildings. Although ambient conditions are, in reality, always time-dependent, agricultural buildings are usually sufficiently lightweight in a thermal sense that steady-state (or step-wise steady-state) conditions may be assumed.

One way to estimate whether a building is lightweight in a thermal sense is to consider how rapidly the building might respond to a sudden change of outdoor temperature. Air temperature in a ventilated barn, for example, would be expected to follow outdoor temperature with a lag of less than 1 hr. The same could be expected for temperature changes of other components of the barn such as the walls. Compare this to the period of the diurnal cycle of outdoor temperature which is 24 hrs. The response time of the barn is much less than the period of the outdoor temperature changes, thus, the barn may be considered thermally lightweight and steady-state analysis would be suitable as an approximation of the actual conduction heat transfer process. (Thermal lag is associated with the conduction process; radiation heat transfer is nearly instantaneous, and convective heat transfer occurs within seconds.) As a rough rule, if the maximum air temperature within a building is reached within 2 hrs of the maximum outside temperature, the building may be classed as a lightweight structure.

3-2.2. Temperature Fields. If two boundary conditions for Equation 3-7 are known, the equation can be solved to determine the temperature field. Example 3-1 shows the solution of Equation 3-7 for one-dimensional heat transfer by conduction through a planar solid, a situation resembling heat loss through a homogeneous wall of a building.

Example 3-1

Problem: Determine the temperature field within a homogeneous wall of thickness, L, when the temperature at one boundary is t_1, and the temperature at the other boundary is t_2.

Solution: This example can be considered a case of one-dimensional heat transfer in cartesian coordinates and Equation 3-7 reduces to

$$d^2t / dx^2 = 0, \qquad (3\text{-}8)$$

which can be integrated twice to yield

$$t = c_1 x + c_2. \qquad (3\text{-}9)$$

For $0 < x < L$ the temperature field is

$$t = t_1 + (t_2 - t_1)(x / L). \qquad (3\text{-}10)$$

Equation 3-10 represents a temperature field linear in distance for the situation of steady-state heat transfer in one dimension in a homogeneous solid with no internal heat sources. Other coordinate systems do not necessarily yield temperature profiles linear in distance.

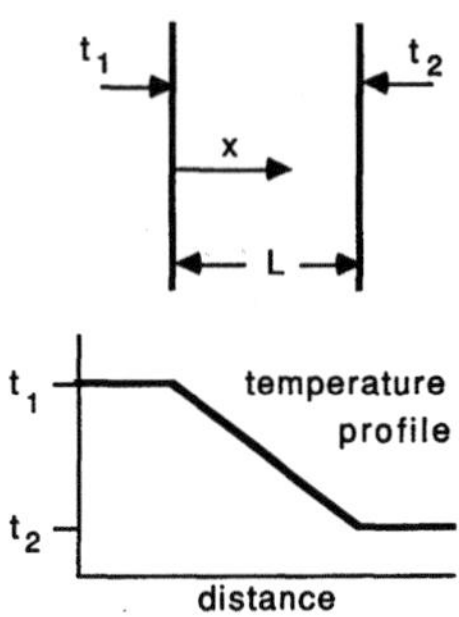

Conduction heat transfer through cylindrical walls can also be important in building environment analysis and design. For example, a forced hot air furnace may be used to heat a nursery for young animals or a greenhouse. If a duct to carry heated air is round, and is insulated so the wall is thick relative to the duct diameter, conduction heat transfer in cylindrical coordinates must be considered. Example 3-2 demonstrates the solution of such a situation, and a derivation of the expression for temperature in one-dimensional cylindrical coordinates. Heat conduction through insulation on water pipes almost always must be analyzed in cylindrical coordinates.

Example 3-2

Problem: Determine the temperature field within a homogeneous cylindrical shell of inner radius r_i and outer radius r_o when the temperature of the inner surface is t_i and the temperature of the outer surface is t_o. Assume steady state conditions.

Solution: Equation 3-7 applies again in the form

$$\frac{d^2t}{dr^2} + \frac{1}{r}\frac{dt}{dr} = 0, \qquad (3\text{-}11)$$

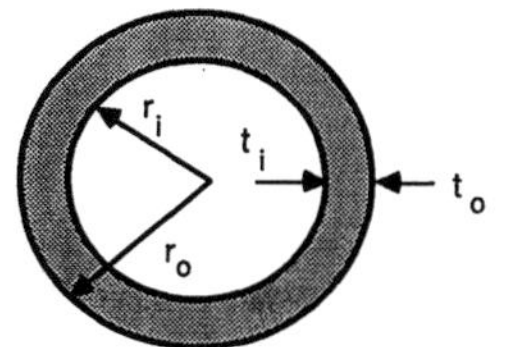
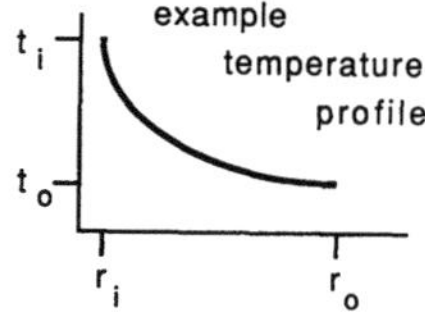

which can be solved readily by substituting $S = dt/dr$. This reduces Equation 3-11 to

$$\frac{dS}{dr} + \frac{S}{r} = 0. \qquad (3\text{-}12)$$

If Equation 3-12 is multiplied by rdr, it becomes

$$r\, dS + S\, dr = 0, \qquad (3\text{-}13)$$

which is, by definition,

$$d(rS) = 0. \qquad (3\text{-}14)$$

Equation 3-14 can be integrated once to

$$rS = c_1 \qquad (3\text{-}15)$$

and a second time to

$$t = c_1 \ln r + c_2. \qquad (3\text{-}16)$$

Applying the boundary conditions leads to

$$t_i = c_1 \ln r_1 + c_2, \text{ and} \qquad (3\text{-}17)$$

$$t_o = c_1 \ln r_o + c_2. \qquad (3\text{-}18)$$

The constants of integration are determined using Equations 3-17 and 3-18, and the temperature field can be expressed as

$$t = \frac{t_o - t_i}{\ln(r_o/r_i)} \ln r + \frac{t_i \ln r_o - t_o \ln r_i}{\ln(r_o/r_i)} \qquad (3\text{-}19)$$

Equation 3-19 shows temperature in this steady state case is not linear in distance, but rather shows a logarithmic increase (or decrease, depending on the boundary conditions). Linearity would not be expected intuitively, of course, because the cross-sectional area of heat transfer is a function of radius, thus the temperature gradient must also be a function of radius for steady state heat transfer to occur.

3-2.3. Conduction Heat Transfer. Thermal flux due to conduction heat transfer can be determined by

$$q'' = - k \, dt / dn, \qquad (3\text{-}20)$$

in a restatement of Equation 3-2 where dt/dn is the temperature gradient and q" is heat flux by conduction. Equation 3-20 expresses the <u>Fourier law of heat conduction</u>. When integrated over the area of heat flow (area normal to the direction of flux) heat flux becomes heat flow (watts, for example).

Under steady-state conditions, Equation 3-20 may be integrated along the path and over the area of heat flow to yield heat flow, q

$$q = kA\Delta t / L; \; \Delta t = t_1 - t_2 \qquad (3\text{-}21)$$

Equation 3-21 is frequently rearranged and the terms k/L grouped into a single term, U, the <u>unit area thermal conductance</u>,

$$q = UA\Delta t; \; U = k / L. \qquad (3\text{-}22)$$

Thermal conductance is an extensive property, whereas, thermal conductivity is intensive.

A final rearrangement is frequently made for convenience. The inverse of unit area thermal conductance, termed <u>unit area thermal resistance</u>, R, is used in a restatement of Equation 3-22,

$$q = A\Delta t / R; \; R = L/k. \qquad (3\text{-}23)$$

Example 3-3 is a calculation of heat transfer through a single layer wall in cartesian coordinates, Example 3-4 provides a similar calculation for cylindrical coordinates.

Example 3-3

<u>Problem:</u> A wall has been built of concrete and is 200 mm thick with a cross-sectional area of 10 m². The temperature of one face is 20 C, the temperature of the other is - 10 C. Assume thermal conductivity of concrete is 0.56 W/mK.

Determine the temperature field and estimate the rate of conduction heat transfer through the wall.

Solution: The temperature field is independent of thermal properties for one-dimensional, steady-state heat transfer by conduction, and properties are uniform and isotropic. Thus Equation 3-10 applies, where $t_1 = -10$ C, $t_2 = 20$ C and $x = 0$ where $t = -10$ C.

$$t = -10 + (20 - (-10))(x/0.2)$$
$$= -10 + 150 \ (x \text{ in mm})$$

and by Equation 3-23,

$$R = L/k = 0.2/0.56$$
$$= 0.357 \ m^2 K/W$$
$$q = A\Delta t/R = (10)(30)/0.357$$
$$= 840 \ W.$$

Example 3-4

Problem: A circular sheet metal duct carries refrigerated air to a cold storage room for apples. The duct itself is 250 mm in diameter (outside), and is covered with a 50 mm layer of insulation, making the outer diameter of the insulated duct 350 mm. The duct wall thickness is 1 mm.

The inner surface temperature of the duct is 0 C, and the outer surface temperature of the insulation is 25 C. Thermal conductivity of sheet metal is 60 W/mK, and of the insulation is 0.04 W/mK.

Calculate the rate of conduction heat gain through the insulation per meter length of the duct. Assume steady-state conditions.

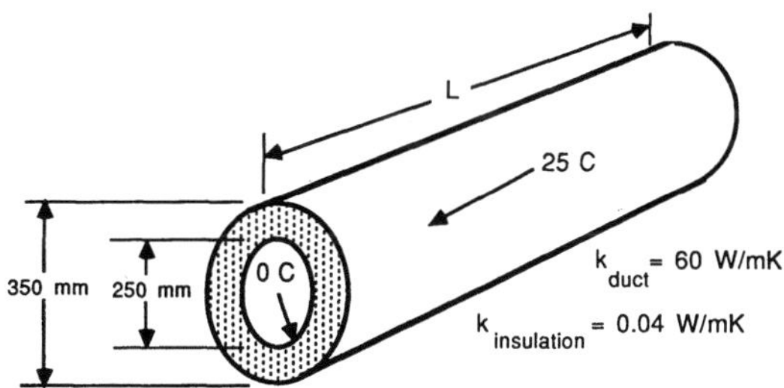

Solution: In cylindrical coordinates, thermal resistance to conductive heat transfer is

$$R = \frac{\ln(r_o/r_i)}{2\pi kL} \tag{3-24}$$

where L is the length of the cylinder, and r_o and r_i are the outer and inner radii, respectively. It can be a useful exercise to develop this expression using the Fourier law of heat conduction, the definition of heat transfer using the resistance analogy (Equation 3-23), and the equation for the temperature field in cylindrical coordinates (Equation 3-19).

Note that Equation 3-24 presents thermal resistance differently than is the typical case for cartesian coordinates. For the cylinder, resistance is total thermal resistance for the cylindrical wall, not unit area or unit length thermal resistance (unless L = 1). This is an important distinction which will be emphasized later.

This example is to calculate heat gain per meter of duct, thus, L is 1 m in Equation 3-24, k is 0.04 W/mK for the insulation, and r_o and r_i are 175 and 125 mm, respectively. For now, consider just the insulation layer; it is likely to be the limiting resistance along the path of heat transfer. The duct wall itself will have relatively little resistance.

The resistance of one meter length of insulation is

$$R = \ln(175/125)/(2\pi)(0.04)(1.0) = 1.34 \text{ mK/W}$$

and the heat gain per unit length is

$$q = \Delta t/R = (25 \text{ K})/(1.34 \text{ mK/W}) = 18.7 \text{ W/m}.$$

3-2.4 Resistances in Series. If total resistance, R', is considered instead of unit area thermal resistance, conduction heat transfer in steady state through more complex geometries resembles electrical current flow through an electrical circuit of resistors, and

current = difference in potential/total resistance or

$$q = \Delta t / R'. \tag{3-25}$$

The electrical analogy is useful for realistic conduction heat transfer problems; standard electrical circuit equations can be used to simplify thermal circuits. In an electrical circuit with resistors in series, the total resistance of the circuit equals the sum of the individual resistances,

$$R' = \Sigma R_{individual}. \tag{3-26}$$

A typical wood frame wall in a building has several layers. There is inside sheathing, a wall cavity filled with insulation, exterior sheathing, and siding. It is likely no two layers are made of the same material, and each is a different thickness.

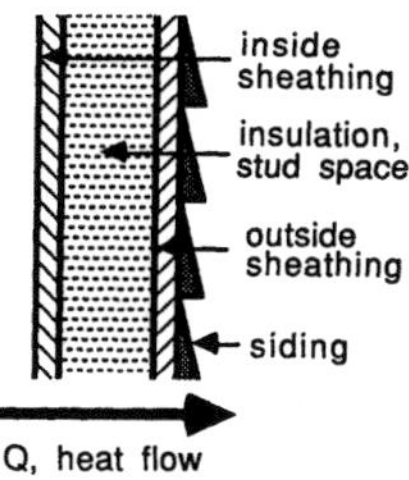

Conduction heat transfer is through each of the layers in series. This is a <u>series thermal circuit</u> and the total resistance of the wall equals the sum of resistances of the layers. A typical calculation is in Example 3-5.

Example 3-5

<u>Problem:</u> A wall is constructed as shown in Figure 3-3. Calculate the unit area thermal resistance of the wall section, and the heat flux which would result if one side of the wall were held at 20 C and the other at - 5 C.

<u>Solution:</u> This is a series thermal circuit; Equation 3-26 can be used to determine the total unit area thermal resistance of the wall, and Equation 3-25 can then be used to determine thermal flux. The first step is to calculate the unit area thermal resistance of each layer:

$$R_1 = L_1/k_1 = (0.020m)/(0.25W/mK) = 0.08 \ m^2K/W,$$

$$R_2 = L_2/k_2 = (0.100m)/(0.15W/mK) = 0.67 \ m^2K/W,$$

$$R_3 = L_3/k_3 = (0.030m)/(0.20W/mK) = 0.15 \ m^2K/W.$$

The total unit area thermal resistance is

$$R' = 0.08 + 0.67 + 0.15 = 0.90 \ m^2K/W.$$

The heat flux is

$$q' = \Delta t / R' = (20 \ C - (- 5 \ C))/0.90 \ m^2K/W = 27.8 \ W/m^2.$$

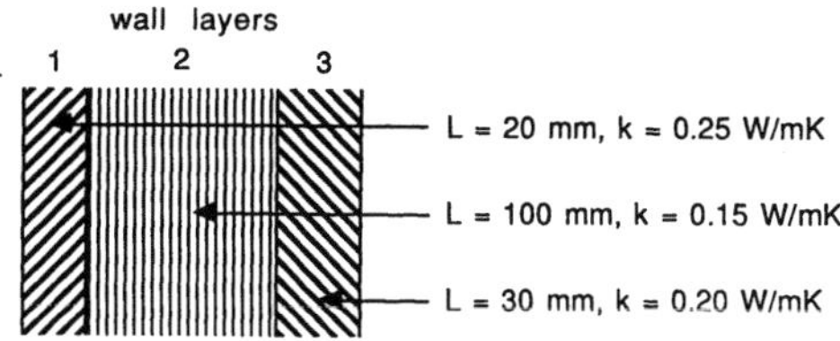

Figure 3-3. Cross-section of the wall considered in Example 3-5.

At this point it would be a useful exercise to return to Example 3-4, calculate the thermal resistance of the sheet metal duct wall, compare that value to the resistance of the insulation and sheet metal in series, and judge whether it was a reasonable assumption to neglect the effect of the duct wall on heat gain into the duct.

3-2.5. Resistances in Parallel. When electrical resistors are in parallel, the total resistance, R', is found by the inverse rule,

$$1/R' = \Sigma\,(1/R_{individual}). \qquad (3\text{-}27)$$

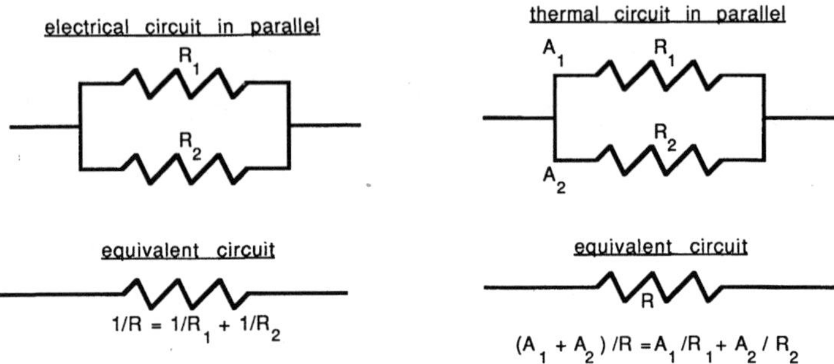

A typical building has numerous heat loss paths operating in parallel. Heat transfer through the walls, windows, ceiling, doors, and floor are heat transfer paths in parallel. Each path bridges the same temperature difference – indoor air temperature to outdoor air temperature.

Note: When the electrical analogy is applied to heat transfer paths in parallel, the resistances used in Equation 3-27 must be the total resistances of each path not the unit area resistances. The parallel heat transfer resistance relationship is normally stated as

$$A_{total}/R' = \Sigma\,(A_{individual}/R_{individual}) \qquad (3\text{-}28)$$

where A_{total} is the sum of areas of the heat loss paths,

$$A_{total} = \Sigma\,A_{individual}, \qquad (3\text{-}28a)$$

and $R_{individual}$ is the unit area thermal resistance of each path. In Equation 3-28, R' is the unit area thermal resistance averaged over all heat transfer paths. Example 3-6 demonstrates an application of parallel heat transfer calculations.

Example 3-6

Problem: A wood-framed wall in a building (see Figure 3-4) is well insulated. However, framing occupies 20% of the wall area, and framing does not have the insulative effect of insulation.

Heat loss through such a wall is a situation of conductive heat transfer paths in parallel. One path of heat loss is through the framed part of the wall, the other path is through the insulated part. For this example, the framed part of the wall has a unit area thermal resistance of 2.3 m^{2}K/W, and the insulated part has a unit area thermal resistance of 4.1 m^{2}K/W.

If the wall is 3 m high and 10 m long, what is the average unit area thermal resistance of the wall, and what will be the heat loss through the wall when it is 20 C indoors and - 5 C outdoors?

<u>Solution:</u> Equation 3-28 is used to calculate the average unit area thermal resistance. There are two paths of heat loss. The total area of heat loss is 30 m^2 (3 m x 10 m). Framing occupies 20% of the wall, a heat loss area of 6 m^2 (0.20% of 30 m^2). The insulated part of the wall is the rest, 24 m^2. Equation 3-28 becomes

$$30 \text{ m}^2 / \text{R}' = (6 \text{ m}^2 / 2.3 \text{ m}^2\text{K/W}) + (24 \text{ m}^2 / 4.1 \text{ m}^2\text{K/W}).$$

The average unit area thermal resistance is

$$\text{R}' = 3.55 \text{ m}^2\text{K/W}.$$

The heat loss through the wall can be calculated using Equation 3-23,

$$\begin{aligned} q &= (30 \text{ m}^2)(20 \text{ C} - (-5 \text{ C})) / 3.55 \text{ m}^2\text{K/W}, \\ &= 212 \text{ W}. \end{aligned}$$

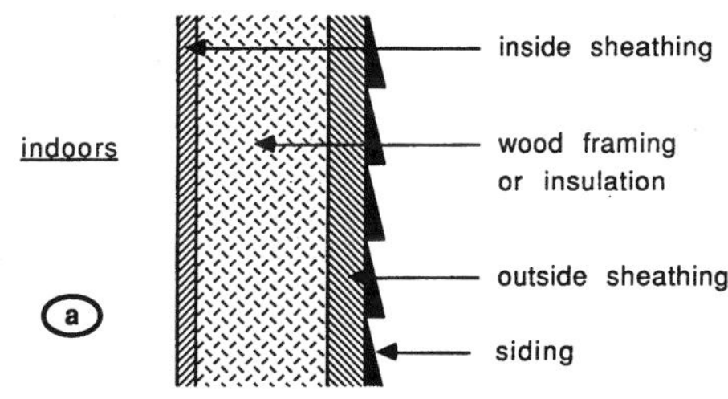

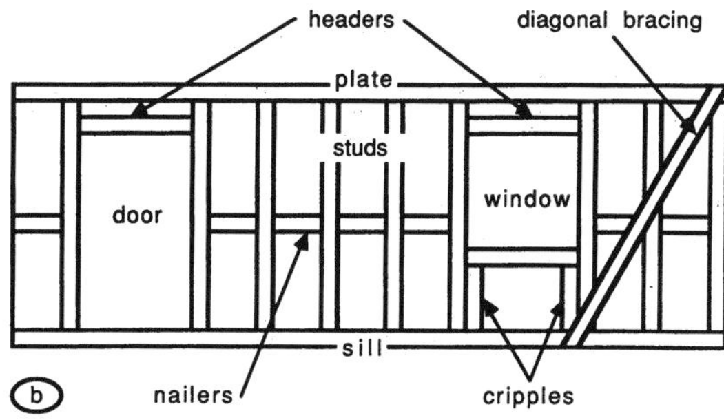

Figure 3-4. Cross-section of a wood-framed wall considered in Example 3-6. Wall cross-section is in (a), and the elevation is in (b).

Note the influence of a small area of relatively low unit area thermal resistance. The insulated wall section unit area thermal resistance of 4.1 m^2K/W is reduced more than 13% to 3.55 m^2K/W by framing which comprises only 20% of the wall.

Examples 3-5 and 3-6 have demonstrated calculations for heat flow paths in series and parallel. In real situations, both calculations are required. A strategy to approach such problems, in a cartesian coordinate system, is first to examine each series heat flow path, calculate its unit area thermal resistance, and then complete the calculation for all the parallel paths. For example, in a building with walls, ceiling, doors, and windows, the areas of each path and the unit area thermal resistance of each path are calculated first. Then the average unit area thermal resistance and total heat loss are calculated. In cylindrical coordinate systems, total thermal resistance must be used rather than unit area thermal resistances. Modifications of this strategy work, also, but should be attempted only after this straight forward sequence is well understood.

Data to calculate conduction heat transfer exist in two forms. In many engineering applications, thermal conductivity values are used along with Equation 3-21. However, in buildings, many materials are used in standard modules and unit area thermal resistances (R-values) of the modules appear in standard tables. Appendix Table A3-2 contains such a collection of mixed data. Where a material is not used in standard thickness (e.g., concrete) resistance per meter thickness is presented. Where a standard module thickness is typical, the R-value of that module is listed.

The R-value of a desired thickness of material for which an R-value per meter is given can be obtained by scaling the unit thickness R-value. For example, concrete with sand and gravel aggregate, not oven dried, has an R-value of 0.56 m^2K/W per m thickness. A wall 0.2 m thick would, therefore, have an R-value of 0.112 m^2K/W (20% of 0.56 m^2K/W). This calculation, in effect, invokes the concept of resistances in series.

On the other hand, the R-value for Douglas fir plywood, 15.88 mm thick, is given directly as 0.145 m^2K/W. However, if a thickness of material is to be used that is not listed in the table, for example, 10 mm thick Douglas fir plywood, its R-value can be estimated by scaling from given data. The data state 15.88 mm of the plywood has an R-value of 0.145 m^2K/W, thus, 10 mm must have an R-value of 0.091 m^2K/W (= (10 / 15.88) (0. $145m^2K/W$)) .

Rounding differences may give slightly different values depending on which database one starts with. If a range of values is given, the midpoint should be used unless manufacturer's data suggest otherwise.

3-3.1. Natural Convection. Convective heat transfer occurs due to a temperature difference between a solid object and its surrounding fluid. In natural convection, fluid moves because of density differences caused by temperature, humidity, or other air constituent gradients. Many engineering studies have been presented which provide empirical correlations between convective heat transfer rates and various thermal and geometric properties of the fluid and solid object.

As an example of natural convection, consider the air duct described in Example 3-4. Cold air within the duct causes the exterior of the duct to be colder than the surrounding room air (if the duct is inside a heated space). Air in the vicinity of the duct cools because of contact with the duct surface, becomes slightly more dense than the surrounding air, and begins to fall. When the cooler air falls away from the duct, it is replaced by room air which is still warm. That air cools, falls, and the cycle is repeated as long as the temperature difference between the air and duct is maintained.

Density differences enhance fluid movement, viscosity retards it. Fluid properties which influence the rate of convective heat transfer are: thermal conductivity, the coefficient of thermal expansion, viscosity, density, and specific heat. The gravitational constant, g, as well as the size and shape of the solid object involved in the heat exchange also influence the rate of convective heat transfer.

In heat transfer, as in fluid mechanics, analyses of physical phenomena are described in terms of dimensionless ratios of the properties and parameters which influence and describe the phenomena. In natural convection, three dimensionless numbers are important.

The first is the Nusselt number, Nu, calculated as

$$Nu = hL / k, \qquad (3\text{-}29)$$

where L is a characteristic dimension of the solid object involved, k is the thermal conductivity of the fluid, and h is the coefficient of convective heat transfer, also called a film coefficient or a surface coefficient. The characteristic dimension could be the diameter of a horizontal, circular duct, the height of a wall, or the equivalent diameter of a non-circular duct. The Nusselt number can be interpreted as the ratio of the ease with which heat is transferred by convection to the ease with which heat is transferred by conduction, and thus is a measure of the intensity of the convective heat transfer process.

The coefficient, h, is a conductance defined as the flux of heat transferred convectively divided by the temperature difference between the fluid and solid,

$$h = q'' / \Delta t, \text{ or } h = (q / A)\Delta t, \qquad (3\text{-}30)$$

where q is total heat transferred over an area, A, between a solid object and a fluid due to a temperature difference, Δt.

If the Nusselt number can be determined, convective heat transfer can be quantitatively predicted.

The second dimensionless member, the Prandtl number, Pr, expresses the ratio of the diffusion of momentum to the diffusion of heat from a solid surface through a boundary layer into a surrounding fluid. It is calculated as

$$Pr = \mu c_p / k \tag{3-31}$$

where, μ is the dynamic viscosity of air, c_p is its specific heat, and k is its thermal conductivity. Example air properties, and the corresponding Prandtl numbers as functions of temperature, are in Table 3-1.

The third dimensionless ratio, important in natural convective heat transfer, is the Grashof number, Gr. This ratio can be interpreted as the ratio of buoyancy forces to viscous drag forces within the fluid,

$$Gr = g\rho^2 \beta L^3 \Delta t / \mu^2, \tag{3-32}$$

where β is the coefficient of thermal expansion, g is the gravitational constant, and other terms are as previously defined.

In general, natural convective heat transfer processes have been found to follow the relationship

$$Nu = c(GrPr)^n. \tag{3-33}$$

Convective heat transfer mirrors fluid mechanics, where there is a laminar domain, a fully turbulent domain, and between laminarity and turbulence is a domain of mixed laminar and turbulent flow. The coefficient and exponent in Equation 3-33 depend on the fluid flow domain. When flow is laminar, n = 0.33. When flow is turbulent, n = 0.25. When flow is in the transition region between laminar and turbulent, n is between 0.33 and 0.25. In this text only laminar and turbulent flow will be considered, and the transition range will not be covered.

Table 3-1. Properties of dry air at standard atmospheric pressure, 101.325 kPa.

Temperature, K	μ, kg/ms	k, W/mK	Pr	ρ, kg/m^3	c_p, kJ/kgK
200	1.329E-5	0.01809	0.739	1.7684	1.0061
250	1.488E-5	0.02227	0.722	1.4128	1.0053
300	1.983E-5	0.02624	0.708	1.1774	1.0057
350	2.075E-5	0.03003	0.697	0.9980	1.0090
400	2.286E-5	0.03365	0.689	0.8826	1.0140

(Adapted from Sucec, 1985.)

Air is the fluid of interest in environmental control problems, and the temperature range involved is usually quite narrow. This has permitted simplified forms of Equation 3-33 to be developed for specific applications, forms in which fluid property values are used at standard conditions. More general relationships can be found in the *ASHRAE Handbook of Fundamentals*, for example. Table 3-2 contains equations for natural convection useful in environmental control applications, for dry air at standard atmospheric pressure and 20 C, and have been adapted from the *ASHRAE Handbook of Fundamentals*.

The determination whether flow is laminar or turbulent is based on the GrPr product. Laminar range equations apply for GrPr between 10^4 and 10^8, and the turbulent equations apply for GrPr between 10^8 and 10^{12}. For standard air conditions, GrPr can be approximated by

$$GrPr = 10^8 L^3 \Delta t, \qquad (3\text{-}34)$$

where L is in meters and Δt in K.

Examples 3-7, 3-8, and 3-9 demonstrate uses of convective heat transfer calculations. The examples apply to air ducts and pipes. Convective heat transfer to and from walls of buildings is treated differently, as we will see later in the text.

Table 3-2. Natural convention heat transfer coefficients.

Vertical plates

| laminar range | $h = 1.42(\Delta t/L)^{0.25}$ | (3-35) |

| turbulent range | $h = 1.31(\Delta t)^{0.33}$ | (3-36) |

Horizontal plates facing upward when cooled or downward when heated (always laminar convective heat transfer)

$$h = 0.59(\Delta t/L)^{0.25} \qquad (3\text{-}37)$$

Horizontal plates facing upward when heated or downward when cooled

| laminar range | $h = 1.32(\Delta t/L)^{0.25}$ | (3-38) |

| turbulent range | $h = 1.52(\Delta t)^{0.33}$ | (3-39) |

Horizontal cylinders

| laminar range | $h = 1.32(\Delta t/L)^{0.25}$ | (3-40) |

| turbulent range | $h = 1.24(\Delta t)^{0.33}$ | (3-41) |

Vertical cylinders can be treated as vertical plates.

(Δt in K, L in m, and h in W/m^2K)

Example 3-7

Problem: Consider a horizontal heating pipe for a greenhouse. The pipe carries warm water and releases heat into the greenhouse air by convective heat transfer. The outside diameter of the pipe is 50 mm, and its surface temperature is 80 C. The greenhouse air temperature is 20 C. What will be the surface convective heat transfer coefficient, and what will be the rate of convective heat loss from the pipe?

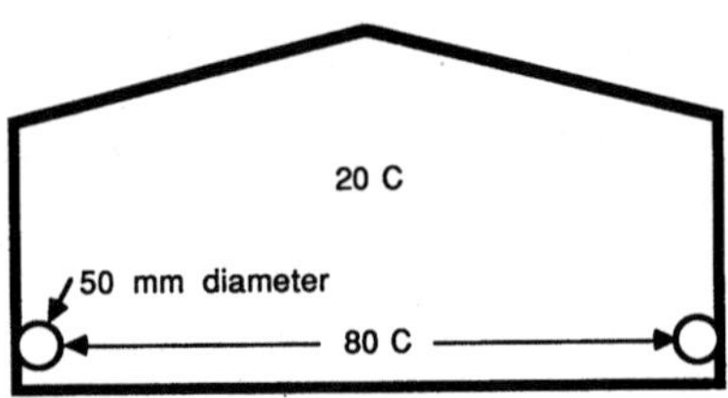

Solution: The pipe can be considered a horizontal cylinder (no fins on the pipe were mentioned). However, to know which of the equations in Table 3-2 to use, we must determine whether the convection will be laminar or turbulent. From Equation 3-34,

$$\text{GrPr} = 10^8 (0.050 \text{ m})^3 (80 \text{ C} - 20 \text{ C})$$
$$= 0.75\text{E} + 6$$

which is within the laminar range. The convective heat transfer coefficient can thus be calculated as (using Equation 3-40)

$$h = 1.32(60 \text{ K} / 0.050 \text{ m})^{0.25}$$
$$= 7.8 \text{ W/m}^2\text{K}.$$

The rate of heat loss can be determined by rearranging Equation 3-30 as follows:

$$q'' = h\Delta t \qquad\qquad (3\text{-}42)$$
$$= (7.8 \text{ W/m}^2\text{K})(60 \text{ K}) = 470 \text{ W/m}^2.$$

Heat loss from pipes is often expressed per unit length of pipe rather than unit area. Each meter of pipe length corresponds to

$$A = 2\pi rL = \pi(0.050 \text{ m})(1.0 \text{ m}) = 0.157 \text{ m}^2$$

thus,

$$q = (470 \text{ W/m}^2 \text{ K})(0.157 \text{ m}^2/\text{m}) = 74 \text{ W/m}.$$

(This has been a calculation based on basic considerations of convective heat transfer. It should be noted that heat losses from greenhouse heating pipes are usually calculated based on data provided by manufacturers of heating systems, data which have been obtained from experiments, which apply to the specific arrangement of pipes being considered in a design, and which include both convective and radiative heat transfer.)

Example 3-8

<u>Problem:</u> Consider a horizontal sheet metal duct which carries heated air from a furnace to a heated space. The outside diameter of the duct is 0.5 m, and its outside surface temperature is 80 C. The air temperature in the space through which the duct passes is 20 C. What will be the surface convective heat transfer coefficient and the rate of heat loss by convection from the duct?

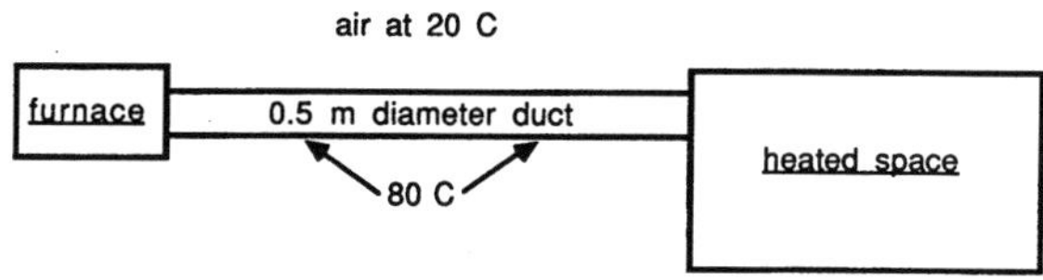

<u>Solution:</u> This example is similar to Example 3-7, only the dimensions have changed. Again, first check to determine whether laminar or turbulent conditions apply,

$$GrPr = 10^8 (0.5 \text{ m})^3 (80 \text{ C} - 20 \text{ C})$$
$$= 3.0E+9,$$

which is in the turbulent range. The convective heat transfer coefficient can, thus, be calculated by

$$h = 1.24(60 \text{ K})^{0.33}$$
$$= 4.79 \text{ W/m}^2\text{K}.$$

The rate of convective heat transfer can be calculated as in Equation 3-42,

$$q'' = (4.79 \text{ W/m}^2\text{K}) (60 \text{ K})$$
$$= 287 \text{ W/m}^2.$$

Each meter of duct has a surface area of

$$A = \pi(0.5 \text{ m})(1.0 \text{ m}) = 1.57 \text{ m}^2,$$

thus, the heat loss by convection is

$$q = (287 \text{ W/m}^2)(1.57 \text{ m}^2/\text{m}) = 451 \text{ W/m}.$$

Compare this example to Example 3-7. Airflow has changed from laminar to turbulent, but the convective heat transfer coefficient has decreased. It is not necessarily true that the average (which is what we have calculated) convective heat transfer coefficient will increase with the onset of turbulence, if the turbulence is caused simply by having a larger solid object. Of course, the much larger size of the air duct leads to a significantly larger rate of heat loss per meter length if not per unit area of duct.

Example 3-9

<u>Problem:</u> A hot water pipe is insulated to prevent heat loss from hot water as it flows between the water heater and the point of use.

The pipe's outside diameter is 100 mm, the insulation thickness is 50 mm, the pipe temperature is essentially that of the hot water, 90 C, and air temperature surrounding the pipe is 10 C.

The thermal conductivity of the insulation on the pipe is 0.05 W/mK.

Calculate the heat flux from the surface of the insulation, and the heat loss per meter of pipe.

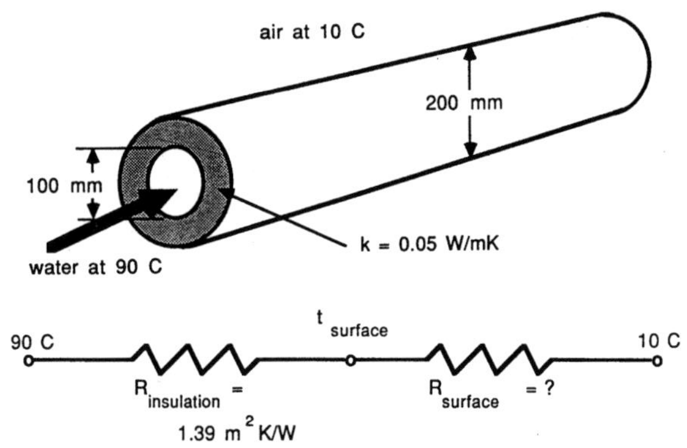

<u>Solution:</u> Inherent in this problem are assumptions that: (a) the pipe wall does not contribute a significant thermal resistance to the path of heat loss from the water to the ambient air, (b) there is little thermal resistance due to the convective heat transfer process from the water to the inside surface of the pipe, and (c) contact between the pipe and insulation is sufficiently good that there is little contact thermal resistance.

Checking the first assumption is an exercise left for practice. Hint: Use the thermal resistance equations for cylindrical coordinates. Convective resistances from water to a surface are approximately two orders of magnitude less than from air to a surface. Thus, in this problem, the second assumption is reasonable. The third assumption must be based on knowledge of how the insulation will be applied to the pipe; we will assume it will be wrapped tightly and taped.

This problem can be viewed as a thermal circuit with two resistances in series — the insulation, and the convective heat transfer resistance between the surface of the insulation and the air (a <u>surface resistance</u>). The resistance of the insulation can be calculated using Equation 3-24

$$R_{insulation} = (\ln(0.10 \text{ m} / 0.05 \text{ m}))/(2\,\pi)(0.05 \text{ W/mK})(1.0 \text{ m})$$
$$= 2.21 \text{ m K/W}.$$

Recall this thermal resistance is based on a unit length of pipe. Convective thermal resistance is based on unit area. We must decide which basis to use. Either will work, but for this example use unit area. The convective thermal resistance is based on the area of the outside of the insulation layer; the insulation resistance will be based on the same area.

Each meter length of pipe has a surface area of

$$A = \pi(0.20 \text{ m})(1.0 \text{ m}) = 0.628 \text{ m}^2/\text{m}$$

Therefore, the unit area thermal resistance (based on the surface area) is

$$R_{insulation} = (2.21 \text{ m K/W})(0.628 \text{ m}^2/\text{m}) = 1.39 \text{ m}^2\text{K/W}.$$

The thermal circuit is expressed as an electrical analog, and the surface resistance can be determined as the inverse of the convective heat transfer coefficient found using one of the equations in Table 3-2. The question is, which equation? That depends on whether the flow will be laminar or turbulent, and that is a function of the insulation surface temperature. But the insulation surface temperature is a function of the surface resistance. We are in a circle. To approach such a problem, <u>assume</u> a condition, solve the problem using that assumption, and then <u>check the assumption</u>. If the assumption is found to have been incorrect, it is changed and the problem is solved again.

To begin the solution, assume conditions are laminar. An energy balance written for the outside surface of the insulation is

$$h_{surface}(t_{surface} - 10 \text{ C}) = (90 \text{ C} - t_{surface})/ R_{insulation}$$

With the assumption of laminar airflow,

$$h_{surface} = 1.32((t_{surface} - 10 \text{ C})/0.2 \text{ m})^{0.25}$$
$$= 1.97(t_{surface} - 10 \text{ C})^{0.25}.$$

The energy balance can be rewritten as

$$1.97(t_{surface} - 10)^{1.25} = (90 - t_{surface})/1.39.$$

Although this is a single equation in one unknown, it is nonlinear. A simple solution technique is to use trial and error, searching for a value of surface temperature which balances the energy balance. Computers and programmable calculators are well suited to this type of search.

This equation is sufficiently simple that it is most easily solved using a calculator. One approach is to rewrite the energy balance as

$$2.74(t_{surface} - 10)^{1.25} + t_{surface} = 90$$

and search for values of surface temperature until one is found such that the left hand side (LHS) of the equation equals 90. Such a search sequence is

$t_{surface}$	LHS
20 C	68.7
25	105.9
23	90.6
22.9	89.9
22.92	90.0

Considering significant digits, a surface temperature of 23 C is estimated. Now check the assumption of laminar flow.

$$\begin{aligned} GrPr &= 10^8(0.2 \text{ m})^3(23 \text{ C} - 10 \text{ C}) \\ &= 1.1E + 7 \end{aligned}$$

This is within the laminar range; the initial assumption was correct. Several procedures can now be used to calculate heat flux from the pipe. One way is to calculate the surface convective thermal resistance, the total thermal resistance, and the heat flux.

$$\begin{aligned} h_{surface} &= 1.32((23 \text{ C} - 10 \text{ C}) / 0.2 \text{ m})^{0.25} \\ &= 3.75 \text{ W/m}^2\text{K}, \end{aligned}$$

and $R_{surface} = 1/3.75 \text{ W/m}^2\text{K} = 0.267 \text{ m}^2\text{K/W}$.

The total series resistance is

$$R_{total} = 1.39 \text{ m}^2\text{K/W} + 0.267 \text{ m}^2\text{K/W} = 1.66 \text{ m}^2\text{K/W},$$

and the heat flux is

$$q'' = (90 \text{ C} - 10 \text{ C}) / 1.66 \text{ m}^2\text{K/W} = 48 \text{ W/m}^2.$$

Each meter length of pipe has 0.628 m^2 surface area, thus, heat loss per meter is

$$q = (48 \text{ W/m}^2)(0.628 \text{ m}^2/\text{m}) = 30 \text{ W/m}.$$

It would be a useful exercise to rework this problem using unit length thermal resistances rather than unit area thermal resistances.

3-3.2. Forced Convection. Forced convective heat transfer is usually greater on a unit area basis than is natural convective heat transfer. Fluid motion caused by a fan or pump is normally more rapid than the rate of motion which can be achieved by thermal buoyancy. As an example of forced convection, air in the heating duct of Example 3-8 moves because of a fan and exchanges thermal energy with the duct wall by forced convective heat transfer. Many forced convection situations are described in heat transfer texts, only a few apply frequently to environmental control calculations in agricultural buildings.

The convective heat transfer equation for forced convection is the same as for natural convection,

$$q'' = h\Delta t,$$

however, the equations which provide values for h differ from those for natural convective heat transfer. The Nusselt number and Reynolds number, Re, are the dimensionless ratios important in forced convective heat transfer. The Reynolds number can be interpreted as the ratio of momentum forces to viscous forces and expresses the level of turbulence,

$$Re = \rho VL / \mu, \tag{3-43}$$

where ρ is mass density, μ is dynamic viscosity, L is a characteristic length, and V is the averaged velocity of fluid flow, $V = V/A$.

Environment control applications usually involve air, and forced convective heat transfer applies to airflow inside ducts, for example. The flow is invariably turbulent for realistic situations. For airflow inside a duct at standard atmospheric pressure, the following simplified equation applies,

$$h = c\, G^{0.8} / D^{0.2}. \tag{3-44}$$

In Equation 3-44, c is a coefficient computed from the thermal properties of air and is a function of temperature. Table 3-3 can be used to estimate its value. The parameter G is the mass flow of air in the duct, per unit cross-sectional area of duct, kg/m^2s,

$$G = \rho V. \tag{3-45}$$

The parameter D is the hydraulic diameter of the duct, m,

$$D = 4(\text{area}) / (\text{perimeter}). \tag{3-46}$$

The hydraulic diameter of a round duct equals its physical diameter. Example 3-10 illustrates an application of Equation 3-44, and Example 3-11 is an extended problem which combines forced convective heat transfer and heat conduction.

Generalized correlations, and equations for other fluids and flow situations, can be found, for example, in the *ASHRAE Handbook of Fundamentals*.

Example 3-10

<u>Problem:</u> Air flows through a rectangular heating duct, 0.3 m by 0.6 m. The air velocity is 5 m/s, air density is 1.3 kg/m³, and air temperature is 35 C.

What is the convective heat transfer coefficient between the heated air and the duct wall?

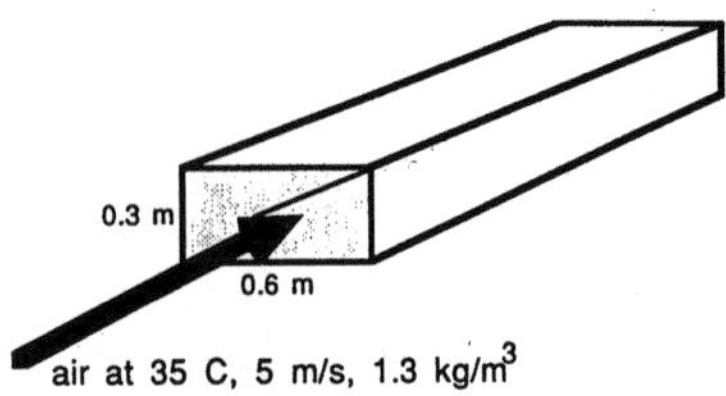

<u>Solution:</u> To use Equation 3-44, three parameters are needed. The first is c, and its value can be obtained by interpolation from Table 3-3. For a temperature of 35 C, c = 3.23.

The mass flow rate is

$$G = (1.3 \text{ kg/m}^3)(5 \text{ m/s}) = 6.5 \text{ kg/m}^2\text{s},$$

and the hydraulic diameter is

$$D = 4(0.18 \text{ m}^2)/(1.8 \text{ m}) = 0.4 \text{ m}.$$

The convective heat transfer coefficient is

$$h = (3.23) (6.5 \text{ kg/m}^2\text{s})^{0.8}/(0.4 \text{ m})^{0.2}$$
$$= 17.3 \text{ W/m}^2\text{K}.$$

Table 3-3. Coefficient c in Equation 3-44 (SI units).

Temperature C	c
-18	3.09
4	3.18
27	3.21
49	3.26
71	3.32
93	3.37

Adapted from the *ASHRAE Handbook of Fundamentals* : for h in W/m²K; G in kg/m² s; D in m. An approximation of the data is the equation c = 3.14783 + 0.00240267t.

Example 3-11

<u>Problem:</u> Air is heated in a furnace and distributed to a heated space through a round sheet metal duct at a volumetric flow rate of 4 m³/s. The duct diameter is 0.8 m and the outer surface of the duct is covered with 10 mm of expanded polyurethane having a thermal conductivity of 0.023 W/mK.

The duct is 50 m long and passes through an unheated space where air temperature is 5 C. The surface resistance outside the duct insulation is 0.1 m²K/W and includes both convective and radiation heat transfer. Density of the heated air is expected to be 0.9 kg/ m³.

If air leaves the furnace at 60 C, what will be its temperature at the end of the 50 m long duct? At what rate will heat be lost from the heated air?

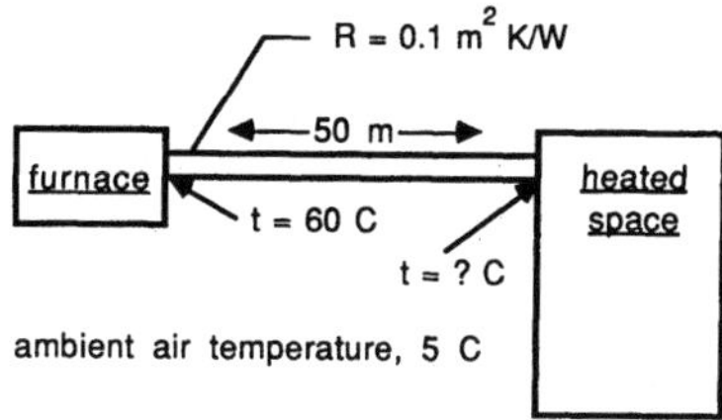

<u>Solution:</u> The solution of this problem is not simple, but the example can be solved using only the heat transfer principles covered thus far.

Two steps are required. First, the thermal resistance between air inside the duct and the air outside must be calculated. Then air temperature change along the length of the duct must be determined.

The series thermal circuit from inside the duct to outside is

60 C ——WW——WW——WW—— 5 C
R_{inside}   R_{wall}   R_{outside}

and $R_{outside}$ is given as 0.1 m²K/W.

The unit area convective heat transfer coefficient at the inside surface of the duct may be determined using Equation 3-44. At 60 C, c in Equation 3-44 is found in Table 3-3 by interpolation, c = 3.29, and

$$D = 0.8 \text{ m},$$
$$G = (0.9 \text{ kg/m}^3) (4 \text{ m}^3/\text{s})/(\pi) (0.4^2)$$
$$= 7.16 \text{ kg/m}^2\text{s, and}$$

$$h = (3.29)(7.16 \text{ kg/m}^2\text{s})^{0.8}/(0.8 \text{ m})^{0.2}$$
$$= 16.6 \text{ W/m}^2\text{K}.$$

The unit area convective resistance is the inverse,

$$R_{inside} = 1 / 16.6 \text{ W/m}^2\text{K}$$
$$= 0.060 \text{ m}^2\text{K/W}.$$

The third resistance in the series thermal circuit is that of the insulation and duct wall. It was stated earlier that a sheet metal wall contributes little to the thermal resistance of an insulated duct, thus, only the insulation will be considered.

The two convective resistances have been calculated on a unit area basis. The conductive resistance must have the same basis to be comparable, but the conductive resistance equation (Equation 3-24) is based on length. A unit area of duct (outside surface area) has a length of

$$L = 1.0/(0.8\pi) = 0.398 \text{ m}.$$

Equation 3-24 is used to calculate the thermal resistance of the wall (insulation) for a duct length of 0.398 m (r_o= 0.41 m, r_i = 0.40 m)

$$R_{wall} = (\ln(0.41 / 0.40))/(2\pi)(0.023)(0.398)$$
$$= 0.429 \text{ m}^2\text{K/W}.$$

(It could be anticipated that a calculation of thermal resistance of the wall in cartesian coordinates might be adequate for this example because the insulation thickness is much less than the radius of the duct. In fact, $R = L/k = 0.01$ m/ 0.023 W/mK = 0.435 m^{2}K/W is very close.)

A second correction (although small in this example) will be made. The outside surface convective resistance and the wall thermal resistance have been based on the outside surface area. The convective resistance of the inside surface has been based on the inside area. The two areas are not equal. Although the difference in this example is small, it can be significant in other problems.

The correction is

$$R_{inside}(\text{corrected}) = (0.060 \text{ m}^2\text{K/W})(0.82 \text{ m} / 0.80 \text{ m})^2$$
$$= 0.062 \text{ m}^2\text{K/W}.$$

One would expect this increase intuitively. There is slightly less than a unit area of surface inside the duct for each unit area outside. The resistance for an area less than one unit should be greater than the resistance for a unit area, and areas scale by the square of the ratio of their diameters.

The total resistance of the series thermal circuit is

$$R_{total} = 0.062 + 0.429 + 0.10 = 0.591 \text{ m}^2\text{K/W}$$

based on surface area outside the insulation.

The next step, to find the temperature change of the air, requires an application of calculus. The temperature difference between air inside the duct to outside changes continuously along the length of the duct. A simple heat loss equation will not apply (unless the change is very slight, a situation we do not know at this point).

Consider an elemental length of the duct. Heat loss from air as it traverses the elemental length must equal the transfer of heat through the wall to the outside air. This simple energy balance can be written in integral equation form, solved, and used to determine air temperature at any point along the duct.

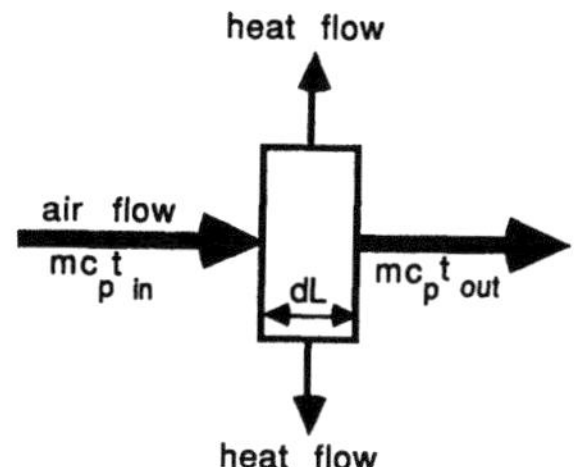

Heat transfer through the wall of the elemental length can be written (Equation 3-23) as

$$q = A\Delta t / R; \quad A = \pi DdL = 0.82\pi dL = 2.58 dL.$$

We have calculated the unit area thermal resistance, $R_{total} = 0.591 \text{ m}^2\text{K/W}$, thus,

$$q = 4.36(t_{air} - 5 \text{ C})dL.$$

This thermal exchange must be balanced by heat loss from the mass of air flowing through the element, m,

$$q = - mc_p dt_{air}; \quad m = (0.9 \text{ kg/m}^3)(4 \text{ m}^3\text{/s}) = 3.6 \text{ kg/s}.$$

The negative sign is introduced so a positive heat loss is associated with a negative temperature change.

If the specific heat of air within the duct is approximated as 1006 J/kgK, heat loss can be written as

$$q = - 3620 dt_{air}.$$

Heat loss must equal heat gain, thus,

$$4.36\ (t_{air} - 5)\ dL = -\ 3620\,dt_{air}$$

which can be rearranged and integrated along the 50 m length of the duct in the form

$$\int_{60\,C}^{t\,exit} \frac{dt_{air}}{(t_{air} - 5)} = -\int_{0\,m}^{50\,m} 0.112E\text{-}2\ dL$$

and solved as

$$t_{exit} = 5 + 55\ \exp(-0.056) = 57.0\ C.$$

The energy loss equation can be used again to estimate the rate of heat loss from the air where Δt is now a temperature change not a temperature difference.

$$\begin{aligned}
q &= mc_p\Delta t \\
&= (3.6\ kg/s)(1006\ J/kgK)(60\ C - 57.0\ C) \\
&= 10{,}900\ W\ (or\ 10.9\ kW).
\end{aligned}$$

(A natural next question is whether the cost of added insulation would be balanced by the value of heat energy saved.)

In general terms, air temperature in a process such as this example can be calculated from

$$t_{exit} = t_{ambient} + (t_{initial} - t_{ambient})\ \exp(-A/mc_pR)$$

where $t_{ambient}$ is ambient air temperature, $t_{initial}$ is the temperature of air entering the duct, A is the surface area of the duct, m is the mass flow rate of air through the duct, c_p is the specific heat of air, and R is the unit area series thermal resistance of the heat transfer path from inside the duct to outside.

Because of the small temperature change of the warm air as it traverses the duct, we could have avoided the differential equation approach and used a constant temperature difference of (60 C - 5 C = 55K) to calculate heat loss from the air. However, it is useful to see the general approach for use in other applications where temperature changes might be greater.

The simpler approach provides a check on the accuracy of the first approach. The unit area series thermal resistance is 0.591 m^2K/W, and the total area of heat transfer is $\pi(0.82\ m)(50\ m) = 129\ m^2$.

Thus
$$\begin{aligned}
q &= A\Delta t\,/\,R = (129\ m^2)(55\ K)\,/\,0.591\ m^2K/W \\
&= 12{,}000\ W
\end{aligned}$$

which is somewhat greater than the first estimate because air has not been permitted to cool along the length of the duct.

3-4. Radiation Heat Transfer

3-4.1. General. As has been stated, all objects at temperatures above absolute zero emit thermal radiation. Emission is over a wavelength band, as shown in Figure 3-5. The emissive power shown in Figure 3-5, W/micron, is the thermal energy emitted within each micron waveband. The total emitted thermal energy is the integral or the area under the curve. Approximately a quarter of the emitted thermal energy is at wavelengths shorter than the maximum; three-quarters is at wavelengths above the maximum. As temperature increases, the peak becomes more sharply defined.

The wavelength for peak emission intensity is found using Wien's law,

$$\lambda_{max} = 2898 \, / \, T, \tag{3-47}$$

where λ_{max} is wavelength in microns and T is the surface temperature, K, of the emitting object. In Figure 3-5, the curve for a temperature of 300 K peaks at a wavelength of 9.66 microns, for example. Thermal radiation from objects at earth temperature is loosely referred to as having a wavelength of approximately 10 microns. It is often termed "long-wave" or "low-temperature" thermal radiation. The curves in Figure 3-5 show the peak emission of objects at earth temperature is at approximately 10 microns, but significant emission occurs well below and well above this wavelength. Solar radiation peaks at 0.6 microns which represents an effective emitting temperature of 4800 K.

The radiant fluxes emitted from objects at temperatures shown in Figure 3-5 equal the integrals of the curves shown, and can be calculated using the Stephan-Boltzmann relationship,

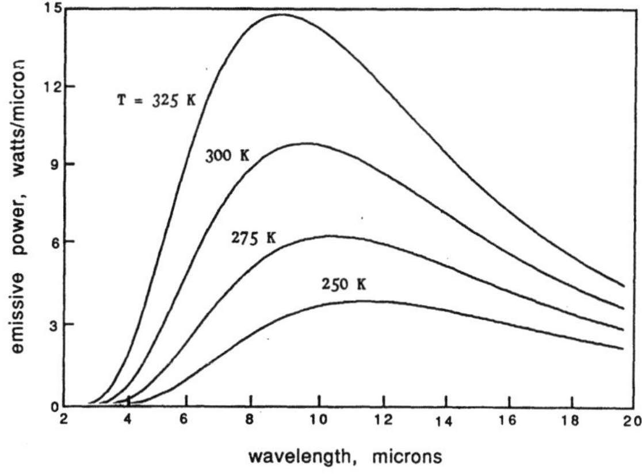

Figure 3-5. Monochromatic emissive power of black bodies at 250, 275, 300, and 325 K.

$$q'' = \sigma T^4. \tag{3-48}$$

The constant, σ, is the Stephan-Boltzmann constant, 5.6697E - 8 W/m^{2}K^4.

The intensity with which earth-temperature thermal radiation is emitted is frequently overlooked. Consider Figure 3-6, in which is graphed thermal flux emitted according to Equation 3-48. Also graphed is the solar constant, which is the intensity of direct-normal solar radiation just outside the earth's atmosphere. The intensity of direct normal radiation on the earth's surface is seldom more than 70% of the solar constant, and then only at high altitudes on very clear, dry days. One-half the solar constant is more typically received on a sunny day, and then only at midday. Figure 3-6 demonstrates that, at temperatures frequently dealt with in building environments (280 to 300 K), the intensity of emitted thermal radiation is approximately one-third the solar constant and compares to solar intensity at sea level on all but the clearest days. So why doesn't everyone freeze to death by loss of thermal radiation? The answer lies in radiation heat transfer calculations, and the exchange of thermal radiation, not just the loss.

3-4.2. Emitted Thermal Radiation. Equation 3-48 applies to a perfect emitter, frequently called a <u>black body</u>. However, no object is a perfect emitter; real substances are characterized by an efficacy of radiant emission, the emittance ε. Radiant energy flux from real objects is calculated from

$$q'' = \varepsilon \sigma T^4. \tag{3-49}$$

Emittance values are normally a function of wavelength. That is, an object will radiate thermal energy with a different efficacy at one temperature than at another. If emittance is not a function of wavelength, the object is termed a <u>gray body</u>. Emittance does not change rapidly as a function of wavelength, thus gray body radiation is assumed to apply in environmental control calculations involving thermal radiation, i.e., calculations over a relatively narrow waveband.

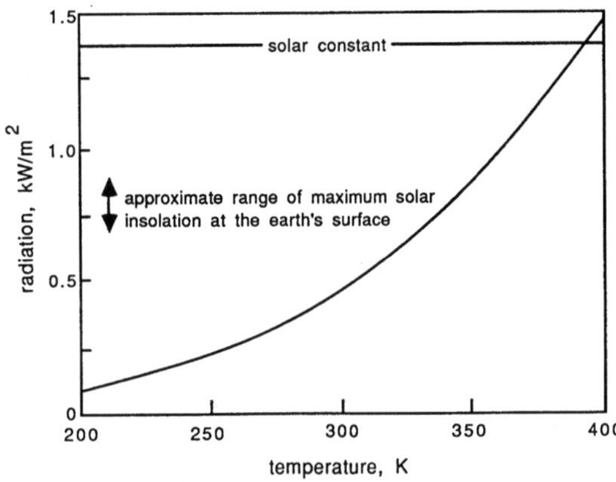

Figure 3-6. Thermal radiation emitted from a black body as a function of its temperature.

It should be emphasized that thermal radiation is a phenomenon determined by the surfaces of objects. Radiation properties of an object are determined by its surface to a depth of only several wavelengths of the radiation involved. As a consequence, thermal radiation properties of a surface can be changed by simply painting over it. Thermal radiation emittance data for common materials can be found in Appendix 3-3. An application of thermal radiation calculations is in Example 3-12.

Example 3-12

Problem: A steam heating pipe in a greenhouse has a surface temperature of 90 C. The surface has been painted with aluminized paint, $\varepsilon = 0.45$. What is the radiant flux leaving the surface, and by how much would the flux change if the pipe were repainted with an oil base or latex paint having an emittance of 0.95?

Solution: The Stephan-Boltzmann equation, 3-49, applies. The pipe surface temperature is 90 C + 273.15 = 363.15 K. The heat flux with the aluminized paint is

$$q'' = (0.45)\,(5.6697E - 8)\,(363.15)^4$$
$$= 444 \text{ W/m}^2.$$

When the surface emittance changes to 0.95, the flux increases to

$$q'' = (0.95)\,(5.6697E - 8)\,(363.15)^4$$
$$= 937 \text{ W/m}^2,$$

which is an additional 493 W/m^2.

The heating pipe also loses thermal energy by convection, but this comparison demonstrates the importance of emittance in determining the effectiveness of steam pipe heating systems.

Field experiments have demonstrated a 15 to 20% improvement of steam heating system heat delivery when aluminized paint is covered with ordinary oil or latex paint. Why are so many radiators in homes and steam pipes in greenhouses painted with aluminum paint? There seem to be no reasons other than aesthetics and custom.

3-4.3. Reflected and Transmitted Thermal Radiation. Reflectance, another thermal radiation property, is defined as the ratio of thermal radiation which irradiates a surface and is reflected to the total irradiation upon the surface (irradiation is defined as the total radiation incident on a surface per unit area and unit time).

A third property, transmittance, is defined as the fraction which is transmitted through the receiving object. Absorptance, a fourth property, describes the fraction absorbed by the surface (and which is converted to thermal energy). Conservation principles dictate the sum of reflectance, transmittance, and absorptance must be 1.0.

3-4.4. Absorptance and Emittance. Absorptance and emittance are opposites, physically, but they are numerically equal for a given surface and for radiation at the same wavelength. A surface which has an emittance of 0.6 for the wavelength which is characterized by that surface's temperature will also have an absorptance of 0.6 for radiation having the same wavelength. For example, white paint has a solar absorptance which is low – most solar radiation striking white paint is reflected. However, the table in Appendix 3-3 shows white paint has a thermal radiation emittance greater than 0.9, it emits nearly as efficiently as does a black body when it is at earth temperatures. This value of emittance applies to wavelengths near 10 microns.

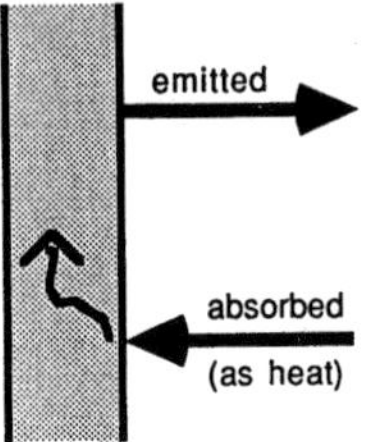

The equality of emittance and absorptance at the same wavelength means white paint also has an absorptance greater than 0.9 for thermal radiation at approximately 10 microns wavelength.

If human eyes were sensitive to wavelengths only between 9 and 10 microns what we now call white paint would appear nearly as black as carbon. In fact, almost all objects would be black – all objects except those with metallic surfaces. This is because, as can be seen in Appendix 3-3, only metals have thermal emittances and thereby thermal absorptances less than approximately 0.9 and a surface with an absorptance that high appears black.

3-4.5. Angle Factors. When two objects exchange <u>diffuse</u> thermal radiation, the net exchange is determined by radiation flux leaving each object as well as the radiation <u>angle factor</u> between the two objects. The angle factor is also termed a shape factor or a configuration factor. The angle factor from one object to another can be interpreted as the fraction of radiation leaving the first object intercepted directly by the second. Another way to interpret the angle factor from, for example, object 1 to object 2, is to imagine how much thermal radiation "universe" of object 1 is occupied by object 2.

Any object exchanging thermal radiation with a second must have a non-zero angle factor with the second object. Angle factor values range from 0.0 (no exchange) to 1.0 (exchange only with the second object). An object can have an angle factor to itself if part of the radiation leaving the surface of that object is intercepted directly by another part of the same object (for example, the inside surface of a hemisphere radiates in part to itself).

Each of any two objects involved in a thermal radiation exchange has an angle factor for the other object, and the two angle factors are not equal. However, a reciprocity relationship can be used to relate them,

$$F_{1\text{-}2}A_1 = F_{2\text{-}1}A_2, \tag{3-50}$$

where $F_{1\text{-}2}$ is the angle factor from object 1 to object 2, A_1 is the surface area of object 1, etc.

An angle factor frequently encountered in agricultural engineering environmental analysis applies to thermal radiation exchange between a small object and large surroundings. For example, a cow in a barn and a plant in a greenhouse are small relative to their radiation surroundings. The angle factor from the cow to the walls, ceiling, and floor of the barn is almost 1.0 unless other objects in the barn exchange a significant amount of thermal radiation with the cow. The angle factor from a greenhouse plant to the structural cover of the greenhouse may be approximately 0.5; one-half the radiation "universe" of the plant is the cover (roof and walls). The other half is the greenhouse bench and surrounding plants. By Equation 3-50, the angle factors of the barn to the cow and greenhouse structural cover to the plant are small but not zero.

Note: The sum of angle factors from an object to all other objects (possibly including itself) must equal 1.0. This is an expression of the conservation of thermal radiation. All thermal radiation which leaves an object must be directly intercepted by the other objects with which the first object exchanges thermal radiation.

Extensive tables of angle factors can be found in many heat transfer texts. In Appendix 3-4 are graphs of angle factors for several common thermal radiation exchange situations. Configuration 1 could represent a ceiling and floor exchanging thermal radiation, configuration 2 could represent adjacent walls, a wall and a floor, etc. Configurations 3, 4, and 7 could be assumed to represent a small area exchanging thermal radiation with a ceiling or a specific wall. For example, an animal in a barn could be approximated in shape as a box, with each face of the box exchanging thermal radiation with the surfaces of the barn which it "sees" in a thermal radiation sense, or it could be represented as a sphere. Configurations 5 and 6 could, for example, represent a heating pipe in a greenhouse exchanging thermal radiation with a wall or other large surface.

Examples 3-13 and 3-14 demonstrate angle factor calculations.

Example 3-13

<u>Problem:</u> One pig is in a barn. The pig has a surface area of 2 m^2. The barn is 10 m by 20 m with 3 m high walls. The pig exchanges thermal radiation with the inside walls of the barn. Estimate the angle factor from the pig to the barn and the barn back to the pig.

<u>Solution:</u> A pig in a barn qualifies as a small object in a large space. The radiation temperature of the animal's surroundings is assumed to be everywhere uniform, thus, it is not necessary to consider thermal radiation exchange from each side of the animal with the barn, and calculate separate angle factors for each part of the animal's surface to each part of the inside of the barn. We will neglect the angle factor of the pig's surface to itself. For example, while the pig stands, each leg will exchange thermal radiation with the other legs, the under side of the abdomen, etc. However, we will assume for the sake of the example that this is a relatively insignificant part of the total thermal radiation exchange of the pig.

The angle factor from the pig to the barn (F_{1-2}) can be immediately estimated as 1.0. In return, the surface area of the walls, floor, and ceiling of the barn must be calculated and is 580 m^2. By Equation 3-50

$$\begin{aligned} F_{2-1} &= F_{1-2}A_1/A_2, \\ &= (1.0)(2 \text{ m}^2)/(580 \text{ m}^2), \\ &= 0.00345. \end{aligned}$$

The other object with which the barn exchanges thermal radiation is itself. Because the sum of angle factors must equal unity,

$$F_{2-2} = 1.0 - 0.00345 = 0.99655.$$

Example 3-14

<u>Problem:</u> For the barn described in Example 3-13, determine the angle factor for thermal radiation exchange between the ceiling and floor, and between the 3 m by 10 m end wall and an adjacent 3 m by 20 m sidewall. First use the angle factor graphs and then the angle factor equations in Appendix 3-4 (the angle factor from the floor to ceiling is configuration 1, for example) .

<u>Solution:</u> Appendix 3-4 can be used to determine the angle factor between the ceiling and floor. The ratio b/c is 20/3 or 6.67. The ratio a/c is 10/3 or 3.33. The graph shows the angle factor is approximately 0.65 and is the same from the ceiling to floor as from the floor to ceiling.

The equation corresponding to this angle factor calculation, $X = a/c = 6.67$ and $Y = b/c = 3.33$, is

$$F_{1\text{-}2}(\pi XY/2) = \ln\left(\frac{(1+X^2)(1+Y^2)}{1+X^2+Y^2}\right)^{1/2} + Y(1+X^2)^{1/2}\tan^{-1}\left(\frac{Y}{(1+X^2)^{1/2}}\right)$$

$$+X(1+Y^2)^{1/2}\tan^{-1}\left(\frac{X}{(1+Y^2)^{1/2}}\right) - Y\tan^{-1}Y - X\tan^{-1}X$$

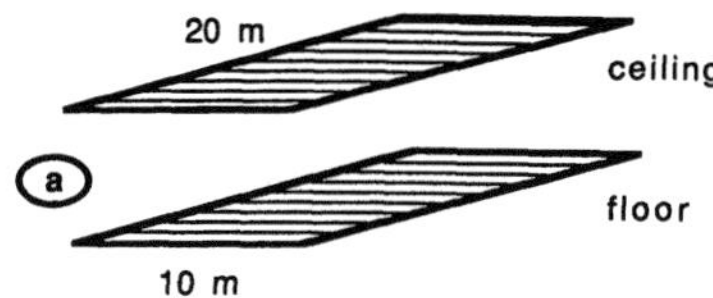

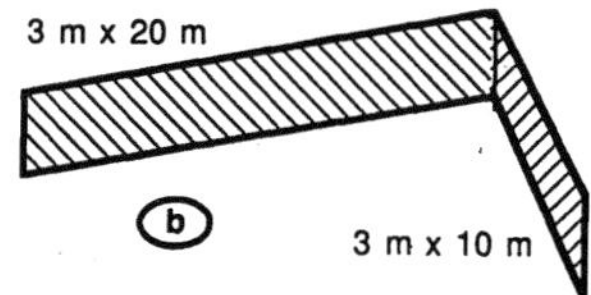

The solution for X and Y values as above is $F_{1\text{-}2} = 0.66$, approximately the same as was determined from the graph. The symmetry of the problem indicates $F_{2\text{-}1} = 0.66$ also.

The angle factor for thermal radiation exchange between adjacent walls can be found from Appendix 3-4 also.

If A_1 is the 10 m long wall, the ratio Y/X is $10/3 = 3.33$ and the ratio Z/X is $20/3 = 6.67$. The angle factor from A_1 to A_2 is approximately 0.13.

If A_1 is the 20 m long wall, the ratio Y/X is 6.67 and the ratio Z/X is 3.33. The angle factor from A_1 to A_2 is approximately 0.06.

Note: There are small inaccuracies from reading the graphs, for the two angle factors should relate through reciprocity according to Equation 3-50,

$$F_{1\text{-}2}/F_{2\text{-}1} = A_2/A_1.$$

This relationship was approximately confirmed (0.13/0.06 approximately equals 60/30).

As an exercise, use the equation for this geometry as given in Appendix 3-4, simplify it for walls at right angles, and calculate the actual value of the angle factor.

It would also be a useful exercise to determine the angle factors from one wall to all other walls and the ceiling and floor and to check whether they add to unity.

Angle factor algebra can be used to determine angle factors for more complex situations than in Examples 3-13 and 3-14. Calculations are based on conservation of thermal radiation. If diffuse thermal radiation leaves surface 1 and is intercepted by surface 2, and if surface 2 is divided into two areas, 2a and 2b, then

$$A_1 F_{1\text{-}2} = A_1 F_{1\text{-}2a} + A_1 F_{1\text{-}2b}. \tag{3-51}$$

The concept embodied in Equation 3-51, along with inventiveness, permits one to determine angle factors for geometries which are seemingly quite different from those listed in angle factor catalogs.

As a simple example of angle factor algebra, consider the situation as shown where the angle factor is desired between areas A_{1b} and A_{2b}.

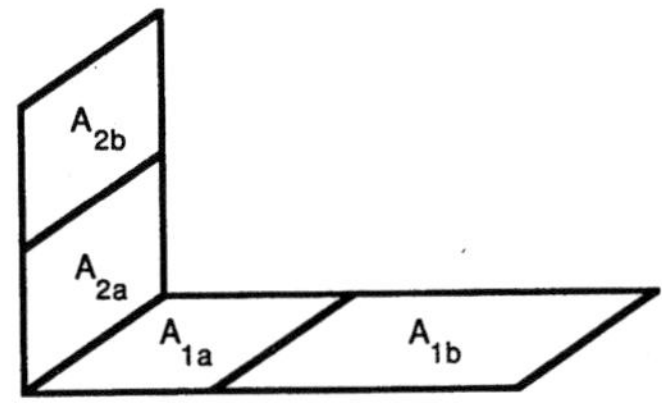

Applying Equation 3-51,

$$A_{1b} F_{1b\text{-}2b} = A_1 F_{1\text{-}2b} - A_{1a} F_{1a\text{-}2b}, \tag{3-52}$$

where $A_1 = A_{1a} + A_{1b}$. Further application of conservation of thermal radiation leads to

$$A_1 F_{1\text{-}2b} = A_1 F_{1\text{-}2} - A_1 F_{1\text{-}2a}, \text{ and} \tag{3-53}$$

$$A_{1a} F_{1a\text{-}2b} = A_{1a} F_{1a\text{-}2} - A_{1a} F_{1a\text{-}2a}. \tag{3-54}$$

Configuration 2 in Appendix 3-4 can be used to determine all angle factors on the right hand sides of Equations 3-53 and 3-54. The areas will be known, thus the two equations can be solved, and in turn, Equation 3-52 used to calculate $F_{1b\text{-}2b}$.

When infinitesimal areas must be integrated to determine finite area angle factors (see configurations 3-6 in Appendix 3-4), the following relations apply:

$$F_{1\text{-}2} = \frac{1}{A_1} \int_{A_1} F_{dA1\text{-}A2} \, dA_1, \text{ and} \qquad (3\text{-}55)$$

$$F_{2\text{-}1} = \int_{A_1} dF_{A2\text{-}dA1}. \qquad (3\text{-}56)$$

The angle factor differential in Equation 3-56 is determined using the equation form of the appropriate angle factor in Appendix 3-4.

Thermal radiation exchanges are frequently ignored in environmental analyses or are incorporated into empirical coefficients as we will see later. However, when situations arise where thermal radiation is important, and the assumptions built into the empirical coefficients do not apply, a radiation exchange analysis can begin using the information provided here. More detailed thermal radiation exchange information and procedures can be found for example, in Sparrow, and Cess (1978).

3-4.6. Thermal Radiation Exchange. Thermal radiation exchange can be calculated once angle factors are known. A simple situation is the case of a small object in large surroundings. This is simple because once thermal radiation leaves the small object (surface number 1) it will be absorbed by the large surroundings (surface number 2). In other words, surface 2 is thermally black. Even if the absorptance of surface 2 is low, multiple reflections from place to place on surface 2 will eventually absorb all the thermal radiation. Very little will be reflected back to surface 1 because of the very small angle factor from 2 to 1. For this situation, the net exchange of thermal radiation can be calculated,

$$q_{1\text{-}2} = A_1 \varepsilon_1 \, \sigma \, (T_1^4 - T_2^4). \qquad (3\text{-}57)$$

Example 3-15 applies Equation 3-57 to an environmental analysis situation.

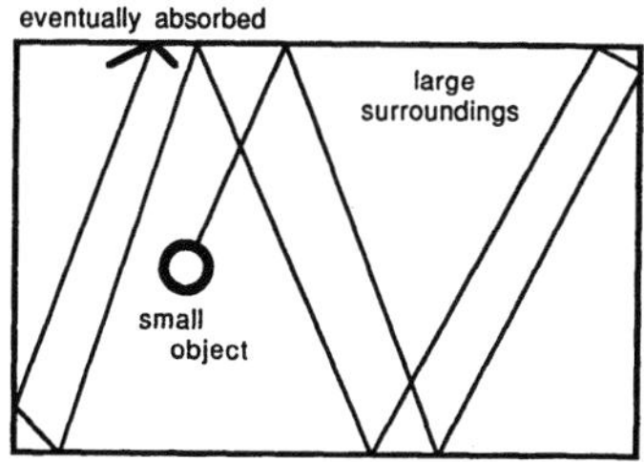

Example 3-15

<u>Problem:</u> Consider again Example 3-13. If the pig's skin temperature is 35 C and the average temperature of the radiant surroundings (walls, etc.) is 10 C,

calculate the rate of heat loss by radiation from the pig. The emittance of skin can be assumed to be 0.90.

<u>Solution:</u> This is a direct application of Equation 3-57. The skin area of the pig is 2 m^2, and the pig will be termed surface 1. Surface temperatures are:

$$T_1 = 35 + 273.15 = 308.15,$$
$$T_2 = 10 + 273.15 = 283.15.$$

The heat loss is

$$q = (2 \text{ m}^2)(0.90)(5.6697E - 8)(308.15^4 - 283.15^4)$$
$$= 264 \text{ W.}$$

The total thermal radiation leaving the pig is 920W and is found from Equation 3-49. The difference (920 W - 264 W = 656 W) is what returns from the surroundings. This example shows the importance of the thermal radiation environment of an object and even if indoor air temperature is adequate, explains why people feel cold when sitting near a large window in cold weather. The glass is cold and net thermal radiation exchange with it increases heat loss.

More complicated thermal radiation exchange situations require more sophisticated analyses and are most easily implemented on a computer.

To begin, define radiosity, B, as the total radiation which leaves a surface. It is the sum of what is emitted and what is reflected (plus that possibly transmitted)

$$B = \varepsilon W_b + \rho H_2 (+ \tau H_1) \tag{3-58}$$

where W_b is black body emissive power calculated using Equation 3-47, H is irradiation, ρ is reflectance, and τ is transmittance.

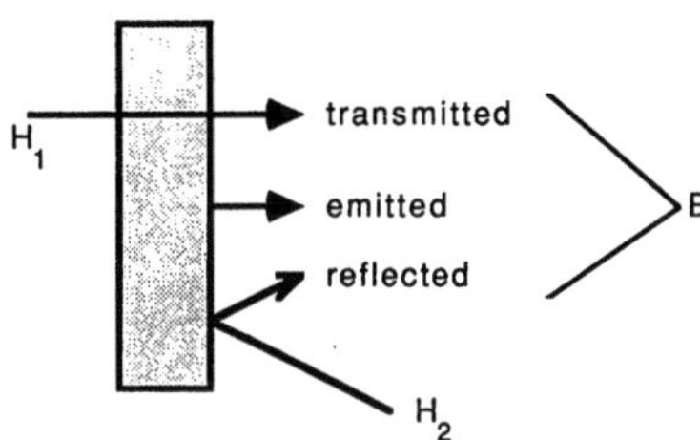

When the surface is opaque, $\rho = (1 - \alpha)$ or $(1 - \varepsilon)$, $\tau = 0$, and

$$B = \varepsilon W_b + (1 - \varepsilon)H. \tag{3-59}$$

The difference between radiosity and irradiation is the net energy flux lost by an object by thermal radiation (defined as positive if leaving the surface),

$$q/A = B - H$$
$$= \varepsilon W_b + (1 - \varepsilon)H - H. \qquad (3\text{-}60)$$

From Equation 3-59,

$$H - (B - \varepsilon W_b)/(1 - \varepsilon), \qquad (3\text{-}61)$$

which can be combined with Equation 3-60 to yield

$$q = \varepsilon A(W_b - B)/(1 - \varepsilon), \qquad (3\text{-}62)$$

Equation 3-62 applies to an object exchanging radiation with all objects in its radiation surroundings. Consider the situation with n isothermal surfaces exchanging radiation as would be the case for inner surfaces of walls in a room. For any one surface, i, the irradiation on i is the sum of radiation received from all other surfaces,

$$H_i A_i = \sum_j F_{j\text{-}i} B_j A_j = \sum_j F_{i\text{-}j} B_j A_i. \qquad (3\text{-}63)$$

$$\text{or } H_i = \sum_j F_{i\text{-}j} B_j. \qquad (3\text{-}64)$$

Summations are taken over all n surfaces participating in the radiation exchange.

This expression for irradiation is substituted into Equation 3-59 to obtain the following set of simultaneous equations relating irradiation on each of the n surfaces.

$$B_1 = \varepsilon_1 \sigma T_1^4 + (1 - \varepsilon_1)\sum_j F_{1\text{-}j} B_j,$$

$$B_2 = \varepsilon_2 \sigma T_2^4 + (1 - \varepsilon_2)\sum_j F_{2\text{-}j} B_j,\text{...and} \qquad (3\text{-}65)$$

$$B_n = \varepsilon_n \sigma T_n^4 + (1 - \varepsilon_n)\sum_j F_{n\text{-}j} B_j.$$

Equation 3-65 can be rearranged as a matrix equation for convenience,

$$
\begin{bmatrix}
1 - (1-\varepsilon_1)F_{1\text{-}1} & -(1-\varepsilon_1)F_{1\text{-}2} & -(1-\varepsilon_1)F_{1\text{-}3}\cdots & -(1-\varepsilon_1)F_{1\text{-}n} \\
-(1-\varepsilon_2)F_{2\text{-}1} & 1 - (1-\varepsilon_2)F_{2\text{-}2} & -(1-\varepsilon_2)F_{2\text{-}3}\cdots & -(1-\varepsilon_2)F_{2\text{-}n} \\
\vdots & & & \\
-(1-\varepsilon_n)F_{n\text{-}1} & -(1-\varepsilon_n)F_{n\text{-}2} & -(1-\varepsilon_n)F_{n\text{-}3}\cdots & 1 - (1-\varepsilon_n)F_{n\text{-}n}
\end{bmatrix}
\begin{bmatrix}
B_1 \\ B_2 \\ \vdots \\ B_n
\end{bmatrix}
$$

$$
=
\begin{bmatrix}
\varepsilon_1 \sigma T_1^4 \\
\varepsilon_2 \sigma T_2^4 \\
\vdots \\
\varepsilon_n \sigma T_n^4
\end{bmatrix}
\qquad (3\text{-}66)
$$

Equation 3-66 is a set of n equations in n unknowns, $B_1, B_2,..B_n$.

Before they can be solved to determine radiosities, all emittances, angle factors, and surface temperatures must be known. Emittances will be known from the choice of materials in the building design. Angle factors can be determined using tables such as in Appendix 3-4 or calculated from the relevant equations. Surface temperatures can be determined using energy balances which will be a later topic in this text. (For now, assume they are known or given.)

Simple matrix analysis procedures suited for computer implementation can be used to solve Equation 3-66. One candidate solution method is Gaussian elimination, perhaps with pivotal condensation to eliminate the possibility of a pivot near zero, and the resulting inaccuracies.

3-5. Mixed Mode Heat Transfer

Situations frequently arise in environmental analyses where conduction, convective, and radiation heat transfer occur simultaneously. Examples are: (a) heat loss through insulation on a heating duct, where loss on the outside surface is both convective and radiative, and (b) heat gain at a roof's upper surface with loss by conduction to the underside and loss on the upper side by convective and radiative heat transfer. Such problems can be solved in a straightforward manner if a proper energy balance is developed.

Typically, an energy balance is written for the surface with an unknown temperature and solved for that temperature. Ambient conditions are completely specified. The energy balance is usually nonlinear and may be solved most readily by trial and error. Trial and error or iterative solutions are not elegant, but they work.

Examples 3-16 and 3-17 illustrate mixed process heat transfer problems which can arise in environmental analysis.

Example 3-16

<u>Problem:</u> The sun shines on the roof of a barn with an insolation of 600 W/m². The absorptance of the roof for solar energy is 0.6.

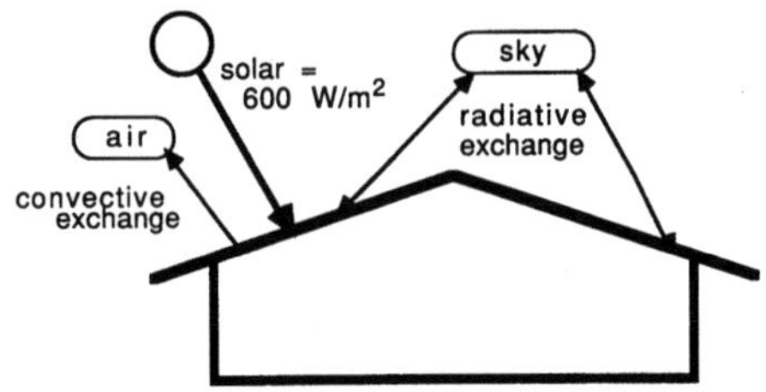

The roof exchanges thermal radiation with the sky. The surface emittance is 0.90, and sky temperature can be approximated by the Swinbank model,

$$T_{sky} = 0.0552 T_{air}^{1.5}, \qquad (3\text{-}67)$$

where temperatures are expressed in Kelvin degrees.

The roof also exchanges thermal energy by convective heat transfer with the outside air, the convective coefficient is 30 W/m^{2}K. Ambient air temperature is 25 C.

The roof is insulated, having an R-value of 2 m^{2}K/W. Air temperature under the roof is 30 C, and the surface resistance between the lower surface of the roof and air inside the barn is 0.2 m^{2}K/W. This surface resistance includes both convective and radiative resistance.

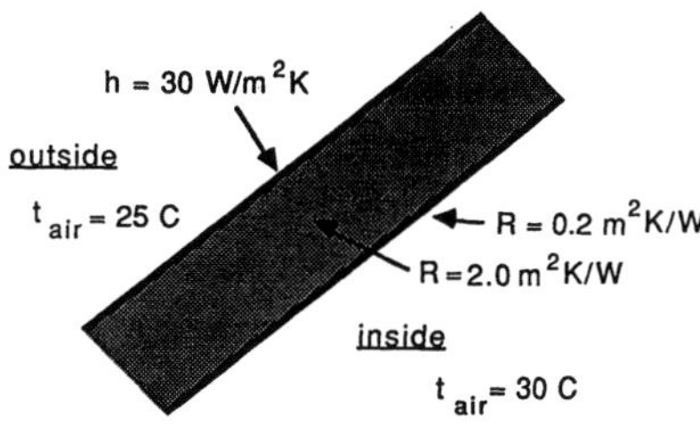

The temperature of the lower surface of the roof, t_{ls}, is important in determining the comfort of animals within the barn; they exchange thermal radiation with the roof. What will be the temperature of the lower surface of the roof with the given conditions?

Solution: All boundary conditions are known except for sky temperature, and a model is available to calculate that,

$$\begin{aligned} T_{sky} &= 0.0552(25\ C + 273.15)^{1.5}, \\ &= 284.19\ K\ (= 11\ C,\ \text{or } 14\ K \text{ below air temperature}). \end{aligned}$$

The temperature of the lower surface of the roof is desired, but before that can be found, the temperature of the upper surface, t_{us}, must be calculated. To determine t_{us} an energy balance on the upper surface of the roof is formed as follows:

$$\begin{aligned} \text{gains} &= \text{losses}, \\ q''_{solar} &= q''_{convective} + q''_{radiative} + q''_{conductive}. \end{aligned}$$

In this energy balance, solar radiation is considered the only gain, other fluxes are losses. The assumptions as to which are gains and which losses are not fixed. The energy balance is correct as long as temperature differences are assigned to agree with the assumed directions of heat transfer.

Absorbed solar flux is calculated as

$$q''_{solar} = (0.60)(600 \text{ W/m}^2) = 360 \text{ W/m}^2.$$

Convective heat loss is

$$q''_{convective} = h\Delta T = 30 \text{ W/m}^2\text{K}(T_{us} - 298.15 \text{ K}).$$

Absolute temperatures will be used in all terms of the energy balance because they are required for radiative heat transfer calculations.

Conductive heat transfer is ($R = 2.2$ from T_{us} to the inside air)

$$q''_{conductive} = \Delta T/R = (T_{us} - 303.15 \text{ K})/2.2 \text{ m}^2\text{K/W}.$$

Radiation heat transfer between the barn's roof and the sky can be considered a situation of a relatively small object in large surroundings. Thus thermal radiation loss to the sky can be written

$$q''_{radiation} = \varepsilon_{us}\,\sigma\,(T_{us}^4 - 284.19^4)$$

$$= (0.9)(5.6697\text{E-}8)(T_{us}^4 - 65.228\text{E} + 8),$$

$$= 5.1\text{E-}8\,T_{us}^4 - 332.7.$$

All the terms can be substituted into the energy balance and rearranged to solve for T_{us},

$$360 \text{ W/m}^2 = 30(T_{us} - 298.15)$$

$$+ (T_{us} - 303.15)/2.2 + 5.1(T_{us}/100)^4 - 332.7$$

or $5.1(T_{us}/100)^4 + 30.4545\,T_{us} = 9775.3.$

Note: For convenience, T_{us} is divided by 100 to eliminate the need to carry exponents of 10 in the equations. A trial and error solution will be used to determine a value for T_{us} such that the left hand side (LHS) of the last equation equals 9775.3.

T_{us}, K	LHS
310	9911.89
305	9729.96
306	9766.23
306.5	9784.39
306.25	9775.31

We have reached a solution, $t_{us} = 306.25 \text{ K} - 273.15 = 33.1 \text{ C}.$

Heat transfer through the roof to air within the barn forms a series thermal circuit from which the temperature of the lower surface of the roof, t_{ls}, can be found.

This is a series thermal circuit, thus, temperature differences scale linearly with resistances. The temperature of the lower surface of the roof is

$$t_{ls} = 33.1\ C + (2.0/2.2)(30\ C - 33.1\ C)$$
$$= 30.3\ C.$$

The lower surface temperature is only slightly above inside air temperature and solar heating has little effect. This is because the roof is well insulated. As an exercise, do the example again using a roof thermal resistance of 0.3 m^2K/W, an uninsulated case.

A second useful exercise would be to calculate the magnitude of each term of the energy balance on the upper surface of the roof and determine whether conductive, convective, and radiative heat losses equal the absorbed solar energy when added together, and compare them as to each one's importance as a means of heat loss from the upper surface of the roof.

Example 3-17

<u>Problem:</u> Greenhouses are frequently heated by steam circulated through iron pipes. The condensing steam contains a great deal of energy which is added from the outer surface of the pipe to the greenhouse by convection to the air and radiation to the interior parts of the greenhouse.

Consider a case where heating pipes with outside diameters of 56 mm are used.

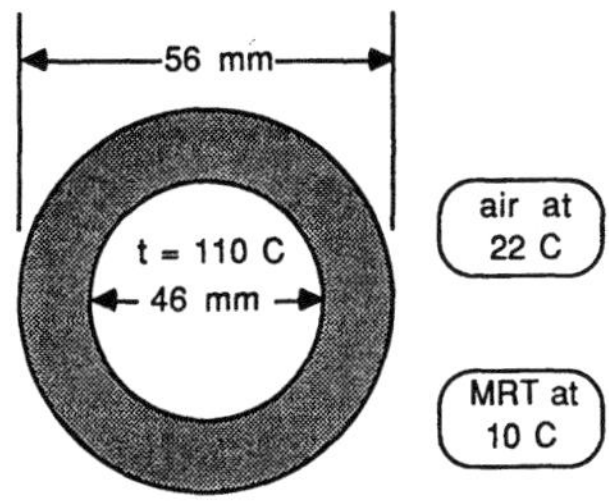

The inside diameter of each pipe is 46 mm. The iron in the pipe has a thermal conductivity of 52 W/mK.

The steam is pressurized and has a condensation temperature of 110° C. Condensation is such a vigorous process that we can assume the inside surface of the pipe equals steam temperature.

The greenhouse air temperature is 22 C and the mean radiant temperature of the surroundings of the heating pipe is 10 C (the pipe is next to a cold outside wall). Mean radiant temperature is the hypothetical temperature which would cause the same net radiative heat transfer with the heating pipe as if the surroundings were all at this mean radiant temperature instead of their many different actual temperatures.

If the heating pipe is painted black and has a surface emittance of 0.95, what is the net thermal exchange between the pipe and the greenhouse? How much of the heat transfer is convective and how much radiative?

Solution: This is a situation of both series and parallel heat transfer involving conductive, convective, and radiative heat transfer. The convective and radiative transfers from the outer surface are in parallel, and the two together are in series with conductive transfer through the pipe wall. The parallel transfer is somewhat different from what we have seen before. Normally, parallel heat transfer is defined as being between the same temperature difference, but here convective heat transfer is to air at 22 C and radiative heat transfer is to surroundings at 10 C. However, the problem is readily solved using energy balance analysis.

The heat transfer network is as shown (with temperature expressed in Kelvin).

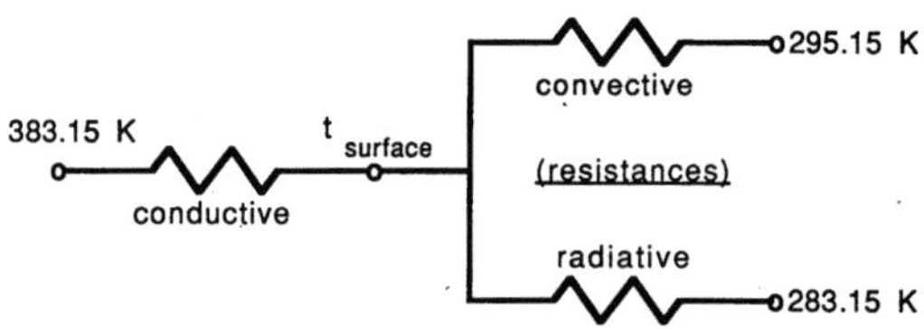

Radiative heat transfer will not be considered a thermal resistance for solution of heat transfer, although in principle it could be. Instead, an energy balance will be developed for the outside surface of the pipe, with the goal of determining the pipe's surface temperature. After that, heat transfer magnitudes will be determined. This strategy is useful to solve many heat transfer problems of this sort, as it was in Example 3-16.

At the pipe's surface,

conductive gain = radiative loss + convective loss.

The conduction heat transfer thermal resistance is found in cylindrical coordinates from

$$R_{conductive} = (\ln(r_o / r_i))/2\pi kL$$
$$= (\ln(28/23))/2\pi(52) \ (1) \quad \text{(per unit length)}$$
$$= 0.000602 \text{ mK/W}.$$

Each meter of pipe has 0.1759 m^2 of surface area. For now, calculations will be on a unit area basis. The unit area (based on the outside surface area) resistance of the pipe wall to conduction heat transfer is

$$R_{conductive} = (0.000602 \text{ mK/W}) (0.1759 \text{ m}^2/\text{m})$$
$$= 0.000106 \text{ m}^2\text{K/W},$$

and heat gain to the surface by conduction is

$$q''_{conductive} = \Delta T / R$$
$$= (383.15 \text{ K} - T_{surface}) / 0.000106 \text{ m}^2\text{K/W}.$$
$$= 9441(383.15 - T_{surface})$$

The pipe will be assumed to radiate as a small object in large surroundings; the angle factor will be assumed to be 1.0, and Equation 3-51 applies. Loss of heat from the surface by radiation is

$$q''_{radiative} = (0.95) (5.6697\text{E} - 8) (T^4_{surface} - 283.15^4)$$
$$= 5.386(T_{surface} / 100)^4 - 346.3$$

Convective heat transfer is by natural means, but it is not clear whether it will be by laminar or turbulent transfer. We can be guided in deciding whether to assume laminar or turbulent conditions by the criterion of Equation 3-34. The laminar and turbulent ranges divide at a GrPr value of 10^8, thus, at the dividing value

$$L^3\Delta T = 1; L = 0.056 \text{ m.} \tag{3-68}$$

If flow is laminar, the temperature difference between the pipe's surface and the air must be no more than

$$\Delta T = 1 / L^3 = 5694 \text{ K.}$$

We can be confident airflow around the pipe, and convective heat transfer, will be laminar.

With laminar heat transfer, the convective coefficient is calculated using Equation 3-40 from Table 3-2,

$$h = 1.32 \ ((T_{surface} - 295.15 \text{ K}) / 0.056 \text{ m})^{0.25},$$
$$= 2.713 \ (T_{surface} - 295.15)^{0.25}.$$

Convective heat transfer from the surface is

$$q''_{convective} = h\Delta T = 2.713(T_{surface} - 295.15)^{1.25}.$$

The energy balance can now be written (for the surface temperature expressed as T_s)

$$9441(383.15 - T_s) = 5.386(T_s/100)^4 - 346.3 + 2.713(T_s - 295.15)^{1.25}$$

and rearranged as

$$5.386(T_s/100)^4 + 2.713(T_s - 295.15)^{1.25} + 9441 T_s - 3{,}617{,}760 = 0$$

which is a single equation with one unknown. As before, a trial and error solution will be used, searching for a value of surface temperature such that the left hand side (LHS) of the rearranged energy balance equals zero. A sequence which leads to the solution is

T_s, K	LHS
382	-9432
383	32
382.99	-63
382.995	-16
382.997	3
382.9967	0.3

In this solution, the value of LHS changes rapidly with small changes of T_s, thus, the value $T_s = 382.9967$ K ($= 109.8467$ C) is very close, and convergence need be pursued no further.

In terms of significant digits, as determined by good engineering practice, 382.9967 has too many, but for the purpose of illustration, all decimal places will be carried to check how well the energy balance is satisfied. Note how in this example the surface temperature must be found with considerable care because of the large influence small errors have on the calculated value of conductive heat gain. To obtain accuracy, the constant in the energy balance (3,617,760) must be carried to this number of significant digits because the convergence criterion in the solution of the rearranged energy balance is of the order of unity.

With the surface temperature found, heat transfer magnitudes can be calculated,

$$q''_{conductive} = 9441(383.15 - 382.9967) = 1542 \text{ W/m}^2,$$
$$q''_{radiative} = 5.386(382.9967/100)^4 - 346.3 = 813 \text{ W/m}^2,$$

and

$$q''_{convective} = 2.713(382.9967 - 295.15)^{1.25} = 730 \text{ W/m}^2.$$

The energy balance is essentially satisfied. Conductive gain to the surface is

1542 W/m^2, and the sum of losses is $813 + 730 = 1543$ W/m^2. This magnitude of error is to be expected, for the conduction heat transfer term is extremely sensitive to small rounding errors. A check to be sure the balance is satisfied is important to prevent inadvertent errors.

Heating per unit length of pipe is frequently desired; such values can be obtained from the above by multiplying each flux by the surface area per meter of pipe, 0.1759 m^2/m.

The temperature difference between the pipe surface and ambient air is much less than that required for turbulent convective heat transfer; that assumption was correct.

Note the loss of heat from the pipe is balanced between convective and radiative. The large radiative component is important in determining the environment of plants near the pipe. They will be subjected to a significantly warmer environment than will those plants shaded from thermal radiation and may exhibit different growth and timing.

The large radiative term is also significant in leading to a means to save heating energy. If the heating pipe is along the outside foundation wall, a reflective (foil-faced) insulation placed on the inside surface of the wall will reflect much of the radiant energy back into the greenhouse instead of allowing the wall to absorb and lose it to the outdoors.

It would be a useful exercise to analyze the heating situation again using a different value of surface emittance for the pipe (0.40 to represent aluminum paint, for example). Would you expect the convective component to change significantly?

A second useful exercise would be to solve the example using the assumption that the pipe wall's thermal resistance is insignificant and the pipe's outside surface temperature equals the steam temperature. Would you expect much difference in the calculated radiative and convective heat transfer rates?

3-6. Program BALANCE

Repeated solutions of energy balances, such as in Examples 3-16 and 3-17, are tedious. Program BALANCE, which is an executable file containing built-in instructions for its use, provides a quicker means to solve the generalized equation contained in the two examples:

$$A1(T/100)^4 + A2(T - A3)^{A4} + A5T + A6 = 0. \qquad (3\text{-}69)$$

For example, in Example 3-17,
   A1 = 5.386,
   A2 = 2.713,

A3 = 295.15,
A4 = 1.25,
A5 = 9441, and
A6 = -3,617,760.

In Example 3-16,
Al = 5.1,
A2 = 0.0,
A3 = 0.0,
A4 = 0.0,
A5 = 30.4545, and
A6 = -9775.3.

Example 3-18

Problem: Repeat Example 3-17 and develop a graph to show the effect of the thermal conductivity of the pipe wall on the surface temperature of the pipe.

Solution: The thermal conductivity of the pipe wall influences terms A5 and A6 of Equation 3-69. This can be seen by returning to Example 3-17 and reviewing the derivation of the energy balance. For the same mean radiant temperature of the surroundings,

$$A5 = 1/((0.1759 \text{ m}^2/\text{m})(\ln(r_o/r_i))/2\pi k) = 181.56k, \text{ and}$$

$$A6 = -346.3 - (383.15)\left(\frac{1}{(0.1759 \text{ m}^2/\text{m})(\ln(r_o/r_i)/2\pi k)}\right),$$

$$= -346.3 - 69{,}566k.$$

The thermal conductivity used in Example 3-17 was 52 W/mK. From Appendix 3-1, thermal conductivities span the range from approximately 400 down to approximately 0.05 W/mK. The following values of conductivity lead to the A5 and A6 values shown, and BALANCE provides the surface temperature values.

k, W/mK	A5	A6	surface temperature, K
0.1	18.16	-7,302	342.6
0.5	90.78	-35,129	369.4
5.0	907.8	-348,176	381.5
50	9,078	-3,478,646	383.0
500	90,779	-34,783,346	383.1
0.05	9.078	-3,825	328.41
0.01	1.816	-1,042	303.16

The data, when graphed, provide the following on a semilogarithmic scale. It is obvious that a very low value of thermal conductivity of the pipe wall would be necessary to affect the pipe's surface temperature significantly, a value much lower than typifies metals.

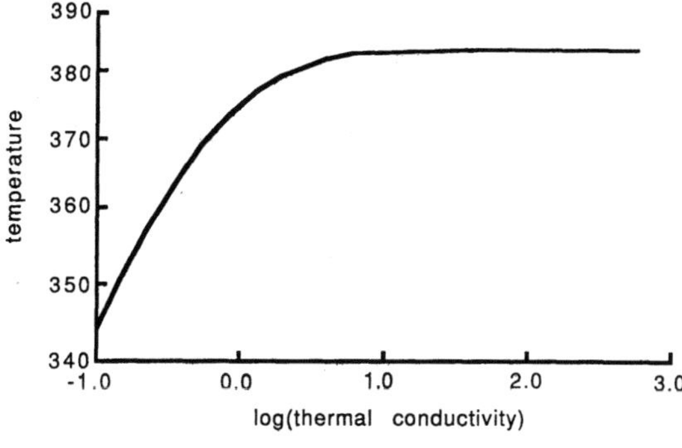

3-7. Combined Convective and Radiation Surface Coefficients

Calculations for radiation heat transfer differ markedly in form from calculations for convective and conductive heat transfer. A desire to attain greater similarity among the three has led to the concept of a radiation surface coefficient of heat transfer. Although radiation heat transfer is properly calculated as described in previous sections, one can, in concept, propose to estimate radiation heat transfer as

$$q'' = h_r \ A(T_s - T_a), \qquad (3\text{-}70)$$

where T_s and T_a are the temperatures of the surface under consideration and the ambient air, respectively. The coefficient, h_r, is defined as the radiation surface coefficient, and is used to calculate the total convective and radiative heat loss from the surface by

$$q'' = (h_r + h_c)(T_s - T_a). \qquad (3\text{-}71)$$

In concept, this is correct for the form of h_r has not been restricted. When heat loss from a surface is expressed in this form, analysis is made easier because radiation heat transfer is linearized and problems of solving nonlinear equations (e.g., Examples 3-16 and 3-17) are avoided. Prior to accessibility to programmable calculators and small computers, avoiding nonlinear problems was a useful goal to achieve.

Linearizing radiation heat transfer can be visualized as an application of a Taylor's series expansion of the Stephan-Boltzmann expression for thermal radiation emission. If the temperature difference between two objects involved in a thermal radiation exchange is small, the expansion may be usefully truncated after the first term leaving an approximating expression for the net exchange between two objects (assume for now only two objects are involved in the exchange, such as a small object within a large enclosure, or one wall of a room exchanging thermal radiation with the other surfaces of the enclosure).

Radiation surface coefficients are obtained from knowledge of actual thermal radiation exchange. For example, consider the previously considered case of a small object in a large room,

$$q'' = \varepsilon_1 \, \sigma \, (T_1{}^4 - T_2{}^4),$$

where surface 1 is the small object in the large room, surface 2.

Thus, the radiation surface coefficient defined in Equation 3-70 for this situation is

$$h_r = \varepsilon_1 \, \sigma \, (T_1{}^4 - T_2{}^4)/(T_1 - T_a). \qquad (3\text{-}72)$$

Unfortunately, this linearization still involves a nonlinear equation to determine h_r, and h_r is a function of temperatures of the surfaces involved in the heat exchange. If T_2 does not equal T_a, there is little advantage in using a radiation surface coefficient. However, if T_2 and T_a are equal, $A_1 \ll A_2$, and $(T_1 - T_a) \ll T_1$, the following approximation may be made:

$$h_r = 4\varepsilon_1 \, \sigma \, T_{ave}{}^3, \qquad (3\text{-}73)$$

where $T_{ave} = 1/2(T_1 + T_a)$.

It is frequently assumed that h_r is a constant which can be determined, but in strict terms this is not true. Since nonlinear equations in the forms we have seen can be solved using computers, the need to linearize heat transfer equations is less strong and radiation heat transfer equations can be used directly to avoid the approximation of the radiation surface coefficient.

One situation is still universally used wherein radiation heat transfer is treated using a surface coefficient, and convective and radiative heat transfer are combined as in Equation 3-71. This involves determining the surface coefficients for heat transfers through walls, etc., of buildings. For this situation, convective heat transfer equations such as in Section 3-3 and radiative heat transfer equations such as in Section 3-4 are not used. Instead, surface coefficients which have been determined empirically to represent average conditions for buildings are accepted, radiation is obscured in the process, and heat gain or loss from the surface of the wall, etc., is determined using Equation 3-71 with the empirical coefficient used in place of $(h_r + h_c)$. Appendix 3-5 contains surface coefficient data as presented, for example, in the *ASHRAE Handbook of Fundamentals* (1989).

These combined coefficients will be used repeatedly throughout this text to calculate building heat transfer gains and losses. It is useful at this point, however, to highlight an important assumption inherent in the data of Appendix 3-5. The data apply in a strict sense only when the air temperature involved in the heat exchange (e.g., the air inside a barn exchanging heat with the inside surface of an outside wall) <u>is the same as the mean radiation temperature of the surroundings of the surface in question.</u>

This is obviously a crude assumption for many applications to agricultural buildings where the temperatures of walls, for example, can differ significantly from air temperature. This would be an especially rough approximation for greenhouses where glass temperature is cold during winter. The data was developed for commercial and industrial buildings with heavy and insulated walls, and in those applications the assumption is more suitable. However, this difference has been frequently overlooked in heat loss calculations for poorly insulated walls, and the errors introduced by the assumption have been accepted.

3-8. Thermal Resistance of Plane Airspaces

A situation which can arise in environmental analysis, and which involves both convective and radiative heat exchange, is the transfer of heat across plane airspaces. One example is loss of heat through an uninsulated wall where the wall cavity is hollow. Another is loss of heat through a night curtain in a greenhouse, where the curtain is composed of at least two layers of fabric or sheet plastic and separated by airspaces.

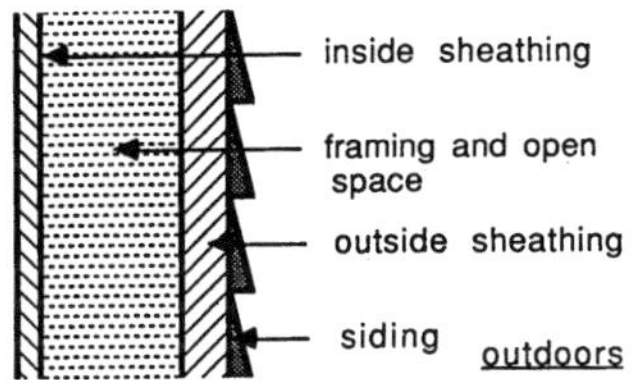

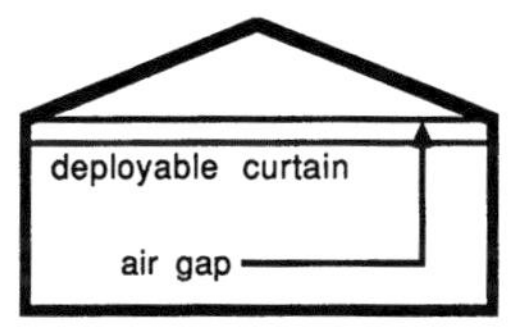

Values for the thermal resistance of such airspaces have been determined experimentally and are published, for example, in the *ASHRAE Handbook of Fundamentals* (1989). Appendix 3-6 gives thermal resistances of plane airspaces.

The thermal resistance depends on several factors. The mean temperature and the thickness of the airspace, the temperature difference from one side to the other, and the direction of heat transfer are four factors which influence convective heat transfer. Radiative heat transfer is affected by the temperatures of the two surfaces which form the airspace and the emittances of the two surfaces.

It is unlikely both surfaces bounding the airspace will be thermally black, thus an effective emittance, E, of the cavity must be determined.

$$E = (\varepsilon_1^{-1} + \varepsilon_2^{-1} - 1)^{-1} \qquad (3\text{-}74)$$

When the emittances of the two sides of the space are estimated, the cavity emittance is calculated, and the temperatures are estimated, the effective resistance can be calculated for a given direction of heat transfer.

Example 3-19

Problem: Estimate the R-value of just the cavity in a hollow wall framed with lumber which leaves approximately a 90 mm airspace. Consider a situation of winter heat loss from the building where the air temperature inside the building is approximately 20 C and it is - 10 C outdoors.

Solution: Several assumptions are required to obtain a solution. First, it will be assumed the mean airspace temperature is the average of inside and outside temperatures. If we knew more about the wall construction we might be able to obtain a better estimate of the mean temperature, but for now we will use a simple average. We will also assume the temperature difference across the cavity is 10 K. This, again, will depend on the wall construction and how much insulation value is in the inside and outside sheathing, and outside siding, compared to the R-value of the cavity. Finally, assume the materials to sheath the wall are thermally black, their effective emittances for thermal radiation exchange are 0.9.

By Equation 3-74, the effective emittance of the airspace is

$$E = [(1/0.9) + (1/0.9) - 1]^{-1} = 0.82.$$

If this is a building wall, the orientation of the airspace is vertical and heat transfer is horizontal. The tabulated data in Appendix 3-6, for a mean airspace temperature of 5 C, an airspace thickness of 88.9 mm, an effective cavity emittance of 0.82, and a temperature difference of 16.7 K shows a thermal resistance of 0.16 m^{2}K/W. When the temperature difference is 5.6 K, the resistance is 0.18 m^{2}K/W. Our assumption of a 10 K temperature difference places us approximately in the middle, thus, we can estimate the actual thermal resistance will be 0.17 m^{2}K/W.

3-9. Thermal Radiation Exchange with Gases

To this point all thermal radiation exchange has been assumed to occur among solid objects. Gases also emit and absorb thermal energy. The model for sky temperature in Example 3-16, Equation 3-67, is actually a model for the mean radiation temperature of the lower part of the atmosphere. Unless the sky is very clear and dry, thermal radiation exchange with the sky is exchanged only with the lowest few hundred meters of air.

The effective emittance of air depends on both humidity and the carbon dioxide content of the air, both of which can be high in agricultural buildings. The following data are emittance values for carbon dioxide and water vapor at 24 C:

Path length, m	Carbon Dioxide, ppm			Relative Humidity, %		
	1,000	3,000	10,000	10	50	100
3	0.03	0.06	0.09	0.06	0.17	0.22
33	0.09	0.12	0.16	0.22	0.39	0.47

In a barn during times of minimum ventilation, for example, the carbon dioxide level may be greater than 3000 ppm and the relative humidity may be higher than 70%. For an animal in the center of a barn, the effective path length of air surrounding it may be greater than 3 m, thus, thermal radiation exchange with the air may be significant. As a rough rule, the effective path length for air radiating to the walls in an enclosure equals four times the mean hydraulic radius of the enclosure. However, in environment control analyses this factor is usually ignored. With the advent of computers and their processing power, it is more feasible to include radiation exchange calculations in determining the environment provided to animals and plants. However, radiation exchange calculations which involve gases are quite complicated, and thermal radiation exchange with the air is neglected. It is not clear, however, this is a reasonable assumption. Sparrow and Cess (1978), for example, develop analyses for radiation exchange between solid objects and gases, methods which can be useful in exploring the thermal radiation interaction of solid surfaces and air.

SYMBOLS

A	area, m^2
B	radiosity, W/m^2
c	coefficient in Equations 3-33 and 3-44
c_1, c_2	integration constants
c_p	specific heat, kJ/kgK
D	hydraulic diameter, m
E	effective emittance
F	radiation angle factor
g	gravitational constant, m/s^2
G	unit area mass flow rate, kg/m^2s
Gr	Grashof number, see Equation 3-32
h	coefficient of convective heat transfer, W/m^2K
h_r	radiation surface coefficient of heat transfer, W/m^2K
H	irradiation, W/m^2
k	thermal conductivity, W/mK
L	thickness or length, m
n	generalized spacial variable, m
n	exponent in Equation 3-33
Nu	Nusselt number, see Equation 3-29
Pr	Prandtl number, see Equation 3-31

q   heat transferred, W

q'' heat flux, W/m^2

q_{gen} internal heat generation, W/m^3

r   radial dimension, m

R   unit area thermal resistance, $m^2 K/W$

Re   Reynolds number, see Equation 3-43

t   temperature, C

T   absolute temperature, K

U   unit area thermal conductance, $W/m^2 K$

V   velocity, m/s

$\dot{V}$ volumetric flow rate, m^3/s

W   thermal radiation emissive power, W/m^2

x   cartesian dimension, m

α thermal diffusivity, s/m^2

β coefficient of thermal expansion, K^{-1}

ε thermal radiation emittance

λ wavelength, microns

μ dynamic viscosity, kg/ms

ρ density, kg/m^3

ρ reflectance

σ Stephan-Boltzmann constant, $W/m^2 K^4$

τ time, s

τ transmittance

EXERCISES

1. Consider a wall made of three layers. The outer layer is four inches of concrete, the inner layer is the same, and the center layer is made of two inches of expanded polystyrene board stock (molded beads, approximately 20 kg/m^3). If the temperatures of the inner and outer surfaces of the wall are 25 and 11 C, respectively, what temperature exists at the interface between the inside and center layers and the center and outside layers? What is the flux of heat through the wall?

2. A hot water heating system is used to heat a calf nursery. Heated water at 70 C flows from the boiler to the nursery through a pipe with an outside diameter of 60 mm. Estimate the convective heat transfer coefficient (based on unit area, and then on unit length) between the pipe surface and the surrounding, still air when the air is at 20 C.

3. Atmospheric air at 60 C enters a sheet metal duct with a cross-section of 0.3 m by 0.5 m. The duct wall is at a temperature of 10 C. During operation, the airflow rate varies between 0.1 and 0.3 m^3/s. Develop a graph of the convective heat transfer coefficient between the air and the duct wall for the range of expected airflow rates.

4. Consider the glass glazing on a greenhouse. The greenhouse is single-

glazed and glass thickness is 8.5 mm. Temperatures of the inside and outside surfaces of the glass are -5 C and -6 C, respectively. Solar insolation with an intensity of 300 W/m^2 strikes the glass (direct normal intensity). Eight percent of the insolation (or 24 W/m^2) is absorbed within the glass. Determine the rate of heat loss from inside the greenhouse to the glazing (W/m^2) when the sun shines, and compare that rate to the rate of loss if there were no absorption of solar energy within the glass.

5. A significant problem related to convective heat transfer in environment control is heat gain or loss from air ducts carrying cold or warm air to conditioned spaces. It is often of interest to determine whether the heat gain or loss is sufficient to warrant concern or a sufficient reason to add insulation.

A refrigeration system provides cooled air for an onion storage room at a food processing plant. For reasons no one can explain, the refrigeration system is located at the opposite end of the building from the storage room – a distance of 50 m. The duct passes through the building on the way to the storage room, and air in the building (surrounding the duct) is at 20 C. The cooled air should arrive at the storage room at 1 C. The question is, what should be the temperature of the air when it leaves the refrigerator?

The problem is one of forced and natural convection – forced inside the duct and natural outside. Radiation and conduction complicate the problem, but for now make the following assumptions:

(a) the thermal resistance and thickness of the duct wall are negligible (it is made of sheet metal and has no insulation),

(b) the combined natural convection and radiation surface coefficient on the outside surface of the duct is 11 W/m^{2}K,

(c) although the duct will be cold, there will be no condensation on its outside surface. This is obviously not realistic, for there likely will be condensation and associated additional heat exchange, but for now ignore this complicating factor, and

(d) assume the duct is round with a diameter of 0.5 m and airflow at a rate of 0.8 m^3/s. The density of the air inside the duct is approximately 1.3 kg/m^3.

Do the following:

(a) Determine the temperature at which air should be discharged from the refrigeration system to provide 1 C air at the storage room.

(b) Determine the actual natural convection coefficient for heat transfer to the duct from the surrounding still air.

6. A small temperature sensor is used to measure air temperature in an air duct in which hot air flows. The actual air temperature is 95 C, and the actual duct wall temperature is 45 C. The convective coefficient between the air and sensor is estimated to be 150 W/m^{2}K. The surface emittance of the sensor is 0.95.

   The sensor can measure only its own temperature, and the goal of using a sensor is for its temperature to equal the temperature of the medium being sensed. What will be the temperature indicated by the temperature sensor — the apparent air temperature?

7. In a food processing plant, cold water at 3 C is pumped through a 20 mm (outside diameter) galvanized steel pipe. The pipe passes through a room where the air temperature can be as high as 30 C and the relative humidity can be as high as 90%. To prevent condensation on the pipe surface, it is insulated with a sleeve of foam rubber (thermal conductivity of 0.03 W/mK). What thickness of insulation is needed to prevent condensation? (Begin by assuming laminar airflow around the pipe.)

8. Consider a 300 mm diameter, round sheet metal duct insulated on the outside with 50 mm of cellular polyurethane (R-11 exp.). Air moves through the duct at a velocity of 7 m/s, and is at 50 C and 30% relative humidity. The heat transfer coefficient (convective plus radiative) at the outer surface of the insulation is 20 W/m^{2}K; ambient air temperature is 10 C. Calculate the outside surface temperature of the insulation.

9. A single layer night curtain is being used in a large greenhouse. The material of the curtain has no significant thermal resistance in itself; all the resistance is in the convective and radiative properties of the two surfaces. Air temperatures below and above the curtain are 17 C and 8 C, respectively. Radiation temperatures of the greenhouse above and below the curtain are -5 C and 15 C. Emittances of the two sides of the curtain material are 0.20 and 0.90 (one side is aluminum foil, the other is cloth). The greenhouse above and below the curtain can be assumed thermally black ($\varepsilon = 1.0$). For maximum thermal benefit, should the foil side of the curtain be on the upper or lower side of the curtain?

REFERENCES

ASHRAE. 1989. Handbook of Fundamentals. American Society of Heating, Refrigerating, and Air Conditioning Engineers, Atlanta, GA.

Holman, J.P. 1981. Heat Transfer. McGraw-Hill Book Co., New York. 569 pp.

Sparrow, E.M. and R.D. Cess. 1978. Radiation Heat Transfer. Hemisphere Publishing Company, Washington DC. 366 pp.

Swinbank, W.F. 1963. Long wave radiation from clear skies. Royal Meteorological Society Quarterly Journal 89(381):339-348.

Sucec, J. 1985. Heat Transfer. Wm. C. Brown Publishers, Dubuque, IA. 837 pp. plus appendices.

CHAPTER 4
STEADY-STATE THERMAL ANALYSIS

4-1. Introduction

The mechanisms of conductive, convective, and radiative heat transfer affect thermal exchanges between a building and its ambient surroundings. Heat is transferred through walls, the ceiling and roof, doors, windows, and the floor. Walls may be entirely above ground, entirely below ground, or a combination.

Models to describe heat flow along these paths have been developed and are widely used. We will focus on steady-state models. Numerous time-dependent models can be found in the research literature, and some are being adopted in engineering practice. That is happening in the design of thermally massive commercial and industrial buildings and some residences, but has not yet been widely adopted for most agricultural buildings.

Because heat transfer basics have been presented in Chapter 3, this chapter will concentrate on examples to demonstrate calculations of steady state heat transfer through the components of a building's structure.

4-2. Heat Transfer Through Walls

Design and construction practices can make walls of buildings complex to analyze if heat transfer through them is considered in fine detail. Fortunately, various simplifying assumptions have been found adequate and provide sufficiently accurate calculations for most design purposes.

<u>Types.</u> Many wall types are encountered in building practice. Appendix 4-1 shows several common designs. Perhaps the most common is the first shown in the appendix, a wood-framed wall with an inside sheathing to protect the wall, an outside sheathing also for strength and mechanical protection, and siding to protect the wall from weather.

The cavity between studs in a wall can be left empty, if no added insulation value is desired, or can be partially or totally filled with insulation. Insulation is commonly used even when energy savings are not a concern. If a wall is insulated, its inner surface will be warmer during cold weather and condensation on the inner surface will not be as severe a problem as for an uninsulated wall. In agricultural buildings, insulation may be limited to that provided by outside sheathing rather than added insulation in the wall cavity.

<u>Insulation.</u> Insulation can be in several forms. Mineral fiber (glass wool, mineral wool) is commonly used. The strands of such insulation suppress air circulation within the pore spaces so that heat moves primarily by conduction.

Air is a good insulator against thermal conduction.

When mineral fiber insulation is used, care must be taken to prevent moisture condensation within the wall cavity. If excessive free water forms, insulation will become compact and its insulation value lost forever. Care must be taken to prevent wind or an exhaust ventilation system from forcing outside air slowly through small leaks and through the stud cavity. That is, air infiltration must be avoided. If air infiltrates through mineral fiber insulation, insulation acts as an effective air filter but not as insulation.

Foamed insulation is also frequently used in agricultural buildings, either as foam boards or formed in place using foam. Foam has the advantage of a high insulation value. If wet conditions are expected, closed-cell foam is recommended to prevent moisture saturation and serious loss of insulation value. If foam is to be used where sunlight can strike it, it must be covered for sunlight slowly degrades foam by accelerating the oxidation process. When foam contains a gas other than air (to achieve a higher insulation value), the R-value used in design should be the "aged" value. This means the value after sufficient time has passed to permit a representative fraction of gas to leak out of the foam and be replaced by atmospheric air.

Loose fill such as mineral fiber, cellulose, and vermiculite can also be used as insulation. However, the moist conditions of agricultural buildings can cause rapid and severe damage to such insulations. When an insulation such as cellulose fiber is wetted, it compacts rapidly and drying will not restore it.

<u>Sheathing.</u> The inner sheathing on a wall acts primarily to protect the wall from mechanical damage (and water spray, etc.). The outer sheathing provides mechanical protection and strength, and more recently materials have been used which provide a significant degree of thermal insulation as well.

<u>R-values.</u> A wood-framed wall is a combination of series and parallel thermal circuits. We have seen in Chapter 3 how to combine series and parallel thermal circuits into an overall R-value. These concepts will now be applied to engineering design problems. The thermal bridging effect of framing can be a significant factor in reducing the effective R value of a building shell.

The parallel thermal circuit feature of wood-framed walls arises because heat can transfer either through the wall by way of the framing or through insulation. Framing comprises a significant fraction of a wall. Not only wall studs, but sills and plates at the floor and ceiling, bracing, framing for windows and doors, and other uses of wood in the wall can be as much as 20% of a wall's cross-section. A lightly framed wall can be as little as 12% wood. Typical values for the area over wood are: 15% for walls framed from 38 x 140 mm (2 x 6) lumber, and 20% for 38 x 90 mm (2 x 4) lumber. Wall studs of larger lumber can be spaced farther apart. These data can be made more exact by knowledge of (or specifying) the wall framing.

Example 4-1

<u>Problem:</u> The walls of a dairy barn have been constructed of (three oval core) concrete blocks 203.2 mm thick (8 in. blocks). Determine the R-value of the wall and estimate the temperatures of the inside and outside surfaces when it is 15 C indoors and - 5 C outdoors. (Note: Conversion from IP to SI units should not increase the number of significant digits. However, ASHRAE data in Appendix 3-2 will be our database, so we will use the SI dimensions as listed in that table.)

<u>Solution:</u> Appendix 3-2 contains R-values of building materials. Appendix 3-5 has surface resistances to be used in building heat loss calculations. The problem concerns only a series thermal circuit.

In Appendix 3-2, concrete blocks with three oval cores, cast using sand and gravel aggregate, are in the section titled: Masonry Units. Conductance for a block 203.2 mm thick is listed as 5.11 W/m^2 K. (Note that light-weight aggregate and cinder blocks, with more air in pore spaces, have a conductance roughly 40% lower, but are not used as frequently because of their lower strength.) The thermal resistance of the block is the inverse of conductance and is given as 0.20 m^2 K/W.

Surface resistance values (in Appendix 3-5) for a vertical wall apply. Surface emittance will be assumed to be 0.90, and winter conditions with a wind velocity of 6.7 m/s will be assumed. For these conditions, surface resistances of the inside and outside surfaces of the wall are 0.12 and 0.030 m^2 K/W, respectively.

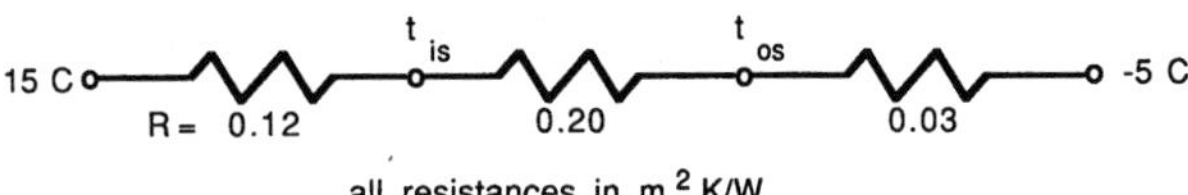

The series thermal circuit is comprised of three resistances with resistance values of 0.12, 0.20 and 0.03 m^2 K/W. The three total a resistance of 0.35 m^2 K/W. The thermal circuit is shown above where t_{is} and t_{os} are the temperatures of the inside and outside surfaces. In steady-state heat transfer, temperature differences scale linearly with resistance differences, thus,

$$t_{is} = 15 - (0.12 / 0.35)(15 - (- 5)) = 8.1 \text{ C, and}$$
$$t_{os} = 15 - ((0.12 + 0.20) / 0.35)(15 - (- 5)) = - 3.3 \text{ C.}$$

Example 4-2

<u>Problem:</u> A wood-framed wall is to be constructed with an inside sheathing of plywood, an outside sheathing of vegetable fiber board, and plywood siding. A

question has arisen whether it will be necessary to specify full-thickness glass wool insulation (which means insulation to fill the wall cavity) or whether a lesser thickness will be sufficient.

The desired wall insulation value is 2 m^2 K/W. Framing members are 88.9 mm thick and 38.1 mm wide (2 x 4 wall studs). Inside sheathing is to be 15.88 mm thick Douglas fir plywood, outside sheathing is to be 12.70 mm thick regular density vegetable fiber board, and the siding is to be 11.11 mm thick hardboard. Winter design conditions are assumed.

<u>Solution:</u> Framing has not been specified in detail, assume wood to fill 20% of the wall's cross section. The desired R-value of 2.0 m^2 K/W includes the effect of framing. Full thickness insulation will be assumed for now. Surface emittances are assumed to equal 0.90 (typical construction materials).

Each of the two parallel heat transfer paths must be analyzed. The R-values can be obtained from Appendix 3-2 for construction materials and Appendix 3-5 for surface coefficients. Values are as shown below.

Heat transfer paths through the insulation and framing are in parallel with each square meter of wall having 0.8 m^2 of insulated area and 0.2 m^2 of wood-framed area. The overall unit area R-value is, thus,

$$R_{overall} = 1.0 / (0.8 / 2.86 + 0.2 / 1.15) = 2.20 \ m^2 \ K/W.$$

The insulation value is slightly greater than desired. To bring the wall to an R-value of 2, the insulated part must have an R-value of

$$R_{insulated} = 0.8 / (1.0 / 2.0 - 0.2 / 1.15) = 2.45 \ m^2 \ K/W.$$

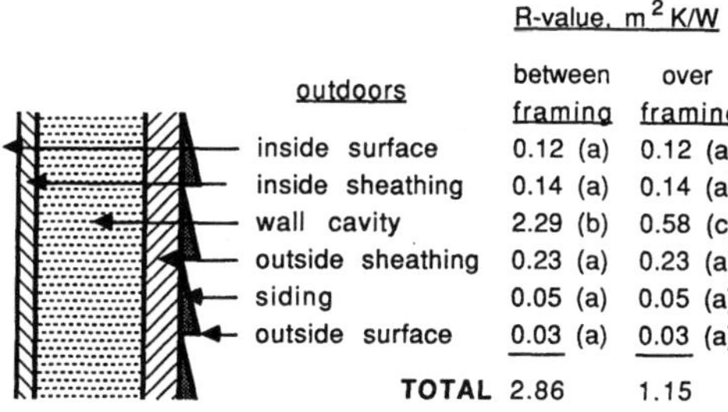

Notes:
(a) obtained directly from data tables
(b) obtained directly from data table for 88.9 mm thickness
(c) assume Dougles fir will be the framing lumber, and
    its R-Value will be an average of data in Appendix 3-2
    (6.5 mK/W)(0.0889 m) = 0.58 m^2 K/W

To reduce the insulated part of the wall to an R-value of 2.45 m^2 K/W, the R-value within the cavity must be reduced from 2.29 to 1.88 m^2 K/W. This would equal the R-value of the insulation plus any resulting airspace created if the insulation were not full thickness.

Blanket insulation is not available in a continuous gradation of thicknesses. The next step below full thickness (88.9 mm, or 3.5 in.) is 50.8 mm (2 in.). This would leave an airspace 38.1 mm thick. The R-value of blanket insulation 50.8 mm thick is approximately 1.31 m^2 K/W. The airspace must, thus, have the equivalent of 1.88 - 1.31 = 0.57 m^2 K/W insulation value.

Appendix 3-6 contains data for R-values of airspaces. The airspace is vertical with horizontal heat flow. The temperature difference across the airspace will likely be between 5.6 and 16.7 K, and the mean temperature will be approximately 10 C in most buildings. Data in the appendix show it is possible to attain an airspace R-value of 0.56 only if the temperature difference across the space is low and the effective emittance of the cavity is very low.

To attain an effective emittance of 0.03 or 0.05 requires one surface forming the airspace to have a surface emittance of nearly zero. This is not practical. The best one can expect is to use foil-backed insulation. The foil has an effective emittance of at least 0.20, making the effective emittance of the cavity approximately 0.20. This does not achieve the required insulation value for the wall.

In conclusion, one would specify full thickness insulation; this meets the required R-value of the wall. To specify less would make the wall slightly less insulated than desired. Of course, the outside sheathing could be changed to obtain more insulation value and permit the wall insulation to be reduced to, perhaps, 50.8 mm thickness. An engineering economic analysis could reveal which would be most cost-effective.

Example 4-3

<u>Problem:</u> A horizontal (eave to eave) thermal curtain for a greenhouse is to be designed using two layers of foil-faced fabric separated by an airspace approximately 20 mm thick.

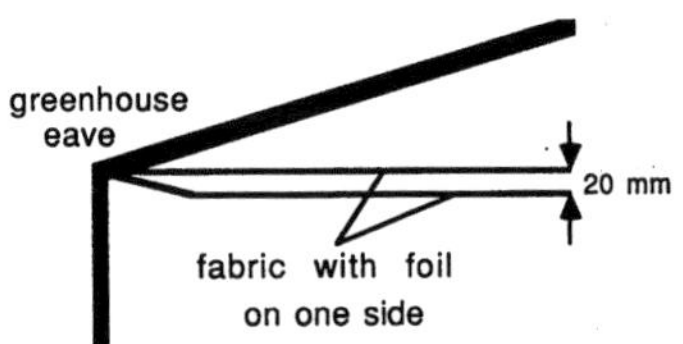

The fabric in the curtain is a loose-weave polyester skrim faced with aluminum foil on one side. The fabric itself has negligible thickness. The surface emittance of the foil side of the fabric is approximately 0.20 and the other side has an effective emittance of approximately 0.60. The two fabric layers can be oriented so

1. each layer has its foil side up,
2. each layer has its foil side down,
3. the top layer has the foil face up and the bottom layer has the foil face down, and
4. the bottom layer has the foil face up and the top layer has the foil face down.

Which of the four orientations provides the largest R-value?

<u>Solution:</u> The fabric was stated to have negligible thickness, thus, all the thermal resistance of the thermal curtain is provided by surface resistances (top and bottom) and the resistance of the airspace between the layers.

Appendices 3-5 and 3-6 contain data for surface resistances and airspace resistances, respectively. A horizontal surface (Appendix 3-5) with heat flow upward (the function of a thermal curtain in a greenhouse is to conserve heating energy) has a surface coefficient which is a function of surface emittance. Data for surface conductance as a function of emittance form a straight line when graphed.

The equation for the line is

$$h_s = 4.02 + 5.82\varepsilon \tag{4-1}$$

For surface emittances, ε, of 0.20 and 0.60, surface conductances, h_s, are 5.18 and 7.51 W/m^2 K, respectively, making surface resistances 0.19 and 0.13 m^2 K/W, respectively.

The airspace can be formed with each surface facing into the space having an emittance of 0.2 or 0.6, or it can be formed with one surface having an emittance of 0.2 and the other 0.6. Resulting effective emittances (see Equation 3-74) of the airspace are shown in the table below.

Appendix 3-6 will be used to obtain airspace R-values. The mean temperature of the airspace will be assumed to be approximately 10 C and the temperature difference across the space will be assumed to be 5.6 K. The airspace is 20 mm thick or approximately 19.1 mm. The airspace is horizontal and heat flow is up.

Interpolation must be used to obtain values (with R-value units of m^2 K/W) and are as follows:

Effective emittance = 0.11:

$$R = 0.39 + (0.11 - 0.05)(0.30 - 0.39)/(0.20 - 0.05)$$
$$= 0.35 \text{ m}^2 \text{ K/W}$$

ε, side one	ε, side two	effective emittance
0.20	0.20	0.11
0.60	0.60	0.43
0.20	0.60	0.18

Effective emittance = 0.43:

$$R = 0.30 + (0.43 - 0.20)(0.20 - 0.30)/(0.50 - 0.20)$$
$$= 0.22 \text{ m}^2 \text{ K/W}$$

Effective emittance = 0.18:

$$R = 0.39 + (0.18 - 0.05)(0.30 - 0.39)/(0.20 - 0.05)$$
$$= 0.31 \text{ m}^2 \text{ K/W}$$

The four possible orientations of the layers and the resulting R-values are found in the table below.

| Orientation of foil face | | R-values | | | |
top layer	bottom layer	top surface	bottom surface	air space	total
up	up	0.19	0.13	0.31	0.63
down	down	0.13	0.19	0.31	0.63
up	down	0.19	0.19	0.22	0.60
down	up	0.13	0.13	0.35	0.61

Differences among the four possibilities are not great and would not be expected to be so. However, the best arrangement is to have both layers facing the same direction. Whether they face up or down does not appear to make a difference based on what has been calculated. However, field experience shows it best if both layers have foil sides up. Why? Think about it. Hint: The data for surface coefficients are based on the assumption the surrounding mean radiant temperature equals the air temperature. In a greenhouse, glazing temperature will be significantly lower than greenhouse air temperature during cold weather.

Note: The above example is based on the assumption the surface characteristics of the thermal curtain material do not degrade (as would happen with dust accumulation). Such degradation may overwhelm any small advantages as were calculated.

Example 4-4

<u>Problem:</u> A refrigerated apple storage building is to be designed. The walls will be constructed as a composite of 101.1 mm thick concrete blocks as an inner wall, the same as an outer wall, and polyurethane foam boards will be used between the inner and outer walls to provide insulation. The desired R-value of the wall is 2.5 m^2 K/W. Design the wall using steady-state analysis to meet this goal. (In reality, the wall's mass may cause transient analysis to be used in practice.)

<u>Solution:</u> The R-value contributions of the surfaces and concrete blocks cannot

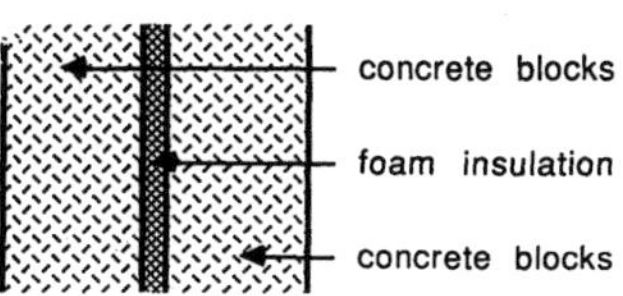

be changed. The thickness of foam boards is the one variable over which there is control in the design.

Winter conditions will be assumed and the concrete blocks will be specified in the design (by you, the design engineer) to be three oval core blocks, sand, and gravel aggregate. Component R-values are available in Appendix 3-2 and surface resistances are in Appendix 3-5. Only a series thermal circuit is involved for the concrete block walls and foam boards form a homogeneous plane.

The concrete blocks have R-values of 0.12 m^2 K/W apiece. The inside surface will be assumed to have a surface emittance of 0.90. The inside and outside wall surface resistances are thereby 0.12 and 0.03 m^2 K/W, respectively. The series thermal circuit consists of the foam boards, two concrete blocks, the inside surface, and the outside surface. The contributions of the blocks and surfaces add to

$$0.12 + 0.12 + 0.12 + 0.03 = 0.39 \text{ m}^2 \text{ K/W}$$

Note how the thermal resistance of the inside surface is as great as that of each block. Masonry has very little thermal resistance.

An insulation value of 2.5 m^2 K/W is desired; 2.5 - 0.39 = 2.11 m^2 K/W must be provided by the insulation. Appendix 3-2 lists the R-value of unfaced cellular polyurethane which has been expanded (foamed) using R-11 refrigerant; the value is 43.38 m^2 K/W per meter thickness. Note R-11 does not refer to the insulation value of the foam board. Refrigerant R-11 can be used to create the foam and provides better insulation value than would air.

The thickness of foam board which is required is 49 mm (2.11 m^2 K/W)/(43.38 mK/W). As the engineer responsible for the design, you would choose and specify a standard thickness of unfaced polyurethane board with at least this much R-value.

Note in this calculation no thermal resistance was assigned to the areas of contact between the foam boards and concrete blocks. A mastic would likely be used to hold the foam board onto one of the walls during construction and contact would be continuous with both concrete block walls after the wall is completed. The exact construction practice would dictate the extent of contact resistance; good practices would result in very small values. Thus, it is conservative design to ignore contact resistances in most situations.

Finally, note that polyurethane foam has a thermal resistance greater than has polystyrene. However, it is also likely to be more expensive. In a complete engineering design, the cost of each to attain the desired R-value could be determined, and the least cost design chosen.

4-3. Heat Transfer Through Ceilings

Ceiling construction is generally simpler than walls. A sheathing forms the ceiling, and insulation (if any) is placed above the sheathing. In modern agricultural buildings, attic spaces are generally unused and sheathing (attic flooring) is not placed above the insulation.

Less framing lumber is used in a ceiling. Generally, the ceiling is attached to the underside of roof trusses. The fraction of framing is generally assumed to be 0.1 when trusses are spaced at 0.4 m and 0.07 when spaced at 0.6 m.

The effect of the roof is frequently assumed to be negligible when calculating heat loss though a ceiling. Standard engineering practice calls for attics to be well ventilated to prevent moisture accumulation. This is especially important in barns where the animal housing space is frequently very humid during cold weather. If the attic is well ventilated, attic temperature will be near outdoor air temperature. When heat passes through the ceiling into the attic air, it is carried to the outdoors by ventilation rather than conduction through the roof. In this case, the roof effect is insignificant.

During hot weather, much of the heat transmission down through a ceiling results from thermal radiation emitted by the under side of the roof and absorbed by the upper side of the ceiling. It has been estimated that as much as three-quarters of the heat gain through a ceiling is due to this radiation effect. Data and procedures to estimate this effect may be found in the *ASHRAE Handbook of Fundamentals*.

Data to calculate the thermal resistance of ceilings is found in Appendix 3-2, and the procedure to calculate resistances is identical to calculations for walls, as in Section 4-2. Note how the direction of heat flow affects the calculated R-value of a ceiling.

4-4. Heat Transfer Through Glazings

4-4.1. General. The technical term for a light-transmitting opening in a building is fenestration. Greenhouses, obviously, are almost entirely light transmitting. Barns, such as for dairy cows and swine, usually have some windows to provide light during the day. Other agricultural buildings such as poultry houses and refrigerated horticultural storages have no windows.

Common glazing materials are glass or plastic sheets. Double-walled rigid plastic sheets and air inflated double polyethylene sheets are frequently used for greenhouse glazing. Windows in barns are usually glass. Double glass for insulation can be used to control condensation on the inside surface, if desired. In barns, the windows need not be openable for the ventilation system should be designed to provide all necessary fresh air. In fact, opened windows usually defeat the ability of a well designed ventilation system to provide good air distribution.

Glazings act both to admit solar radiation and to transmit heat between the inside and outside of the building. Procedures to calculate solar transmittance are described in more detail in Chapter 5.

4-4.2. R-Values. Overall coefficients of heat transfer have been determined for several common glazing materials. Appendix 4-2 contains glazing data for windows and greenhouse glazings.

It should be noted the data for windows apply only to conditions listed in footnotes of the appendix. The conditions are relatively extreme and, thus, should be used to determine peak heating or cooling loads. If used to determine season-long heating or cooling needs, the needs will be overestimated. The data include the effects of both convection and radiation from surfaces and radiation effects grow nonlinearly with increasing temperature differences.

It should also be noted that heat loss coefficients for glazings typically include the effects of inside and outside surface resistances. Single glazings have little thermal resistance in the glazing itself; almost all the resistance is provided by the surfaces.

Example 4-5

<u>Problem:</u> The wood-framed wall described in Example 4-2 is to be constructed with the full thickness of insulation and have an average R-value of 2.20 m^2K/W. The wall will be 3 m high and 60 m long. Double glazed windows of 3 mm glass separated by 6 mm airspaces will be installed. The windows will be in wood frames and will be 1.0 m wide and 0.6 m high. There will be 20 of the windows in the wall. What will be the expected R-value of the windows and when the windows are installed what will be the average R-value of the wall with its windows? Assume cold weather conditions.

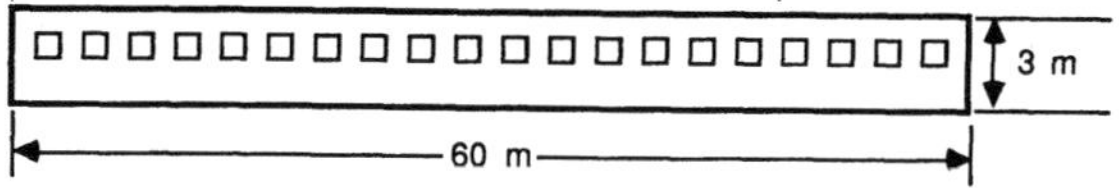

<u>Solution:</u> From Appendix 4-2, for the windows as described, the unit area thermal resistance is 0.30 m^2 K/W. Wood frames during winter provide an

116

adjustment factor between 1.00 and 1.11. We do not have exact data from the window manufacturer, so assume an average adjustment factor of 1.05. This brings the unit area thermal resistance to 0.32 m^2 K/W.

Total window area is

$$A_{windows} = 20(1.0 \text{ x } 0.6) = 12 \text{ } m^2,$$

and total wall area, less windows, is

$$A_{wall} = (3 \text{ x } 60) - 12 = 168 \text{ } m^2.$$

Heat loss through the windows and wall forms a parallel thermal circuit and Equation 3-28 applies. The sum of wall and window area is 180 m^2, and

$$180 / R_{average} = 12 / 0.32 + 168 / 2.20$$

from which the average unit area thermal resistance, $R_{average}$, is calculated as 1.58 m^2 K/W.

This is a significant decrease of the wall's insulation value even though the window area is relatively small. This effect is one of the problems with windows when energy conservation is a goal. Even double glazing or storm windows do not compare well with an insulated wall section in terms of R-value.

It would be a useful exercise to repeat this calculation assuming only single glazed windows in aluminum frames are to be used.

4-5. Heat Transfer Through Doors

Doors are often a significant factor in calculating heat loss from agricultural buildings, although they may comprise only a small fraction of total wall area. Flush doors are used for permitting people to enter and leave and overhead (usually panel) doors permit entry of equipment and animals. Unit area thermal resistances of doors are listed in Appendix 4-3. The data include surface resistances and interpolation and moderate extrapolation are reasonably accurate for conditions different from those listed in the footnotes of the appendix.

Example 4-6

<u>Problem:</u> The wall described in Example 4-5 is to be modified to include an overhead door 2.5 m high and 3 m wide and a flush door 2.5 m high and 1 m wide. The overhead door will be a panel door with 11 mm thick panels, 35 mm thick frame, with no glazing, and the flush door will be a 35 mm thick solid-

core door with no glazing. Calculate the effect of the doors on the average R-value of the wall for winter conditions with windows and doors included. No storm doors are to be installed.

Solution: Appendix 4-3 contains the data needed to solve this problem, a further example of parallel heat transmission.

The unit area thermal resistance of the flush door and panel door are 0.45 and 0.31 m^2 K/W, respectively. The respective areas are 2.5 and 7.5 m^2. Adding the doors reduces the framed wall area to 168 m^2 - 10 m^2 = 158 m^2. Windows are not affected. The average R-value can be calculated using Equation 3-28.

$$180 / R_{average} = 12 / 0.32 + 158 / 2.20 + 2.5 / 0.45 + 7.5 / 0.31$$
$$= 37.5 + 71.8 + 5.6 + 24.2$$
$$= 139.1 \text{ W/K.}$$

The average unit area thermal resistance, $R_{average}$, is 1.29 m^2 K/W.

This shows the potentially significant effect of doors. The average R-value of the wall has changed from 1.58 to 1.29 m^2 K/W – an 18% decrease. What was originally an insulated wall (Example 4-2) with an R-value of 2.86 through the insulation, reduced to an R-value of 2.20 by the effect of the framing, reduced again to an R-value of 1.58 because of the windows, and then dropped to 1.29 when doors were added. The insulation value of the wall when averaged with its windows and doors is only 45% of the insulated section. If energy conservation were a goal in the design, it would be logical to search first for means to increase the averaged R-value by changing the windows and doors rather than by choosing the more expensive approach of redesigning the wall to add more insulation.

The heat loss factor calculated above for the wall (139.1 W/K) is a sum of four contributions. The relative magnitude of each contribution is an indication of its importance. In this example, windows contribute 27% of the heat transmission though the wall (37.5/139.1), the wall contributes 52%, and the flush and panel doors contribute 4% and 17%, respectively.

4-6. Heat Transfer Through Floors on Grade

Many agricultural buildings are built on concrete slabs and have no basement. The perimeter of the slab is likely to have a concrete foundation wall extending down to a concrete footing below frost line. Insulation may or may not be a part of the perimeter.

Heat transmission through the perimeter is likely to be an important factor in barns, especially those with well insulated walls, but will be less significant in greenhouses. Perimeter heat loss is also likely to be significant in horticultural product storages where above-ground walls and the ceiling are well insulated.

Thermal comfort, however, may be a factor in some barns and greenhouses. Small animals, such as calves and pigs, penned along an outside wall may suffer significantly from chilling when lying on a cold floor. Plants along the outside wall in a greenhouse, if on the floor rather than on benches or in ground beds, may suffer from low root temperatures during cold weather. When this factor is expected to exist, perimeter insulation may be recommended.

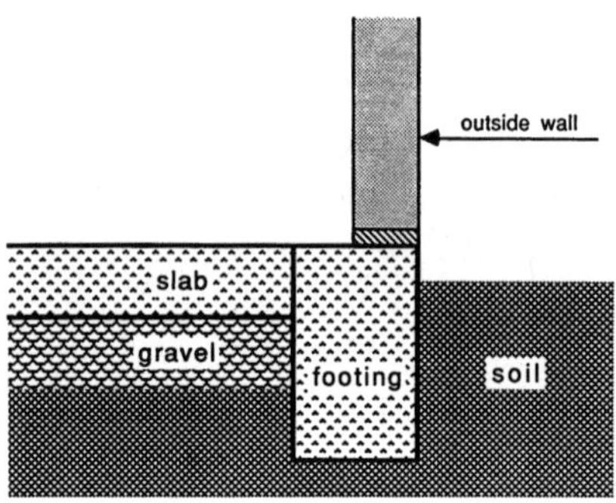

An unheated floor slab has been found to lose heat primarily from its perimeter rather than from the interior part of the building. If a building on a slab is heated, soil under the interior parts of the building is warmed by heat loss from the building. The thermal conduction path to cold soil surrounding the building is long and the heat does not transfer quickly away from the central regions. In contrast, heat transmitted through the floor near the perimeter can move readily to the cold soil surrounding the building. In terms of heat conduction, the outer two meters of perimeter are the only parts of the floor which are important.

Experimental evidence indicates heat loss from perimeters of buildings is proportional to the length of the perimeter and the temperature difference between air inside the building and out. If steady-state conditions are assumed, heat loss from the perimeter can be estimated by

$$q_{floor} = FP(t_{inside} - t_{outdoors}), \qquad (4\text{-}2)$$

where q_{floor} is heat loss in W; P is the building perimeter, m; t is air temperature; and F is an experimentally determined perimeter heat loss factor, W/m K.

Values of F for an uninsulated and unheated slab floor on grade range between 1.4 and 1.6 W/m K, depending on the severity of the winter. The factor is reduced to between 0.8 and 0.9 W/m K when insulation of R = 0.95 m^2 K/W is installed from under the edge of the slab down to the top of the footing (applied to the outside surface of the foundation wall, or the inside surface before the slab is poured, see Figure 4-1 for the first case). Within the ranges of the perimeter factor, the large values apply when winters are less severe and air temperature differences are smaller.

More complete data for perimeter heat losses in agricultural buildings should be based on highly detailed computer analysis; finite element analysis works well.

Computer analysis has shown the integrity of insulation around a perimeter is critical in determining the effectiveness of the insulation. Thermal bridges (Figure 4-2) formed by concrete in the slab or foundation wall seriously detract from energy conservation efforts. If, for example, the slab is in physical contact with the foundation wall, the conduction of heat through that bridge may reduce heat savings by 20 to 30%, compared to construction where the joint between the slab and the foundation is insulated.

4-7. Heat Transfer Through Basement Walls and Floors

Heat loss from heated basements is complicated by the two- and three-dimensional nature of the conduction heat transfer process. Heat lost from the upper parts of a basement wall has a much shorter path length to the outside air than does heat lost from near the floor. Heat loss from the floor is influenced more by soil temperature than by outdoor air temperature, except indirectly through the long-term influence of air temperature on ground temperature.

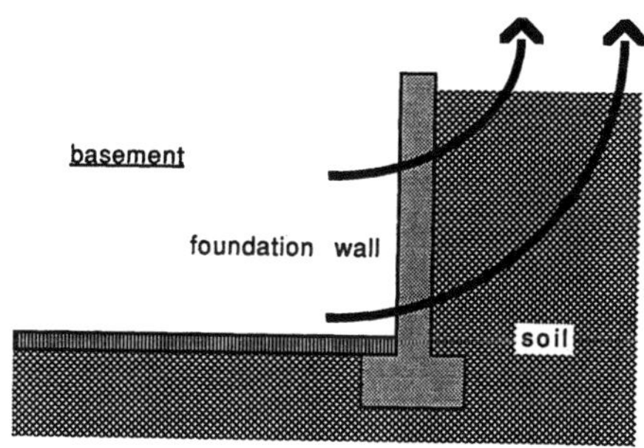

To permit estimates of basement heat loss, empirical factors have been determined to be used in calculations. The factors are used in a heat loss equation

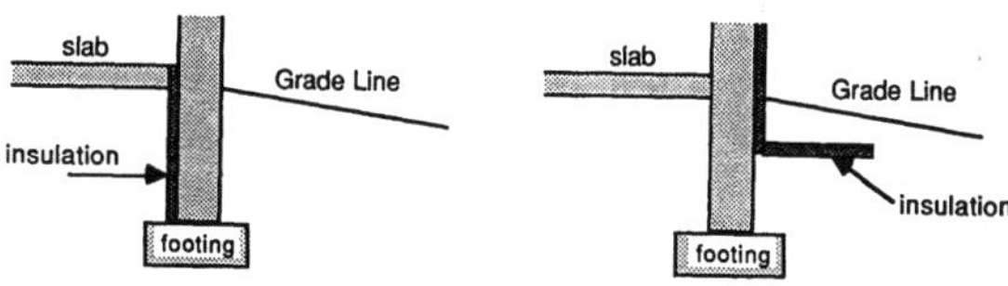

Figure 4-1. Examples of perimeter insulation placement.

$$q_{ground} = U'(t_{inside} - t_{ground}) \qquad (4\text{-}3)$$

where q is heat loss in W, t is temperature in C, and U' is an integrated conductance over the area of heat loss, having units of W/K. Note that Equation 4-3 is written to express heat loss in terms of ground temperature rather than outdoor air temperature.

Basements are not deep enough to be surrounded by soil at constant temperature. Essentially constant temperatures are not reached until 10 m below ground surface, thus, t_{ground} must be estimated and chosen to represent the heating season.

Yearly average soil temperature approximately equals yearly average air temperature. Winter soil temperature is offset from yearly average temperature by a decrement. Figure 4-3 can be used to determine the magnitude of that decrement. The decrement, subtracted from the yearly average, is the heating season soil temperature.

Yearly average soil temperature in a region frequently is the same as the temperature of water from deep wells in the region. For example, locate the region of Central New York State on the map in Figure 4-3. It is just southeast of Lake Ontario at approximately 75° west longitude and 43° north latitude. The average yearly temperature in the region is approximately 10 C. The map in Figure 4-3 shows the decrement which represents winter soil temperatures is 10 C. Thus, the winter design soil temperature is 0 C, which is used as t_{ground} in Equation 4-3.

Table 4-1 provides data to estimate the value of U' for both basement walls and floors. The data for walls is both for uninsulated construction and for three levels of possible insulation. The floor is assumed to be not insulated.

Example 4-7

Problem: A packing shed for a fruit farm is to have a basement for storage. The basement is to be 8.5 m by 12 m, and the basement floor will be 1.8 m below grade. The shed is to be located in the state of Washington at 120° west

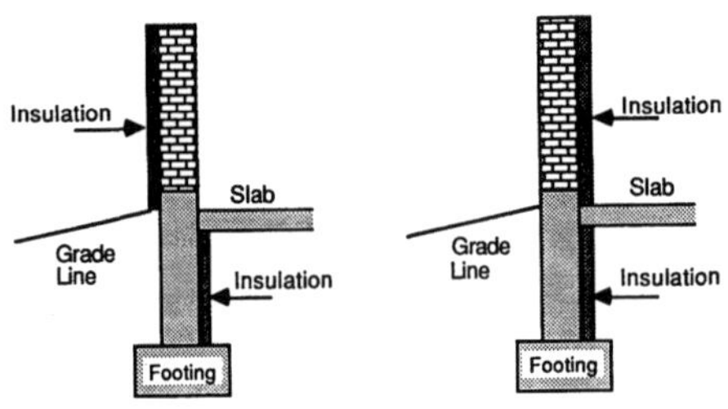

Figure 4-2. Examples of insulation added to slab floors and footings which form thermal bridges.

longitude and 45° north latitude. The average yearly temperature is 12 C. The basement will be maintained at 15 C and the walls will not be insulated. Estimate the rate at which heat will be lost from the basement by conduction through the walls and floor to the soil.

Solution: Heat loss through the walls and floor can be estimated using the data in Table 4-1. Conductances for walls between 0 and 1.8 m can be averaged to obtain the average conductance per unit length of wall perimeter. The average of the six values is

$$U_{average} = (2.33 + 1.26 + 0.88 + 0.67 + 0.54 + 0.45)/6$$
$$= 1.02 \text{ W/m}^2\text{K} \times 1.8 \text{ m}^2/\text{m} = 1.84 \text{ W/mK.}$$

The perimeter of the basement is 41 m, thus, the wall's contribution to the total conductance, U'' to be used in Equation 4-3, is

$$U'_{wall} = (1.84 \text{ W/mK})(41 \text{ m}) = 75 \text{ W/K.}$$

The shortest width of the basement is 8.5 m, thus, for a depth below grade of 1.8 m the unit area conductance is 0.14 W/m^{2}K. The floor area is 102 m^2 and the floor's contribution to the total conductance is

$$U'_{floor} = (0.14 \text{ W/m}^2\text{K}) (102 \text{ m}^2) = 14 \text{ W/K.}$$

Therefore, the total conductance, $U'_{basement}$, is $75 + 14 = 89$ W/K.

From the map in Figure 4-3, the decrement of ground temperature below

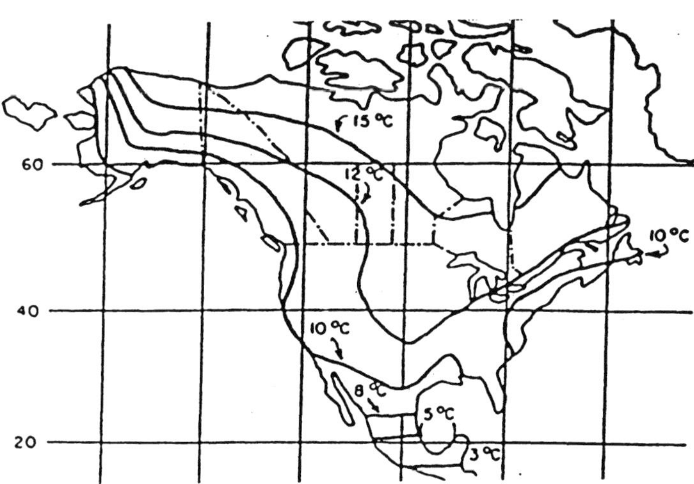

Figure 4-3. Isotherms of the decrement of soil temperature below average yearly soil temperature during the heating season. Adapted from the 1989 *ASHRAE Handbook of Fundamentals.*

average yearly temperature is approximately 10 C, thus, ground temperature is estimated to be 2 C during the heating season.

The rate of heat loss to the ground, computed using Equation 4-3, is

$$q_{ground} = (89 \text{ W/K})(15 \text{ C} - 2 \text{ C}) = 1157 \text{ W}.$$

The heat loss from the part of the basement wall above grade must be added to obtain the total basement heat loss, of course.

4-8. Program RVALUE

The total conductance of a building's shell, ΣUA, summed over the walls, ceiling, windows and doors, is important in determining thermal behavior of the building. Sections of this chapter have shown how to obtain the components needed to calculate this sum.

Good design should include investigating many possible options and selecting the best based on overall economic considerations and other factors. Many options exist when designing a building to meet a thermal standard. Can the ceiling be heavily insulated (which is less expensive to do than heavily insulating walls), and the walls lightly insulated and achieve the same thermal standard? Can insulative outer sheathing be substituted for insulation within a wall cavity and achieve the same thermal standard at a lower cost? Or vice-

Table 4-1. Heat loss conductance for basements

Through walls

Depth below grade	Conductance, W/m^2K, for wall insulated by			
	uninsulated	R = 0.73	1.47	2.20 m^2K/W
0 - 0.3 m	2.33	0.86	0.53	0.38
0.3 - 0.6	1.26	0.66	0.45	0.36
0.6 - 0.9	0.88	0.53	0.38	0.30
0.9 - 1.2	0.67	0.45	0.34	0.27
1.2 - 1.5	0.54	0.39	0.30	0.25
1.5 - 1.8	0.45	0.34	0.27	0.23
1.8 - 2.1	0.39	0.30	0.25	0.21

Through the floor, conductance in W/m^2K

Depth of floor below grade	Minimum width of basement, m			
	6.0	7.3	8.5	9.7
1.5 m	0.18	0.16	0.15	0.13
1.8	0.17	0.15	0.14	0.12
2.1	0.16	0.15	0.13	0.12

Notes:
a. Assumes thermal conductivity of the soil is 1.38 W/mK.
b. Uninsulated wall assumed to be masonry.
c. Data adapted from the *ASHRAE Handbook of Fundamentals*.

versa? If window area is significant, can the cost of double glazing be avoided by adding more or different insulation to the walls or ceiling and still have the same overall building U value? Or, can windows be double glazed and walls insulated less for lower total cost and the same overall building thermal characteristics? These are only some of the questions a designer might ask while deciding the best design for a specific application.

The effort involved in calculating ΣUA for a building is significant, and precludes repeated calculations to answer the "what if" questions so important to good design. In addition, a lack of time to complete a wide selection of designs works against obtaining an intuitive feeling for the importance of various factors in determining building thermal behavior. Numerous thermal analysis programs for buildings have been written for small computers and are becoming commercially available. They can be invaluable tools for saving time in thermal analysis. If less time is required for analysis, more time is available for design, which should be the main focus of engineering.

Program RVALUE is an executable file provided as an example of the utility of a building thermal analysis program. It is not a general purpose program intended for general building design; it is too limited. However, RVALUE can be used to obtain some understanding of how parts of a building interrelate thermally to yield an overall thermal resistance, R, and an overall thermal conductance, ΣUA. Instructions for using the program are included within the menus provided the user.

Example 4-8

<u>Problem:</u> A building (10 by 50 m, with 3 m high sidewalls) is to be built. Windows will be single glass, normal emittance. The building will have two, 2.5 by 4 m, 44 mm thick, panel doors with 11 mm thick panels. The ceiling will be sheathed on the underside with 15.88 mm plywood and insulated with 150 mm of mineral fiber blanket. The walls will have 12.7 mm plywood as inside sheathing, 88.9 mm of mineral fiber blanket as the cavity insulation, 19.8 mm regular density vegetable fiber board as outside sheathing, and 15.88 mm plywood as siding. Wall framing is estimated to constitute 20% of the wall area.

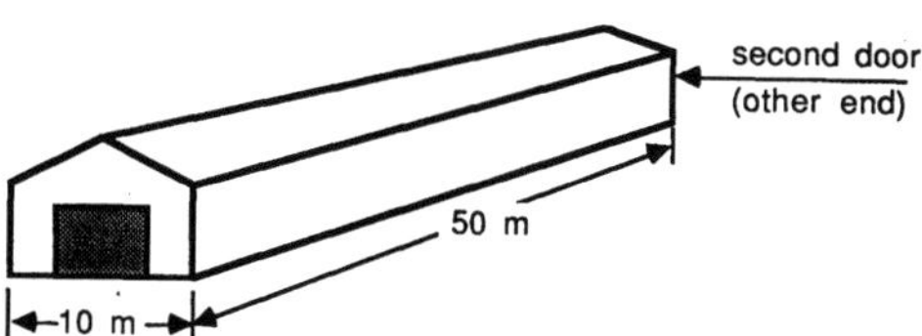

Develop a graph to show the effect of glazing area on the overall R-value of the building. Graph the <u>building R-value</u> as a function of <u>glazing area</u> for a glazing area ranging from 0% to 30% of the gross wall area. Assume winter conditions.

<u>Solution:</u> Program RVALUE may be used to accumulate data for the required graph. Door data (the door area is 20 m^2), window data, ceiling data, and wall data are entered as requested, and the window data is repeatedly changed to obtain the overall R-value as a function of glazing area from 0 m^2 to 30% of the gross wall area, which is

gross wall area = 3 m x 120 m = 360 m^2.

The R-value data is:

Glass Area, %	Glass Area, m^2	Building R-value
0	0	2.41 m^{2}K/W
5	18	1.86
10	36	1.51
15	54	1.28
20	72	1.10
25	90	0.97
30	108	0.87

Note: In using RVALUE, one must be careful not to enter impossible conditions. For example, in this problem, to enter a window area of 100 m^2 and a door area of 300 m^2 (by mistake) results in a net wall area of - 40 m^2, clearly impossible.

The data for building R-value is graphed in Figure 4-4. For comparison, the extreme condition of the walls being entirely glazed (100% glazing) results in an overall R-Value for the building of 0.37 m^{2}K/W. The graph in Figure 4-4 shows the relatively rapid change of the building's R-value as glazing area is added. When the glazing area changes from 0% to 5% of the wall area (5% is a modest amount of glazing), the overall building R-value drops from 2.41 m^2 K/W to 1.86 m^2 K/W. This is a 23% decrease with a corresponding 30% increase of heat loss.

It would be a useful exercise to repeat this example, using RVALUE, to ensure its use is understood.

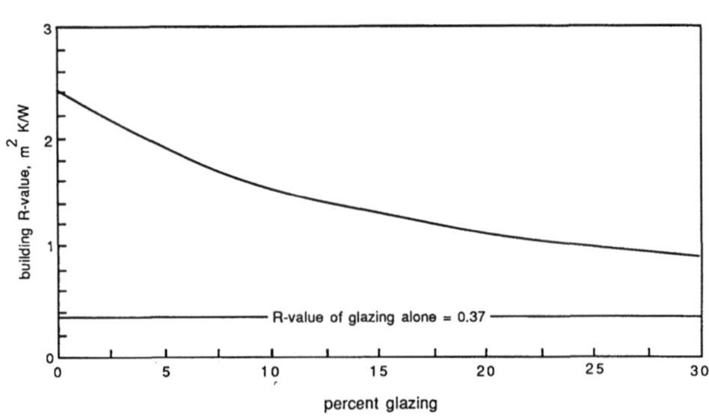

Figure 4-4. Average R-Value of the structural cover of the building described in Example 4-8 as a function of the percentage of the wall devoted to glazing.

4-9. Sol-Air Temperature

All the analyses of heat gains and losses from buildings have, to this point, assumed thermal exchange is between regions at the temperature of the outside air and the temperature of the inside air. This may be a valid assumption at night, or when it is cloudy, but during times of significant solar insolation, solar heating is an additional factor which can significantly affect the magnitudes of heat exchange. A concept which permits consideration of solar effects and long-wave thermal reradiation, without introducing the nonlinearities of radiative heat transfer, is the sol-air temperature. In preview, the sol-air temperature, in the absence of solar heating and long-wave exchange, acts as an equivalent air temperature which would cause heat to be exchanged in the same magnitude that it is exchanged when actual air temperature, thermal radiation, and solar heating are considered.

Consider a building surface irradiated by solar insolation, and exchanging thermal energy with the ambient air by convection, and with the radiant surroundings by long-wave thermal radiation. An energy balance on the surface contains, as fluxes on the upper surface,

$$h(t_a - t_w) + \alpha I + \varepsilon_w \sigma (T_s^4 - T_w^4). \qquad (4\text{-}4)$$

Simplify Equation 4-4 by treating long-wave exchange as a single term, $\varepsilon \Delta R$, the difference between thermal radiation leaving the surface, and that striking it, to modify the thermal gains terms to

$$h(t_a - t_w) + \alpha I - \varepsilon \Delta R. \qquad (4\text{-}5)$$

Next hypothesize a situation where there is no solar irradiation or long-wave thermal exchange, only convective exchange between the surface and air at a temperature, t_{sa}, which produces exactly the same net heat exchange with the surface. The gains in the thermal energy balance for this situation are only convective,

$$h(t_{sa} - t_w). \qquad (4\text{-}6)$$

Heat exchanges in the two cases are the same, thus, the gains on the upper surface must be equivalent. Equating the two gains yields

$$t_{sa} = t_a + (\alpha I - \varepsilon \Delta R) / h. \qquad (4\text{-}7)$$

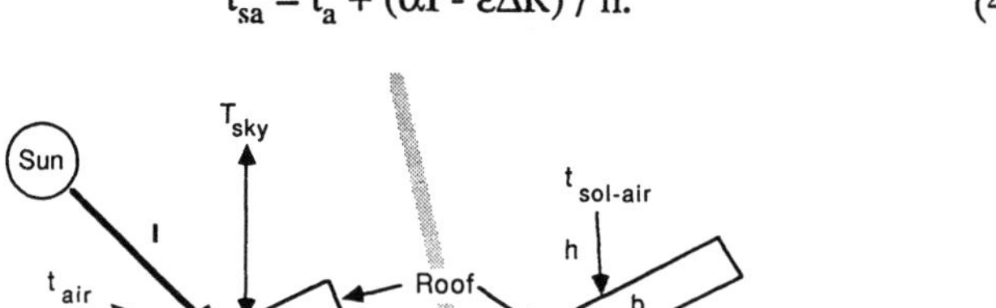

Same heat conduction through the roof

If this equivalent temperature, the "sol-air temperature", is known for each building surface, thermal exchange between the building and its surroundings can be calculated using only linear equations. Fortunately, data exists to estimate the sol-air temperature.

Methods to determine solar gain will be covered later, in Section 5-3. For now, assume a value of I can be determined for each surface. Means to determine the reradiation term are less obvious. It has been observed (Parmelee and Aubele, 1952) that ΔR is approximately 0 for vertical building surfaces. Frequently, the sky is colder than a vertical surface, but the ground is warmer because of solar heating and the net effect is to balance the loss and gains. The same researchers found ΔR to be approximately 60 W/m^2 (loss from the surface) for a horizontal surface facing a clear sky. Later researchers (Lokmanhekim, 1971, and Kusuda, 1976) modified ΔR further to account for cloudiness,

$$t_{sa} = t_a + (\alpha I - 6\varepsilon \cos\phi(10 - \Omega))/ h, \qquad (4\text{-}8)$$

where Ω is the cloudiness factor, ranging from 0 for a clear sky, to 10 for complete cloud cover. The angle ϕ is the tilt angle of the surface under consideration, being 0° for a horizontal surface and 90° for a vertical surface.

Example 4-9

<u>Problem:</u> A building roof having a 1:3 slope is irradiated with solar heat at a rate of 500 W/m^2. Air temperature is 28 C and the sky is clear. The roof's solar absorptance is 0.8, and its thermal emittance is 0.95. The surface convective coefficient is approximately 30 W/m^2 K ($R_{os} = 0.03$ m^2 K/W). Calculate the sol-air temperature which applies and compare it to the actual air temperature.

<u>Solution:</u> Equation 4-8 can be used to calculate the sol-air temperature directly. The roof slope is 1:3, or 18.43°, and

$$
\begin{aligned}
t_{sa} &= 28\ C + [(0.80)\ (500\ W/m^2) \\
&\quad - (6\ W/m^2)\ (0.95)\ (\cos(18.43°))\ (10 - 0)]/30\ W/m^2\ K \\
&= 39.5\ C
\end{aligned}
$$

This is a significant increase above the actual air temperature. In effect, the solar (primarily) effect is to raise the effective outdoor temperature by more than 11 C. The solar effect is more than 13° (400 Wm^{-2} / 30 Wm^{-2} K^{-1}), while reradiation is nearly a 2° reduction, making the net effect 11.5 C. Of course, this example is for a building surface receiving a high amount of solar radiation. However, on any building surface irradiated by the sun, the net flow of heat is likely to be into the building (in the steady state) instead of out of the building, as we have assumed for example in Section 4-2.

4-10. Heat Exchangers

Engineers frequently must design equipment to transfer thermal energy from one fluid stream to another. Heat exchangers are used for this task. Standard design techniques are available which can be applied to most design problems and which are well suited to computerized analysis as a tool for design.

Heat exchangers find wide application in the food processing industry to heat and cool food products. In environmental control, heat exchangers are used, for example, to prewarm fresh air by recovering heat from ventilation air being exhausted from a building. Heat exchangers can be used to transfer heat from steam or water (heated in a boiler) to air in air circulators such as fan-jet units.

4-10.1. Overall Heat Transfer. The concept of series thermal circuits has already been discussed in Section 3-2.4. In heat exchangers, heat is transferred from one stream of fluid to another stream, with transfer occurring across a wall separating the two fluid streams. The series thermal circuit contains two convective resistances and one conductive resistance.

The thermal circuit is as shown in Figure 4-5, and unit area thermal flux, q'', can be expressed as

$$q'' = \frac{t_b - t_a}{1/h_a + x/k + 1/h_b},\qquad(4\text{-}9)$$

where k is the thermal conductivity of the wall material and x is its thickness.

However, a plane wall with uniform temperatures on its two sides is not a practical concern with heat exchangers. The goal is to transfer heat, thus, the two fluid streams continuously vary in temperature. The total heat transfer can be expressed in terms of an overall heat transfer coefficient, U, a total area for heat exchange, A, and some average temperature difference, Δt_m,

$$q = UA\Delta t_m,\qquad(4\text{-}10)$$

The determination of Δt_m depends on the type of heat exchanger being designed. For example, consider the double pipe exchangers shown in Figure 4-6. With parallel flow heat exchangers, the temperatures of the two fluid streams approach a single, intermediate value and Δt_m varies from the initial difference

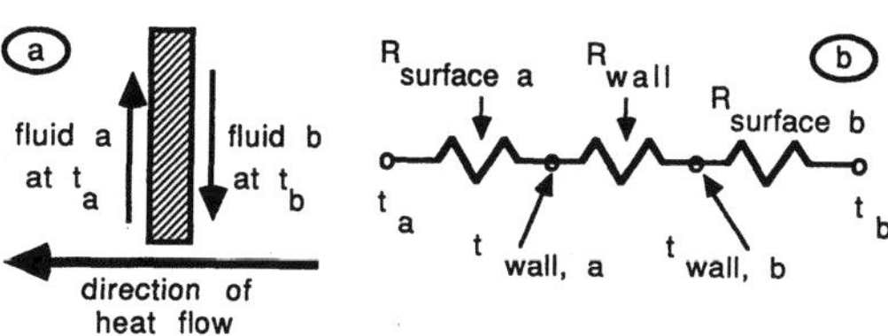

Figure 4-5. Thermal circuit for heat exchange: (a) fluid flow past the heat exchanger wall and (b) the series thermal circuit showing resistances (R) and temperatures (t).

between the entering streams to no difference if the exchanger is large enough to be completely effective. With the counter flow exchanger, the temperature difference between the streams may or may not change along the length of the exchanger but the difference is not known a priori at any point along the exchanger.

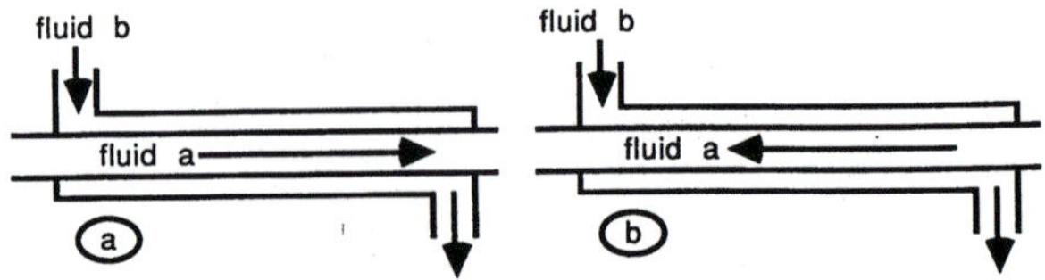

Figure 4-6. Double pipe heat exchangers: (a) parallel flow and (b) counter flow.

As a practical matter, counter flow heat exchangers are more typically used for environmental control applications. Temperature changes on each side of the exchanger can be greater than would be the case for parallel flow exchangers. Parallel flow heat exchangers are useful when a rapid initial temperature change is desired in one or both of the fluid streams.

4-10.2. Overall Heat Transfer Coefficient. When commercially designed heat exchangers are to be used, the manufacturer's data should be available to determine the overall UA value of the exchanger, perhaps as a function of fluid properties and flow rates, or perhaps as a relatively constant value. Such data is obtained by experiment and is the most accurate as long as the exchanger is used as tested. However, if a heat exchanger is to be designed and constructed for local use, heat transfer analysis procedures covered previously can be used to estimate the UA value of the exchanger.

Example 4-10

<u>Problem:</u> A heat exchanger is to be designed and locally fabricated to prewarm ventilation air for a calf nursery. The exchanger will be built of parallel plates of 0.5 mm thick galvanized sheet metal. Dimensions are as shown and the airflow rate in each channel is to be approximately 0.25 m³/s. Determine the average unit area thermal conductance (U) for the exchanger.

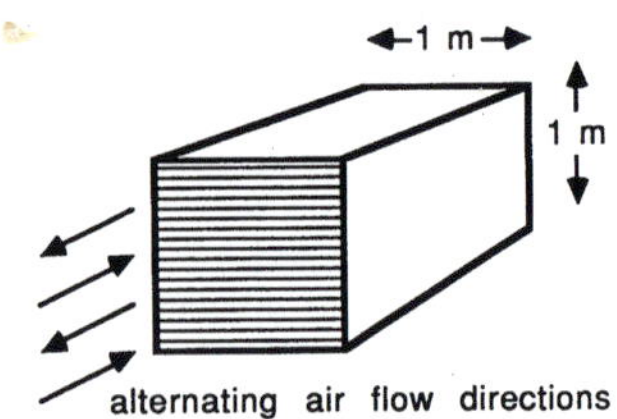

<u>Solution:</u> The unit area thermal conductance can be determined using

$$U = 1 / (1 / h_1 + x / k + 1 / h_2),$$

where x = 0.5 mm and k = 45.3 W/mK (from Appendix 3-1 for mild steel); and the convective coefficients, h_1 and h_2, must be determined. Each is a forced convective heat transfer coefficient, which is discussed in Section 3-3.2.

Equation 3-44 applies,

$$h = cG^{0.8} D^{-0.2},$$

where G and D are as defined in Equations 3-45 and 3-46. From the sketch of the exchanger and the dimensions given, each airflow channel is approximately 0.125 m by 1 m, and

$$D = 4(0.125 \ m^2)/(2.25 \ m) = 0.222 \ m.$$

To determine G, air density is required. Operating conditions are not known, but we can assume if the unit is used to temper outdoor air, air temperature will be between outdoor air temperature and room air temperature. Temperatures will vary from day to day and within the exchanger but often will be below freezing. As an approximation, assume air density to be approximately 1.10 kg/m^3 (cool air somewhat above sea level).

Density variations within the exchanger will be ignored and it will be assumed the convective coefficient remains constant from place to place within the exchanger. Variations should be small and insignificant compared to other unknowns such as variation of local air velocity. Airflow nonuniformity from top to bottom in each channel, and from channel to channel, will likely contribute a greater degree of uncertainty than air density differences. Such practical matters lead to the need to confirm overall U values by experimental means. However, heat transfer analysis can be used as a first approximation to evaluate designs. The mass rate airflow, G, is thereby

$$G = (1.10 \ kg/m^3)(0.25 \ m^3/s)/(0.125 \ m^2) = 2.2 \ kg/m^2 \ s.$$

The coefficient c required to determine h is estimated (Table 3-3) to be approximately 3.10 at temperatures at which the exchanger would be expected to operate. The convective heat transfer coefficient is, thus,

$$h = (3.10)(2.2 \ kg/m^2 \ s)^{0.8}(0.222 \ m)^{-0.2} = 7.87 \ W/m^2 \ K.$$

This value applies to each side of the exchanger, thus, the overall unit area thermal conductance is

$$\begin{aligned} U &= 1/[(1 \ / \ 7.87 \ W/m^2 \ K) + (0.0005 \ m \ / \ 45.3 \ W/mK) + (1 \ / \ 7.87 \ W/m^2 \ K)] \\ &= 3.94 \ W/m^2 \ K. \end{aligned}$$

4-10.3. Logarithmic Mean Temperature Difference. The average temperature difference from one end of the heat exchanger to the other is needed to determine the total heat exchanged. Heat transfer texts show how differential calculus can be applied to derive equations for the types of heat exchangers in single pass counter flow and parallel flow exchangers, Figure 4-6. Texts such as Holman (1981) present correction factors to determine the average temperature difference for many other types of heat exchangers.

The average, or mean, temperature difference is called the "log mean temperature difference" (LMTD) defined as the temperature difference at one end of the exchanger, less the temperature difference at the other end of the exchanger, divided by the natural logarithm of the ratio of the two temperature differences. In symbol form,

$$\Delta t_m = \frac{(t_{h2} - t_{c2}) - (t_{h1} - t_{c1})}{\ln[(t_{h2} - t_{c2})/(t_{h1} - t_{c1})]}. \tag{4-11}$$

The entering and exiting fluid stream temperatures must be known to determine the LMTD. The concept is, thus, most useful in determining design parameters to achieve certain heat exchange goals such as shown in Example 4-11.

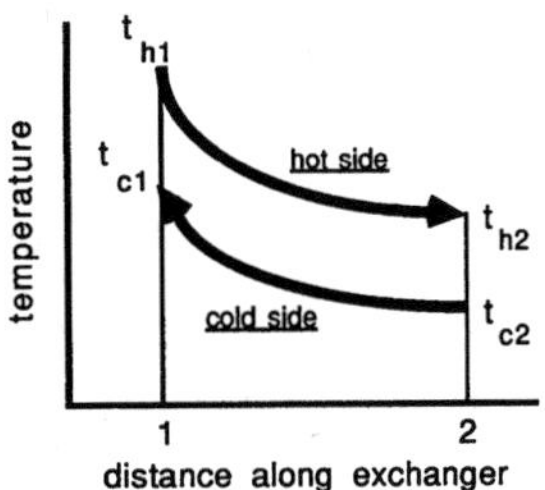

Example 4-11

Problem: A heat exchanger is used to transfer heat from hot water to recirculated air within a greenhouse. Hot water arrives at the exchanger at 80 C, and leaves at 60 C, and flows at a rate of 0.05 kg/s. Air flows through the exchanger at a rate of 0.4165 kg/s, and is heated from 20 C to 30 C. The measured unit area thermal conductance of the heat exchanger is 25 W/m^2 K (this high rate is achieved using fins on the air side of the exchange surface).

What area of heat exchange is needed to achieve these conditions?

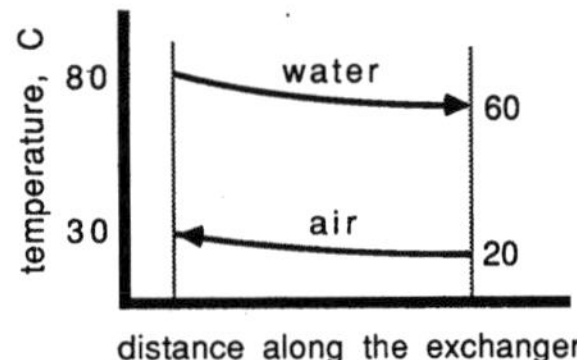

Solution: The specific heats of air and water are 1006 and 4190 J/kgK, respectively (approximately). The rate of heat exchange can be found from

$$q = mc_p \Delta t$$

$$= (0.05 \text{ kg/s})(4190 \text{ J/kgK})(20 \text{ K}) \text{ on the water side, or}$$
$$= (0.4165 \text{ kg/s})(1006 \text{ J/kgK})(10\text{K}) \text{ on the air side, thus,}$$
$$= 4{,}190 \text{ J/s.}$$

The LMTD is

$$\text{LMTD} = \frac{(80 - 30) - (60 - 20)}{1n(80 - 30)/(60 - 20)]} = 44.81 \text{ K.}$$

Note the LMTD is intermediate between the 50 K difference at one end of the heat exchanger and the 40 K difference at the other end, but is not the simple average. Equation 4-10 can be restated as

$$q = UA(\text{LMTD}),$$

thus,

$$A = q / U(\text{LMTD})$$
$$= (4{,}190 \text{ J/s})/(25 \text{ W/m}^2\text{K})(44.81 \text{ K})$$
$$= 3.74 \text{ m}^2.$$

4-10.4. Heat Exchanger Effectiveness and the NTU Method. As used in Example 4-11, the LMTD approach to heat exchanger design is useful when all fluid flow temperatures are known and the heat exchange area, or thermal conductance, is to be determined. When inlet or outlet fluid temperatures are to be determined for a given exchanger, the LMTD method becomes cumbersome.

Environmental control design frequently involves using a commercially available heat exchanger and designing fluid temperatures and flow rates to achieve certain goals, rather than designing the exchanger itself. The NTU method is more useful for such design problems.

The effectiveness, ε, of a heat exchanger is defined as

$$\varepsilon = \frac{\text{actual rate of heat transfer}}{\text{maximum possible rate of heat transfer}}. \qquad (4\text{-}12)$$

The actual rate of heat transfer can be determined by calculating either the rate heat is lost from the hot fluid or the rate heat is gained by the cold fluid.

$$q_{actual} = m_h c_h \Delta t_h, \text{ or} \qquad (4\text{-}13a)$$

$$= m_c c_c \Delta t_c. \qquad (4\text{-}13b)$$

where m is mass flow rate, c is specific heat, Δt is temperature change; the subscripts h and c refer to the hot and cold sides of the exchanger, respectively.

The maximum possible rate of heat exchange will be achieved when fluid on one side of the exchanger attains the entering temperature of the other fluid stream. The stream in which this is possible is the one with the minimum value of the (mc) product, because an energy balance shows the heat gained by the cold side must equal the heat lost by the hot side. Thus,

$$q_{max} = (mc)_{min}(t_{h, \, inlet} - t_{c, \, inlet}). \tag{4-14}$$

When entering fluid temperatures, mass flow rates and specific heats are known, the maximum possible rate of heat exchange can be determined and if the effectiveness can be predicted, the actual rate of heat transfer can be found.

Equations to calculate heat exchanger effectiveness can be found in heat transfer texts (such as Holman, 1981). For simple parallel and counter flow heat exchangers, effectiveness values can be determined from

$$\text{parallel flow: } \varepsilon = \frac{1 - \exp[-\, NTU(1 + C)]}{1 + C} \tag{4-15}$$

$$\text{counter flow: } \varepsilon = \frac{1 - \exp[-\, NTU(1 - C)]}{1 - C \exp[-\, NTU(1 - C)]} \tag{4-16}$$

$$\text{counter flow, } C = 1: \varepsilon = \frac{NTU}{1 + NTU} \tag{4-17}$$

$$\text{all exchangers for } C = 0: \varepsilon = 1 - \exp(-\, NTU) \tag{4-18}$$

$$\text{where } NTU = UA/(mc)_{min}, \text{ and} \tag{4-19}$$

$$C = (mc)_{min}/(mc)_{max}. \tag{4-20}$$

The grouping of terms, $UA/(mc)_{min}$, is termed the "Number of Transfer Units", or NTU. This term arises because the grouping can be interpreted to indicate the relative size of the heat exchanger. Equations to determine NTU are:

$$\text{parallel flow: } NTU = \frac{-\, \ln[1 - (1 + C)\varepsilon]}{1 + C} \tag{4-21}$$

$$\text{counter flow: } NTU = (C - 1)^{-1} \ln\left(\frac{\varepsilon - 1}{\varepsilon C - 1}\right) \tag{4-22}$$

$$\text{counter flow, } C = 1: NTU = \varepsilon/(1 - \varepsilon) \tag{4-23}$$

$$\text{all exchangers, } C = 0: NTU = -\, \ln(1 - \varepsilon) \tag{4-24}$$

Heat transfer texts provide graphs to determine ε and NTU, but equations are more useful for computerized analyses.

Applications of the NTU method will be seen through two examples. The basic method of solution is based on trial and error.

Example 4-12

<u>Problem:</u> Warm water from the cooling condenser of a power generating station is to provide heat for a large greenhouse range. The warm water flows through counter flow heat exchangers (many exchangers operating in parallel) and greenhouse air is heated by being circulated through the exchangers. Warm water is available at 30 C and greenhouse air is to be heated from 18 C to 27 C. Air flows through the exchanger at a mass flow rate of 2.5 kg/s. The heat exchanger surface area is 100 m^2 and the average unit area thermal conductance (U) is 50 W/m^2 K.

At what rate must water be pumped through the exchanger to provide this amount of air heating?

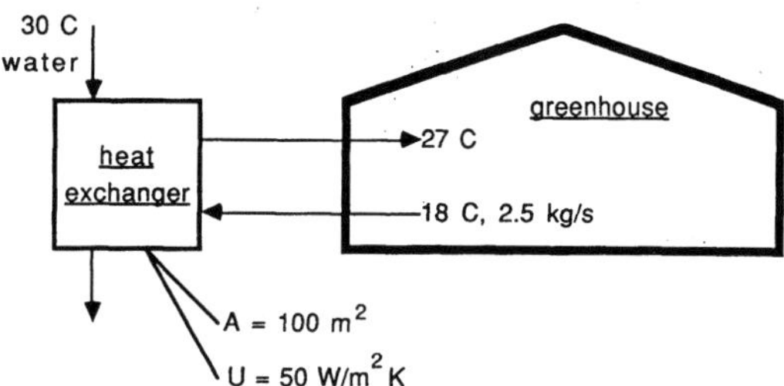

<u>Solution:</u> The limiting side of the heat exchanger is not known at this point. The trial and error approach of the NTU method assumes one of the sides is limiting, completes the analysis, and then checks to determine whether the assumption was correct. The check is whether the heat gained and lost by the fluid streams is the same as the heat transferred across the heat exchange surface.

Begin by assuming the air side is limiting. (It is easier to start using the air side because both m and c are known for the air side.)

$$(mc)_{min} = (2.5 \text{ kg/s}) (1006 \text{ J/kg K}) = 2515 \text{ J/s K.}$$

Air enters at a temperature of 18 C and exits at 27 C, thus, the heat gain is

$$q_{air} = (2515 \text{ J/s K})(27 \text{ C} - 18 \text{ C}) = 22{,}635 \text{ W.}$$

The Number of Transfer Units can be calculated,

$$\text{NTU} = UA / (mc)_{min} = (100 \text{ m}^2) (50 \text{ W/m}^2 \text{ K})/(2515 \text{ J/s K})$$
$$= 1.99,$$

and the effectiveness is

$$\varepsilon = \frac{\text{actual temperature change}}{\text{maximum possible temperature change}}$$
$$= (27\ C - 18\ C)/(30\ C - 18\ C) = 0.75,$$

assuming the specific heat is not a function of temperature and the best that can be done is to heat the air to the entering water temperature (from 18 C to 30 C).

Now the question is: Are the NTU and ε values compatible for a value of $0 < C < 1$? If so, the original assumption of air as the limiting side is correct ($C < 1$). If $C < 1$ the minimum is less than the maximum, as it should be.

Equation 4-16 expresses effectiveness of a counter flow heat exchanger, and everything is known but C. Unfortunately, the equation is nonlinear and a solution for C is not straightforward. One method to approach a solution for C is by trial and error, assuming values for C, solving the equation, and iterating until the value of effectiveness from the equation approximates the known value of $\varepsilon = 0.75$. This is an ad hoc method of numerical solution which could be readily implemented on a computer. Sample results of iterations for NTU = 1.99 are:

C	ε	
0.5	0.773	(already close to 0.75)
0.4	0.793	(wrong direction)
0.6	0.753	(looking better)
0.65	0.742	(too far)
0.62	0.748	(still too far)
0.61	0.750	(good enough for me)

So what has been learned? The initial assumption was correct. Air is the limiting side and $C = 0.61$. From the definition of C,

$$(mc)_{max} = (mc)_{min} / C = 2515\ \text{J/s K} / 0.61 = 4123\ \text{J/s K}.$$

In the temperature range from 18 C to 27 C the specific heat of water is approximately 4180 J/kg K, thus,

$$m_{water} = (4123\ \text{J/s K})/(4180\ \text{J/kg K}) = 0.986\ \text{kg/s}.$$

The density of water in the temperature range is approximately 0.998 kg/L, thus, the pumping rate should be approximately 1 L/s.

The exit temperature of water can also be determined. The rate of heat exchange is 22,635 W, water flows at 0.986 kg/s with a specific heat of approximately 4180 J/kg K, thus, its temperature change must be

$$\Delta t_{water} = (22{,}635\ \text{J/s})/(0.986\ \text{kg/s})(4180\ \text{J/kg K})$$
$$= 5.5\ K.$$

Water exits the exchanger at 30 C - 5.5 K = 24.5 C.

Example 4-13

Problem: Reconsider Example 4-12 and determine the required water pumping rate when water enters the heat exchanger at a temperature of 35 C instead of 30 C.

Solution: The same approach as before can be used; air can be assumed to be the limiting side. The NTU is unchanged and equals 1.99. The effectiveness is now

$$\varepsilon = (27\ C - 18\ C)/(35\ C - 18\ C) = 0.53.$$

The same iterative approach may be used as before,

C	ε	(as before)
0.5	0.773	
0.7	0.731	
0.9	0.688	
0.99	0.668	(not even close)

To reach $\varepsilon = 0.53$ would require $C > 1$, which is clearly impossible. What is the conclusion? Air must not be the limiting side. It may not be intuitively obvious that warmer water suddenly causes the water side to be limiting, but the equations are sufficiently complicated that intuition should not be trusted.

So what can be done? A solution is possible using a trial and error solution for mc $_{water}$. A beginning point is known, mc $_{water}$ must be less than 2515 J/sK (the mc of the air side) for the air side is not limiting.

Values of mc $_{water}$ can be assumed, and the effectiveness determined both by Equation 4-12 and Equation 4-16. When the two agree, a solution has been attained.

mc $_{water}$	C	NTU = UA/C_{min}	Δt_{water}	$\varepsilon = \Delta t_{water}/17$	ε by Eq. 4-16
2000	0.795	2.500	11.32 K	0.666	0.766
1500	0.596	3.333	15.09	0.888	0.876
1600	0.636	3.125	14.15	0.832	0.853
1550	0.616	3.226	14.60	0.859	0.865
1530	0.608	3.268	14.79	0.870	0.869
1533	0.610	3.262	14.77	0.869	0.868
1534	0.610	3.259	14.76	0.868	0.868

Values in the table are determined from the following:

$$C = C_{water}/2515 \text{ J/s K}$$
$$NTU = (50 \text{ W/m}^2 \text{ K})(100 \text{ m}^2)/mc_{water}$$
$$\Delta t_{water} = (22{,}635 \text{ J/s}) / mc_{water}$$
$$\varepsilon = \Delta t_{water} / 17 \text{ is based on a maximum possible water temperature change of 17 K (from 35 C to 18 C).}$$

The effectiveness value calculated in two ways agrees when mc_{water} is 1534 J/s K, which is less than mc_{air}, thus, water is indeed the limiting side.

With mc_{water} = 1534 J/s K and the specific heat of water = 4180 J/kg K, the mass rate flow of water must be

$$m_{water} = (1534 \text{ J/s K})/(4180 \text{ J/kg K}) = 0.367 \text{ kg/s,}$$

and with a water density of 0.998 kg/L the flow rate of water should be (0.367 kg/s)/(0.998 kg/L) = 0.368 L/s. The exiting water will be cooled by 14.76 K (already calculated in the table). The exiting water temperature will thus be 35 C - 14.76 K = 20.24 C, which is more than 4 K colder than the exiting water temperature in Example 4-12 where entering water temperature was 5 K colder.

A careful comparison of Examples 4-12 and 4-13 can help in gaining an understanding of the action of heat exchangers. In these two examples, when the water enters at a higher temperature, heat is transferred more quickly at the beginning and less heat must be transferred at the end, thus, a colder water exit temperature can be tolerated. The greater temperature change in the water permits a significantly lower water flow rate, which is important for design. One could conclude that when reject heat sources are used for heating, slight source temperature increases can lead to significant changes of operation.

4-10.5. Program XCHANGER. As is obvious from Examples 4-12 and 4-13, designing heat exchangers can be a tedious exercise if done using hand methods. Program XCHANGER is provided as a tool useful for exploring heat exchanger designs. The program uses the NTU method and can calculate any of four temperatures: the warm side entering or exiting temperature, or the cold side entering or exiting temperature.

Examples 4-12 and 4-13 demonstrated how the NTU method can be used to find required mass flow rates and exiting temperatures of the flow streams. The capability of calculating required entering temperatures is included in XCHANGER through iterative searching for an entering temperature which satisfies an energy balance (in the manner demonstrated in Examples 4-12 and 4-13). The program also determines when a solution is not possible – which occurs more frequently than one might intuitively expect!

Example 4-14

<u>Problem:</u> Reconsider Example 4-13. The heat exchanger manufacturer markets units in many sizes, ranging from a heat exchange surface area of 20 m^2 to an area of 200 m^2 in increments of 20 m^2. The choice of 100 m^2 in Examples 4-12 and 4-13 was somewhat arbitrary. In this example, explore the effects of using exchangers of different areas for the design conditions of Example 4-13.

<u>Solution:</u> Program XCHANGER was used to determine all data provided below. The program was run repeatedly, changing the UA value for each iteration.

Convergence is very sensitive to the UA value. At UA = 1895 W/K there is not convergence, while at UA = 1896 W/K there is a solution for a water flow rate of 1308 kg/s, which is clearly not suitable.

Area, m^2	Limiting	UA, W/K	m_{water}	t_{exit}^{water}	NTU	ε
20	solution does not converge					
40	cold side	2000	3.4837	33.446	0.80	0.5294
60	warm side	3000	0.5474	25.107	1.31	0.5819
80	warm side	4000	0.4110	21.826	2.33	0.7749
100	warm side	5000	0.3669	20.243	3.26	0.8681
120	warm side	6000	0.3468	19.386	4.14	0.9185
140	warm side	7000	0.3361	18.887	4.98	0.9479
160	warm side	8000	0.3298	18.581	5.80	0.9658
180	warm side	9000	0.3260	18.387	6.61	0.9772
200	warm side	10000	0.3235	18.262	7.39	0.9846

Data such as calculated in this example could be used to select the most appropriate size of exchanger. For example, an area of 60 m^2 results in need for much less water than would be required by an exchanger with an area of 40 m^2, yet increasing the area to 80 m^2 does not result in a comparable further savings. If water supply and pumping costs are a major consideration, a larger area might be cost-effective. If not, the exchanger with 60 m^2 of heat exchange area could be best. However, even if water flow is a large concern, an exchanger area larger than 100 m^2 will not provide significant savings in the need for water.

Note that all the exchangers except the first are acceptable in that they solve the design problem of providing heat to the greenhouse. However, some are more cost-effective than others and the financial aspects of engineering design are always important.

SYMBOLS

A	area, m^2
C	ratio of mc products, see Equation 4-15
c	coefficient, see Equation 3-44
c_c, c_h	specific heat, kJ/kg K
D	hydraulic diameter, m
F	perimeter heat loss factor, W/m K
G	unit area mass flow rate, kg/m^2 s
h	convective heat transfer coefficient, W/m^2 K
I	solar irradiation, W/m^2
k	thermal conductivity, W/m K
m	mass flow rate, kg/s
P	perimeter, m
q	heat transferred, W
R	unit area thermal resistance, m^2 K/W
ΔR	net long-wave thermal radiation exchange, W/m^2
t	temperature, C
T	absolute temperature, K
U	unit area or unit length thermal conductance, W/m^2 K or W/m K
x	spacial variable, m
α	solar absorptance
ϵ	heat exchanger effectiveness
ϵ	surface emittance for radiation heat exchange
ϕ	surface tilt angle
Ω	degree of cloudiness, from 0 (none) to 10 (total)

EXERCISES

1. You are designing a building to be 14 m wide, 60 m long, and 3 m high. It will have a well ventilated attic space.

   The ceiling is hung from the roof trusses and will be sheathed on the lower side with 19.05 mm plywood and insulated with 88.9 mm of mineral fiber insulation above the sheathing. The walls are framed using 38.1 mm x 88.9 mm lumber, sheathed on the inside with 15.88 mm thick high density particleboard, insulated with full thickness cellulosic insulation, sheathed on the outside with 25.4 mm cellular polyurethane (surfaced) and sided with 9.53 mm lapped plywood. There are two panel doors in the building, each 2.5 m x 3.5 m, and each panel door is 44 mm thick with 11 mm thick panels. Single glazed windows (no storm windows) comprise 4% of the gross wall area.

   The owners decided a window area of 4% of gross wall area will not permit enough light to enter, yet they do not want simply to increase window area at the expense of energy conservation. Calculate the window area which would provide the same overall R-value of the shell of the building if double glass ("thermopane") with 3 mm thick glass and a 6 mm thick airspace were used in place of single glazed windows.

2. Determine the effective unit area R-value of a framed wall made of 2 x 6s (38.1 mm x 139.7 mm) which are 24 in. (610 mm) on center. The wall has 88.9 mm of batt insulation (mineral fiber) on the inside (warm) wall side of the wall cavity. The inside wall sheathing is 12.7 mm plywood and the outside of the wall is sided with 15.88 mm plywood (there is no outside sheathing under the siding). Include effects of framing on the R-value. Assume framing is 18% of the gross wall area.

3. Consider the barn wall in Example 4-5 constructed as outlined in Exercise 2 above. The windows will be single glazed with wood frames. Determine the average unit area R-value of the wall. Assume window area is 8% of the gross wall area.

4. Continue with the wall described in Exercise 3 above. Add a panel door (44 mm thick with 11 mm panels) which is 2.8 m high and 4.2 m wide and determine the unit area R-value of the wall. Assume doors are equivalent to 3% of the gross wall area.

5. A wall has been built which has an average unit area R-value of 1.8 m^2 K/W. The wall is built on an unheated slab floor insulated along the edge by an insulation with R = 0.95 m^2 K/W. The wall is 2.8 m high. Of all the heat lost through the wall and perimeter, what percent is through the wall and what percent is through the perimeter?

6. For the wall described in Exercise 5 above, by how much would heat loss through the wall and perimeter increase if the perimeter were not insulated?

7. The heat exchanger described in Example 4-10 is to be built with a total heat exchange area of 50 m^2. If outdoor air is - 5 C and indoor air is 18 C, what will be the rate (watts) of heat exchange between the two airstreams? What will be the temperatures of the two airstreams as they exit the exchanger? Assume there is no condensation within the exchanger.

8. A forced air heating unit (counter flow heat exchanger) is to be used for a greenhouse. The unit, which has a UA value of 800 W/K, receives water from a boiler at 85 C. During night the heater must warm the greenhouse air from 16 C to 32 C (at a mass airflow rate of 2.3 kg/s). What water flow rate is required to do this?

During the day, greenhouse temperature rises from 16 C to 23 C and air enters the exchanger at 23 C. What will be the exit temperature of the greenhouse air? What will be the heat delivery rate (watts) during the night, and what will it be during the day?

REFERENCES

ASHRAE. 1976. Procedures for determining heating and cooling loads for computerizing energy calculations. American Society of Heating, Refrigerating and Air Conditioning Engineers, Atlanta, GA. 182 pp.

ASHRAE. 1989. Handbook of Fundamentals. American Society of Heating, Refrigerating and Air Conditioning Engineers, Atlanta, GA.

ASHRAE. 1985. Environmental Control Principles, a textbook supplement to the ASHRAE Handbook 1985 Fundamentals Volume. American Society of Heating, Refrigerating, and Air Conditioning Engineers, Atlanta, GA.

Esmay, M.L. and J.E. Dixon. 1986. Environmental Control for Agricultural Buildings. The AVI Publishing Co., Westport, CT. 287 pp.

Holman, J.P. 1981. Heat Transfer. McGraw-Hill Book Co., New York. 570 pp.

Kusuda, T. 1976. The computer program for heating and cooling loads in buildings. National Bureau of Standards Load Determination Program. U.S. Department of Commerce/National Bureau of Standards. BSS No. 69. Washington D.C.

Lokmanhekim, M., editor. 1971. Algorithms for building heat transfer subroutines. American Society of Heating, Refrigerating, and Air Conditioning Engineers, Atlanta, GA.

Mackay, C.O. and L.T. Wright. 1946. The sol-air thermometer – A new instrument. Transactions ASHVE 52:271.

Midwest Plan Service. 1978. Professional design supplement. Midwest Plan Service, Ames, IA. 218 pp.

Midwest Plan Service. 1983. Structures and Environment Handbook. Midwest Plan Service, Ames, IA.

Parmelee, G.V. and W.W. Aubele. 1952. Radiant energy emission of atmosphere and ground. Transactions ASHVE 58:85-106.

Stoecker, W.F. and J.W. Jones. 1982. Refrigeration and Air Conditioning. McGraw-Hill Book Company, New York. 443 pp.

Threlkeld, J.L. 1970. Thermal Environmental Engineering. Prentice-Hall, Englewood Cliffs, NJ. 493 pp.

Timmons, M.B. and L.D. Albright. 1978. Wind directional dependence of sol-air temperatures. Transactions of the ASAE 21 (4) :742-746.

CHAPTER 5
STEADY-STATE ENERGY AND MASS BALANCES

5-1. Introduction

Three important concepts underlie environmental analysis of buildings: (a) control volumes, (b) conservation of energy, and (c) conservation of mass. The concept of energy conservation is applied to sensible heat, and mass conservation is applied to latent heat (humidity) and gaseous (or other) contaminants. Both conservation concepts rely on use of control volumes.

This chapter will begin with descriptions of control volumes and energy and mass balances. It will then focus on the components of each balance in turn, and, finally, meld the components together into solutions to answer questions such as: What will be the inside air temperature? How much heat will be needed? How many fans will be needed for ventilation? What size should the cooling system be?

Agricultural buildings are typically characterized by having single airspaces. This simplifies the analysis computationally, but in concept everything to be covered also applies to multiple airspace buildings.

<u>Control volumes.</u> Processes in thermodynamics, fluid mechanics, and heat transfer can be viewed in two ways. One way is to examine a process with a focus on its internal features. The other is to enclose the process with an imaginary boundary and examine only what passes across the boundary. The second is the control volume approach and is commonly used in environmental analyses and system design.

Although, at first, the control volume approach may appear to simplify processes excessively (a black box approach), in practice it is a very powerful tool. Many processes can be analyzed without knowing the details of what actually happens within the control volume. In some ways, this is like what a depositor knows of a bank account.

A person with a bank account has little concern for the internal matters of postings, proofings, and other recordkeeping functions within a bank. Knowing if the same dollar which was deposited is later the one withdrawn is immaterial. Everything the customer needs to know about the account can be determined by knowing what goes into the account and what comes out. The balance is obtained knowing the data. The current balance figure which the bank provides is redundant information (and, of course, a check on the accuracy of the account owner's and the bank's records). Internal workings of the banking process are sufficiently complicated to prevent ordinary customers from determining their bank balances if the only information available is complete detail of the internal processes.

The same is true of certain thermal processes in buildings. Some processes can be analyzed knowing only what enters the airspace and what leaves it. Obviously a bank account is not a perfect analogy to building environmental control, but the comparison illustrates how processes can be described sufficiently by what crosses an imaginary boundary surrounding the process.

A simple example of applying control volume concepts is the following. Consider a forced hot air heating system in an older, single family house. Some of the heating ducts pass through the attic; the attic is otherwise unheated. Should the heating ducts be insulated? If they are, the attic will be colder and the lower attic temperature will increase heat loss through the attic floor from the heated rooms below. Will the increase of heat loss by conduction from the rooms below be greater than the savings from insulating the ducts? Will insulation be a wise investment or a mistake?

It would be a significant exercise to answer this question in detail by analyzing the heat transfer processes involved, that is, by concentrating on heat transfers from place to place within the house. Instead, mentally place a control volume around the entire house. When it is cold outdoors, heat escapes from all parts of the house and the rate of heat loss depends on the temperature difference from inside the house to out. If the attic is colder when the ducts are insulated, there will be less heat loss from the attic and, thus, less from the house. This can be stated with confidence although nothing is known about the rate of heat loss from the ducts before or after insulating or the insulation value of the floor between the attic and the rooms beneath.

<u>Sensible energy balance.</u> A simple set of sensible energy transfers for a single airspace building is shown in the sketch of Figure 5-1.

The energy flow terms are:

q_s: sensible heat gain from animals (or people) within the airspace.

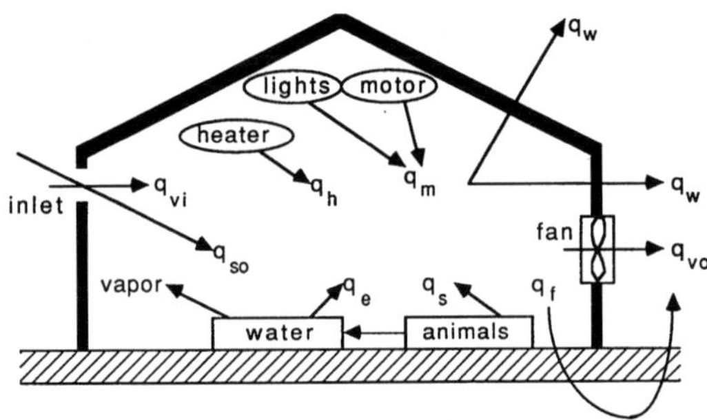

Figure 5-1. Contributors to a sensible energy balance in an animal housing building.

144

q_m: sensible heat gain from "mechanical" sources such as motors and lights. Such sources are usually electrical devices, and the heat gain is from conversion of electrical energy to sensible heat.

q_{so}: sensible heat gain from the sun. This can be gain through windows in a barn, and be relatively small, or it can be solar gain into a greenhouse and dominate all other heat gains.

q_h: sensible heat gain from a heating system.

q_{vi}: the sensible heat contained in the ventilation air entering the space referenced to some temperature datum. The datum is immaterial as, eventually, only the difference of sensible heat contents of the entering and exiting airstreams will be considered.

q_w: the transfer of sensible heat through the structural cover of the building; i.e., walls, ceiling, windows, doors, etc.

q_f: sensible heat transfer to the floor of the building primarily at the perimeter. It will be assumed that heat exchange with the floor in the interior of the building is relatively insignificant.

q_e: the rate of conversion of sensible heat to latent heat within the airspace. For example, evaporation of water from the floor of a barn or transpiration and evaporation of water from plants in a greenhouse are conversions of sensible to latent heat.

q_{vo}: the sensible heat contained in the ventilation air leaving the space referenced to the same temperature datum as q_{vi}.

The control volume for the energy balance is the air within the space bounded by the walls, floor, ceiling, and imaginary planes at the ventilation inlets and outlets. We need not specify the type of ventilation, or if total ventilation has a component of air infiltration through cracks. The arrows in Figure 5-1 indicate the assumed directions of heat transfers. The assumed directions need not be fixed, as long as the resulting energy balance is written to agree with the directions.

The general form of an energy balance for a control volume is

Gains - Losses = Change of Storage,

and if conditions are steady-state, there is no change of storage.

The steady-state sensible energy balance for Figure 5-1 rearranged in the form

Gains = Losses   is

$$q_s + q_m + q_{so} + q_h + q_{vi} = q_w + q_f + q_e + q_{vo} \qquad (5\text{-}1)$$

<u>Mass balance.</u> A simple mass balance for the same airspace is shown in the sketch of Figure 5-2. The same control volume is used as for the sensible heat balance.

The mass flow terms are:

m_p: the rate the material of interest (water vapor, carbon dioxide, etc.) is produced within the space.

m_{vi}:   the rate at which the material of interest is carried into the airspace by ventilation air.

m_{vo}:   the rate at which the material of interest is carried out of the airspace by ventilation air.

Several assumptions are contained within the sketch in Figure 5-2. Mass transfer by diffusion through the structural cover and floor is assumed to be sufficiently slow as to be negligible. The sources of the material of interest may be many but are shown as a collective whole. The material of interest is assumed not to undergo a transformation within the space to another material [such as the conversion of latent heat (in humidity) to sensible heat by condensation to form fog]. The only removal mechanism is ventilation; for example, dust in the air does not settle out onto the floor or adhere to the walls. Air in the space is well mixed.

The steady-state mass balance for Figure 5-2 is

$$m_p + m_{vi} = m_{vo}. \qquad (5\text{-}2)$$

Individual terms of the energy and mass balances will now be examined in detail.

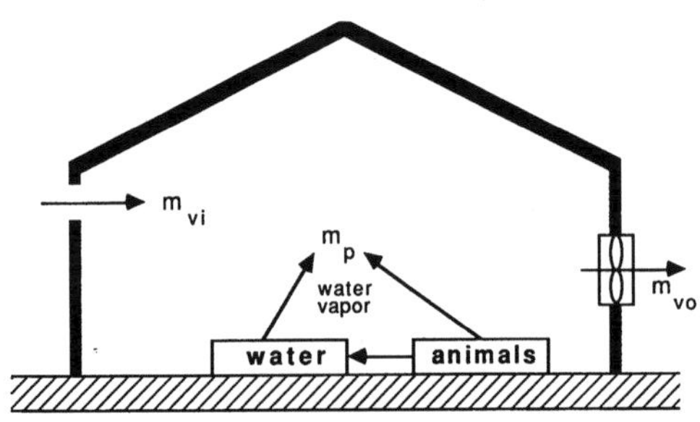

Figure 5-2. Mass balance for a single airspace.

5-2. Components of the Sensible Energy Balance

The energy balance can be used for many purposes. Heating or cooling required to maintain a predetermined air temperature can be calculated. The ventilation rate required to maintain design conditions can be found. The indoor air temperature in response to imposed conditions can be determined. The amount of insulation needed to limit heating may be of interest. However, before those calculations are possible, the components of the energy balance must be quantified. Some of the components have been discussed previously – some have not.

5-2.1. Sensible Heat Produced by Animals, q_s. Mammals are homeothermic creatures. That is, they attempt to maintain constant body temperatures (a state of homeothermy). Body temperatures are generally above those of the ambient air.

While poultry body temperatures are approximately 41 C, mammals of commercial importance in agriculture maintain body temperatures in the vicinity of 38 C. Internal physiological processes which maintain constant conditions are extremely complex and together are a condition termed "homeostasis". More complete descriptions of animal physiology and homeothermy can be found, for example, in Hellickson and Walker (1983), Esmay and Dixon (1986), and especially Clark (1981) and Curtis (1983).

The first priority of homeostasis is to maintain body temperature. If there is not sufficient feed to support body temperature, growth, and production, then production and growth slow. During cold weather when more heat is needed to maintain body temperature, more feed is eaten if available. In extreme cold, growth may become negative when fat is metabolized to maintain body temperature.

Sensible heat production is a by-product of maintenance, growth, and production; if conditions are extremely hot, production (e.g., milk or eggs) and feed intake are reduced to limit the quantity of sensible heat which must be expelled to the environment.

However, good agricultural practices mean ambient conditions should be regulated and sufficient food provided to the animals so all three functions can be supported with something approaching an optimal level. The temperature zone where this is possible is called the "comfort zone"; conditions outside the zone characterize "thermal stress".

Ambient thermal environment significantly affects animals, and the animals housed within a building significantly affect the environment within the building. A focus on the second effect is required to solve the energy balance in Equation 5-1.

Animals lose heat to their surroundings by radiative and convective transfer and, in certain circumstances, by conductive heat transfer (e.g., from a pig lying on the floor to the floor). Evaporative heat loss is a major component of total heat exchange, for expired air is nearly saturated with water vapor at body temperature. In cold weather, this means each breath expels a great deal of latent heat compared to the latent heat content of the ambient air.

Sensible heat is lost primarily from the outer surfaces of animals; latent heat is lost primarily from respiratory tracts. In general, farm animals do not have a sufficient number of sweat glands to make evaporation from the skin a highly significant factor.

Calorimetric studies have provided extensive data which can be used to estimate sensible heat production from commercially important farm animals. Such data is used widely, but limitations of the data should always be observed.

Calorimetric data does not contain the effect of evaporation of animal waste products from floors, etc., which is a conversion of sensible to latent heat. Thus, calorimetric data applied to a barn does not correctly estimate the amounts of sensible and latent heat actually added to the airspace.

Genetic differences of new breeds may affect heat and moisture production from animals. The best example of this is chickens, where the breeding process is ongoing and birds of a (human) generation ago were very different from today's birds.

The production level of the animals used to obtain the calorimetric data may have been very different from the production levels of the animals to be housed in the building being designed (for example, milk production from dairy cows). Different feeding and production levels lead to different sensible heat generation levels.

Finally, the ambient conditions used when the calorimetric data were obtained may have been significantly different from conditions expected in the animal housing being designed. For example, it is common to operate animal growth chambers and calorimeters at approximately 50% relative humidity. In winter, animal barns are frequently more humid than 50% and can be during summer drier. Total heat production from an animal may not be significantly affected by relative humidity (within reasonable ranges) but partitioning between sensible and latent heats will be.

However, much engineering design is accomplished using "best available" rather than "perfect" information. Such designs are adequate – adequate if sufficient safety margins are incorporated to guard against reasonable variations of material properties, operating conditions, and biological responses.

Fortunately, animals and plants have elastic responses to their environments. A column overloaded by a small amount for a short time may collapse, but a plant or animal kept a few degrees too warm or cold for a day will survive and respond to the improper conditions with only a temporary change of growth or production. Of course, if conditions become too extreme, the plant or animal will also "collapse" and may die or suffer long-lasting effects. Environmental control and alarms must be installed to prevent such extreme conditions and provide warnings of potential disasters.

Appendix 5-1 contains animal heat and moisture production data, abstracted from data standard D270.4 of The American Society of Agricultural Engineers (ASAE). Data are generally for a standard size of animal – an animal unit. For example, one animal unit for dairy cattle is a 500 kg cow. Many dairy cows are larger than 500 kg and some are smaller, so a means must be found to extrapolate heat production data to other animals of various sizes.

It has been determined that mammalian sensible heat production is linearly correlated more closely to body surface area than body weight, and surface area and weight are not linearly correlated. The relationship between heat production and weight is such that heat production is proportional to weight raised to the power of 0.734. Thus, for example, a cow which weighs 20% more than another will produce $(1.2)^{0.734} = 1.14$ times as much heat.

Example 5-1

Problem: Determine the sensible heat production from 100 (590 kg) dairy cows housed in a barn at 12 C.

Solution: Data in Appendix 5-1 for dairy cattle provide the needed information after interpolation and extrapolation. At 10 C a 500 kg dairy cow loses 1.5 W/kg of sensible heat and at 15 C she loses 1.2 W/kg. Interpolating linearly between these two values for heat production from one 500 kg cow at 12 C,

$$(q_s)_{500\,kg} = 1.2 + (2/5)(1.5 - 1.2) = 1.32 \text{ W/kg}.$$

A cow with a mass of 590 kg will lose heat at a rate greater than will a 500 kg cow, by the ratio $(590/500)^{0.734}$, the heat loss is

$$(q_s)_{590\,kg} = 1.32(500)(590/500)^{0.734} = 745 \text{ W/cow}.$$

With 100 cows in the barn, total sensible heat addition to the air will be

$$q_s = (745 \text{ W/cow}) (100 \text{ cows}) = 74,500 \text{ W} = 74.5 \text{ kW}.$$

Note: The heat production data do not state the relative humidity at which the data was obtained. The original reference (Yeck and Stewart, 1959) states that humidities between 50% and 70% were maintained during their tests. We can be confident of our calculations only if expected relative humidities are within this range. In practice, barns are between 50% and 70% relative humidity much of the time.

For computerized analysis of building thermal environment, it is easier to use equations than tabled data. One means to express data is in polynomial form; the animal heat production data in Appendix 5-1 can be expressed in such a form. For example, sensible heat loss from 500 kg dairy cows is given at five air temperatures. The data can be expressed as a polynomial function of air temperature, and the function can have an order as high as four. However, it is best not to use high order expressions because of the possible lack of smoothness between data points. Generally, a second order polynomial suffices.

Example 5-2

Problem: Determine a second order polynomial which can be used to estimate sensible heat production from 500 kg dairy cows as a function of ambient air temperature.

Solution: The enclosed program POLYNOM was used for five data points and second order. A second order polynomial fit of the five heat production data points for dairy cows yielded the following expression:

$$(q_s)_{500\ kg} = 1.86 - 3.074E - 2t_{air} - 5.268E - 4(t_{air})^2$$

where t_{air} is air temperature in C and q_s is in W/kg.

To assess adequacy of the fit, the observed and predicted sensible heat production data can be compared.

Observed	Predicted
1.9 W/kg	1.9 W/kg
1.5	1.5
1.2	1.3
1.1	1.0
0.6	0.6

Agreement is not perfect but is likely to be adequate within our knowledge of other design parameters in a typical engineering design problem.

5-2.2. Mechanically Produced Heat, q_m. Lighting is the largest source of electrically generated heat in most agricultural buildings. Unless motors operate for a significant fraction of the time, their contributions are negligible.

To estimate heat production from lights, total electrical input must be determined. That is, incandescent lights produce heat approximately equal to the wattage of the installed lights. Fluorescent, metal halide, mercury, and sodium vapor lights add more heat than the installed wattage because of the power required to operate the ballasts (unless the ballasts are remote, as might be the case in a growth chamber, for example).

Motors are generally rated based on their outputs not their inputs. Small, single-phase motors have efficiencies typically in the range from 55% to 75%. Motors may not be used to their full rated capacity; thus, to estimate the heat addition from a motor, its expected load and efficiency must be known. Although some of the power delivered by a motor will ultimately be in the form of mechanical energy (for example, potential energy in feed raised from a storage to the birds in a poultry house), most of the electrical power input appears as heat from the motor and frictional heat along the mechanical linkages driven by the motor. Of course, heat from motors on exhaust fans is generally carried immediately outdoors and does not contribute to the building energy balance.

In many designs of environmental control systems for buildings, the addition of heat from lights and motors is assumed negligible as a worst case for winter designs and, at a maximum, as a worst case for summer designs (if not ignored entirely). In growth chambers and growth rooms in greenhouse operations, lights are the major source of heat within the airspace and must be carefully accounted for. In greenhouses where artificial lights are used, the lights may be sufficient to provide even a majority of the heat needed during cold weather, especially if movable insulation is used.

Example 5-3

<u>Problem:</u> A poultry building is lighted by fluorescent lights, designed for approximately 40 W of installed capacity per square meter of floor area. The barn is 12 m wide and 40 m long. Lights are the only significant source of electrically generated heat within the barn. Estimate the sensible heat production rate from this source.

<u>Solution:</u> With a floor area of 480 m^2 and lighting installed at 40 W/m^2, a total lighting capacity of 19,200 W is present. However, fluorescent lights have ballasts which typically draw approximately an additional 20% power. Thus, the heat released into the space will be 1.2(19,200 W) = 23,000 W = 23 kW.

5-2.3. Solar Heat Gain, q_{so}. When sunlight irradiates a glazing, three things may happen. The solar energy may be transmitted, it may be reflected, or it may be absorbed. If I is the intensity of solar insolation at a transparent surface, the three components are:

absorbed = αI, where α is absorptance,
reflected = ρI, where ρ is reflectance, and
transmitted = τI, where τ is transmittance.

<u>Absorption:</u> A simple model for absorption within a transparent material uses the extinction coefficient, K, which characterizes the material. The extinction coefficient is a measure of absorption,

$$- dI_\lambda = I_\lambda K_\lambda dx, \qquad (5\text{-}3)$$

where I_λ is the monochromatic intensity at a point along the insolation's path through the glazing, K_λ is the monochromatic extinction coefficient, dx is the differential path length, and dI_λ is the differential change of monochromatic intensity. Fortunately, most common glazings show little change of extinction coefficient over the solar spectrum, a wavelength range of 0.3 to 3 microns. The major error in this assumption occurs in the ultraviolet wavelength band which contributes only a little to the total energy within the insolation. (Note that many glazings have extinction coefficients for longwave thermal radiation which are very different for solar insolation. For example, there is the "greenhouse effect".)

If L is path length,

$$\alpha = 1 - \exp(- KL), \qquad (5\text{-}4)$$

where K is not a function of wavelength within the solar spectrum, and represents an average over the spectrum. Extinction coefficients of typical materials are in Table 5-1.

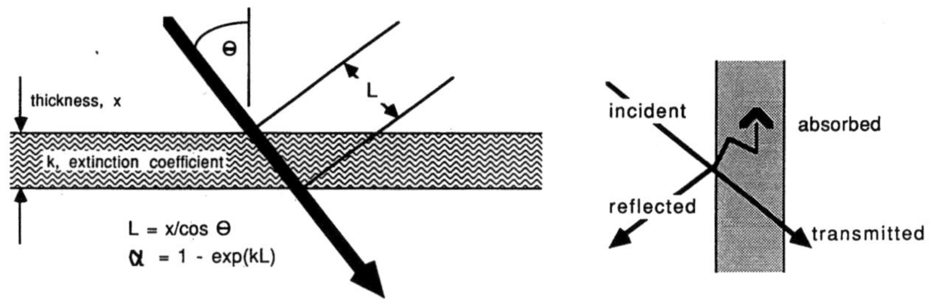

Table 5-1. Extinction coefficients for transparent materials.

Material	Extinction Coefficient, mm^{-1}
ordinary window glass	0.03 (approx.)
polyethylene	0.165
low-iron glass ($<0.01\%\,Fe_2O_3$)	0.004 (approx.)
heat-absorbing glass	0.13 to 0.27
Tedlar[a] (polyvinyl fluoride)	0.14
Mylar[a] (polyethylene terephthalate)	0.205
Teflon[a] (fluorinated ethylene propylene)	0.06

[a]trademark of E. I. DuPont de Nemours, Wilmington, DE.

If the angle of solar irradiation and thickness of the glazing are known and the extinction coefficient is estimated, absorptance can be calculated using Equation 5-4.

Example 5-4

<u>Problem:</u> Calculate the solar absorptance at normal incident angle for insolation passing through ordinary window glass 3 mm thick.

<u>Solution:</u> The extinction coefficient (from Table 5-1) is assumed to be approximately 0.03 mm^{-1}. Using Equation 5-4,

$$\alpha = 1 - \exp(-(0.03)(3 \text{ mm})) = 0.0861.$$

Nearly 9% of the insolation is absorbed as it passes through the glass. This is in addition to insolation lost due to reflection.

<u>Reflection.</u> The index of refraction, n, determines the speed of light through a material, and also determines reflection from the surface. As shown in Figure 5-3, when light strikes a reflective surface of a transparent material at an angle of incidence, ϕ, specular reflection will also be at the angle ϕ. Light passing through the transparent material will be bent to an angle ϕ. The refractive index, n, is

$$n = \sin\phi/\sin\theta. \tag{5-5}$$

Snell's law relates the refractive index and surface reflectance, ρ. Reflectance has two components: one polarized parallel to the plane of incidence and one polarized perpendicularly. The two components are

$$\rho_{parallel} = \sin^2(\phi - \theta)/\sin^2(\phi + \theta), \text{ and} \tag{5-6a}$$

$$\rho_{perpendicular} = \tan^2(\phi - \theta)/\tan^2(\phi + \theta). \tag{5-6b}$$

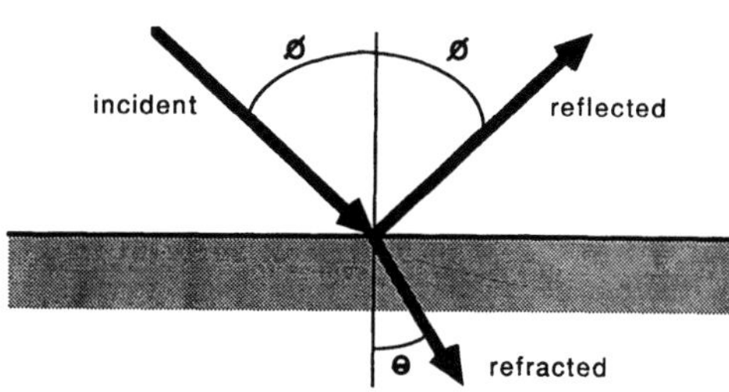

Figure 5-3. Angles of incident, reflected (specular), and refracted light on striking a transparent material.

When the incident radiation is perpendicular to the surface, the two components are equal and when the incident angle approaches 90 degrees each component of reflectance approaches 1.0. Table 5-2 contains refractive indices for common glazing materials. Example 5-5 illustrates a calculation of specular reflectance.

As an alternative means to estimate specular reflectance, an approximation of solar reflectance may be estimated using Figure 5-4. The graph applies to specular reflectance from a smooth glass surface.

Example 5-5

<u>Problem:</u> Determine the two components (parallel and perpendicular) of reflectance of light irradiating a glass surface at an angle of 50 degrees.

<u>Solution:</u> We will assume, based on Table 5-2, the refractive index is 1.50. The incident angle is 50°, thus, the refractive angle is

$$\phi = \text{arcsin} (\sin 50° / 1.50)$$
$$= 30.7°$$

Using Equations 5-6a and b,

$$\rho_{parallel} = \sin^2 (50 - 30.7) / \sin^2 (50 + 30.7)$$
$$= 0.112, \text{ and}$$

$$\rho_{perpendicular} = \tan^2 (50 - 30.7) / \tan^2 (50 + 30.7)$$
$$= 0.003.$$

Sunlight is relatively unpolarized, thus,

$$\rho_{average} = (0.112 + 0.003) / 2 = 0.058.$$

Table 5-2. Refractive indices for light in the visible waveband.

Material	Refractive Index
air	1.00
window glass	1.50 to 1.55
Tedlar[a] (polyvinyl fluoride)	1.45
Mylar[a] (polyethylene terephthalate)	1.64
Teflon[a] (fluorinated ethylene propylene)	1.34

[a]trademark of E.I. DuPont de Nemours, Willmington, DE.

Transmittance. Transmittance, reflectance, and absorptance sum to unity. When absorptance and reflectance are known, transmittance may be readily calculated. However, the methods outlined above to determine absorptance and reflectance through a glazing are relatively crude. Multiple interreflections within the glazing change absorptance and reflectance values; Stokes' equations may be used to estimate the effects of these reflections,

$$\rho_{actual} = \rho_s(1 + \frac{\tau_s^2(1-\rho_s)^2}{1-\rho_s^2\tau_s^2}), \text{ and} \qquad (5\text{-}7a)$$

$$\tau_{actual} = \tau_s(\frac{(1-\rho_s)^2}{1-\rho_s^2\tau_s^2}). \qquad (5\text{-}7b)$$

The subscript, s, indicates values obtained for single passes of radiation (values obtained using Equations 5-6a and b). Absorptance for this situation is calculated using the fact that absorptance, reflectance, and transmittance sum to unity.

$$\alpha_{actual} = 1 - \tau_{actual} - \rho_{actual} \qquad (5\text{-}7c)$$

Example 5-6

Problem: Calculate the absorptance, reflectance, and transmittance of direct beam solar radiation striking 3 mm thick window glass at an angle of incidence of 50°. Estimate the amount of radiation reflected, absorbed, and transmitted if the intensity of solar radiation is 500 W/m^2.

Solution: Assume the window glass has a refractive index of 1.50 and an extinction coefficient of 0.03 mm^{-1}. The angle of incidence is 50°. The angle after refraction is 30.7° as calculated in Example 5-5. Thus, the path length of travel through the glass is

L = 3 mm/cos (30.7°) = 3.49 mm.

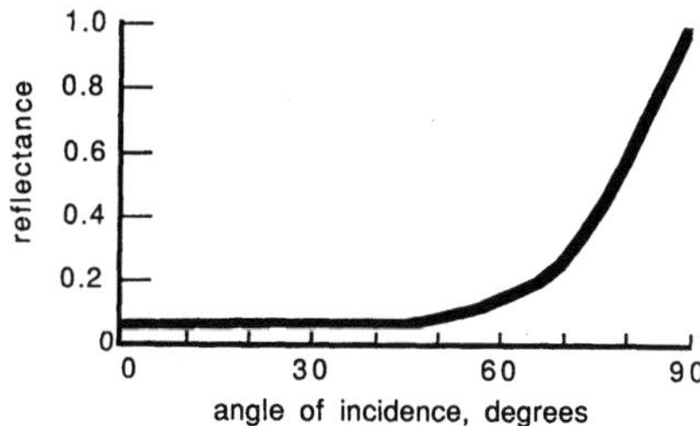

Figure 5-4. Reflectance (specular) for light incident on glass as a function of the angle of incidence.

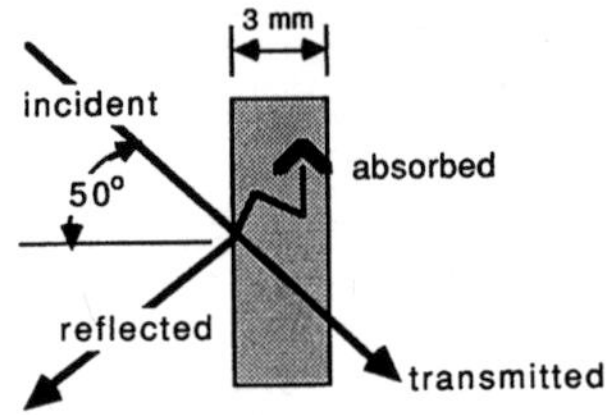

Absorptance for a single pass through the glass is

$$\alpha_s = 1 - \exp\left(- (0.03 \text{ mm}^{-1})(3.49 \text{ mm})\right) = 0.099$$
$$\text{(based on that which passes into the glass).}$$

Reflectance for irradiation at this angle of incidence is 0.058 as calculated in Example 5-5 (assuming the insolation is unpolarized). By conservation of energy, the transmittance for a single pass once the radiation passes into the glass is

$$\tau_s = 1.0 - 0.099 = 0.901.$$

Corrected for interreflections,

$$\rho_{actual} = 0.058\left(1 + \frac{0.901^2(1 - 0.058)^2}{1 - 0.058^2\,0.901^2}\right)$$
$$= 0.100, \text{ and}$$
$$\tau_{actual} = 0.901\left(\frac{(1 - 0.058)^2}{1 - 0.058^2\,0.901^2}\right)$$
$$= 0.802.$$

The overall solar absorptance is

$$\alpha_{actual} = 1.0 - 0.100 - 0.802 = 0.098 \text{ (based on irradiation).}$$

Solar insolation is 500 W/m^2, thus,

reflected radiation $= (0.100)\,(500 \text{ W/m}^2) = 50 \text{ W/m}^2$,
absorbed radiation $= (0.098)\,(500 \text{ W/m}^2) = 48 \text{ W/m}^2$, and
transmitted radiation $= (0.802)\,(500 \text{ W/m}^2) = 401 \text{ W/m}^2$.

The three sum to 499 W/m^2; the single W/m^2 which was lost in the process can be attributed to rounding effects. The transmitted radiation passes into the airspace and is the contribution to the energy balance of Equation 5-1, q_{so}.

5-2.4. Heating System, q_h. This energy term is often zero in many agricultural buildings, or it is the term in the energy balance to be determined and is unknown. If there is a predetermined heating contribution, its magnitude must be determined from the rating of the heating device, the expected level of average operation, and the heat addition is then used in the energy balance.

5-2.5. Ventilation, q_{vi} and q_{vo}. The two ventilation energy terms, q_{vi} and q_{vo}, will be considered together.

The mass flow rates of air at the inlet and exhaust are usually assumed to be equal. This is not strictly true because as air passes through a ventilated space moisture will likely be added and the mass flow rate of moist air exiting will be slightly greater than that entering. However, the difference is generally so small that a constant mass flow rate is usually assumed.

The change of sensible heat content of ventilation air is measured by its change of temperature. As we have seen before, the specific heat of air is approximately 1006 J/kg K and the heat added to the air is

$$q_{vo} - q_{vi} = 1006\rho \dot{V}(t_i - t_o), \tag{5-8}$$

where ρ, air density, is typically based on conditions at the inlet of the fan, $\dot{V}$ is the volumetric flow rate of air (also typically measured at the fan), and t_i and t_o are temperatures inside and outside the building, respectively. If conditions inside the building are known, specific heat corrected for humidity ratio (as described in Chapter 2, Psychrometrics) can be used in place of the value 1006 J/kg K.

5-2.6. Structural Heat Loss, q_w. Several of the concepts needed to calculate the heat transferred through walls, roofs, etc., have been seen in the chapter on heat transfer (Chapter 3) and the chapter on steady-state heat transmission through building boundaries (Chapter 4). We have seen the importance of R-values and the analogy to electrical resistance networks in calculating the overall conductive heat transfer through a thermal circuit.

The equation to calculate structural heat loss is

$$q_w = \sum_n (A / R)_n (t_i - t_o) \tag{5-9}$$

where there are n paths of transfer, each path is (most likely) a series thermal circuit.

Unit area thermal resistances are used in Equation 5-9. The factor $\sum(A / R)$ characterizes the overall conductance of the building shell and includes the effects of framing, windows, and doors. Procedures to calculate the overall conductance are shown in Examples 4-1, 4-2, 4-4, 4-5, and 4-6.

5-2.7. Heat Exchange with the Floor, q_f. The means to calculate heat exchange with a slab floor was developed in Chapter 4. The exchange can be calculated by

$$q_f = FP(t_i - t_o)$$

as presented in Equation 4-2.

Heat exchange with a basement wall and floor does not follow this model but such situations are rare in agricultural buildings and will be ignored for now. If such a situation exists, the heat exchange is treated as a constant value to be added to an energy balance. It is constant because soil temperature is viewed as unchanging.

5-2.8. Evaporation, q_e. Evaporation has two primary components in barns: evaporation from wash water and animal wastes, and evaporation from the animals (primarily from their respiratory systems). In greenhouses, evaporation is from the floors and benches when water is spilled, from the surface of the potting medium (or soil), and from transpiration by plants.

Data for heat produced from animals may or may not incorporate and obscure the conversion of heat from sensible to latent within the airspace. If the data are based on housing studies, they include conversion of sensible heat to latent (such as evaporation from the floor) within the space. If presented as calorimetric data, moisture production within the airspace must be estimated and sensible heat production decreased accordingly. The conversion of sensible to latent heat will depend on the housing system, waste handling system, and air temperature, and is not easy to estimate.

Latent heat production in greenhouses depends to a large extent on solar insolation, for transpiration is the major component of evaporation. As a rough rule, insolation which passes through the greenhouse cover can be partitioned - one half is converted immediately to sensible heat added to the air, one quarter is added to the air as latent heat, and one quarter is either reflected back to the outside (approximately 10%), used in photosynthesis (2% to 3%) or stored in the intrinsic thermal mass to be released later.

5-3. Uses of the Sensible Energy Balance

A sensible energy balance may be used for several design purposes. The ventilation rate to maintain specified inside conditions may be calculated. The inside air temperature which would result from ambient conditions and other factors may be determined. The rate of heat addition needed to maintain inside air temperatures during cold weather may be of interest. These can be found starting with the sensible energy balance.

Equation 5-1 can be rewritten in the following form:

$$q_s + q_m + q_{so} + q_h = \Sigma UA(t_i - t_o) + FP(t_i - t_o)$$
$$+ 1006\rho\dot{V}(t_i - t_o) + q_e. \qquad (5\text{-}10)$$

The addition of solar heat, q_{so}, is kept in symbol form and would be calculated separately as the product of transmittance, glazing area, and solar intensity; less the fraction reflected back outside, used for photosynthesis if the building of interest is a greenhouse, and stored in the intrinsic thermal mass inside the building.

In common practice, Equation 5-10 is simplified for applications in barns and greenhouses.

<u>Animal housing.</u> When animal heat data are presented as net sensible heat production, the terms q_s and q_e are combined into one, which will be written as q_s and understood to be a net sensible heat addition. Transmitted solar input to barns is usually neglected. Heat from mechanical sources is frequently assumed to be negligible in barns. Finally, supplemental heat in barns (except when young animals are housed) is not frequently used. When these simplifying assumptions are accepted, the energy balance reduces to

$$q_s = (\Sigma UA + FP + 1006\rho\dot{V})(t_i - t_o). \qquad (5\text{-}11)$$

The inside air temperature can be determined if other parameters are known,

$$t_i = t_o + q_s / (\Sigma UA + FP + 1006\rho\dot{V}). \qquad (5\text{-}12)$$

and the energy balance can be rearranged to calculate the required ventilation rate to maintain desired conditions,

$$\dot{V} = \frac{q_s - (\Sigma UA + FP)(t_i - t_o)}{1006\rho \, (t_i - t_o)}. \qquad (5\text{-}13)$$

Example 5-7:

<u>Problem:</u> Determine the ventilation rate (m^3/s) required to maintain a dairy barn at 15 C, given the following conditions:

elevation: 500 m
number of cows: 60
average weight: 580 kg
wall area: 274 m^2
wall R-value: 2.05 m^2 K/W
ceiling area: 520 m^2
ceiling R-value: 1.97 m^2 K/W
window area: 12 m^2

window R-value: 0.30 m^2 K/W
door area: 15 m^2
door R-value: 0.49 m^2 K/W
perimeter length: 110 m
perimeter heat loss factor: 1.5 W/m K
outside air temperature: - 5 C
inside air relative humidity: 70%
an exhaust ventilation system

Assume there is no significant heat addition from lights and motors and little solar heating of the barn can be expected. The attic is well ventilated. The animal heat data for dairy cows in Appendix 5-1 reflect net sensible heat after latent heat conversion has been deducted.

<u>Solution:</u> Equation 5-13 can be used, but several parameters are needed in addition to the data given.

A 500 kg dairy cow at 15 C can be expected to add 1.2 W/kg of sensible heat to the air in a barn (see Appendix 5-1). Cows in the example have a mass of 580 kg and there are 60 of them. Total sensible animal heat production is, thus,

$$q_s = (60)(1.2)(500)(580 / 500)^{0.734} = 40,000 \text{ W} = 40 \text{ kW.}$$

Air inside the barn is 15 C, with a relative humidity of 70%. The barn is at 500 m elevation; standard atmospheric pressure for that elevation is assumed (95.461 kPa). The program PLUS can be used to estimate air density, which is 1.14 kg/m^3.

Data given in the statement of the problem can be used directly to calculate structural heat loss; perimeter heat loss conductance is

$$FP = (1.5 \text{ W/m K})(110 \text{ m}) = 165 \text{ W/K.}$$

Heat loss conductances from the walls, ceiling, windows, and doors are:

$$
\begin{aligned}
\text{walls:} \quad & UA = A / R = 274 \text{ m}^2 / 2.05 \text{ m}^2 \text{ K/W} = 134 \text{ W/K,} \\
\text{ceiling:} \quad & UA = A / R = 520 \text{ m}^2 / 1.97 \text{m}^2 \text{ K/W} = 265 \text{ W/K,} \\
\text{windows:} \quad & UA = A / R = 12 \text{ m}^2 / 0.30 \text{ m}^2 \text{ K/W} = 40 \text{ W/K, and} \\
\text{doors:} \quad & UA = A / R = 15 \text{ m}^2 / 0.49 \text{ m}^2 \text{ K/W} = 31 \text{ W/K.}
\end{aligned}
$$

The sum, ΣUA, is 470 W/K.

Equation 5-13 can now be applied to calculate the required ventilation rate, where the indoor to outdoor air temperature difference is 20 K,

$$\dot{V} = \frac{40,000 \text{ W} - (470 \text{ W/K} + 165 \text{ W/K})(20 \text{ K})}{(1006 \text{ J/kg K})(1.14 \text{ kg/m}^3)(20 \text{ K})}$$

$$= 1.2 \text{ m}^3/\text{s.}$$

Ventilation is by an exhaust system and this ventilation rate was calculated at the indoor conditions, thus, the fans must ventilate at a rate of 1.2 m³/s (or 1.14 kg/m³ x 1.2 m³/s = 1.37 kg/s). Fans are normally specified according to their volumetric capacity.

Example 5-8

Problem: Assume the same barn and conditions as in Example 5-7. A minimum barn ventilation rate of 0.9 m³/s has been determined as necessary to maintain healthy conditions for the animals. How cold must the outdoor air temperature be before it will no longer be possible to maintain 15 C inside the barn?

Solution: Equation 5-12 can be rearranged to solve this problem. Air density will be 1.14 kg/m³ and sensible heat from the animals will be 40 kW at 15 C. Outdoor air temperature to satisfy the energy balance is

$$t_o = t_i - q_s / (\Sigma AU + FP + 1006 \rho \dot{V}),$$

$$= 15\,C - \frac{40{,}000\ W}{470\,W/K + 165\,W/K + (1006\ J/kg\ K)(1.14\,kg/m^3)(0.9\,m^3/s)}$$

$$= -9\,C.$$

If outdoor air temperature is below - 9 C and the ventilation rate of 0.9 m³/s is maintained, the indoor air temperature will fall below 15 C. If the outdoor air temperature is extremely low and the minimum ventilation rate continues, the barn could freeze. Fortunately, sensible animal heat production increases at lower air temperatures. This effect acts as a negative feedback to help prevent freezing. Of course, freezing is always possible if weather is sufficiently cold, thus, alarms or other freeze protection must be incorporated into the environmental control system. This problem will be explored more fully later.

Example 5-9

Problem: For the same conditions as in Example 5-7 (except the indoor air temperature), but with a ventilation rate of 2 m³/s, determine the expected indoor air temperature.

Solution: Equation 5-12 applies. However, two factors prevent direct application of the equation: the dependence of air density on indoor conditions and the effect of air temperature on sensible heat production from cows. One way to approach the problem is to assume values for air density and heat production, determine the resulting air temperature and density, and iterate until the temperature, relative humidity, heat production, and air density are internally consistent. For now, the effect of ventilation on relative humidity within the airspace will be neglected and a constant relative humidity of 70% will be assumed.

The ventilation rate is higher than the value calculated in Example 5-7, thus it is reasonable to expect the indoor air temperature to be lower, air density to be higher, and more sensible heat to be available from the cows. Assume air density is 1.15 kg/m^3 and sensible heat production is 45 kW. A candidate value for indoor air temperature, from Equation 5-12, is

$$t_i = -5\,C + \frac{45,000\ W}{470\,W/K + 165\,W/K + (1006\ J/kg\ K)(1.15\ kg/m^3)(2m^3/s)}$$

$$= 10.3\ C$$

Now the assumptions must be checked. At 10.3 C, air density will be 1.16 kg/m^3 and animal heat production (per animal unit) will be

$$q'_s = 1.5 + (0.3/5.0)(1.2 - 1.5) = 1.48\ W/kg.$$

Total sensible heat production from the herd will be

$$q_s = (60\ cows)(1.48\ W/kg)(500\ kg/cow)(580\ kg/500\ kg)^{0.734}$$
$$= 49,500\ W.$$

Air density and animal heat production were estimated slightly too low. Final air temperature will be above 10 C, thus, animal heat production will not be quite as high as 49.5 kW. For the next iteration, assume air density is 1.16 kg/m^3 and sensible heat production is 48 kW. The next candidate air temperature is, thus,

$$t_i = -5\,C + \frac{48,000\ W}{470\,W/K + 165\,W/K + (1006\ J/kg\ K)(1.16kg/m^3)(2m^3/s)}$$

$$= 11.2\ C$$

At 11.2 C, air density is 1.16 kg/m^3 and heat production per kg for a 500 kg cow is

$$q'_s = 1.5 + (1.2/5.0)(1.2 - 1.5) = 1.4\ W/kg.$$

Total animal sensible heat production is

$$q_s = (60)(1.4)(500)(580/500)^{0.734} = 47,000\ W.$$

This is sufficiently close to the final environment. Further iterations would converge at an indoor air temperature of approximately 11 C, an air density of 1.16 kg/m^3, and a sensible heat production of approximately 47.5 kW. However, data for building thermal parameters and animal heat are not known with sufficient accuracy to make further precision meaningful. We can conclude that indoor air temperature with the new ventilation rate will be 11 C.

5-4. Components of the Mass Balance, Humidity

Although the mass balance of Equation 5-2 applies in principle to any component of air within a control volume, in practice it can be used only when the production rate of the component can be quantified. Further, in the mass balance the only environmental control parameter which can be determined is the ventilation rate, for ventilation is assumed to be the only removal mechanism for the component. These limitations simplify applications of the mass balance.

In animal housing, only data for humidity and carbon dioxide production are known with some certainty. Each has been found proportional to the number and type of animals and their sizes. Rates of production of other components such as dust, ammonia, or other gases are functions more of the waste handling system than of the animals and their sizes. Such rates have not been quantified in a way which applies in general and is useful for design.

Fortunately, it has been found that design to remove moisture and carbon dioxide usually provides sufficient ventilation to maintain the other components below levels of concern. In certain situations where another component rises to a level which is too high, the reason can often be traced to improper handling of the animal waste rather than improper design of the ventilation system and often can be cured by changing the way waste is managed.

Greenhouses usually are ventilated only for temperature control. High humidity during cold weather is tolerated; relative humidity often will be 70% to 80%. The cold surfaces of the glazing are dehumidifiers which prevent humidity from reaching saturation.

Moisture production within a greenhouse is a function of the plant population, light intensity, and water management practices of the grower and has not been quantified in a way useful for environmental design. For these reasons, although mass balance relations obviously apply to greenhouses, they are seldom used to design greenhouse environmental control systems. Only temperature is controlled.

5-4.1. Moisture Production From Animals.

Data are presented in Appendix 5-1 for moisture production of various animals. Moisture production rates for dairy cows must be extrapolated if other than the 500 kg standard animal unit is anticipated (see Example 5-1). Data for broilers and swine are presented for various animal sizes and extrapolation should not be needed. Data for laying hens apply to 1.8 kg white leghorns, but that is the standard size of a laying hen. If different sizes are expected, the data may be scaled.

Example 5-10

Problem: Determine the rate at which moisture is generated and must be removed from a swine barn housing 100, 40 kg; 100, 60 kg; and 100, 80 kg pigs. Air temperature is expected to be 20 C.

Solution: Data from Appendix 5-1 may be used. At 20 C, moisture production is

Size, kg	Moisture Production	
	mg/kg-s	mg/pig-s
40	0.61	24
60	0.47	28
80	0.39	31

which is, for 100 of each size pig,

$$\text{moisture} = 100(24 + 28 + 31)(1.0E - 6 \text{ kg/mg}) = 0.0083 \text{ kg/s}.$$

Example 5-11

Problem: Determine the rate at which moisture is generated and must be removed from a dairy barn housing 100 cows averaging 570 kg, when the indoor air temperature is 10 C.

Solution: By the data in Appendix 5-1, at 10 C moisture production of a 500 kg dairy cow is 0.28 mg/kg-s. Moisture production rate from a 570 kg dairy cow can be estimated as

$$\text{moisture} = 0.28(570 / 500)^{0.734} = 0.31 \text{ mg/kg-s}.$$

Note: This formulation for moisture production is assumed to mirror the form which has been found to represent sensible heat production from mammals.

Moisture production from the herd will be

$$m_p = 0.31 \text{ mg/kg-s}(570 \text{ kg})(100 \text{ cows})(1.0E - 6 \text{ kg/mg})$$
$$= 0.018 \text{ kg/s}.$$

5-4.2. Ventilation, m_{vi} and m_{vo}. As was done for energy transport by ventilation in the sensible energy balance, m_{vi} and m_{vo} will be considered together. The moisture content of air equals the product of the humidity ratio and the mass of air. Entering air, carrying m_{vi}, has a moisture content of

$$m_{vi} = \rho_o \dot{V}_o W_o, \qquad (5\text{-}14)$$

and expelled air, carrying m_{vo}, has a moisture content of

$$m_{vo} = \rho_i \dot{V}_i W_i, \qquad (5\text{-}15)$$

where ρ_o and ρ_i are air densities at the inlets and outlets, $\dot{V}_o$ and $\dot{V}_i$ are the volumetric rates of airflow at the inlets and outlets, and W_o and W_i are humidity ratios at the inlet and outlet, respectively. This assumes well-mixed air.

The products of air densities and volumetric flow rates are mass flow rates at the inlets and outlets, $(m_{air})_i$ and $(m_{air})_o$, and can be assumed to be equal. This is not strictly true, for the moisture content changes as air moves from the inlet to outlet, but the approximation is close. Ideally, entering air flows through planned inlets or fans and exhausted air through planned outlets or fans. Air leaks and wind and thermal buoyancy effects distort this idealization so Equations 5-14 and 5-15 are based on total entering and total leaving air, not just that entering and leaving through planned inlets and outlets.

5-5. Uses of the Mass Balance, Moisture

In most design situations, outdoor conditions are chosen based on design weather data and indoor conditions are specified based on the needs of the animals or plants within the space. If the moisture production rate, m_{water}, is known, the ventilation rate required to maintain indoor humidity at its design value may be calculated using a rearrangement of the mass balance, Equation 5-2.

In practical terms, this calculation provides an estimate of the minimum ventilation rate. If a higher ventilation rate is necessary to maintain conditions at the design temperature, a humidity lower than design conditions will result.

$$m_{air} = m_{water} / (W_i - W_o). \qquad (5\text{-}16)$$

This presents no problem, for design humidity conditions are usually chosen to be the maximum desired. If an exact humidity level is desired, irrespective of the ventilation needed for temperature control, moisture must be added to or removed from the air by mechanical means (a humidifier or a dehumidifier).

Example 5-12

Problem: A poultry house is being designed for 30,000 leghorn laying hens (1.8 kg, average). The poultry house is located in a region with an elevation of 1000 m. Determine the ventilation rate required to maintain the indoor air at 23 C and 70% relative humidity when it is - 20 C and 55% relative humidity outdoors.

Solution: Equation 5-16 applies. Data to quantify moisture production from the birds is in Appendix 5-1 and program PLUS may be used to determine the relevant humidity ratios.

At 23 C moisture production from 1.8 kg leghorn laying hens is estimated as

$$(m_{water})_{kg} = 0.97 + (5/10)\,(1.19 - 0.97) = 1.08 \text{ mg/kg-s.}$$

There are 30,000 hens, averaging 1.8 kg, thus, the total moisture production within the airspace is estimated as

$$m_{water} = (30,000)\,(1.8)\,(1.08) = 58,320 \text{ mg/s} = 0.05832 \text{ kg/s.}$$

The humidity ratios needed for Equation 5-16 are:

$$W_o = 0.000393 \text{ kg/kg, and}$$
$$W_i = 0.013919 \text{ kg/kg.}$$

The required mass flow rate of air is

$$m_{air} = (0.05832 \text{ kg/s})/(0.013919 \text{ kg/kg} - 0.000393 \text{ kg/kg})$$
$$= 4.3 \text{ kg/s (dry air).}$$

Correcting the required air mass flow rate for humidity is a sufficiently small change as to be disregarded considering the lack of precision in other design parameters, but will be included here for illustration.

Each kg of dry air holds 0.013919 kg of water, thus, the exhaust ventilation rate must be $4.3 + 0.013919 = 4.314$ kg/s. If an exhaust ventilation system is used, the applicable air density is at indoor conditions and is 1.03 kg/m^3. The volumetric ventilation is, therefore,

$$\dot{V}_{air} = (4.314 \text{ kg/s})/(1.03 \text{ kg/m}^3) = 4.44 \text{ m}^3/\text{s,}$$

or 0.14 L/s-bird.

It would be a useful exercise to develop a graph for the required ventilation rate to maintain the indoor design conditions as a function of the outdoor temperature at several outdoor relative humidity values. For example, develop the graph for the required ventilation rate for outdoor temperatures between - 30 and 20 C, at outdoor relative humidities of 0%, 50% and 100%. Explain any difficulties which might arise.

5-6. Components of the Mass Balance, Carbon Dioxide

Carbon dioxide is frequently ignored as a design parameter for animal housing in general engineering practice. Experience has shown that ventilation to control temperature and moisture is usually sufficient to control carbon dioxide (and other gaseous contaminants of the air). However, carbon dioxide is an asphyxiant and should be of concern at high levels. The level for concern, however, is not clear. Rules exist for human exposure to carbon dioxide. For example, long exposure to levels over 10,000 ppm should be avoided. Some engineering recommendations for animal housing suggest a maximum continuous exposure level of 2500 ppm. The carbon dioxide level where health and productivity are affected is not known; the level of 2500 ppm was chosen to be sufficiently low that effects would not be anticipated. Further research is required before safe levels will be known with certainty.

Carbon dioxide is frequently added in greenhouses when light levels are high and there is little or no venting. At low light levels, light is limiting and carbon dioxide is not; ambient carbon dioxide suffices for growth.

The amount of carbon dioxide required for optimal photosynthesis is open to question. Common practice is to supplement the level to somewhere in the range from 800 to 1500 ppm. Sensing carbon dioxide within the plant canopy is important. If air movement is limited within the canopy, carbon dioxide depletion may be serious even though the level measured outside the canopy seems sufficient. For example, carbon dioxide levels as low as 150 to 200 ppm have been detected within dense plant canopies on bright days. Photosynthesis is seriously affected in this range. Fans to mix the greenhouse air have been found to help prevent depletion within canopies.

5-6.1. Carbon Dioxide Produced By Animals. Carbon dioxide is a by-product of metabolism, as is heat and moisture. The ratio between carbon dioxide production and total animal heat production is fixed by the biochemical processes of metabolism. Additional carbon dioxide may be produced from decomposition of wastes and digestive processes of ruminants, but such additional production is small compared to that respired by the animals.

One liter of carbon dioxide is produced, on the average, for every 24.6 kJ of total heat added to the environment by an animal. Total heat production of animals can be estimated using data in Appendix 5-1.

5-7. Uses of the Mass Balance, Carbon Dioxide

The minimum ventilation rate required to maintain a specified carbon dioxide level in animal housing is constant over the range of possible outdoor conditions. The ambient carbon dioxide level is approximately 345 ppm.

Example 5-13

<u>Problem:</u> Consider the poultry house described in Example 5-11. Calculate the carbon dioxide level when ventilation is 4.3 m³/s.

<u>Solution:</u> The mass balance of Equation 5-2 applies when rearranged to solve for the carbon dioxide concentration within the airspace. Well-mixed conditions are assumed and the concentration of carbon dioxide in the exhaust air will be the same as the concentration within the space.

The mass of carbon dioxide in air can be calculated by

$$M_{CO2} = M_{air}(C_{CO2})_{mass},$$

where M is mass and $(C_{CO2})_{mass}$ is the mass concentration, kg/kg, of carbon dioxide.

The carbon dioxide partial volume in air can be calculated similarly by

$$V_{CO2} = V_{air}(C_{CO2})_{volume}.$$

where V is volume and $(C_{CO2})_{volume}$ is the volumetric concentration, m^3/m^3, of carbon dioxide.

Carbon dioxide production depends on total heat production, which at 23 C is

$$q_{kg\ of\ bird} = 6.8 + (5/10)\ (6.6\text{-}6.8) = 6.7\ W/kg.$$

The 30,000 birds average 1.8 kg, thus, total heat production within the building is

$$q_{total} = (30,000)(1.8)(6.7) = 361,800\ W.$$

The conversion factor from total heat to carbon dioxide production leads to a carbon dioxide production rate of

$$\dot{V}_p = (361.8\ kW)/(24.6\ kJ/l) = 14.7\ L/s = 0.0147\ m^3/s.$$

The carbon dioxide balance, when expressed on a volumetric basis, is (for constant pressure)

$$\dot{V}_p + \dot{V}_{vo} = \dot{V}_{vi},$$

where

$$\dot{V}_{vi} = \dot{V}_{air}(C_{CO2})_{volume},$$

$$= (4.3\ m^3/s)\ (C_{CO2})_{volume},\ and$$

$$\dot{V}_{vo} = \dot{V}_{air}(C_{CO2})_{volume} = (4.3\ m^3/s)\ (0.000345),$$

$$= 0.001484\ m^3/s,$$

for ambient air at 345 ppm carbon dioxide concentration.

The balance for carbon dioxide is, thus,

$$0.0147 \text{ m}^3/\text{s} + 0.001484 \text{ m}^3/\text{s} = (4.3 \text{ m}^3/\text{s})(C_{CO_2})_{volume},$$

and

$$(C_{CO_2})_{volume} = 0.003764 = 3764 \text{ (approximately 3800) ppm.}$$

This concentration of carbon dioxide is well below the 5000 ppm limit used by OSHA for human occupation. However, if research were to discover that a lower limit is preferred, a minimum ventilation rate might be based on carbon dioxide rather than moisture control.

This example could also be approached using mass rather than volume and it would be a useful exercise to do so. The factor to convert carbon dioxide volumetric to mass concentration is the ratio of molecular weights of carbon dioxide to air, which is 1.519. The mass concentration of carbon dioxide in ambient air is, thus, approximately 524 mg/kg.

Example 5-14

Problem: A greenhouse at sea level with a volume of 3000 m³ experiences an air infiltration rate of 0.75 air changes (ac) per hour. Carbon dioxide is added to enhance plant growth. At what rate (kg/s) is carbon dioxide lost from the greenhouse by exfiltration when the greenhouse carbon dioxide level is 1000 ppm?

Solution: For contrast to Example 5-12, this example will be solved using mass rather than volume. Assumptions are needed, as is usually the case in engineering design. Assume the greenhouse air will be at approximately 20 C and 70% relative humidity (typical), and the air will be well-mixed.

Air enters the greenhouse with a carbon dioxide concentration of 524 mg/kg, and leaves with a concentration of 1519 mg/kg. The approximate density of air at 20 C and 70% relative humidity at sea level is 1.18 kg/m³. The mass flow rate of air through the greenhouse is, thus,

$$m_{air} = (1.18 \text{ kg/m}^3)(3000 \text{ m}^3)(0.75 \text{ ac/hr})(3600 \text{ s/hr})^{-1},$$
$$= 0.7375 \text{ kg/s.}$$

The change of carbon dioxide concentration is 1519 mg/kg - 524 mg/kg, or 995 mg/kg. Thus, the loss of carbon dioxide, in terms of mass, is

$$m_{vo} = (0.7375 \text{ kg/s})(995 \text{ mg/kg}) = 733.8 \text{ mg/s} = 0.0007338 \text{ kg/s,}$$
$$\text{or } 2.64 \text{ kg/hr.}$$

By comparison, carbon dioxide uptake by a greenhouse crop with a full canopy in good sun should be approximately 0.7 mg/m^{2}s. A greenhouse with a volume of 3000 m^3 will have a floor area of approximately 1000 m^2. The carbon dioxide uptake by the crop will, thus, be approximately 0.0007 kg/s, which is approximately the rate at which carbon dioxide is lost by exfiltration.

SYMBOLS

A	area, m^2
C	concentration, ppm
F	perimeter heat loss factor, W/mK
I	irradiation, W/m^2
K	extinction coefficient, mm^{-1}
L	thickness, m or mm
m	mass flow rate, kg/s
m	rate of mass production, kg/s
M	mass, kg
n	index of refraction
P	perimeter, m
q	heat transferred, or produced, W
R	unit area thermal resistance, m^{2}K/W
t	temperature, C
U	unit area thermal conductance, W/m^{2}K
V	volume, m
$\dot{V}$	volumetric flow rate or volumetric production rate, m^3/s
W	humidity ratio, kg/kg
x	spacial variable, m
α	absorptance
ϕ	angle of refraction
λ	wavelength, microns
ρ	density, kg/m^3
ρ	reflectance
τ	transmittance
ϕ	angle of incidence

EXERCISES

1. Determine the sensible, latent, and total heat production from a flock of 30,000 white leghorn laying hens (averaging 1.9 kg body mass) housed at 24 C.

2. Determine the coefficients for a second order polynomial to represent the sensible heat production of the laying flock described in Exercise 1 above, including the heat added by lights. Lighting within the housing space is

designed at 40 W/m^2, and the building has a floor area of 1200 m^2. Compare predicted heat production to the data used to determine the coefficients of the polynomial.

3. Determine the coefficients of a second order polynomial to describe the carbon dioxide production from a herd of 100 dairy cows (averaging 560 kg body mass) as a function of air temperature.

4. Direct beam solar radiation strikes a single glazed window (ordinary window glass 3 mm thick). What is the greatest angle of incidence at which at least half the light passes through the glass?

5. A swine barn housing 1000 growing pigs (averaging 55 kg body mass) is located at an elevation of 800 m. Desired indoor conditions are 18 C and 60% relative humidity. Develop a graph of the required ventilation rate for moisture control (m^3/s, based on inside conditions) for outdoor conditions of 80% relative humidity and a temperature range of - 20 to + 5 C.

6. For the conditions described in Example 5 above, develop a similar graph of ventilation rate as a function of outdoor temperature when ventilation is calculated based on temperature control. Assume the (UA+FP) value for the building is 800 W/K.

7. For the barn described in Examples 5 and 6 above, will supplementary heat be required to maintain the barn at 18 C indoors when it is - 20 C outdoors? If so, how much?

8. For the conditions described in Example 5 above, develop a graph of supplemental heat required as a function of outdoor air temperature if ventilation is for moisture control at all times.

REFERENCES

ASAE Standards, 34th ed. 1987. Standard D270.4. Design of ventilation systems for poultry and livestock shelters. American Society of Agricultural Engineers, St. Joseph, MI.

ASHRAE. 1989. Handbook of Fundamentals. American Society of Heating, Refrigerating and Air Conditioning Engineers, Atlanta, GA.

Clark, J.A. 1981. Environmental Aspects of Housing for Animal Production. Butterworths, London. 511 pp.

Curtis, S.E. 1983. Environmental Management in Animal Agriculture. The Iowa State University Press, Ames, IA. 410 pp.

Esmay, M.L. and J.E. Dixon. 1986. Environmental Control for Agricultural Buildings. The AVI Publishing Co., Inc. Westport, CT. 287 pp.

Hellickson, M.A. and J.N. Walker. 1983. Ventilation of Agricultural Structures. American Society of Agricultural Engineers, St. Joseph, MI.

Midwest Plan Service. 1983. Structures and Environment Handbook. Midwest Plan Service, University of Iowa, Ames IA.

Yeck, R.G. and R.E. Stewart. 1959. A ten year summary of the psychroenergetic laboratory dairy cattle research. Transactions of the ASAE 2(1):71-77.

CHAPTER 6
VENTILATION RATES

6-1. Introduction

Frequently, the first step in designing environmental control for agricultural buildings is to determine required ventilation rates. Ventilation in both animal housing and greenhouses is critical. In animal housing, ventilation is frequently the only means of environmental modification. In greenhouses, both heating and ventilation are used for environmental modification, and ventilation is critical when dealing with the large heating effect of the sun.

A properly designed and installed ventilation system must provide sufficient ventilation for summer conditions, adequate minimum ventilation rates for winter, proper staging between the minimum and maximum ventilation rates, and safety alarms to alert the operator should the system fail to maintain the environment within prescribed limits.

Sensible energy and mass balances are the tools used to calculate required minimum and maximum ventilation rates. This chapter will focus on <u>calculating</u> ventilation rates, not on applying field rules which can differ from region to region. However, field recommendations and calculated values should not differ greatly, and if they do, the differences should be explainable (otherwise one must be in error).

Strategies to calculate minimum and maximum ventilation rates for greenhouses differ significantly from strategies for animal housing. Greenhouses usually are ventilated only for temperature control, but may be ventilated for a short time during the late afternoon in cold weather to "dry out the greenhouse". However, for most of a cold day, the only ventilation will be infiltration. During warm weather, greenhouses are ventilated at rates sufficient to limit the temperature rise of the ventilation air as it passes through the greenhouse, or may be evaporatively cooled, with the temperature rise of the evaporatively cooled ventilation air limited to a prescribed amount between inlet and exhaust. A typical limit of temperature rise of ventilation air is 4 K during warm weather.

In contrast, animal housing is usually ventilated during all weather, with the ventilation rate modulated depending on inside air temperature. Fresh air is needed during even the coldest weather to remove moisture, odors, and aerosols from the aerial environment within the building. Oxygen depletion is not normally a concern. The major danger of inadequate ventilation arises from stress and disease potential created by high levels of humidity, noxious gases, carbon dioxide and aerosols; and excessively high temperatures during warm weather.

6-2. Design Weather Data

Weather data has been collected in the United States and worldwide for many years. Air temperature data is always collected; precipitation, wind and solar data may be available. For animal housing design air temperatures usually suffice, but proper greenhouse design requires solar data in addition to air temperatures.

Mean coincident wet-bulb temperature is also important for agricultural building design, especially with regard to evaporative cooling system design. Mean coincident wet-bulb temperatures are the expected wet-bulb temperatures at the design dry-bulb temperature values.

Weather data may be presented in one of two ways. The traditional method has been to compile "design temperatures". Design temperatures are based (usually) on averaged data collected for at least 15 years, and frequently for 40 years or more. For example, a 97.5% summer design temperature for a location is the temperature not exceeded for more than 2.5% of the time during summer (June, July, and August). Outdoor air temperatures will be above the 5% winter design temperature for all but 5% of December through February.

Extensive weather design data is available, for example, in the *ASHRAE Handbook of Fundamentals*. If this information is not available in reference tables, it can usually be obtained from local utility companies, heating system contractors, and Cooperative Extension offices. Sample design temperatures are available in Appendix 6-1 and Figure 6-1, which contains maps of the 48

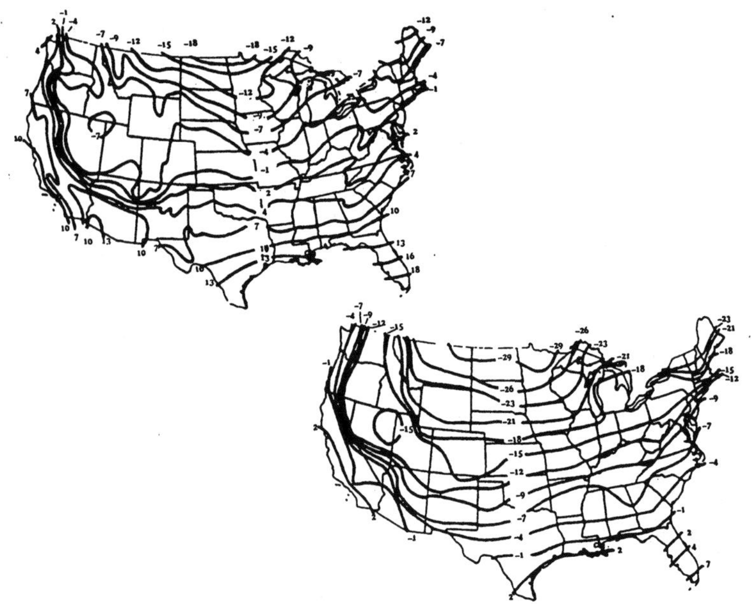

Figure 6-1. Average daily air temperature, C, during January and 97.5% winter design temperature in the contiguous United States.

contiguous states of the United States with averaged and 97.5% winter design temperatures.

More recently, with the advent of computerized data collection methods, air temperatures have been presented as "bin data". In this technique, the number of hours during an average year that outdoor air temperature is within a certain range is determined, for temperature ranges spanning the local climate. For example, the number of hours the temperature is between - 20 and - 15 C, between - 15 and - 10 C, etc., up to the range where maximum summer temperatures are found. Bin data have been found useful for energy use calculations and are available from several sources: ASHRAE, The Government Printing Office (AFM 88-29), The National Climatic Data Center of NOAA, and engineering handbooks as examples. Much of this data, for example the ASHRAE and NOAA databases, are available on magnetic media format. Sample bin data can be found in Appendix 6-2.

6-3. Animal Housing Ventilation

6-3.1. General. Mature animals normally produce sufficient body heat to permit ventilation with no need for supplemental heat during typical cold weather, provided stocking density is at or near full capacity. For example, when weather is cold, a large dairy cow is approximately equivalent to a 1 kW heater in terms of her sensible heat production. If a building design does not result in adequate temperature at the minimum ventilation rate without supplemental heat, a design change should be considered with insulation added to make ventilation possible during cold weather. Supplemental heat should be considered as a final resort when housing mature animals because of its cost. Of course, young animals often require higher temperatures for comfort and health than do mature animals and do not produce sufficient body heat to maintain that temperature during cold weather. Supplemental heat is frequently necessary in those cases.

During warm weather, animal housing ventilation is normally designed to limit indoor air temperature rise above outdoor air temperature and evaporative cooling may be used where the outside air temperature is high and midday relative humidity is low. The choice of a maximum temperature rise must be tempered with a realization of the cost to keep the rise very small. As a close approximation, the temperature rise is halved when the ventilation rate doubles. Thus, the ventilation rate grows geometrically while the added benefit diminishes at each step. If the ventilation rate is X at an 8 K rise, it is 2 X for a 4 K rise, 4X for a 2 K rise, 8 X for a 1 K rise, etc. To increase ventilation from 4X to 8X to attain a benefit of only one degree may be very costly. Field experience has shown the practical minimum air temperature rise is approximately 1.5 to 2 K during the warmest weather and likely should be no more than 4 K unless the ventilation air is cooled such as by evaporative cooling.

6-3.2. Maximum Ventilation Rate. The situation of no evaporative cooling will be examined first. The sensible energy balance, Equation 5-10, applies,

$$q_s + q_m + q_{so} + q_h = \Sigma UA(t_i - t_o) + FP(t_i - t_o) + 1006\rho \dot{V}(t_i - t_o) + q_e.$$

Maximum ventilation rates are based on warm weather conditions, as a consequence, q_h will be zero. Also, for simplicity and to agree with animal heat production data in Appendix 5-1, the terms q_s and q_e will be combined into a single term, q_s. This assumes animal heat production data will be based on confinement conditions. If not, a correction must be made. As will be demonstrated in a later section, temperature control dictates the maximum ventilation rate during warm weather. If ventilation is adequate to control temperature during warm weather, it is adequate to control moisture and other aerial contaminants.

The sensible energy balance can be rearranged to calculate the ventilation rate, $\dot{V}$, m³/s,

$$\dot{V} = \frac{q_s + q_m + q_{so} - (\Sigma UA + FP)(t_i - t_o)}{1006\rho\,(t_i - t_o)}. \qquad (6\text{-}1)$$

Equation 6-1 has the form shown in Figure 6-2 (the line for temperature control) where the ventilation rate is a function of outside air temperature and other parameters are unchanged.

During early afternoon, when outside air temperature is highest, the sun is high in the sky and solar entry through glazing is often limited and usually is neglected. Heat from mechanical sources is often small during the middle of the day. Lights, for example, would not be on. Thus, it is common practice to eliminate q_m and q_{so} from the sensible energy balance. However, a designer should always be aware of the assumptions contained in that approach, and be prepared to challenge the validity of the assumptions for each specific design situation. Solar heat, heat from lights, etc., if significant, must be calculated and added to the energy balance.

The air temperature at which to determine animal sensible heat production is not immediately obvious. As indoor air temperature rises, sensible heat

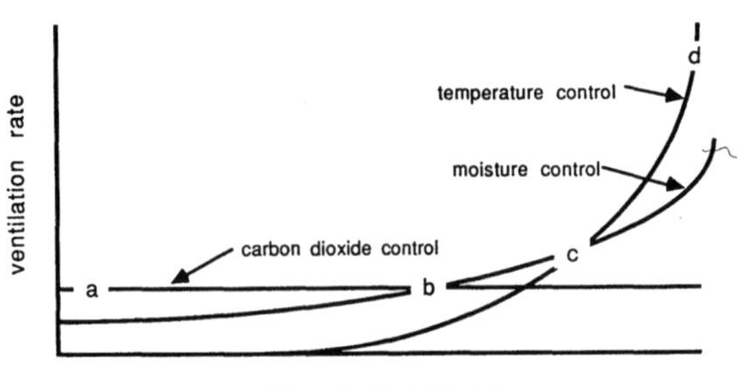

Figure 6-2. Example ventilation graph for temperature (c-d), moisture (b-c), and carbon dioxide (a-b) control in animal housing.

176

production from animals drops as metabolic heat production is reduced in response to heat stress, and a greater fraction is dissipated as latent heat. This is a negative feedback mechanism which assists in limiting the temperature rise of ventilating air during the hottest weather. An implication is that it may be more logical to calculate maximum ventilation rates based on warm summer conditions, rather than the hottest weather (not summer high temperature design conditions).

A criterion to choose a design temperature is to use the temperature at which production from the animals begins to decrease significantly. However, there is a rational basis for arguing that the design temperature should equal the set-point temperature which activates the highest ventilation stage. In this way, the temperature rise chosen to calculate the maximum ventilation rate will not be exceeded at any time during which the highest ventilation stage is activated. (As the air temperature rises, the sensible heat production decreases.) There is a symmetry in calculating the maximum ventilation rate based on the highest thermostat setpoint, and having that occur before the onset of heat stress.

Example 6-1

<u>Problem:</u> Determine the maximum ventilation capacity to be installed in the following mechanically ventilated dairy barn. The barn is to house 100 milking cows (averaging 570 kg weight) in a zero pasture operation. The barn is 13 m wide and 80 m long with 3 m high sidewalls. It is to be located at an elevation of 1000 m.

The ceiling is sheathed on the lower side with 15.88 mm plywood and is insulated with 88.9 mm of mineral fiber blanket. The walls are sheathed on the inside with 12.7 mm plywood, insulated with 88.9 mm of mineral fiber blanket, sheathed on the outside with 19.8 mm thick vegetable fiberboard, and sided with 15.88 mm plywood.

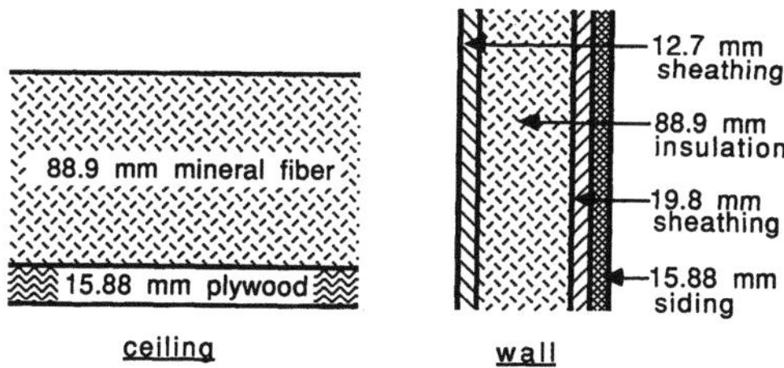

There are two overhead panel doors (44 mm thick with 11 mm thick panels) and each door is 2.5 m high and 3.5 m wide. Double glazed windows are used (6 mm thick airspace, no storm sash) and total window area is 20 m^2.

<u>Solution:</u> The first design decision is to determine an indoor air temperature on which to base the design. Appendix 5-1 contains animal heat production data

and for dairy cows the temperature range for which data are available ends at 27 C. Milk production drops rapidly at temperatures above 27 C (ASAE Data D270.4) although summer weather design conditions in much of the United States can be significantly above 27 C. For convenience and to follow the concept discussed above of designing summer ventilation rates for moderate, not extreme, temperature, the barn will be assumed at 21 C. A moderate relative humidity of 50% will be assumed (and can be adjusted by knowledge of the climate for which the barn is being designed). Program PLUS can be used to determine air density, which is 1.04 kg/m^3 (assuming an exhaust ventilation system and air density measured at indoor conditions).

Sensible heat production from dairy cows at 21 C is 1.1 W/kg, thus, total animal heat production will be

$$q_s = (100 \text{ cows})(500 \text{ kg/cow})(1.1 \text{ W/kg})(570 \text{ kg } / 500 \text{ kg})^{0.734}$$
$$= 60.6 \text{ kW.}$$

The total thermal conductance, ΣUA, of the barn will be obtained using program RVALUE for convenience. Adjustment factors for wood framed windows (assumed to be used) range from 0.90 to 1.0 for double glass; 0.95 will be assumed. This changes the R-value for the windows to (0.29 m^2 K/W)/0.95 (see Appendix 4-2 for window data), or 0.31 m^2 K/W. An "Other" type of window will be specified in RVALUE with this R-value. Door area is 17.5 m^2. The wall framing factor is not specified but the wall appears to be framed with 38 x 90 mm (2 x 4) lumber, thus, the factor will be approximately 20%, the number which should be used in RVALUE.

When these data are used in the program, a building ΣUA value of 784. 8 W/K is obtained. (It would be a useful exercise to use RVALUE to obtain ΣUA and even more useful to try the calculation by hand.)

No perimeter insulation was specified and construction of dairy barn floors is generally slab on grade. From the data in Section 4-6, a perimeter factor of 1.5 W/mK will be chosen (moderate winter severity). The perimeter is 186 m, thus, FP is 279.0 W/K. The temperature difference, t_i - t_o, is chosen to be 3 K, thus, the ventilation rate (Equation 6-1) is

$$\dot{V} = \frac{60,600 \text{ W} - (784.8 \text{ W/K} + 279.0 \text{ W/K})(3 \text{ K})}{(1006 \text{ J/kgK})(1.04 \text{ kg/m}^3)(3 \text{ K})}$$
$$= 18.3 \text{ m}^3\text{/s.}$$

Note an important implication of this calculation. Nothing has been stated about the actual summer weather. Whether the region where the barn is to be built has cool, moderate, or hot summers is not known. However, by the reasoning stated above, this should not matter unless the outdoor air temperature remains always above 18 C(21 C-3 K).

If the summer design temperature is greater than 18 C, the temperature difference from inside the barn to outside will be less than 3 K, but ventilation will not increase no matter how hot the weather becomes. Should it? Not for temperature control, for if the ventilation rate were even to double the gain would be less than 1.5 K at outside air temperatures above 18 C.

This approach does not agree totally with field recommendations. The calculated ventilation rate of 18.3 m^3/s agrees with field recommendations (on a per-cow basis) for climate zones having moderately hot summer weather, but the lack of ventilation rate dependence on climate does not.

The ventilation rate calculation also demonstrates the importance of ventilation for temperature control. Animal heat production is 60.6kW, and (ΣUA + FP) is 1063.8 W/K. When the temperature difference is 3 K, heat loss through the building's structural cover is (1063.8 W/K)(3 K) = 3.2 kW, which is less than 5.3% of the animal heat production. The remainder of the heat is removed by ventilation. As a final comment, there is precedent for ignoring the perimeter heat loss for warm weather calculations, for the soil outside buildings is normally quite warm due to solar heating.

When evaporative cooling is used for environmental modification, the sensible energy balance as already presented applies. However, ventilation air no longer enters at the outside air temperature. It enters at the cooled temperature, t_{oe}, the evaporatively modified outdoor air temperature. The calculation of ventilation rate changes to

$$\dot{V} = \frac{q_s + q_m + q_{so} + (\Sigma UA + FP)(t_i - t_o)}{1006\rho\,(t_i - t_{oe})} \tag{6-2}$$

Evaporatively cooled air is significantly cooler than outside air, thus, the maximum ventilation rate will be lower.

The choice of a design criterion is not limited to maintaining the rise of temperature of the ventilating air to some prescribed limit. That may be the criterion, or the indoor air temperature itself may be prescribed and analysis completed to determine whether the limit can be reasonably attained. Obviously, the design indoor air temperature must be above the temperature to which the ventilating air can be cooled.

In many cases, using a design criterion of building air temperature rather than air temperature rise would be preferred. Using the criterion of a temperature rise is an admission that conditions best for the animals cannot be maintained and the best that can be done is to maintain air temperature near outdoor conditions. When air can be cooled, the possibility is greater that air temperatures which promote health and productivity may be possible.

Example 6-2

<u>Problem:</u> The dairy barn described in Example 6-1 is located in a region of hot summer temperatures and a dry climate. Evaporative cooling has been determined to be desirable to maintain high milk production during the summer. (Note: It is not necessarily true that a zero-pasture housing system would be best for such a climate; the example is used for illustration only. In warm climates, dairy cattle are often kept outdoors in pens or corrals, with sun shades provided to limit heat stress.)

The summer design weather condition (97.5% probability) is 36 C dry bulb temperature, with a mean coincident wet bulb temperature of 18 C. The 97.5% probability wet bulb temperature is 20 C. The evaporative cooling system is expected to be 75% efficient.

Determine the maximum ventilation capacity so indoor air temperature will not go over 27 C (the beginning of heat stress) for more than a few hours per year.

<u>Solution:</u> The decision of which weather condition to use for design is complicated by the need to choose two conditions, dry bulb and wet bulb temperature. The stipulation that the design temperature not be exceeded for more than a few hours per year will be assumed to mean design for the 97.5% probability level. It appears the 97.5% condition for dry bulb temperature has a coincident wet bulb temperature which is close to the 97.5% probability wet bulb temperature. When the wet bulb temperature is 20 C, the dry bulb temperature is below 36 C, thus, less cooling can be expected but less will be needed. Lacking other weather information, the conditions of 36 C dry bulb and 18 C wet bulb temperatures will be chosen, which is an 18 K wet bulb depression.

If the evaporative cooling system is 75% efficient, actual cooling will be 0.75(18) = 13.5 K. At design conditions, ventilating air will enter the building at 22.5 C, while the outdoor air temperature is 36 C. The relative humidity inside the building at this condition is not known but will be assumed for now to be approximately 60%. Thus, the density of indoor air will be 1.03 kg/m^3 (at 27 C, 60% relative humidity, 1000 m elevation).

It will be assumed that solar entry through windows is small enough to be ignored, and during the day there will be little heat generated from lights, etc. The ventilation rate to achieve design conditions with these assumptions is

$$\dot{V} = \frac{q_s - (\Sigma UA + FP)(t_i - t_o)}{1006\rho\,(t_i - t_{oe})}$$

The herd's sensible heat production at 27 C will be 33 kW, thus,

$$\dot{V} = \frac{33{,}000 \text{ W} - (1063.8 \text{ W/K})(27 \text{ C} - 36 \text{ C})}{(1006 \text{ J/kgK})(1.03 \text{ kg/m}^3)(27 \text{ C} - 22.5 \text{ C})}$$

$$= 9.1 \text{ m}^3/\text{s}.$$

Evaporative cooling has permitted nearly a 40% decrease of the maximum ventilation rate.

Excessively high humidity may be a concern when using evaporative cooling. The relative humidity expected in the barn can be determined. Ventilation air enters the space with a dry bulb temperature of 20 C, a wet bulb temperature of 18 C, a relative humidity of 66% and a humidity ratio of 0.012924 kg/kg. At 27 C, a 500 kg cow produces 0.50 mg/kg-s (see Appendix 5-1) of water vapor. The herd will produce

$$m_p = (100 \text{ cows})(0.50 \text{ mg/kg-s})(500 \text{ kg/cow})(570 / 500)^{0.734}$$
$$= 0.0275 \text{ kg/s}.$$

The ventilation rate is 9.1 m^3/s, air density is 1.03 kg/m^3, thus, the change of humidity ratio is

$$\Delta W = m_p / m_{air}$$
$$= 0.0275 \text{ kg/s} / (9.1 \text{ m}^3/\text{s}) \, (1.03 \text{ kg/m}^3)$$
$$= 0.002936 \text{ kg/kg}.$$

Air at a mixed condition within the barn will have a humidity ratio of 0.012924 + 0.002936 = 0.015860 kg/kg. At 27 C, the relative humidity will be 62%. This should not stress the animals and is close to the assumption of 60% relative humidity which was made to determine air density. Although the animals add moisture to the air, their addition of sensible heat lowers the relative humidity of the cooled air. Note that, of course, the results and conclusions require the assumption that data for animal heat and moisture production apply at the conditions existing in the barn and the level of feeding and milk production expected.

6-3-3. Minimum Ventilation Rate. To determine minimum ventilation rates for animal housing, factors in addition to air temperature must be considered and criteria arising from the various factors may conflict. Humidity and other air contaminants may increase when the ventilation rate is low to the point where they dictate the minimum ventilation rate. If that happens, it may no longer be possible to maintain air temperature at the desired level without supplemental heat.

The alternative is to permit air temperature to drop. If air temperature is not far below the optimum level for many hours per year, or for long periods of time, production likely will not be affected. Of course, indoor air temperatures below freezing generally are not permitted, a practice which prevents damage to water pipes, etc. A design minimum indoor air temperature several degrees above freezing is typically chosen for safety. Even at 5 C at the thermostat, thermal stratification and other temperature nonuniformities may produce regions in the barn close to freezing.

The technique to determine which design criterion dictates the minimum ventilation rate is to develop a <u>Ventilation Graph</u>. The graph describes the required ventilation rate as a function of outdoor temperature according to several criteria, for example, temperature control, humidity control, and carbon dioxide control. Sensible energy and mass balances are used to determine the relationship between ventilation and outside air temperature based on these criteria.

To calculate the minimum ventilation rate based on temperature control, contributions of solar heat and heat from lights, etc., are usually discounted — design is for the worst condition when only animal heat is available to warm the air. The energy balance is as before

$$\dot{V}_{temp} = \frac{q_s - (\Sigma UA + FP)(t_i - t_o)}{1006\rho\ (t_i - t_o)} \qquad (6\text{-}3)$$

The moisture balance can be used to determine ventilation for moisture control,

$$\dot{V}_{H_2O} = m_p/\rho_{air}(W_i - W_o), \qquad (6\text{-}4)$$

and a mass balance can be used to determine ventilation for carbon dioxide control,

$$\dot{V}_{CO_2} = (CO_2)_p/((CO_2)_i - (CO_2)_o), \qquad (6\text{-}5)$$

where $(CO_2)_p$ is the volumetric rate of carbon dioxide production, and $(CO_2)_i$ and $(CO_2)_o$ are the volumetric concentrations of carbon dioxide in the indoor and outdoor air, respectively.

The ventilation rate for carbon dioxide control does not vary as a function of outdoor air temperature. Ventilation for temperature control obviously will, and ventilation for moisture control will also vary indirectly as a function of outdoor temperature. A frequently chosen ambient condition for design is a constant relative humidity. In this event, W_o will vary as the outdoor air temperature changes, thus, $\dot{V}_{H_2O}$ must also vary.

A ventilation graph is obtained by calculating the required ventilation rate at a series of outdoor temperatures ranging from the coldest expected for the climate, up to nearly the design indoor temperature, and graphing the results as indicated in Figure 6-2. The graph is drawn for a fixed indoor air temperature, the winter indoor design temperature.

Because the carbon dioxide level in ambient air is not a function of temperature, the line for carbon dioxide control is horizontal on the graph. Although very low at low outdoor temperatures, ventilation for moisture control is always positive because cold air is very dry. The shape of the moisture control curve reflects the shape of the psychrometric chart.

Ventilation for temperature control may be calculated to be negative when it is very cold outdoors. This should be interpreted to mean that when it is very cold, heat loss through the structural cover is so great that the design indoor air temperature cannot be maintained even if there is no ventilation or air infiltration. Supplemental heat is needed.

As outdoor air temperature approaches indoor air temperature, the required ventilation rate becomes very large due to the small value of $(t_i - t_o)$ in the energy balance. This does not indicate the ventilation rate must become increasingly great as the temperature difference becomes small. Instead, the indoor air temperature should be permitted to rise above the winter design temperature (which is generally near or at the lowest desirable temperature).

At a given outdoor air temperature, the criterion which dictates the minimum ventilation rate is that which produces the maximum ventilation rate <u>at that outdoor air temperature.</u> In Figure 6-2, carbon dioxide control dictates when it is very cold outdoors (line ab), moisture control dictates for moderately cold temperatures (line bc), and finally temperature control dictates (line cd). Ideal control of the ventilation rate would follow line abcd on the graph. However, typical fan control is not so finely modulated as to permit this. Instead, common practice is to choose the ventilation rate at point c as the minimum ventilation rate.

When outdoor air temperature is below that at point c, moisture and carbon dioxide criteria are satisfied – they are based on <u>maximum</u> levels. The temperature criterion is not satisfied – it is based on a <u>minimum</u> temperature level. However, most mature farm animals are not affected adversely by air temperatures below those considered optimum. The animals may eat more to maintain production and body temperature, but health, vitality, and productivity are not likely to be affected. There are obviously limits where health and production will be affected, but with typical animal stocking densities there will be sufficient body heat to prevent severely cold indoor air temperatures. If there is not sufficient body heat, supplemental heat may be required, the building may be redesigned for better insulation qualities, or management practices may be changed to increase stocking density.

The temperature at which point c occurs is a function of many design parameters, the insulation value of the structural cover of the building being one. At the minimum ventilation rate, ventilation is still a dominant mechanism of heat loss from the airspace but it is the same order of magnitude as loss through the structural cover. Thus, insulation in the walls, ceiling, and perimeter, and the choice of doors and glazings can affect the temperature criterion line and the location of point c. This can be an important design parameter if the outdoor air temperature at point c is sufficiently high that indoor air temperature cannot be maintained for a significant fraction of the winter.

Example 6-3

Problem: Reconsider the dairy barn described in Examples 6-1 and 6-2. Determine the minimum ventilation rate if the indoor air temperature desired for cold weather conditions is at least 10 C, with a relative humidity no more than 70% and a carbon dioxide concentration no more than 5000 ppm. Outdoor air temperature can be as low as - 30 C.

Solution: A ventilation graph will be used to solve this problem. Program PLUS will be used to determine psychrometric properties of air.

Data in Appendix 5-1 for dairy cows at 10 C are: total heat production is 2.2 W/kg, sensible heat production is 1.5 W/kg, and moisture production is 0.28 mg/kg-s, each based on a 500 kg animal unit. Scaled to 100 cows averaging 570 kg, heat and moisture production rates for the entire barn are:

$$q_{total} = (100 \text{ cows})(2.2 \text{ W/kg})(500 \text{ kg/cow})(570 \text{ kg} / 500 \text{ kg})^{0.734}$$
$$= 121 \text{ kW,}$$

$$q_{sensible} = (100 \text{ cows})(1.5 \text{ W/kg})(500 \text{ kg/cow})(570 \text{ kg} / 500 \text{ kg})^{0.734}$$
$$= 83 \text{ kW, and}$$

$$m_{H_2O} = (100 \text{ cows})(0.28 \text{ mg/kg-s})(500 \text{ kg/cow})(570 / 500)^{0.734}$$
$$= 0.0154 \text{ kg/s.}$$

The simplest line to determine on the ventilation graph is for carbon dioxide control. Ambient carbon dioxide will be assumed at 345 ppm. The conversion from total heat production to carbon dioxide production (Section 5-6.1) yields the carbon dioxide production rate,

$$\dot{V}_{CO_2} = (121{,}000 \text{ J/s})/(24{,}600 \text{ J/L}) = 4.92 \text{ L/s}$$
$$= 0.00492 \text{ m}^3\text{/s.}$$

The change permitted in the carbon dioxide concentration is

$$(C_{CO_2})_{in} - (C_{CO_2})_{out} = 0.005 \text{ m}^3\text{/m}^3 - 0.000345 \text{ m}^3\text{/m}^3$$
$$= 0.004655 \text{ m}^3\text{/m}^3.$$

The ventilation rate for carbon dioxide control is

$$\dot{V}_{CO_2} = (0.00492 \text{ m}^3/\text{s})/(0.004655 \text{ m}^3/\text{m}^3)$$

$$= 1.06 \text{ m}^3/\text{s}.$$

The ventilation rate for moisture control will depend on the outdoor relative humidity, especially as the outdoor air temperature approaches that of the indoor air. For design, a condition of high (but reasonable) outdoor relative humidity is more conservative. The climate where the barn is located is stated to be dry, but during cold weather relative humidities are likely to be high at times in most climate zones. For this design, an outdoor relative humidity of 80% will be assumed.

Indoor conditions of 10 C and 70% relative humidity result in a humidity ratio of 0.006076 kg/kg (recall that this is at 1000 m elevation). Moisture production is 0.0154 kg/s, thus, ventilation for moisture control is calculated from

$$\dot{V}_{H_2O} = (0.0154 \text{ kg/s})/\rho_{air}(0.006076 - W_o),$$

where W_o is the humidity ratio of the outdoor air. At 10 C and 70% relative humidity, air density (ρ_{air}) is 1.10 kg/m^3. The following table contains humidity ratio and ventilation rate data for the outdoor temperature range from - 25 C to 5 C.

Outdoor Temperature	Humidity[a] Ratio, kg/kg	Ventilation Rate, m^3/s
-25 C	0.000353	2.45
-20	0.000577	2.55
-15	0.000925	2.71
-10	0.001457	3.03
-5	0.002256	3.66
0	0.003441	5.31
5	0.004924	12.15

[a] at 80% relative humidity, 1000 m elevation

The ventilation rate for temperature control is determined by the sensible energy balance for the barn, as used in Example 6-2, changed to reflect conditions for this example.

$$\dot{V}_{temp} = \frac{83,000 \text{ W} - (1063.8 \text{ W/K})(10 \text{ C} - t_o)}{(1006 \text{ J/kgK})(1.10 \text{ kg/m}^3)(10 \text{ C} - t_o)}$$

The following table contains data for the ventilation rate as a function of outdoor air temperature.

Outdoor Temperature	Ventilation Rate, m^3/s
-25 C	1.18
-20	1.54
-15	2.04
-10	2.79
-5	4.04
0	6.54
5	14.04

The ventilation rates for carbon dioxide, moisture, and temperature control are shown on the ventilation graph in Figure 6-3. Carbon dioxide never governs the minimum ventilation rate in this example. The lines for moisture control and temperature control cross at a ventilation rate of approximately 3.3 m³/s, at an outdoor air temperature of approximately - 8 C. Thus, the installed minimum ventilation rate would be 3.3 m³/s. This is high for a minimum ventilation rate for a dairy barn, due to the design criterion of only 70% relative humidity. It would be a useful exercise to assume 80% relative humidity and determine the resulting minimum ventilation rate.

When outdoor air temperature is below - 8 C, the temperature control criterion will be violated, resulting in a barn temperature lower than the design value. It would be a useful exercise to determine how low the barn temperature would fall without supplemental heat and how much supplemental heat, if any, would be needed to prevent freezing in the barn.

The sensible energy balance can again be used, although the question of what the barn temperature will be must be determined. Animal heat production is a function of air temperature and air temperature is influenced by the heat production. As the temperature falls, heat production increases, which is a negative feedback mechanism acting to help prevent freezing in the barn. If animal heat production is expressed as a second order polynomial (see section 6-4.1), a quadratic equation results which can be solved to determine the indoor air temperature in closed form.

Before completing this example, a note regarding humidistats is in order. Humidity control in this example was achieved by indirect means – based on assumed conditions and assumed heat and moisture production data applied to the cows. Direct measure of humidity was not considered; humidity measurement in agricultural buildings is not a general practice. The reason can be traced to a lack of suitable humidity sensors. Agricultural buildings often are at high relative humidities and provide aerial environments with high dust levels at times and trace amounts of many corrosive gases. The environment is hostile to sensors and a humidity sensor which can retain calibration in these conditions has not yet been developed. A wet and dry bulb sensor is possible, but requires

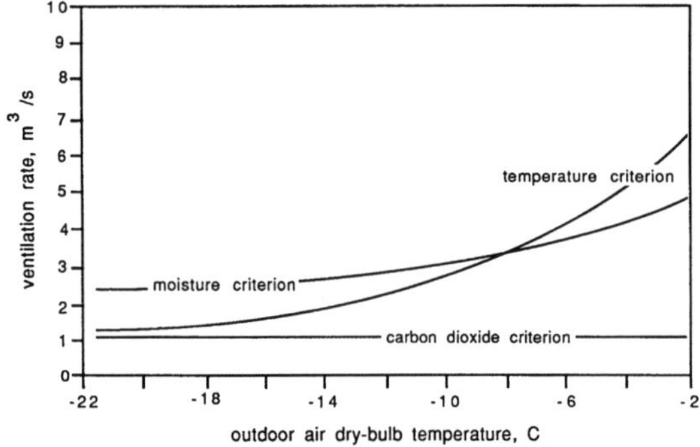

Figure 6-3. Ventilation graph for Example 6-3. Moisture and temperature criteria curves intersect at a ventilation rate of 3.3 m³/s and an outdoor air temperature of -8 C.

186

frequent maintenance and careful use, needs often neglected in the press of daily activities.

6-4. Computerized Procedures to Determine Ventilation Rates

An automated procedure to draw ventilation graphs and determine minimum ventilation rates would be useful for design, for such a program would permit the designer to explore various design options and immediately see the results in terms of their effects on minimum ventilation rates.

The procedures used above to develop a ventilation graph could be implemented in a computer program to calculate and graph data such as in Figure 6-3. The one factor needed for the graphs is a means to calculate heat and moisture production from animals. Data such as in Appendix 5-1 can be used, with interpolation procedures implemented in a computer program to estimate production rates at temperatures intermediate to those in the appendix. Another approach is to approximate the data using a mathematical function which expresses heat or moisture production as a function of air temperature.

6-4.1. Polynomial Regression. Statistical regression procedures are used frequently in engineering and the biological sciences to describe how one parameter varies in response to changes in another. Data to describe the dependence may exist in table form, but for convenience an equation is preferable to express the relationship.

The convective heat transfer equations described in Chapter 3 were developed by the original researchers after they accumulated a large base of experimental data where heat transfer rates by convection were measured, along with other pertinent variables. Numerous regression models were attempted until one was found which provided the "right answer" for the ranges of values of the independent variables which had been considered. Regression analysis is a technique which can be used to express a dependent variable in terms of independent variables thought to influence the response of interest.

Regression models are not based on fundamental principles of the phenomenon under consideration. They are a means to approximate or describe the value of a dependent variable if values for all independent variables are known. Regression models can be expected to have no mathematical resemblance to the true dependence, but they provide adequate answers within a limited range of the independent variables. The limit depends on the width of the original range of data when the equation was developed.

Many forms of regression models are possible–linear, power curves, exponential functions, logarithmic functions, etc. Regression models based on polynomials can also provide suitable means to describe data points. While not best for all situations, polynomials provide an adequate approximation of animal heat and moisture production as a function of air temperature.

Program POLYNOM is provided to calculate the coefficients in a polynomial regression equation for Y as a function of X,

$$Y = a_0 + a_1X + a_2X^2 + a_3X^3 + \ldots \quad (6\text{-}6)$$

The program is limited to no more than a sixth order polynomial, and no more than 100 data points. Two measures of fit are provided. The standard deviation of the data about the regression line is calculated and the data and regression equation can be graphed so the regression adequacy can be evaluated by eye.

Printed copies of the graph of data and the regression equation can be obtained using the PrintScreen key. Note: Before a copy can be obtained, the computer must be in Graphics mode.

Example 6-4

Problem: Develop a polynomial to express sensible heat production (SHP, W/kg) from dairy cows as a function of air temperature (C).

Solution: The basic data are in Appendix 5-1.

Air Temperature	SHP W/kg
-1 C	1.9
10	1.5
15	1.2
21	1.1
27	0.6

Program POLYNOM will be used. There are 5 data points. The maximum possible order of the regression is 4 (the constant term plus 4 coefficients). The coefficients for each order of regression are:

Order	a_0	a_1	a_2	a_3	a_4
0	1.260				
1	1.897	-0.04425			
2	1.860	-0.03074	-5.268E-4		
3	1.845	-0.05708	2.504E-3	-7.693E-5	
4	2.006	0.07884	-2.546E-2	1.540E-3	-2.874E-5

Graphs of the regressions for first through fourth order are in Figure 6-4. The standard deviations for the four orders of regression are: 0:0.483; 1:0.105; 2:0.106; 3:0.124; 4:0.000.

Standard deviations provide a limited measure of goodness of fit. Order 1 (linear regression) is obviously much better than order 0 (only the average). The standard deviation for order 2 shows no improvement (in fact, a very slight increase of the standard deviation), and order 3 is an even poorer fit. Order 4 provides a perfect fit of the five data points, but that should not be taken to mean 4th order is the best means to represent the data. Graphs of the equations for orders 1 to 4 are in Figure 6-4. The fourth order curve, although it passes through every data point, oscillates between the first and second data point in a way inappropriate to describe how the data would be expected to behave. Thus, for this example, a first order equation would be appropriate.

6-4.2. Program VENTGRPH. It is necessary for an engineer designing environmental control systems for barns to know how to construct a ventilation graph. However, to do so repeatedly is laborious, and the effort involved quickly subdues a designer's enthusiasm to explore the many possible options in modifying the initial design of a barn. Program VENTGRPH is provided for such explorations.

All heat and moisture productions from animals are expressed as second order polynomials in VENTGRPH. Although the results of Example 6-4 indicated a first order regression was appropriate to express sensible heat production from dairy cows, a second order polynomial was more appropriate for other data in Appendix 5-1 and was the form adopted for general use in VENTGRPH.

The default data incorporated into VENTGRPH are based on data in Appendix 5-1, but an opportunity is presented to change the data if desired. Also, an "other" category exists to permit ventilation graphs to be constructed for

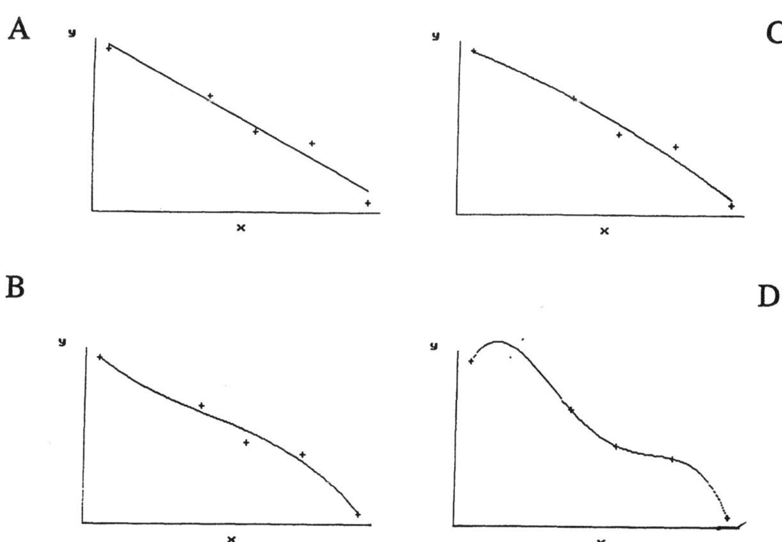

Figure 6-4. Data and regression curves for Example 6-4: (a), first order; (b), second order; (c) third order; and (d), fourth order.

animals other than dairy cows, swine, broilers, and laying hens. To use the "other" category, second order polynomials to approximate sensible and latent heat productions must first be determined.

Example 6-5

Problem: Use program VENTGRPH to develop a ventilation graph for the barn in Example 6-3 and explore the effect of the insulation value of the structural cover on the minimum ventilation rate which should be installed in the barn. Determine the minimum ventilation rate, m^3/s, as a function of the barn's average U value, $W/m^2 K$, over the range of U from 0.1 to 1.0.

Solution: Program VENTGRPH can be used repeatedly to determine the minimum ventilation rate at the various insulation values. Data from previous examples of this chapter will be used as input. The air pressure is 89.874 kPa. Indoor air temperature is 10 C and the relative humidity and carbon dioxide levels are 70% and 5000 ppm, respectively. There are 100 cows averaging 570 kg mass. The perimeter of the barn is 186 m and the perimeter heat loss factor is 1.5 W/mK.

The overall UA value is 784.8 W/K, and the area of the ceiling and walls is: $(13)(80) + (186)(3) = 1598 \ m^2$. Thus, the average U value of the structural cover is: $784.8/1598 = 0.49 \ W/m^2 K$. This is the parameter which will be varied from 0.1 to 1.0, and the result will be a variation in ΣUA from 159.8 to 1598 W/K. A change of this magnitude spans the range from no insulation and very light construction, to a level of insulation which exceeds that normally installed in animal housing.

When program VENTGRPH is used, the minimum ventilation rate (intersection of the moisture/temperature curves) is displayed on the screen. If carbon dioxide controls, its corresponding ventilation rate must be estimated from reading the screen.

Using VENTGRPH repeatedly, the data in the table below can be accumulated where the outdoor air temperature at crossing is the temperature at which the moisture criterion and temperature criterion curves intersect. The carbon dioxide criterion never governs in this example.

As an example of the output from VENTGRPH, in Figure 6-5 is the ventilation graph for the original example, with a ΣUA value of 784.8 W/K. The intersection of the moisture and temperature criteria lines is at approximately 3.5 m^3/s. This is slightly above the result in Figure 6-3. The difference arises from the approximation of animal heat production data by using a polynomial, and graphing the lines more smoothly in Figure 6-5.

The trend for the minimum ventilation rate as a function of the overall building

UA value may appear counter-intuitive at first. Without knowledge of the form of the ventilation graph, one might think that as the ΣUA value increases, more heat will be lost from the building by conduction through the structural cover and less loss can as a consequence be tolerated through ventilation, resulting in a lower minimum ventilation rate.

The data show the opposite is true. Why is this? The reason lies in the shape of the moisture criterion curve. As the ΣUA value increases, the temperature criterion can be met only at a higher outdoor air temperature. At the higher outdoor temperature, the air is less dry (recall the assumption of fixed outdoor relative humidity). When the outdoor air is less dry, more ventilation is needed for moisture control and the moisture control criterion must be satisfied. The result is a higher minimum ventilation rate. What happens to temperature control? The temperature criterion is still met, <u>but at a higher outdoor air temperature.</u>

The crossing of the moisture and temperature control criteria lines at a lower outdoor air temperature is an important, and subtle, result of insulating barns. A first approach to determining the proper insulation level might include only two criteria: prevent condensation on the inside surface of the wall, and conserve sufficient energy to prevent the barn from freezing.

Building U value W/m^{2}K	Building ΣUA value, W/K	Minimum Ventilation Rate, m^3/s	Outdoor Air Temperature at Crossing
0.1	159.8	2.9	-13 C
0.2	319.6	3.1	-11
0.3	479.4	3.2	-9
0.4	639.2	3.4	-8
0.5	799.0	3.6	-6
0.6	958.8	3.8	-5
0.7	1118.6	3.9	-4
0.8	1278.4	4.2	-3
0.9	1438.2	4.4	-2
1.0	1598.0	4.6	-2

The ventilation graph shows two additional advantages of insulation. A lesser minimum ventilation capacity may be installed and this should reduce the electricity cost to ventilate (how the savings can be estimated will be covered later in the text). In addition, the curves cross at a lower outdoor air temperature, thus, the temperature criterion will be violated for fewer hours per year. This could save a little feed and provide an environment within the thermoneutral zone of the animal population for a greater fraction of the year.

6-5. Staging Between Minimum and Maximum Ventilation Rates

To this point, procedures have been developed to calculate the required minimum and maximum ventilation rates for an animal barn. The next design step is to prescribe how the rate shall be varied through the range between.

One means to vary ventilation rates is to use variable speed fans, chosen to provide the maximum ventilation rate at full capacity and the minimum rate when some or all the fans operate at their minimum capacities. To reach the minimum rate, only a few fans may be needed. Variable speed fans have the advantage of continuous variation between their minimum and maximum ventilation rates. They have the disadvantage of losing their ability to resist wind effects when operated at less than full capacity. Fan characteristics will be explored more fully in a later chapter.

Single speed fan systems must be "staged" if they are to modulate between the minimum and maximum ventilation rates. This means, for example, one or more small fans are included to provide the minimum rate (and larger fans not operated). Middle sized and large fans are also installed, and activated as barn temperature rises and the need for ventilation increases. The barn's ventilation rate thereby changes by stages rather than continuously as would be the case with variable speed fans.

Frequently, the small fans continue to operate as the higher stages are needed; each higher stage is an addition of fans. The small fans frequently draw air from near the floor (the coldest air) to conserve a little heat during the coldest weather and the large fans draw air from near the ceiling to remove the warmest air during hot weather.

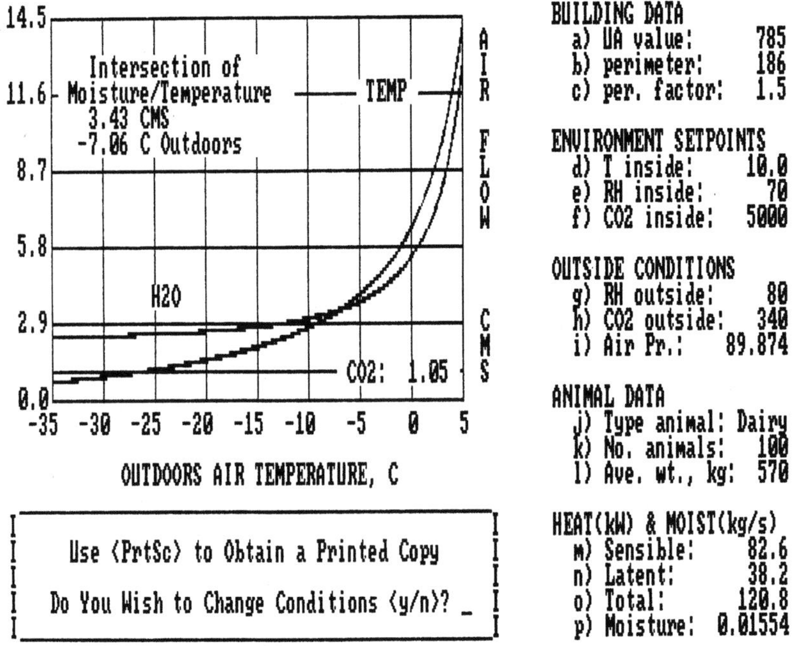

Figure 6-5. Results screen of program VENTGRPH with data and results of Example 6-5, and actual barn design (proposed).

Two factors must be considered when staging a fan system. One is to determine barn temperatures at which control moves from one ventilation stage to another – the control system setpoints. The other is to determine the magnitude of each stage between the minimum and maximum ventilation rates.

The choice of setpoints for ventilation stage changes should reflect the biological needs of the animals. What is best for dairy cows, for example, would be inappropriate for laying hens. A guide to determine what is best are data which reflect production and feed consumption as functions of temperature. For example, the thermoneutral zone for dairy cows centers around the range from 10 to 15 C, but production changes little over a wide temperature range centered about 10 to 15 C. Optimal production and feed consumption for laying hens occurs near 24 C, and is limited to a narrow temperature range (ASAE Data D270.4).

Data in Figure 6-6 can be used to estimate optimum temperature ranges for dairy cows, laying hens, and swine, and the effects of varying from those ranges. Control system setpoints are generally chosen to provide the minimum ventilation rate at the lower end of the optimum production range, and the maximum ventilation rate at the upper end. For example, one might choose 8 C as the lowest thermostat setpoint for Holstein dairy cows. When indoor air temperature is below 8 C, only the minimum ventilation rate for moisture (or perhaps carbon dioxide) control would be used. Above 8 C, the second stage of ventilation would be used. (Of course, a freeze protection would be used in the control sequence; all fans might be off at barn temperatures below 4 C, for example.) The highest stage of ventilation would be activated at the temperature where heat stress could affect production. For dairy cows, this would be approximately 24 C, although common practice is to activate the highest stage several degrees below 24 C.

The number of stages between minimum and maximum ventilation rates is left to the designer's discretion. There should be enough stages so ventilation rate steps from stage to stage do not cause extreme environment changes. For example, a system with only two stages would result in rapid oscillation

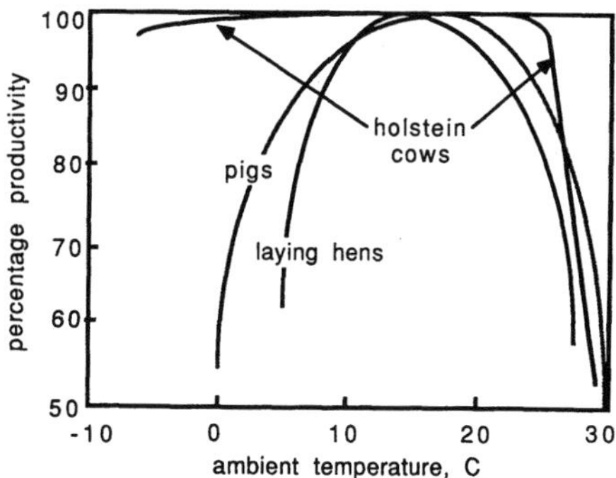

Figure 6-6. Productivity potentials of dairy cows, swine, and laying hens as a function of ambient air temperature (in still air with no solar effect).

between the stages during moderately cold weather and perhaps frequent, sudden chilling of the animals as the ventilation rate changes from the minimum to the maximum. Such a situation would be unsatisfactory. Experience has shown four stages to be a practical minimum. A practical maximum number of stages have been found to be six. More than six stages do not bring significantly better control.

The decision of whether four, five, or six stages are best depends on the total amount of ventilation needed, and availability of fan sizes. If the ventilation system is to be small, fewer stages are more practical to avoid using many small fans. Small fans are generally less energy efficient than large fans and installation costs for many small fans would be greater than for a few large fans.

The second important decision in developing a staging schedule is to specify the magnitude of each stage. The minimum and maximum stages are calculated as shown above. The number of stages can be decided by the size of the total system. Intermediate stages must then be specified.

A traditional method to size the stages is to divide the range between the minimum and maximum rates evenly. Another method is to specify the beginning stages as small steps, and the difference between the next to last and last stages as large.

In Figure 6-7a is a conceptual graph of the need for ventilation as a function of outdoor air temperature. If stages are determined by equal divisions of the ventilation rate range, there will be a large change of outdoor air temperature between stages 1 and 2 (Figure 6-7b). Also the change of ventilation between stages 1 and 2 will be large relative to the magnitude of ventilation at stage 1. When control causes ventilation to oscillate between stages 1 and 2 (outdoor air temperature is too warm to maintain stage 1 continuously and too cold to maintain stage 2), the large changes of ventilation rate could cause localized chilling and rapid changes of temperature within the barn. If stage 2 is only slightly greater than stage 1, these problems can be minimized.

One method to assure the early stages will be separated by small steps is to divide the ventilation rate range using equal divisions of outdoor air temperature, as shown in Figure 6-7c. Experience has shown that equal

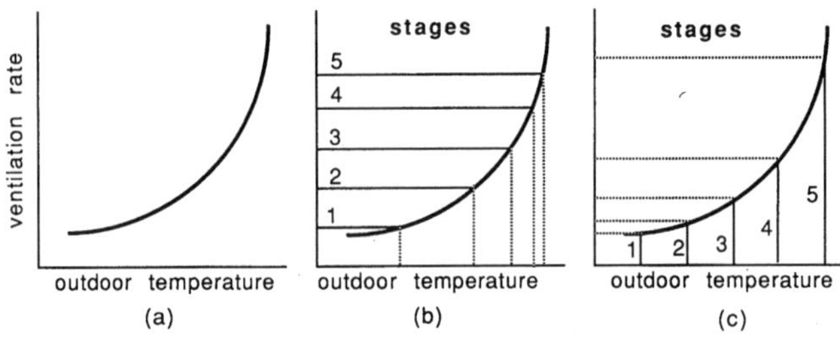

Figure 6-7. Ventilation rate as a function of outdoor temperature: (a) general curve to maintain fixed indoor air temperature, (b) with fan stagings based on equal increments of ventilation rate, and (c) with fan staging based on equal increments of outdoor temperature.

divisions of outdoor air temperature result in stages approximately as listed below in Table 6-1.

Each schedule in Table 6-1 begins with the minimum ventilation rate being 10% of the maximum. Procedures used above to determine minimum and maximum rates will not always result in the 1:10 difference. Depending on assumptions used for design, the ratio may be as small as 1:5. The actual schedule must therefore use the spirit of the data in Table 6-1 rather than always the actual data.

Table 6-1. Staged ventilation rates for 4, 5 and 6 ventilation stages in animal barns*

Number of Stages:	4	5	6
	10	10	10
	15	14	13
	28	19	17
	100	35	25
		100	44
			100

* Expressed as percentages of the maximum ventilation rate.

As an example, suppose a five stage system is desired and calculations have shown the minimum ventilation rate is 15% of the maximum rate. Data for the five stage system show stage 2 is 40% greater than stage 1, and stage 3 is approximately 40% more than stage 2. Stage 4 is nearly twice stage 3. With these as guides, a staging schedule of: 15%, 21%, 29%, 50%, and 100% could be used and fans chosen to provide approximately these stages.

The procedure to choose fans to provide proper staging will be left as a topic for a later chapter. The focus here is on determining required ventilation rates.

Example 6-6

<u>Problem:</u> Reconsider the dairy barn of Example 6-1, et seq. The minimum ventilation rate has been determined to be 3.3 m^3/s and the maximum rate 18.3 m^3/s (no evaporative cooling is to be used). Determine the ventilation rates for a 5 stage ventilation system and specify thermostat setpoints.

<u>Solution:</u> The minimum ventilation rate is 18% of the maximum rate thus the recommended schedule in Table 6-1 must be modified. A candidate is

stage 1: 18% = 3.3 m^3/s
stage 2: 25% = 4.6 m^3/s (stage 1 + approximately 40%)
stage 3: 35% = 6.4 m^3/s (stage 2 + approximately 40%)
stage 4: 60% = 11.0 m^3/s
stage 5: 100% = 18.3 m^3/s

There are many candidate ventilation schedules. The one listed may not be necessarily the optimum, but it reflects the concept contained in Table 6-1. If anything, stages 2, 3, and 4 might be slightly smaller.

There are 5 ventilation stages, there must be 4 thermostat setpoints plus a freeze protection setpoint. A candidate schedule of setpoints is

freeze protection setpoint:	4 C
stages 1 to 2 setpoint:	8 C
stages 2 to 3 setpoint:	12 C
stages 3 to 4 setpoint:	17 C
stages 4 to 5 setpoint:	21 C

The stage 1 to 2 setpoint was chosen to correspond to the previous assumption that the minimum desirable barn temperature will be 8 C. The highest setpoint was chosen to correspond to the temperature at which the maximum ventilation rate was calculated. Intermediate setpoints were rather arbitrarily chosen to divide the range into equal increments.

6-6. Supplemental Heat

Ventilation graphs provide a tool to determine minimum ventilation rates for animal housing. In cold climates, a significant number of hours will find outdoor air temperature below the temperature where moisture and temperature criteria (or carbon dioxide and temperature criteria) curves intersect to determine the minimum ventilation rate. It is possible that freezing temperatures may occur inside the barn under conditions of extreme cold. Freeze protection should be part of the control sequence, but it is useful to determine how frequently freezing conditions could be encountered.

The action of freeze protection is to turn all fans off. In any typical housing for mature animals at normal stocking densities, when there is no ventilation there is more than enough animal heat to maintain the barn at a reasonably high temperature. When fans are off, the barn temperature will rise past the freeze protection point and the fans will turn back on. Operation will oscillate between the lowest stage and off until the extreme cold weather has passed. The effect in the barn will be humidity higher than the design value.

If this occurs for only a few hours during a year the effects are negligible, but if this occurs for many hours, a redesign using more insulation in the structure should be considered, or the barn redesigned to increase stocking density, or both.

Sensible energy balances permit a designer to determine the outdoor air temperature at which the indoor air temperature is at the freeze protection point.

Weather bin data can be used to estimate the number of hours per year that operation might be in this extreme state. If periods with no ventilation are to be avoided completely, weather data and a sensible energy balance can be used to calculate how much supplemental heat will be needed to prevent freezing.

Example 6-7

Problem: Continue to design the dairy barn described in Example 6-1, et seq. The location is near Denver, Colorado. Determine:

(a) how many hours in a typical year the barn will be at the freeze protection temperature (4 C), and

(b) whether, and, if so, how much supplemental heat capacity would be required to prevent the barn from ever being below 10 C, and alternately, ever below the freeze protection temperature if it is desired that the fans never be off.

Assume the minimum ventilation rate is to be constant at 3.3 m^3/s.

Solution, part a: At the freeze protection point, indoor air temperature will be 4 C, the ventilation rate will be 3.3 m^3/s, and animal sensible heat production will be

$$q_{sensible} = (100)(570/500)^{0.734}(500)(1.8603 - 0.03074(4) - 0.0005268(4)^2),$$
$$= 95,200 \text{ W},$$

based on the polynomial for sensible heat production obtained in Example 6-4. The polynomial to express latent heat production is

$$(q_{latent})/kg = 0.521481 + 0.011559t + 0.000575092t^2,$$

which is the default data in VENTGRPH. Note that although in Example 6-4 the linear equation for sensible heat production was best, the second order polynomial is used as it is in VENTGRPH. From the herd,

$$q_{latent} = (100)(570/500)^{0.734}(500)(0.521481 + 0.011559(4)$$
$$+ 0.000575092(4)^2)$$
$$= 31,800 \text{ W}.$$

The heat of vaporization of water is 2460 kJ/kg at body temperature, thus, moisture production is

$$m_p = 31,800/2.46E + 6 = 0.01293 \text{ kg/s}.$$

Compare this to the ventilation graph for 10 C in Figure 6-5. Sensible heat production increased from 83 to 95.2 kW, and moisture production decreased

·from 0.01554 to 0.01293 kg/s.

Air density at 10 C and 70% relative humidity was 1.10 kg/m^3. Density will change at the freeze protection temperature, but for now assume it remains at 1.10 kg/m^3. At this point indoor relative humidity is not known so the new density cannot be determined.

The sensible energy balance is a tool to determine outdoor air temperature when the barn is at 4 C and ventilation is 3.3 m^3/s,

$$t_o = 4\,C - \frac{95,200\ W}{(1006\ J/kgK)(1.10\ kg/m^3)(3.3\ m^3/s) + 1063.8\ W/K}$$
$$= -16.2\ C.$$

At -16 C and 80% relative humidity, the humidity ratio is 0.000843 kg/kg. If the ventilation rate is 3.3 m^3/s at an air density of 1.10 kg/m^3, the mass flow rate is 3.63 kg/s and the humidity ratio change is (0.01293 kg/s)/(3.63 kg/s) = 0.003562 kg/kg. The indoor humidity ratio will, thus, be 0.000577 kg/kg + 0.003562 kg/kg = 0.004139 kg/kg.

At 4 C and a humidity ratio of 0.004139 kg/kg, the relative humidity is 72% and air density is 1.13 kg/m. The initial assumption of air density is in error by a few percent. To be accurate, results to this point will be recalculated.

$$t_o = 4\,C - \frac{95,200\ W}{(1006\ J/kgK)(1.13\ kg/m^3)(3.3\ m^3/s) + 1063.8\ W/K}$$
$$= 15.8\ C.$$

The change is so slight it is safe to assume another iteration is not needed; indoor air density will be 1.13 kg/m^3, with a relative humidity of approximately 72%. You might check this assumption.

Weather bin data for Denver (Appendix 6-2) show outdoor air temperature can be expected to be below - 16 C for

$$1\ hr + 8\ hrs + 35\ hrs + (137\ hrs)((-16\ C - (-17.8\ C))$$
$$/(-12.2\ C - (-17.8\ C)))$$
$$= 88\ hrs/yr.$$

Coincidentally, this number of hours per year is at approximately 4% of a winter, and -16 C is only slightly warmer than the 97.5% winter design temperature for Denver (which is -17 C). Differences among sets of weather data will arise frequently.

The 88 hours per year will accumulate during several short segments of time, probably only for several hours at a time during the coldest nights. For most dairy barns, this should create no problem. The fans would cycle on and off and the relative humidity would reach near 100%, but for only a short time.

<u>Solution, part b:</u> When the barn is held to be no colder than 10 C, supplemental heat will be required any time outdoor air temperature is below - 8 C (the intersection of the moisture and temperature criteria lines on Figure 6-5). However, the heating system must be sized to meet extreme heating needs when weather is at its coldest. From data in Appendix 6-2 for Denver, one hour per year is expected to fall within the range from - 34.4 to - 28.9 C.

The 99% winter design temperature for Denver in Appendix 6-1 is - 21 C but the 99% data can be exceeded for 1% of a winter, which is 21.6 hrs. (Note: the two sets of weather data do not agree exactly — they are from different sources and were obtained from weather data obtained for different periods of time. Such small differences are typical of variations among sets of weather data.)

Conservative design practice would lead to choosing a typical minimum outdoor air temperature for Denver to be - 34.4 C — the low end of the range which occurs 1 hr/yr. A sensible energy balance can be used to calculate supplemental heat required to maintain the barn at 10 C when it is this cold and the ventilation rate is 3.3 m^3/s.

$$q_{suppl.} = (1006 \text{ J/kg K})(1.10 \text{ kg/m}^3)(3.3 \text{ m}^3\text{/s})(10 \text{ C} + 34.4 \text{ C})$$
$$+ (1063.8 \text{ W/K})(10 \text{ C} + 34.4 \text{ C}) - 83{,}000 \text{ W},$$
$$= 162{,}100 \text{ W} + 47{,}200 \text{ W} - 83{,}000 \text{ W} \equiv 126{,}000 \text{ W}.$$

The installed heating capacity must be 126 kW if the barn is never to go below 10 C, the minimum ventilation rate is to be continuous, and the lowest outdoor air temperature is - 34.4 C.

<u>Note:</u> Even at - 34.4 C the structural cover heat loss is only slightly more than half the sensible animal heat production. This would still permit fans to cycle to produce an average ventilation rate slightly less than a quarter of 3.3 m^3/s. Supplemental heat would not absolutely be required, but if it were not used, relative humidity would be above the design value for many hours per year (the hours outdoor air temperature is between - 34.4 C and - 8 C).

If restrictions were relaxed to permit barn temperature to fall to 4 C, sensible animal heat production would rise to 95.2 kW and supplemental heat needs would be lowered to

$$q_{suppl.} = (1006 \text{ J/kg K}) (1.13 \text{ kg/m}^3) (3.3 \text{ m}^3\text{/s}) (4 \text{ C} + 34. 4 \text{ C})$$
$$+ (1063.8 \text{ W/K})(4 \text{ C} + 34.4 \text{ C}) - 95{,}200 \text{ W}$$
$$= 144{,}500 \text{ W} + 40{,}800 \text{ W} - 95{,}200 \text{ W} \equiv 90{,}000 \text{ W}.$$

Permitting barn temperature to be as low as 4 C has reduced the required heating capacity from 126 kW to 90 kW. But, of course, common practice would permit fans to cycle and humidity to rise during those hours of the winter

---

6-7. Greenhouse Ventilation

Greenhouse ventilation systems are sized for temperature control, when the major source of heat is the sun. General practice is to provide sufficient ventilation for 3/4 to 1 air change per minute as the maximum ventilation rate. This rate is sufficient to keep the temperature rise of ventilation air to approximately 5 K. The ventilation rate is sometimes increased for elevations over 600 m and regions with clear skies and high solar insolation. The increase is a field practice and is not based on calculations.

Fans usually are placed on sidewalls, and inlets on sidewalls opposite the fans. Thermal stratification causes the cool, fresh air to travel across the greenhouse and mix very little with hot air near the peak of the greenhouse. The result is an effective maximum ventilation rate larger than 3/4 to 1 air change per minute within the plant growing zone.

Sensible energy balances can be used to calculate the maximum required ventilation rate. As a rough approximation, heat loss by ventilation can be equated with solar gain. A more complete energy balance includes structural heat loss.

Example 6-8

Problem: A gutter-connected, ridge and furrow greenhouse is designed to be 200 m wide and 250 m long. The elevation where it is to be constructed is 500 m. The total UA value has been calculated to be 220,000 W/K.

Weather records show peak solar insolation to be approximately 700 W/m^2 (on a horizontal surface) and peak air temperatures to be 35 C. Calculate the ventilation required to keep the indoor air 5 K above outdoor air temperature.

Solution: Several assumptions are needed to begin this calculation. First, perimeter heat loss will be neglected compared to loss through the structural cover and ventilation. Second, the fraction of solar heat converted to sensible heat inside the greenhouse must be estimated. A rough rule is: a third of ambient solar insolation is reflected back to the outside or converted through photosynthesis, a third contributes through evapotranspiration to latent heat, and a third is sensible heat added to the indoor air. This rule will be used for this problem.

The total sensible heat contribution is estimated to be

$$q_{sensible} = (1/3)\ (700\ W/m^2)\ (50{,}000\ m^2) = 11{,}700{,}000\ W$$

Ventilation is expressed on a volumetric basis, and air density is not known. Outdoor air temperatures of 35 C are expected, and the indoor temperature will be 5 K more, or 40 C. The relative humidity is not known, but will be assumed to be 50%. Program PLUS can be used to determine air density, 1.05 kg/m^3 at indoor conditions.

The required maximum ventilation rate can be calculated as

$$\dot{V} = \frac{q_{sensible} - (\Sigma UA + FP)(t_i - t_o)}{\rho_{air} c_p (t_i - t_o)}$$

The perimeter heat loss is assumed negligible compared with the structural cover heat loss, thus

$$\dot{V} = \frac{11{,}700{,}000\ W - (220{,}000\ W/K)(5K)}{(1.05\ kg/m^3)(1006\ J/kg\ K)(5K)}$$

$$= 2{,}000\ m^3/s.$$

SYMBOLS

A	area, m^2
CO_2	carbon dioxide volumetric concentration, ppm
$\dot{CO_2}$	carbon dioxide production rate, m^3/s
F	perimeter heat loss factor, W/m K
m	mass flow rate, kg/s
m	mass production rate, kg/s
P	perimeter, m
q	heat flow, W
t	temperature, C
U	unit area thermal conductance, W/m^2 K
V	volume, m
$\dot{V}$	volumetric flow rate, m^3/s
W	humidity ratio, kg/kg
ρ	density, kg/m^3

EXERCISES

1. Continue Example 6-3 by determining the barn temperature when outdoor air temperature is - 30 C. If the barn temperature will fall below freezing, calculate the supplemental heat capacity (kW) to be installed to prevent freezing.

2. Repeat the design in Example 6-3, assuming the design criterion for relative humidity inside the barn is 80%.

3. Use VENTGRPH to obtain data for the minimum ventilation rate for the barn in Example 6-3 as a function of design indoor relative humidity. Explore the relative humidity range from 60% to 100%.

4. Determine the minimum required ventilation rate for a poultry laying house with 30,000 white leghorn laying hens averaging 1.8 kg when ΣUA for the building is 650 W/K, the perimeter is 150 m, and the perimeter heat loss factor is 1.2 W/m K. Assume the maximum indoor relative humidity is 80% and carbon dioxide cannot go above 4000 ppm. Identify any assumptions you make in your analysis.

5. Determine the maximum ventilation rate needed for the poultry house described in problem 4.

6. Develop a fan staging schedule for the poultry house described in problems 4 and 5, and the minimum and maximum ventilation rates calculated in the problems.

7. Use VENTGRPH to obtain data for the minimum ventilation rate for the barn in Example 6-3 as a function of design indoor air temperature. Explore the air temperature range from 5 to 25 C. Provide a physical explanation for the trend in your results.

8. Use VENTGRPH to design the ventilation system for the barn described in Example 6-1 for a range of elevations from sea level up to 2000 m. Graph the minimum ventilation rate (m³/s and kg/s) as a function of elevation and interpret the trends of the two curves.

REFERENCES

ASHRAE. 1989. Handbook of Fundamentals. American Society of Heating, Refrigerating, and Air Conditioning Engineers, Atlanta, GA.

ASHRAE. 1985. Bin and degree hour weather data for simplified energy calculations (5 computer disks). American Society of Heating, Refrigerating, and Air Conditioning Engineers, Atlanta, GA.

ASHRAE. 1986. A bibliography of available computer programs in the area of heating, ventilating, air conditioning and refrigeration. American Society of Heating, Refrigerating, and Air Conditioning Engineers, Atlanta, GA.

Ecodyne Cooling Products. 1980. Weather data handbook for HVAC and cooling equipment design. McGraw-Hill Book Co., NY (out of print).

Midwest Plan Service. 1978. Professional Design Supplement, MWPS-17. Midwest Plan Service, Iowa State University, Ames, IA.

Midwest Plan Service. 1983. Structures and Environment Handbook, MWPS- 1. Midwest Plan Service, Iowa State University, Ames, IA.

Randall, J.M. 1979. Efficient ventilation with step control of fans and automatic vents. Farm Building Progress, July. pp.1-5.

U.S. Government Printing Office. 1978 . Engineering Weather Data: Air Force, Army and Navy Manual 88-29. USAF ETAC, Scott Air Force Base, IL. Available through the U.S. Government Printing Office, Washington, DC.

CHAPTER 7
THERMAL INSULATION AND MOISTURE BARRIERS

7-1. Introduction

Insulation is used in agricultural buildings for more than energy conservation. In animal housing, energy conservation may be an insignificant consideration when animal sensible heat production is sufficient.

In Chapter 6, it was demonstrated how insulation in a dairy barn influences the calculated minimum ventilation rate for the barn. Insulation also affects the need for supplemental heat, the thermal radiation environment provided to animals in a barn, and potential moisture condensation problems on the surfaces of and within walls, as examples. These effects will be examined in turn.

7-2. Optimum Insulation Thickness

Insulation often is used for energy conservation. When insulation thickness is small, installed cost is low but the cost to replace lost energy is high. When insulation is thick, installed cost is high but the cost of lost energy is low. The cost of installation is not a linear function of insulation thickness, neither is the cost of lost energy. Thus, an opportunity exists to determine an optimum amount of insulation, a thickness where the ownership cost is minimized.

The installed cost of insulation is normally spread over the expected life of the investment. The annualized cost must be adjusted for the time value of money, tax implications, and possible annual maintenance costs. Those are factors which are the subject of an engineering economics course of study, not this text.

When insulation is installed, the incremental cost of each additional unit of thickness includes both material and labor. Neither of these factors is a linear function of thickness. The relationship between thickness and cost cannot be

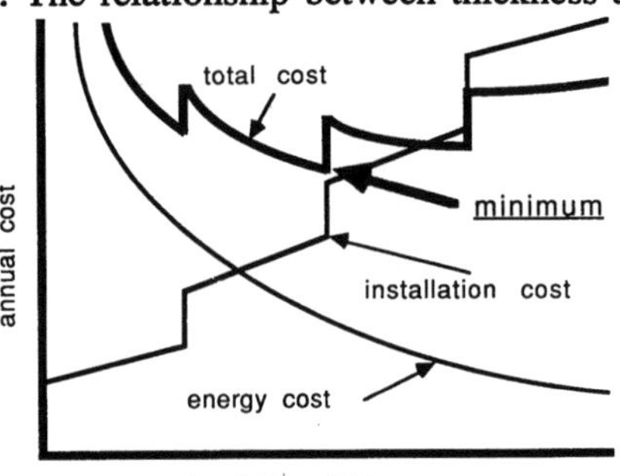

Figure 7-1. Example of factors affecting the economically optimum thickness of insulation used for energy conservation.

generalized. It will depend on construction details necessary to support and then install the additional thickness. For example, when the thickness of a mineral wool blanket increases so the next size of framing lumber is needed, there is in effect a step change of installed cost. A typical relationship between thickness and cost for mineral wool insulation is shown in Figure 7-1.

The cost of lost energy must be determined based on the expected cost of energy over the life of the investment. It is not easy to foresee the future, but a designer can analyze several possibilities to determine how sensitive the design will be to future changes, and then make the best judgment decision. The energy balances described in previous chapters can be used to estimate annual heating energy costs.

One method to estimate annual heating costs relies on bin weather data, such as in Appendix 6-2. This method is best suited for computerized implementation; it is tedious when attempted by hand methods. An energy balance can first be used to determine the outdoor temperature at which heating first will be needed. This balance point will depend on the other heat sources in the building: lights, people, animals, appliances, etc. Then the number of hours in each "bin" of temperature can be used to estimate the total heat needs,

$$q_{heat} = \Sigma UA(t_{in} - t_{out})(\text{hours in the temperature range}), \qquad (7\text{-}1)$$

summed over each bin temperature range. This method is appropriate for many agricultural buildings.

The second method relies on "heating degree day" weather data. Such data generally apply only to small heated buildings such as family homes where there is little heat generated within the heated space compared to the total heat needs, and indoor air temperature is maintained at 24 C. Experience has shown solar and internal heat gains will offset heat loss for such buildings when the average outdoor air temperature is 18.3 C or greater.

(Note: The 18.3 C value arose from a 65 F base used when the concept originally was developed. When house temperatures were held at 75 F, internal heat gains permitted a 10 degree rise above outdoor air temperature.)

Heat loss for a day was found to be proportional to the difference between the average daily outdoor air temperature and 18.3 C. When the average outdoor air temperature was 8.3 C, twice as much heating energy was needed as when the average was 13.3 C. When the average was above 18.3 C, no heat was required. This empirical observation led to the concept of a heating degree day. For example, if the daily high temperature was 10 C, and the daily low was - 4 C, the average was 3 C, and that day contributed (18.3 - 3.0 = 15.3) heating degree days (15.3 degrees difference for one day) .

Weather data for most regions of the world (where heating is required) have been accumulated, and the sum of heating degree days for each day of the year obtained, based on average weather. The United States Weather Bureau, National Climatic Center, has published long-term average value heating degree data to the 18.3 C base, and other base temperatures. The data are based on weather models which use historical records of monthly mean temperatures and their standard deviations. In practice, local heating fuel suppliers and energy companies are sources of heating degree day data for their local region. Sample heating degree day data to base 18.3 are in Appendix 7-1.

To use heating degree day data, the heat loss factor (HLF) for the building being designed must be determined. This is the $\Sigma UA + FP$ value, plus a factor which includes infiltration/ventilation, calculated as $mc_p\Delta t$. From these two factors, the heating need for the period for which the heating degree days (HDD) apply (the heating season, a month, etc.) can be determined from

$$q_{heating} = 24(HLF)\ (HDD)\ (K) \tag{7-2}$$

where $q_{heating}$ is in watt-hours. The empirical factor K applies to single family homes and ranges between approximately 0.8 and 0.6, depending on the number of heating degree days in the heating season. The factor agrees with electric utility company experience, and if used should be obtained from the local utility company for the region under consideration. This factor is an indication that 18.3 may not be an appropriate temperature base for heating degree day compilations. Differing solar climates lead to differing appropriate bases. Also, the base of 18.3 C was chosen approximately 50 years ago when insulation standards for houses were very different from those of today.

Once the heat requirement for a year is known, its dollar value may be calculated if the efficiency of the heating system (η) and the heating value of the fuel (HVF) to be used and its unit cost are known.

$$q_{value} = \frac{(q_{heating})\ (\text{unit cost})}{(\eta)\ (HVF)} \tag{7-3}$$

Fuel heating values and representative heating system efficiencies are:

Fuel	Typical Heating Value [a]	Representative Heating System Efficiencies
electric resistance	3.6 kJ/kWh	1.0
electric heat pump	3.6 kJ/kWh	2.0 - 2.75
oil, #2 heating	38,000 kJ/L	0.65 - 0.75
natural gas	35,000 kJ/m^3	0.70 - 0.85
LP gas	26,000 kJ/L	0.70 - 0.85
coal	28E6 kJ/t	0.60 - 0.70
wood (best stoves)	22E6 kJ/t	0.50 - 0.60
wood (fireplaces)	22E6 kJ/t	0 or negative

[a] only example values; more exact data should be obtained from fuel suppliers.

More complete methods to calculate heating cost are described, for example, in the *ASHRAE Handbook of Fundamentals*. Also, efficiency is more precisely described in several ways: seasonal heating performance factors as found in government regulations, annual fuel use efficiency from the Gas Appliance Manufacturers Association, and seasonal energy efficiency ratio from the Air Conditioning and Refrigeration Institute.

Example 7-1

Problem: Estimate the yearly cost to heat a single family home in a region characterized by 4000 heating degree days (base 18.3 C). The home's heat loss factor (HLF) is 650 W/K, the correction factor K is estimated to be 0.7, and #2 heating oil will be used with an estimated heating system efficiency of 0.75. The oil costs $0.25/L.

Solution: The yearly heat requirement will be calculated,

$$q_{heating} = 24(HLF)(HDD)(K)$$
$$= 24\,(650)\,(4000)\,(0.7) = 4.\,4E7\text{ W-hr}$$
$$= 1.6E8\text{ kJ}.$$

This much heating energy has an estimated value of

$$q_{value} = (1.6E8\text{ kJ})(\$0.25/L)/(0.75)(38{,}000\text{ kJ/L})$$
$$= \$1400/\text{year}.$$

Example 7-2

Problem: A design engineer faces the problem of determining the optimum thickness of foam board insulation to install on the inside surface of the foundation wall of a greenhouse. The wall is concrete block with a basic R-value of 0.35 m^2 K/W (the R-value with no insulation on the wall, but including the surface thermal resistances). The greenhouse is to be located near Lansing, Michigan.

Natural gas is the heating fuel and is expected to cost $0.20/$m^3$ (today's dollar, with fuel inflation equal to the general inflation rate) during the life of the insulation. The annualized cost of installing the insulation has been estimated to be

Thickness	Annual Cost, $/$m^2$	R-Value, m^2K/W
0mm	0	0
12.7	1.75	0.55
25.4	2.50	1.10
50.8	4.50	2.20
76.2	9.50	3.31
101.6	11.50	4.41

The greenhouse will be heated to a constant air temperature of 21.1 C. (Typical practice is to have day and night temperature settings which differ, but for simplicity in this example assume constant indoor air temperature.)

Solution: The optimum insulation thickness for a unit area of the foundation wall will be the same as for the entire foundation wall; for simplicity only 1 m^2 of wall area will be considered.

The R-value of 1 m^2 of foundation wall is 0.35 m^2 K/W, plus the R-value of the installed insulation. The UA value for 1 m^2 is the inverse of the unit area R-value. Total heat loss through the unit area is the sum of UAΔt, summed over the bins of weather data in Appendix 6-2 for Lansing, Michigan.

The bin for data between 15.6 C and 21.1 C will be examined first. There are expected to be 1466 hr when outdoor air temperature is within this range. The average temperature within the range will be assumed to be the average of the end points, or 18.35 C. This assumption is necessary because there is no knowledge of how air temperature actually varies within the range, but the range is narrow enough and the temperature function smooth enough that linearity is a reasonable assumption.

When outdoor air temperature is within the 15.6 C to 21.1 C range, the yearly heat loss per unit area will be

$$q_{15.6-21.1} = UA(21.1 \text{ C} - 18.35 \text{ C})(1466 \text{ hr})$$

with heat expressed in watt-hours. To calculate the heat loss for a year, this computation must be repeated for each bin of weather data lower than the 15.6 C to 21.1 C bin. A means to simplify this calculation is to determine the weighted average of outdoor air temperature when it is below 21.1 C.

$$t_{ave, t < 21.1 \text{ C}} = \frac{1466 (18.35) + 1277(12.8) + ... + 176(-15) + 40(-20.5) + 1(26.1)}{1466 + 1277 + 1153 + 1561 + 1139 + 504 + 176 + 40 + 1}$$
$$= 5.66 \text{ C},$$

for a total of 7317 hr/yr (the sum of the denominator).

The heating value of natural gas is expressed in kJ, watt-hours must be converted to kJ by the factor of (3600 s/hr)(0.001 kJ/J).

$$q_{heating} = UA(21.1 \text{ C} - 5.66 \text{ C}) (7317 \text{ hr}) (3600 \text{ s/hr}) (0.001 \text{ kJ/J})$$

For example, when the wall is uninsulated, R = 0.35 m^{2}K/W; UA = 2.86 W/K, for a unit area. For this situation,

$$q_{heating} = 1.162\text{E6 kJ/m}^2.$$

The cost of heat must also be determined. Equation 7-3 can be used if the heating system efficiency is known. An assumption of 80% efficiency (typical for natural gas) will be made. The heating value of natural gas is 35,000 kJ/m^3 and the unit fuel cost is \$0.20/m^3. When the wall is uninsulated,

$$q = \frac{(1.162E6 \text{ kJ/m}^2)(\$0.20/\text{m}^3)}{(0.80)(35,000 \text{ kJ/m}^3)}$$

$$= \$8.30/\text{m}^2$$

The need for heat has been calculated; the solar contribution must now be introduced. A portion of the required heat will be supplied by the sun during both cloudy and sunny days. Solar data could be obtained and the solar contribution estimated. However, for this example a simple assumption will suffice. Experience has shown, approximately a quarter of a year's need for heat in a greenhouse will be met by the sun in climates which are cold and rather cloudy. This value will be assumed for this example and heating needs reduced by 25%.

Heating costs (corrected for solar gain) for insulation thicknesses listed in the statement of this example are

Insulation Thickness	Wall R-Value	Yearly Heat Loss	Value of Yearly Heat Loss
0 mm	0.35 m^{2}K/W	0.872E6 kJ	\$4.67
12.7	0.90	0.339E6	1.82
25.4	1.45	0.210E6	1.13
50.8	2.55	0.119E6	0.65
76.2	3.66	0.083E6	0.44
101.6	4.76	0.064E6	0.35

The yearly cost of installation and heating can now be determined:

Insulation Thickness	Installation Cost, \$/m^2	Heating Cost, \$/m^2	Total Cost, \$/m^2
0 mm	0	4.67	4.67
12.7	1.75	1.82	3.57
25.4	2.50	1.13	3.63
50.8	4.50	0.65	5.15
76.2	9.50	0.44	9.94
101.6	11.50	0.35	11.85

The optimum thickness can be determined directly from the data; a graph is not needed in this instance because foam boards are available only in fixed thicknesses, the thickness does not vary continuously. The minimum yearly cost is at a foam board thickness of 12.7 mm. Very thick insulation (<u>for this example</u>) is seen not to be economical. It would be a useful exercise to redo this problem to show that, without the benefit of solar gain, the optimal insulation thickness is 25.4 mm.

Greenhouses usually are operated with night temperatures which differ from day temperatures. The analysis procedure used in this problem may still be used in the more complex situation because weather bin data are available for day and night. For example, AFM 88-29, a reference in Chapter 6, divides bin data into three time periods, midnight to 8 A.M., 8 A.M. to 4 P.M., and 4 P.M. to midnight. The time period 8 A.M. to 4 P.M. corresponds approximately to the period when day temperatures are held in greenhouses. When such data are available, day heating and night heating needs can be determined separately, then summed, and the total day heating cost calculated.

Insulating the foundation wall of a greenhouse will have a slight effect on cooling costs also, but since greenhouses are not cooled by refrigeration the cost difference should be slight. This will not be true of, for example, an air-conditioned building where cooling cost savings attributable to insulation may be larger than heating cost savings in warm climates.

7-3. Condensation on Wall Surfaces

When the interior of a building is warmer than outdoor air, the outside surfaces of exterior walls will be warmer than the outdoor air and inside surfaces will be colder than indoor air. If the inside surfaces are colder than the dew point of the indoor air, condensation forms. If the condensation remains for a long time, structural damage may occur and wet conditions will persist which are conducive to the growth of disease organisms.

Sufficient insulation within walls prevents surface condensation. Building design can begin by determining the minimum R-value of the walls and ceiling which will prevent surface condensation. More insulation may later be determined as necessary, based on other criteria, but preventing surface condensation should always be a design consideration.

It may not be practical to prevent surface condensation completely. During extreme outdoor cold, some condensation may be permitted, with the understanding that it will evaporate when the extreme cold passes. Weather design data, or bin data, may be used to determine practical outdoor air temperatures for condensation control design. Doors and windows (especially) will likely never be sufficiently insulated to prevent condensation on their inner surfaces when the interior of the building has a high relative humidity.

It is common in animal barns for windows to become heavily covered with frost during prolonged cold spells. In greenhouses, condensation on glazing during cold weather is a continual factor. Glass houses are designed with channels formed in the framing to carry condensation to drainage. Double polyethylene greenhouses are prone to a significant amount of dripping from the underside of the polyethylene near the peak which can be a factor in disease outbreaks.

The design aerial conditions within a building determine the dew point temperature of the indoor air. Adequate insulation to prevent condensation can be calculated using the series thermal circuit equation for heat conduction as used, for example, in Example 3-16.

The series thermal circuit for a planar section expressed in terms of surface temperature, t_s, follows. The indoor and outdoor air temperatures are t_i and t_o, respectively.

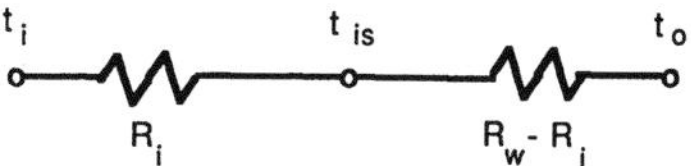

Thermal resistance between indoor air and the wall is R_i, and the thermal resistance of the entire wall (including the inside surface) is R_w. The required thermal resistance of the wall can be calculated in terms of the dew point temperature, t_{dp} of the indoor air, using the requirement that $t_s \geq t_{dp}$,

$$R_w \geq R_i \left(\frac{t_i - t_o}{t_i - t_{dp}}\right) \tag{7-4}$$

Example 7-3

<u>Problem:</u> Determine minimum wall and ceiling thermal resistances to prevent wall and ceiling surface condensation in a poultry house if indoor design conditions are 22 C and 75% relative humidity; the building is to be located near Fort Wayne, Indiana.

<u>Solution:</u> First, the outdoor design air temperature must be decided. In Appendix 6-1, the 99% and 97.5% winter design temperatures are - 20 and - 17 C, respectively. Air temperature is below the 99% and 97.5% air temperature for 22 and 54 hr in a year, respectively. The decision of which to choose as a design condition is not fixed by tradition or typical practice. For this example, the 99% value will be assumed to reduce the number of hours of potential condensation by 32. The 99% and 97.5% outdoor temperatures are sufficiently close that little effect will be seen in the calculated R-values and longer periods of condensation will be avoided if the 99% value is used.

Fort Wayne's elevation is approximately 240 m above sea level, but dew point temperature is independent of atmospheric pressure. Program PLUS, for example, can be used to determine dew point temperature at standard atmospheric pressure. The dew point at 22 C dry bulb temperature and 75% relative humidity is 17.4 C.

The thermal resistance at the inside surfaces of the walls is estimated to be 0.12 m^2K/W (see Appendix 3-5) for common construction materials. At the underside of the ceiling the thermal resistance will be 0.11 m^2K/W for heat flow upward. Thus,

$$R_{wall} \geq (0.12 \text{ m}^2\text{K/W})(\frac{22\text{ C} - (-20\text{ C})}{22\text{ C} - 17.4\text{ C}})$$

$$\geq 1.1 \text{ m}^2\text{K/W, and}$$

$$R_{ceiling} \geq (0.11 \text{ m}^2\text{K/W})(\frac{22\text{ C} - (-20\text{ C})}{22\text{ C} - 17.4\text{ C}})$$

$$\geq 1.0 \text{ m}^2\text{K/W,}$$

assuming the attic will be well ventilated with air temperature approximately equal to that of the outdoor air.

It should be noted that these R-values are modest but higher than provided by an uninsulated wall.

7-4. Condensation Within Walls

Thermal energy migrates through walls due to temperature differences. Moisture migrates due to water vapor partial pressure differences. When it is cold outdoors, vapor partial pressure inside a heated building is significantly higher than outdoors. Water as a consequence will diffuse slowly through the wall. Conversely, water will diffuse into a refrigerated or air-conditioned building during warm, humid weather, which is a problem especially in sub-tropical and tropical climates.

Thermal insulation affects temperatures within a wall. More insulation causes the outer regions of a wall to be colder than would be the case without insulation. Water condenses wherever the partial pressure of water vapor is greater than the saturation partial pressure at the temperature of the region in question.

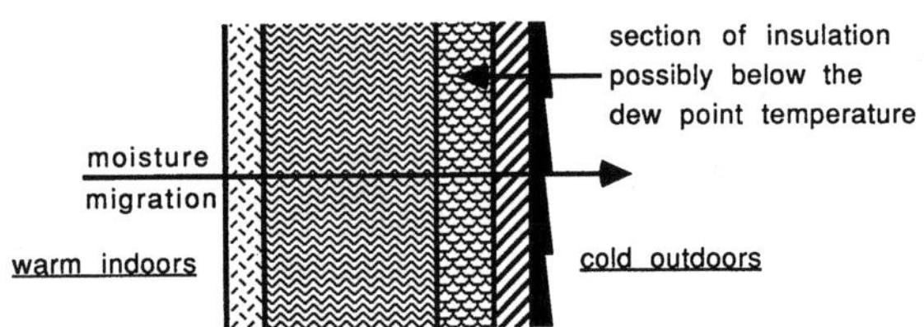

Saturation partial pressure is low at low temperatures. The effect of insulation can be to make condensation more likely near the core of a wall and means must be designed into the wall to prevent such hidden condensation. The means generally involve adding a vapor retarding barrier on the warm and humid side of the wall (where water vapor partial pressure is highest). The barrier retards diffusion of moisture into the wall, thus, moisture which diffuses through the barrier has time to diffuse through the wall and out before enough water vapor accumulates to raise the vapor partial pressure to the condensation point.

7-4.1. Water Vapor Diffusion. The diffusion process is governed by Fick's law,

$$w = -\mu(dp/dx), \qquad (7\text{-}5)$$

where w is the amount of water transmitted per unit area and unit time, μ is the permeability of the material to moisture diffusion (ng water/s-m-Pa) and dp/dx is the water vapor partial pressure gradient (Pa/m). Under conditions of steady-state Equation 7-5 can be solved,

$$w = -\mu(p_1 - p_2)/L, \qquad (7\text{-}6)$$

where L is the path length of diffusion and p is partial pressure.

Permeability, μ, is an intensive property; the combination μ/L is an extensive property termed "permeance". (Permeance can be interpreted as a unit area conductance of water vapor.) Units of permeance are ng/s-m^2-Pa, often shortened to "perms". Resistance to water vapor flow, R_{H_2O} is the inverse of permeance (or L/μ) with units of TPa-m^2-s/kg, frequently designated as "reps", where TPa is terapascals (10^{12} pascals).

Note: The terms "reps" and "perms" are used in both the SI and IP systems of units and have the same physical meaning but very different units. That is, a material with a rating of x reps in the IP system will have a different rep rating in the SI system. Care must therefore be used when manufacturer's data is obtained to be sure the correct system of units is used. The conversion multiplication factor from perm(IP) to perm(SI) is 57.4. That is, 10 perm(IP) = 574 perm(SI).

Note the analogy of this approach to steady-state heat transfer by conduction. Vapor pressure difference is analogous to temperature difference, permeance is analogous to thermal conductance and permeability is analogous to thermal conductivity. The concepts of simple parallel and series resistance circuits apply to water vapor diffusion as they did to heat diffusion.

Appendix 7-2 contains permeance and resistance data for common building materials. More complete data can be found in the *ASHRAE Handbook of Fundamentals*. However, data are more limited than for thermal conductance. Moisture permeability is very sensitive to material density, fiber orientation, moisture content (especially when the material is near saturation), and many other factors. Design specifications should always be based on data from manufacturers which have been obtained in controlled tests. Data in Appendix 7-2 are presented as examples, not as standards for design.

Some data in Appendix 7-2 is based on dry-cup and some on wet-cup test procedures. Because permeance of a material is sensitive to its moisture content, a designer must be careful to use the appropriate value.

Dry-cup data normally is obtained by placing a sample of the material over a container partially filled with a dessicant and measuring how rapidly water diffuses through the material and into the dessicant from air at 50% relative humidity. Wet-cup data are obtained with water in the container instead of a dessicant and measuring how rapidly water diffuses through the material from the container into room air at 50% relative humidity.

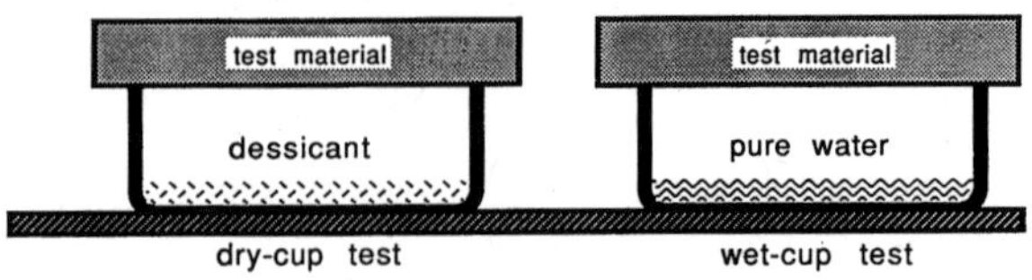

It would be most appropriate to use wet-cup data for applications in moist or humid conditions. Dry-cup data are most appropriate for dry conditions and relative humidities less than 50%, but those are conditions of lesser interest in building design for moisture control.

Example 7-4

Problem: A wood-framed wall is constructed using 15.9 mm thick Douglas fir plywood as outside siding, 12.7 mm thick sheathing quality structural insulating board, 88.9 mm mineral wool backed with asphalt impregnated paper, and 15.9 mm thick gypsum wallboard as inside sheathing. The inside surface of the wall is painted using a latex prime-sealer and one coat of a semi-gloss vinyl acrylic, the outside surface has accumulated three coats of white lead and oil paint.

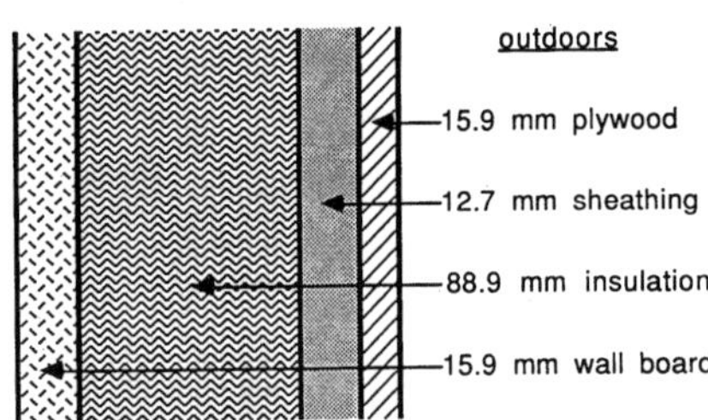

Indoor conditions are expected to be 20 C and 40% relative humidity and design conditions for weather are - 20 C and 60% relative humidity. The building is located at an altitude of approximately 500 m.

Estimate the flux of water vapor diffusing through the wall under these conditions.

Solution: This is a series diffusion circuit. The total resistance to diffusion is the sum of individual resistances, and flux can be calculated using Equation 7-6 where $R_{H2O} = L/\mu$. Because indoor humidity is low, dry-cup data will be used for the paper backing of the insulation, all other materials have only single permeance and resistance values, not wet- and dry-cup data.

When thicknesses of materials differ from values in Appendix 7-2, diffusion resistance values will be scaled linearly. Diffusion resistance values are:

Exterior paint: An average permeance will be assumed, $(17 + 57) / 2 = 37$ perms, thus, the resistance is $1/37 = 0.0270$ reps.

Siding: $(15.9 \text{ mm} / 6.4 \text{ mm})(0.025 \text{ reps}) = 0.062$ reps.

Sheathing: Having no other information, the average permeance will be used, $(2360 + 5900) / 2 = 4130$ perms. The resistance will be $1 / 4130 = 0.00024$ reps.

Insulation: $(88.9 \text{ mm} / 1000 \text{ mm})(0.0041 \text{ reps}) = 0.00036$ reps.

Paper backing: 0.043 reps (the dry-cup value was used because of the design conditions indoors of 40%, and insulation is normally installed with the paper backing nearer the interior side of the wall).

Inside sheathing: $(15.9 \text{ mm} / 9.5 \text{ mm})(0.00035 \text{ reps}) = 0.00059$ reps.

Interior paint: $0.0028 \text{ reps} + 0.0026 \text{ reps} = 0.0054$ reps.

No significant moisture diffusion resistance is assigned to the wall surfaces (as thermal diffusion resistances are). The sum of moisture diffusion resistances is

$$R_{H_2O} = 0.0270 + 0.062 + 0.00024 + 0.00036 + 0.043 + 0.00059 + 0.0054$$
$$= 0.13859, \text{ or approximately } 0.14 \text{ reps.}$$

Program PLUS can be used to estimate partial pressures of water vapor in the air inside and outside the building. At 20 C and 40% relative humidity, the water vapor partial pressure is 0.95 kPa. At 20 C and 60% relative humidity, the partial pressure is 0.06 kPa. The moisture flux is, thus,

$$w = (p_i - p_o) / R_{H_2O}$$
$$= (0.95\text{E} - 9 \text{ TPa} - 0.06\text{E} - 9 \text{ TPa}) / 0.14 \text{ reps}$$
$$= 6.4\text{E} - 9 \text{ kg/s-m}^2$$

7-4.2. Condensation Rates. Unless a perfect vapor diffusion barrier is present in a wall (such as a lining of sheet metal) moisture will always diffuse through the wall when a vapor pressure difference exists. Whether there will be condensation depends on the actual vapor pressure and saturation vapor pressure at every point in the wall. Whether the actual vapor pressure will be greater than the saturation vapor pressure depends on the thermal and moisture diffusion characteristics of the wall materials. A method to predict the possibility of condensation is to graph the actual vapor pressure and saturation vapor pressure values through a cross-section of the wall.

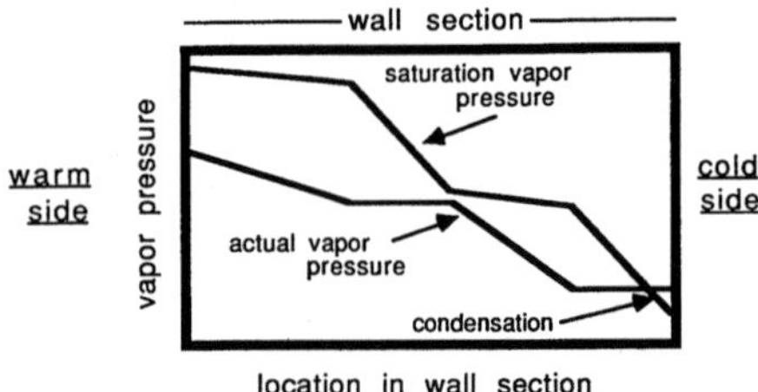

The actual vapor pressure can be determined at the interfaces of adjoining wall layers by applying the electrical resistance circuit analogy to the vapor diffusion circuit. For example, if one starts at the inside surface of a wall, where the vapor pressure of water is p_i, the actual vapor pressure at x inside the wall, p_x depends on the total vapor diffusion resistance of the wall, $R_{H_2O\ total}$; the vapor diffusion resistance between the inside wall surface and x, $R_{H_2O,x}$; and the vapor pressure outside the wall, p_o, as follows:

$$p_x = p_i + (R_{H_2O,\ x} / R_{H_2O,\ total})(p_o - p_i). \qquad (7\text{-}7)$$

The saturation vapor pressure is a function of temperature and the temperature at x is

$$t_x = t_i + (R_{thermal,\ x} / R_{thermal,\ total})(t_o - t_i). \qquad (7\text{-}8)$$

The saturation vapor pressure at t_x can be found using Equation 2-1 in Chapter 2, program PLUS, or psychrometric tables.

The usual approach to determining whether condensation is a possibility is to calculate the actual and saturation vapor pressures at each interface between wall materials and check whether the actual vapor pressure exceeds the saturation vapor pressure at any of the interfaces. Variation of vapor pressure from one side of a wall layer to the other is assumed to be linear.

Example 7-5

Problem: Determine whether condensation can be anticipated within the wall described in Example 7-4, and if so, where.

Solution: The cross-section of the wall is sketched below and interfaces between adjacent layers are numbered.

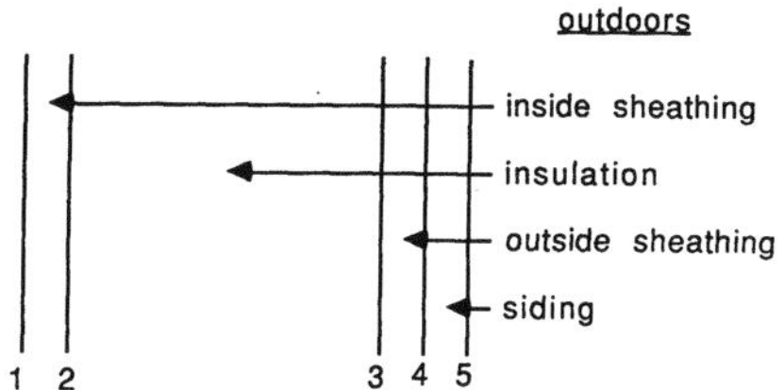

Thermal and moisture diffusion resistances in the thermal and vapor diffusion circuits are:

Item	Thermal Resistance, $m^2 K/W$	Moisture Resistance, reps(SI)
inside surface	0.12	0
interior paint	0	0.0054
gypsum board	0.099	0.00059
backing paper	0	0.043
insulation	1.94	0.00036
sheathing	0.23	0.00024
siding	0.14	0.062
exterior paint	0	0.0270
outside surface	0.03	0
total:	2.559	0.13859

In this example the effect of framing on the wall's insulation value is not considered. Framing will have little influence on moisture migration through the section of wall which is insulated.

Equations 7-7 and 7-8 can be used to determine actual vapor pressures and temperatures existing at interfaces 1 to 5. Program PLUS can be used to determine saturation vapor pressures for temperatures at sections 1 to 5, for an atmospheric pressure of 95.461 kPa. Section 1(a) is on the wall surface, between the indoor air and the paint, and Section 1(b) is between the paint and the gypsum board. Section 2(a) is between the gypsum board and the paper backing on the insulation, and Section 2(b) is between the paper and the insulation. Section 5(a) is underneath the paint on the outside surface of the wall, and Section 5(b) is between the exterior paint and the outdoor air. There is a moisture diffusion resistance between Sections 5(a) and (b), but no significant thermal resistance.

Note the assumptions inherent in the calculations. Paint and paper are assumed to have negligible thermal resistance. Air boundary layers at the wall's surfaces are assumed to have negligible vapor transfer resistance. And, of course, a steady-state is assumed.

The actual temperature at each section was obtained using Equation 7-8 and the actual vapor pressure using Equation 7-7. The saturation vapor pressure was found from program PLUS, using the actual temperatures and 100% relative

Section	$R_{H_2O,x}$	$R_{Thermal, x}$	Temp.	Actual Vapor Pressure	Saturation Vapor Pressure
indoor air	0	0	20.0 C	0.950 kPa	2.37 kPa
1(a)	0	0.12	18.1	0.950	2.10
1(b)	0.0054	0.12	18.1	0.915	2.10
2(a)	0.00599	0.219	16.6	0.912	1.91
2(b)	0.04899	0.219	16.6	0.635	1.91
3	0.04935	2.159	-13.7	0.633	0.19
4	0.04959	2.389	-17.3	0.632	0.13
5(a)	0.11159	2.529	-19.5	0.233	0.11
5(b)	0.13859	2.529	-19.5	0.060	0.11
outdoor air	0.13859	2.559	-20.0	0.060	0.10

humidity (combination c in the program), although the relative humidity was irrelevant because only the saturation vapor pressure was desired.

At section 3 (between the insulation and outside sheathing) a problem arises. The saturation vapor pressure is less than the vapor pressure which would arise from flow continuity (the actual vapor pressure). This cannot be, thus condensation will begin in the colder regions of the insulation. A common practice is to graph the actual and saturation vapor pressure data to visualize where condensation may arise; the data from this example is in Figure 7-2. The lines for actual and saturation vapor pressure cross within the wall section filled with insulation. Wherever the actual vapor pressure is above the saturation vapor pressure, condensation will occur.

Once condensation begins, the data which have been calculated no longer apply. At the onset of condensation the actual vapor pressure at section 3 equals the saturation vapor pressure, which changes actual vapor pressure values elsewhere in the wall section exterior to the condensation. After condensation has begun, water in the insulation destroys the insulation value of the mineral wool, thus temperature values change, which change the actual vapor pressure

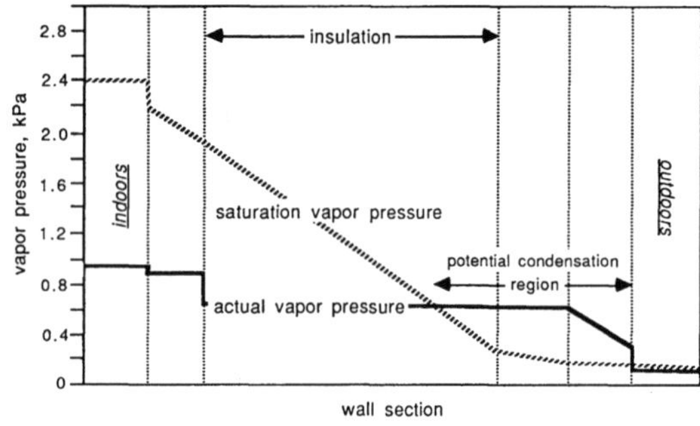

Figure 7-2. Actual vapor pressure and saturation vapor pressure profiles of the wall in Example 7-4.

and saturation vapor pressure values further. The problem then becomes complex and is not readily analyzed. However, it need not be analyzed, for the question is whether there will be condensation, and if so, how can it be prevented. The concern is not what happens after the wall becomes wet.

Before completing this section, it should be noted that after an initial analysis as in Example 7-4, it may be desirable to re-examine the rep values used in the calculations. For example, dry-cup data was used in estimating the rep value of the paper backing the insulation. The data show (for sections 2(a) and 2(b)) the saturation vapor pressure on the two sides of the paper is the same, 1.91 kPa, because the temperature is unchanged. The actual vapor pressure on the 2(a) side is 0.912 kPa, and on the 2(b) side is 0.635 kPa. The relative humidities on the 2(a) and 2(b) sides of the paper are, thus, 48% and 33%. These are both below 50%, the initial assumption was adequate. If the actual humidity in the vicinity of the paper was found to be above 50%, wet-cup data should be considered. This comparison is appropriate for all parts of the wall and highlights the importance of finding complete manufacturer's data (both wet-cup and dry-cup) for all materials used in a wall where condensation may be a problem.

7-4.3. Vapor Retarding Barriers. Vapor retarding barriers are used in buildings to slow the diffusion of water vapor toward cold regions within the building shell where condensation may occur. If diffusion can be slowed sufficiently, the little which diffuses into a wall, for example, will diffuse out before enough vapor accumulates to permit condensation. To be effective, the barrier must have a significant vapor diffusion resistance compared to the resistance of the rest of the wall. The barrier must be placed within the warm section of the wall, for diffusion is generally from warmer to colder sections.

Vapor retarding barriers are usually film materials, most often polyethylene. Mineral fiber insulation is available with an asphalt–impregnated paper as a backing, which provides some resistance to vapor diffusion. The insulation may have a thin aluminum foil applied to the paper backing for more resistance. However, these barriers are typically inadequate because of physical damage to the barriers during installation (tears, holes, etc.) and because of the small gaps at the sides of the insulation at the wall studs. A carefully applied film material such as polyethylene usually provides a more secure barrier.

The required vapor diffusion resistance can be calculated after first compiling data as in Example 7-4. In that example, at section 3 (where insulation and the outside sheathing meet) the saturation vapor pressure was only 0.19 kPa, while the actual vapor pressure arising from flow continuity was 0.635 kPa. The calculation procedure is to determine how much vapor diffusion resistance is needed to cause the actual vapor pressure at 3 to be no higher than 0.19 kPa. Equation 7-7 can be rearranged and solved to determine the required resistance,

$$R_{original, x} + R_{added} = (R_{original, total} + R_{added})(\frac{p_i - p_x}{p_i - p_o}), \qquad (7\text{-}9)$$

$$\text{or } R_{added} = R_{original, total}(\frac{p_i - p_x}{p_x - p_o}) - R_{original, x}(\frac{p_i - p_o}{p_x - p_o}), \qquad (7\text{-}10)$$

where, in this case, p_x is the maximum vapor pressure permitted at the section in question. The vapor diffusion resistance required in the vapor retarding barrier is R_{added}, and $R_{original}$ is the resistance of the wall without the added barrier.

If it is unclear which section will govern condensation, Equation 7-10 can be applied to all sections where condensation is possible and the largest value of R_{added} chosen.

Example 7-6

Problem: Continue with the wall section of Example 7-4 and determine the vapor diffusion resistance (rep value) required for a vapor retarding barrier at 2(a) to prevent condensation within the wall.

Solution: As has been stated, condensation will occur at section 3 unless the actual vapor pressure at 3 can be reduced to 0.19 kPa. Equation 7-10 can be used to calculate the vapor retarding resistance required to accomplish this. The rep value of the original wall is 0.13859, thus,

$$R_{added} = 0.13859 \left(\frac{0.95 - 0.19}{0.19 - 0.06}\right) - 0.04935 \left(\frac{0.95 - 0.06}{0.19 - 0.06}\right)$$
$$= 0.4724 \text{ reps(SI).}$$

The required additional vapor diffusion resistance is 0.4724 rep. This is a large value for a vapor retarding barrier to achieve. Data for films, etc., in Appendix 7-2 show an aluminum foil film of 0.025 mm thickness is required to achieve this resistance.

The factor which has led to the need for such a high resistance in the vapor retarding barrier is the relatively high vapor diffusion resistance located at the outside section of the wall. The plywood siding, with exterior glue, has a resistance of 0.062 reps and the exterior paint (an oil paint) adds another 0.0270 reps. (More than 60% of the wall's vapor diffusion resistance is concentrated at the cold side of the wall.) The temperature at section 3 is low, thus, a very effective vapor retarding barrier is required at 2(a). An alternative for design would be to use a different siding and an exterior paint with less resistance to vapor diffusion (such as a latex paint). With that change, the vapor resistance required at 2(a) should be reduced sufficiently to permit a simple polyethylene film to be effective.

It would be a useful exercise to redesign the wall, choose a new siding and exterior paint, and then design a vapor retarding barrier.

SYMBOLS

A	area, m^2
c_p	specific heat, J/kg K
HDD	heating degree days, see Equation 7-2
HLF	heat loss factor, W/K, see Equation 7-2
HVF	heating value of fuel, see Equation 7-3
K	empirical factor, see Equation 7-2
L	thickness or path length, m
m	mass flow rate, kg/s
p	pressure, Pa
q	heat flow, W
R	unit area thermal resistance, m^2 K/W
R	unit area moisture diffusion resistance, reps
t	temperature, C
U	unit area thermal conductance, W/m^2 K
w	water mass flow rate, ng/m^2 s
x	spacial variable, m
η	efficiency, decimal
μ	permeability, ng/msPa

EXERCISES

1. A poultry house is to be designed and the maximum relative humidity is to be 70%. Develop a graph to show how the required minimum R-value of the walls to prevent condensation on the indoor wall surface varies as a function of the design minimum outdoor air temperature. The outdoor air temperature should be graphed on the x axis, and the range should be from - 40 C to - 10 C.

2. A single family home is to be built in a region characterized by 4800 heating degree days (base 18.3 C). The walls are to be constructed of concrete blocks, with bricks on the outside and insulating foam covered by gypsum wallboard on the inside. Heat will be by #2 fuel oil and a new, high-efficiency furnace will be used having a seasonal efficiency of 90%. Oil prices are expected to be approximately $0.30 per L.

    Assume the R-value of the components of the wall less the foam will be 0.6 m^2 K/W, the foam board has a thermal conductivity of 0.018 W/m K, and its cost will be the same as listed in Example 7-2. Determine the most economical thickness of foam insulation to use.

3. Consider the wall described in Example 7-4. The only change is that the outer surface has been painted with a latex primer-sealer and one coat of a semi-gloss vinyl acrylic. The wall is part of a single family home located in a very hot and humid climate. The indoor air will be conditioned to 24 C and 60% relative humidity, and outdoor design conditions are expected to be 38 C and 70% relative humidity. Graph the temperature profile, the actual vapor pressure profile, and the saturation vapor pressure profile through the wall. Will there be condensation within the wall? If so, determine the rep value of the vapor barrier required to prevent condensation.

REFEkENCES

ACCA. 1986. Manual J, Load calculation for residential winter and summer air conditioning, Seventh edition. Air Conditioning Contractors of America, 1228 17th Street NW, Washington, DC.

ASHRAE. 1976. Energy calculations I - Procedures for determining heating and cooling loads for computerized energy calculations algorithms for building heat transfer subroutines. American Society of Heating, Refrigerating, and Air Conditioning Engineers, Atlanta, GA.

ASHRAE. 1984. Simplified energy analysis using the modified bin method. American Society of Heating, Refrigerating, and Air Conditioning Engineers, Atlanta, GA.

ASHRAE. 1989. Handbook of Fundamentals. American Society of Heating, Refrigerating, and Air Conditioning Engineers, Atlanta, GA.

Degelman, L.A. 1986. A bibliography of available computer programs in the area of heating, ventilating, air conditioning, and refrigeration. American Society of Heating, Refrigerating, and Air Conditioning Engineers, Atlanta, GA.

Midwest Plan Service. 1983. Structures and Environment Handbook. Midwest Plan Service, Iowa State University, Ames, IA.

CHAPTER 8
AIR INLETS AND OUTLETS

8-1. Introduction

Distributing fresh air in an agricultural building presents challenges which differ from those of distributing fresh air in office buildings, for example. Agricultural buildings generally are characterized by being one large airspace rather than many small ones. This difference leads to less complicated air distribution systems (little need for air ducts) but at the same time makes the effects of inlets more important because fresh air must often be distributed to regions of the airspace far removed from the inlet. Thermal stratification is frequently greater in agricultural buildings, a problem which works against good mixing of fresh air with air already in the space.

Agricultural building mechanical ventilation can be accomplished by negative pressure, positive pressure, or neutral pressure methods. The three differ by the air pressure which results within the ventilated space, relative to atmospheric pressure outdoors. In negative pressure (exhaust) ventilation systems, fans remove air from the ventilated space, which creates a slight vacuum within the space and the partial vacuum functions to draw fresh air into the space through inlets. Fans act as outlets. The opposite is true of positive pressure ventilation systems. Fans push fresh air into the ventilated space, and air already within the space is forced out through outlets by the slight overpressure created by the fans. In neutral pressure systems, fans are at both inlets and outlets and are balanced in their characteristics so air pressure within the ventilated space is approximately that of the outdoors air.

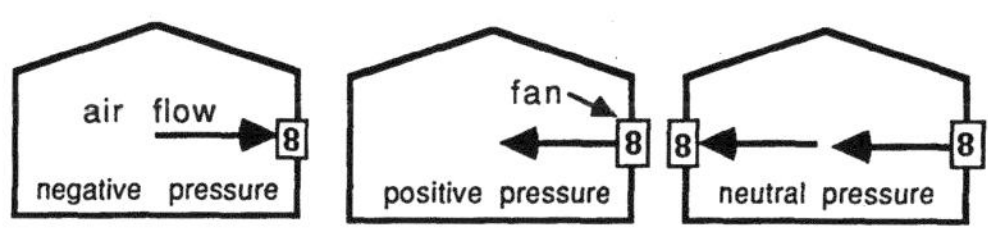

An important underlying principle in ventilation is: <u>a steady-state condition exists, essentially all the time.</u> The mass rate of airflow into a ventilated space, plus any mass added or removed while air is within the space, is balanced by the mass flow rate out of the space. The air pressure which must exist within the ventilated space is that which balances the airflow. As fan or inlet characteristics change, air pressure within the space changes to maintain steady-state. This principle is emphasized here and will be important for this and future chapters.

Overall, negative pressure ventilation systems are preferred for most applications in barns and greenhouses in temperate climates. Although exhaust ventilation systems lead to infiltration through inadvertent openings (cracks in walls, open doors, etc.) they have the advantage of permitting inexpensive air inlets to be easily placed around the building. The most common inlet system today uses a continuous slot around the perimeter of the building, with baffles to direct fresh air either along the ceiling or down the walls. Fans are usually grouped in banks along one wall, the leeward wall referenced to prevailing winter winds.

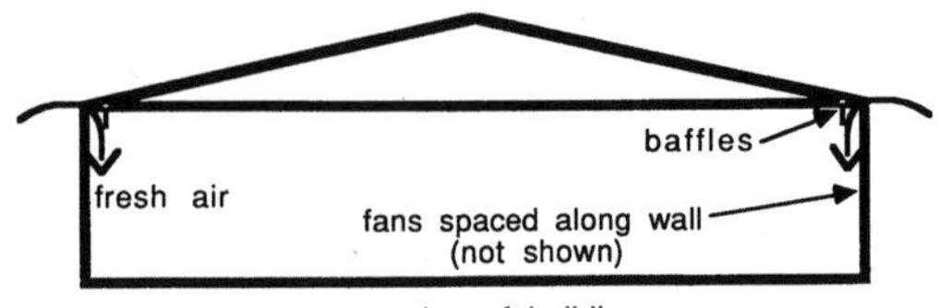

Positive pressure ventilation methods require extensive air distribution methods, sometimes involving a network of air ducts and small, local outlets. Such distribution is necessary to prevent localized chilling near a few large inlets during cold weather. The duct system can be an advantage when carefully controlled air distribution is necessary, such as in horticultural storage rooms, animal laboratories, or a complex of growth rooms for a large greenhouse operation. Positive pressure methods have the additional advantage of preventing unwanted infiltration of outside air which can be a serious problem during times of little ventilation, such as during very cold weather. However, overall the disadvantages of positive pressure systems are generally felt to be greater than their advantages for general application in barns and greenhouses.

Neutral pressure ventilation systems are available commercially and use specially designed fans to supply and exhaust air in a way which maintains air pressure within the ventilated space at approximately atmospheric pressure. Neutral pressure systems often incorporate a method to recirculate room air and mix it with fresh air, which tempers the fresh air and prevents localized chilling conditions. Recirculation can also incorporate filters to remove aerosols and pathogens. These advantages may be important when housing young animals but the method has found only limited application in animal agriculture as a whole, or in greenhouses. Recirculation and mixing can also be incorporated in either negative or positive pressure ventilation systems.

8-2. Fluid Mechanics Basics

Two relationships from fluid mechanics are important for airflow calculations — the continuity equation and the Bernoulli equation.

Flow continuity is a statement of mass conservation. When a fluid (air) with a density of ρ flows through an opening of area A, and has an average velocity of v, the mass flow rate is

$$m_{air} = \rho A v. \qquad (8\text{-}1)$$

When there exist n openings in an airspace, with air flowing in some openings and out others, continuity requires

$$\sum_n (\rho A v)_n = 0. \qquad (8\text{-}2)$$

As a convention for this text, we will identify positive flow and velocity with air entering the airspace (inlets) and negative flow and velocity with air leaving the space (outlets).

The concept of flow continuity will be used many times in this text; a simple application is in Example 8-1.

Example 8-1

<u>Problem:</u> A dairy barn is ventilated using an exhaust system. Conditions within the barn are 18 C and 60% relative humidity and the barn is located at an altitude near sea level. Outdoor conditions are -10 C and 50% relative humidity.

The fans exhaust air at a volumetric capacity of 3 m^3/s (measured at the indoor conditions). Inlets are used having a gross area of 1.2 m^2. Determine the average air velocity within the inlets (averaged over the gross inlet area).

<u>Solution:</u> First the air density indoors and out must be determined:

$$\rho_{in} \;=\; 1.21 \,\text{kg} / \text{m}^3, \text{ and}$$
$$\rho_{out} \;=\; 1.34 \,\text{kg} / \text{m}^3 .$$

Imposing flow continuity,

$$\rho_{in}(3 \text{ m}^3/\text{s}) = \rho_{out} (A_{inlet}) (v_{inlet}),$$

from which v_{inlet}, the inlet air velocity, can be calculated,

$$v_{inlet} = \frac{(1.21 \,\text{kg} / \text{m}^3)(3 \text{ m}^3 / \text{s})}{(1.34 \,\text{kg} / \text{m}^3)(1.2 \text{ m}^2)} = 2.26 \,\text{m} / \text{s}.$$

As used in Equations 8-1 and 8-2, the area of airflow, A, is the effective area. When air approaches an opening, streamlines converge and that convergence acts to reduce the effective area of flow (see Figure 8-1). Turbulence effects and the resulting energy loss also act to reduce flow through the opening. The contraction effect is expressed through the concept of the contraction coefficient,

$$A_{effective} = (C_{contraction})(A_{actual}) \qquad (8\text{-}3a)$$

Friction effects appear as a coefficient of friction, affecting the velocity, and

$$v_{effective} = (C_{friction})(v_{ideal}) . \qquad (8\text{-}3b)$$

The two coefficients combine into a coefficient of discharge to calculate the actual airflow through an inlet ($\dot{V}_{actual}$), given the gross inlet area and ideal velocity,

$$\dot{V}_{actual} = C_{discharge}(A_{inlet})(v_{ideal}) . \qquad (8\text{-}4)$$

From this point on, the coefficient of discharge will be abbreviated to C_d. When the coefficient of discharge is considered, flow continuity is more properly expressed as

$$\sum_n (C_d \rho A v)_n = 0. \qquad (8\text{-}5)$$

Values for discharge coefficients must be determined empirically. The standard for a sharp-edged opening is 0.6. Many inlets which are used to ventilate agricultural buildings have a discharge coefficient equal to 0.6, or greater, and some have coefficients less than 0.6. We will use constant values of discharge coefficients in this text, although some research evidence exists that coefficients are functions of the flow regime of air passing through the openings.

Air velocity through an inlet is a function of the pressure difference across the opening and can be determined using the Bernoulli equation. Whereas the continuity equation is a statement of the conservation of mass, the Bernoulli equation is equivalent to a statement of the conservation of energy.

Along a streamline the energy content of a fluid particle is composed of three parts: kinetic energy, pressure energy, and potential energy. When fluid flows, energy is removed primarily where it contacts a boundary; in the free stream, energy is conserved along streamlines. When written in terms of pressure, Bernoulli's equation is

$$\frac{\rho v^2}{2g_c} + P + \frac{\rho g z}{g_c} = \text{constant}, \qquad (8\text{-}6)$$

where ρ is density, v is velocity, P is static pressure, z is the elevation above an arbitrary but fixed datum, g is the acceleration of gravity, and g_c is a units conversion factor, equaling 1.0 in the SI system and 32.2 when IP units are used. The first term in Equation 8-6 is kinetic energy, the second and third are

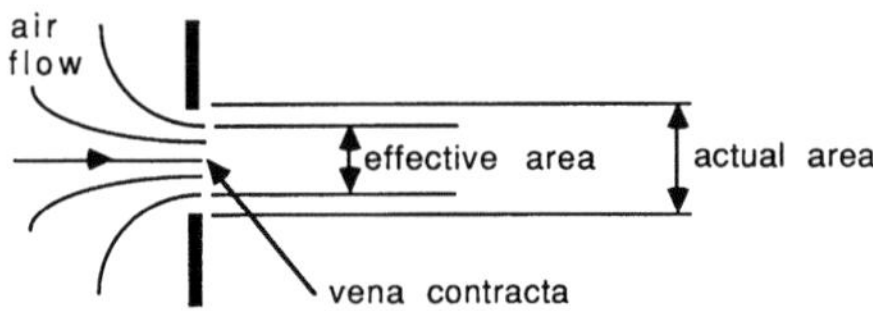

Figure 8-1. Streamlines of air flowing through an opening, showing the vena contracta and effective area of flow.

228

pressure and potential energy, respectively. Each term in the equation has units of N/m² , or pascals, a form of pressure.

The kinetic energy of a fluid is termed the "velocity pressure" and the atmospheric pressure, P, is termed the "static pressure". The sum of the two is termed the "total pressure". When air is considered in ventilation applications, changes of potential energy are small relative to the total pressure (air density has a small value) and are usually neglected. Thus an alternate way to state the Bernoulli equation is to say the total pressure along a streamline remains constant.

Consider an example of air flowing at a speed of 2 m/s through an air duct whose cross-sectional area gradually changes from 2 m² to 1 m². The area is halved – continuity considerations require the velocity to double. Losses are negligible, there is no elevation change, and the Bernoulli equation applies. The total pressure does not change – there are no significant losses. However, the velocity pressure must be different because air velocity changes. To retain a constant total pressure, the static pressure must be lowered by an amount to balance the increase of velocity pressure.

$$\frac{\rho v_1{}^2}{2g_c} + P_1 \left(+ \frac{\rho g z_1}{g_c}\right) = \frac{\rho v_2{}^2}{2g_c} + P_2 \left(+ \frac{\rho g z_2}{g_c}\right) \tag{8-7}$$

$$P_1 - P_2 = \frac{\rho}{2}\left(v_2{}^2 - v_1{}^2\right) \tag{8-8}$$

The sequence of consequences as stated is important to remember. When the cross-sectional area of flow changes, the velocity must change as dictated by the continuity equation, and static pressure then changes as dictated by the Bernoulli equation. Static pressure adjusts so total pressure remains constant.

Example 8-2

<u>Problem:</u> A dairy barn is ventilated using an exhaust system and the fans have created a pressure difference from inside the barn to outside of 15 Pa. There is a square opening in one wall of the barn which is 1 m by 1 m. Outdoor air is at 5 C and 40% relative humidity and the barn is at an elevation of approximately 500 m.

Calculate the velocity of air as it passes through the opening, and estimate the mass rate of airflow.

<u>Solution:</u> The first step is to visualize the way in which Bernoulli's equation applies to this problem. Air first is outside the building and is approximately still (negligible velocity pressure). It is then drawn to the opening where it accelerates to reach a maximum velocity as it passes through the inlet. At the inlet (and just inside it) the static pressure is low enough to balance the velocity

pressure gain. Pressure inside the building is uniform and air velocity in the barn can again be considered nearly zero.

Equation 8-8 can be used to calculate airspeed at the inlet, assuming v_1 applies to the air prior to passage through the opening, and $v_1 = 0$. Air density is 1.19 kg/m^3. When Equation 8-8 is rearranged,

$$v_2 = [2(P_1 - P_2)/\rho]^{0.5}$$
$$= [2(15 \text{ Pa})/1.19 \text{ kg/m}^3]^{0.5} = 5.02 \text{ m/s}.$$

The coefficient of discharge of the opening must be estimated. If the opening is simply cut into the wall, it is reasonably sharp edged and the usual assumption of $C_d = 0.6$ will be made.

$$m_{air} = C_d \rho A v$$
$$= (0.6)(1.19 \text{ kg/m}^3)(1 \text{ m}^2)(5.02 \text{ m/s}) = 3.58 \text{ kg/s}.$$

8-3. Air Jets

An air jet is defined as a region of air moving at a velocity different from the velocity of the surrounding air. When fresh air is drawn into a building through an inlet, it has a significant velocity, while air in the room is approximately still. If the air jet is directed away from all boundaries of the room, it is a free jet. If it is directed to flow along a wall or ceiling, it is a wall jet.

A free jet, if it approaches a solid boundary, will attach to the boundary and become a wall jet. A wall jet will remain a wall jet until it encounters a large solid obstacle or unless thermal buoyancy forces are so great that its flow direction is changed. The phenomenon which keeps a jet attached to a solid boundary is termed the "Coanda effect".

Consider the hypothetical situation shown in Figure 8-2. Fresh air enters through the inlet in a form termed a "plane free jet" (a plane jet because the airspace is long in the z direction and the jet is essentially two-dimensional). The free jet enters the airspace near but not attached to the ceiling. Shear forces at the upper boundary of the jet induce rotation of the fluid at the corner

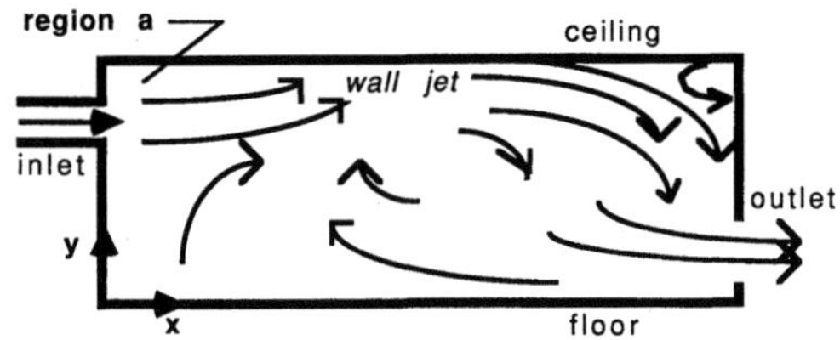

Figure 8-2. A plane jet entering a ventilated space as a free jet and attaching to a nearby surface as a wall jet through action of the Coanda effect.

between the ceiling and the wall with the inlet (region a).

Rotation causes velocity pressure within region a to increase and static pressure to decrease, and lower static pressure causes the jet to be drawn toward the region of low pressure. The result is movement of the jet toward the ceiling and attachment. This is the standard explanation of the Coanda effect. Other motions of air within the airspace will be discussed later in this text.

Free jets find application for ventilating agricultural buildings. A common technique of air distribution, especially if air is to be recirculated and heated, is to attach a length of perforated polyethylene tubing to the discharge side of a fan and distribute air in the form of many small free jets from holes in the polyethylene tube. The fan draws air both from outdoors and from the ventilated space, mixing the two airstreams. Each hole in the polyethylene tube is approximately 50 mm in diameter and the tube may be as long as 50 m. This form of ventilation will be discussed in more detail later in this text.

8-3.1. Air Jet Velocity Decay and Profiles, Wall Jets. When an air jet enters a ventilated space, its character changes along the length of the jet. A plane wall jet of the type used typically to ventilate barns is shown in Figure 8-3.

When the jet enters the airspace, its velocity profile is determined by the inlet characteristics and its inlet speed can be calculated using the Bernoulli equation, if the coefficient of discharge is known. Shear between the jet and the still air is intense (jets are among the most turbulent fluid flow phenomena) but the effect of shear does not penetrate to the core of the jet immediately. The maximum velocity of the jet remains as it was at the inlet for a distance downstream equal to approximately four times the width of the inlet. This is termed zone 1 of the jet flow.

After the effect of shear penetrates to the core of the jet, the maximum velocity within the jet slows as a function of distance along the jet. For a distance extending approximately eight inlet widths downstream the maximum velocity slows proportionally to the inverse of the distance. This is zone 2 and is the zone wherein the jet's velocity profile is established and stabilized.

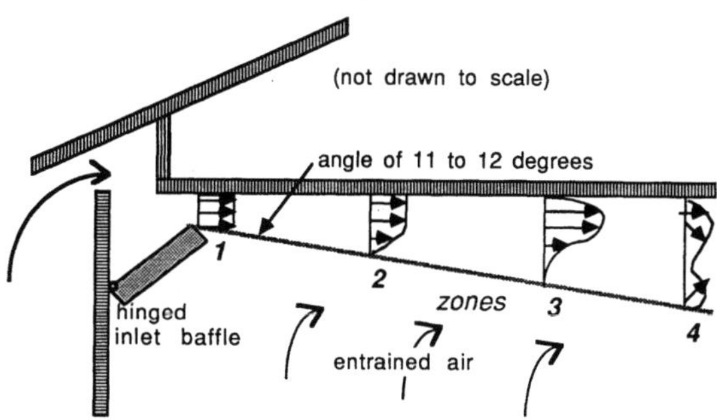

Figure 8-3. A plane wall jet, showing velocity profiles, entrainment and zones of flow.

The next zone, 3, forms the major part of the jet. It is the zone of fully established, turbulent flow where the maximum jet velocity slows as the inverse of the square root of distance from the beginning of the zone. This zone continues until the maximum jet speed is reduced to approximately 0.25 m/s. The zone can be up to 100 jet diameters long and is the zone of primary engineering importance. The distance at which zone 3 ends is termed the "throw" of the jet.

After zone 3, momentum within the jet is sufficiently small that eddies in the air within the ventilated space overwhelm the jet, the velocity profile degenerates, and the jet disappears. This is zone 4 and is limited to a distance of only a few (local) jet diameters.

Detailed equations to calculate the decay of the maximum velocity of wall jets can be found in the *ASHRAE Handbook of Fundamentals*, for example. For practical design calculations related to the type of wall jets used to ventilate agricultural buildings, the following equation has been found to be adequate from the inlet to the throw of a plane wall jet:

$$v_{max} = 2.76 v_{inlet}(x/W)^{-0.5}, \qquad (8\text{-}9)$$

where v_{inlet} is the jet velocity as it passes through the inlet of width W, x is the downstream distance, and v_{max} is the maximum velocity within the jet's velocity profile at x.

Example 8-3

<u>Problem:</u> A poultry house is to be ventilated using an exhaust (negative pressure) system with a continuous slot inlet along the perimeter of the ceiling (similar to that shown in Figure 8-3). The baffle board causes the inlet's coefficient of discharge to be 0.8. Outdoor conditions are 24 C and 60% relative humidity and the elevation is near sea level.

The pressure difference from inside the building to out is to be 10 Pa. How wide must the inlet be so the beginning of zone 4 (where the maximum velocity is 0.25 m/s) is 7 m from the inlet?

<u>Solution:</u> Equation 8-9 applies, but first the inlet jet velocity must be determined using the Bernoulli equation, Equation 8-8. Air density will be 1.18 kg/m^3.

$$v_{inlet} = (2\Delta P / \rho)^{0.5},$$
$$= [2(10\ Pa) / 1.18\ kg/m^3]^{0.5} = 4.12\ m/s.$$

When Equation 8-9 is rearranged,

$$W = \ x \ (v_{max} / 2.76 v_{inlet})^2,$$
$$= (7 \text{ m})[(0.25 \text{ m / s})/(2.76)(4.12 \text{ m / s})]^2 = 0.00483 \text{ m},$$

or 4.8 mm.

The width which has been calculated is at the vena contracta of the jet as it passes through the inlet. The actual inlet width will be equal to the vena contracta width, divided by the coefficient of discharge in this example (because the flow is two-dimensional). Thus the baffle board should be (4.8 mm/0.8) = 6 mm away from the ceiling.

Note: In practical terms this is a very narrow inlet. An inference one might draw is that the throw of the jet will likely be adequate to bring fresh air far into the poultry house if there are no obstacles in the jet's path.

The velocity profile within the established region of a wall jet is well-known and follows flow similarity along the length of zone 3. That is, when graphed in dimensionless form, the profiles remain identical along the entire zone. (Note: In what follows, the terms "wall" and "ceiling" can be used interchangeably.)

The dimensionless distance from the wall, $Y/Y_{0.5}$, is calculated by dividing the actual distance from the wall, Y, by the actual distance at which the velocity equals one half the maximum velocity, termed $Y_{0.5}$. The maximum velocity is at $Y/Y_{0.5} = 0.14$. The line which defines the growth of $Y_{0.5}$ is at an angle of approximately 4°, starting at the beginning of the jet. The equation which defines the profile is sketched in Figure 8-4, and is

$$v / v_{max} = \exp[- 0.937(Y / Y_{0.5} - 0.14)^2]. \qquad (8\text{-}10)$$

Equations 8-8, 8-9, and 8-10 can be used to calculate the actual velocity at any point within the established region of a plane, isothermal wall jet if the inlet air speed or pressure drop, and inlet width, are known.

8-3.2. Entrainment into Wall Jets. Shearing action at the free boundary of wall jets causes still air in the vicinity of the jet to be accelerated. This acceleration of still air is a transfer of momentum from the jet to the still air, a consequence of which is that still air becomes part of the growing jet. The

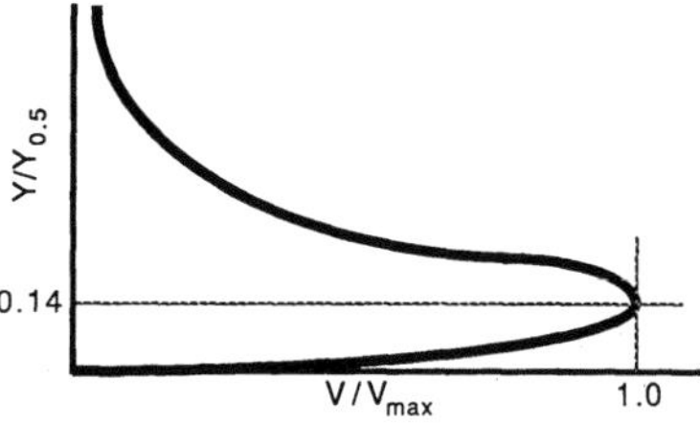

Figure 8-4. Velocity profile of a wall jet within zone 3, graphed in dimensionless form.

boundary between the jet and still air is not a smooth plane – it is an interface characterized by waves of large and small turbulent eddies. The interface ripples as large eddies (long, thin vortices) form and tumble along the boundary.

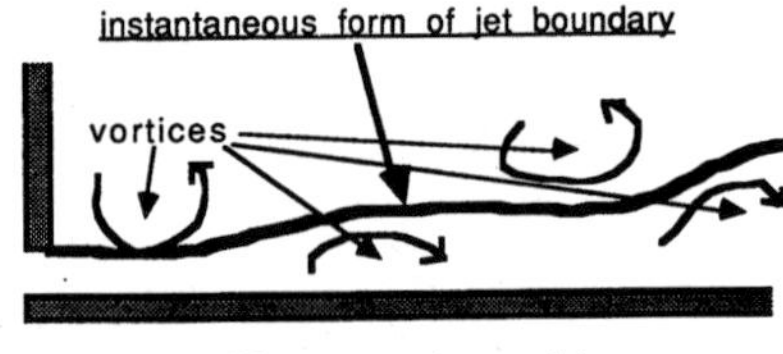

(Note: not to scale)

Growth of a jet is termed "entrainment" – the jet entrains fluid into itself, which slows the jet's velocity and increases its thickness. A fundamental principle in jet theory has been that momentum is conserved within the growing jet; velocity must therefore decrease as the mass increases. Some evidence exists that vorticity is the actual conservative feature, but applications of the concept of momentum conservation have led to numerous experimentally verified predictions of jet behavior.

The entrainment ratio is defined as the ratio of volumetric flow at a distance x along the jet, to the volumetric flow at the inlet,

$$\dot{V}_x / \dot{V}_{inlet} = \sqrt{2}(v_{inlet} / v_{max,x}) \qquad (8\text{-}11)$$

$$= 0.512(x / W)^{0.5}. \qquad (8\text{-}12)$$

Example 8-4

Problem: Reconsider the wall jet described in Example 8-3. What will be

(a) the maximum velocity in the jet velocity profile,

(b) the distance from the wall (ceiling) where the velocity equals 10% of the maximum velocity, and

(c) the entrainment ratio at a downstream distance of 3 m from the inlet?

Solution:

(a) Equation 8-9 can be used to determine the maximum velocity at 3 m. In Example 8-3, the effective inlet width, W, was found to be 0.0048 m. The distance downstream is 3 m and the inlet velocity is 4.12 m/s, thus,

$$v_{max,\,x} = 2.76 v_{inlet}(x / W)^{-0.5}$$
$$= 2.76(4.12 \text{ m / s})(3 \text{ m} / 0.0048 \text{ m})^{-0.5} = 0.455 \text{ m / s}.$$

(b) Equation 8-10 can be used to calculate air velocity at any point within the velocity profile, and $v/v_{max} = 0.10$. However, before the equation can be used, $Y_{0.5}$ must be determined.

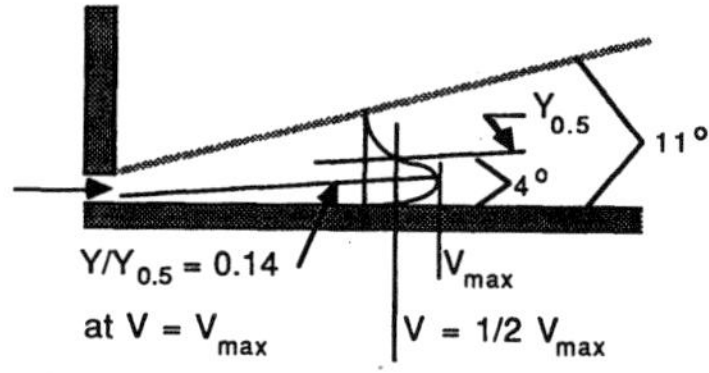

Recall the thickness of a wall jet has been observed to increase at an angle of 11°. The growth of $Y_{0.5}$ is slower and is at an angle of approximately 4°.

At a downstream distance of 3 m from the inlet, $Y_{0.5}$ can be estimated to be

$$Y_{0.5} = 3 \tan(4°) = 0.210 \text{ m},$$

and the distance from the wall where $v/v_{max} = 0.1$ can be found from a rearrangement of Equation 8-10

$$\begin{aligned} Y &= Y_{0.5} [0.14 + 1.033(\ln (v_{max}/v_x))^{0.5}], \\ &= 0.210 [0.14 + 1.033(\ln (10))^{0.5}] \\ &= 0.36 \text{ m}. \end{aligned}$$

Note: The stated wall jet growth angle of 11° leads to a total jet thickness at 3 m of 0.583 m. This corresponds approximately to the point within the jet profile where $v/v_{max} = 0.0015$.

(c) The entrainment ratio can be calculated using Equation 8-11

$$\begin{aligned} \dot{V}_x / \dot{V}_{inlet} &= 0.512(x / W)^{0.5} \\ &= 0.512(3 \text{ m} / 0.0048 \text{ m})^{0.5} = 12.8. \end{aligned}$$

The volumetric airflow rate at the inlet must be determined. The inlet velocity is 4.12 m/s, the area of opening per meter of inlet is 0.006 m², and the coefficient of discharge is stated to equal 0.8. Thus

$$\begin{aligned} \dot{V}_{inlet} &= 0.8 \ (4.12 \text{ m/s}) \ (0.006 \text{ m}^2 / \text{m}) \\ &= 0.0198 \text{ m}^3 / \text{s per m of inlet}. \end{aligned}$$

The volumetric airflow rate at 3 m is 12.8(0.0198 m³/s), or 0.25 m³/s per m of jet width in the z direction (parallel to the axis of the inlet).

8-3.3. Air Jet Velocity Decay and Profiles, Free Round Jets. The concepts of jet development, entrainment, and velocity apply also to free jets. The cross section of a free jet is shown in Figure 8-5. When a free jet issues from a circular opening, the four zones develop and the maximum (centerline) velocity decay may be approximated within zone 3 by the following

$$v_{max,x} = Kv_{inlet}(D_e / x) \qquad (8\text{-}13)$$

where K = 5 for an exit velocity between 2.5 and 5 m/s, and 6.2 for an exit velocity between 10 and 50 m/s. Interpolate for intermediate exit velocities. The variable D_e is the diameter of the jet at the vena contracta at the inlet, not the diameter of the opening through which the jet issues. The diameter at the vena contracta can be determined from the diameter of the opening if the coefficient of contraction is known (frequently assumed equal to the coefficient of discharge, as the coefficient of velocity is typically very close to unity). Other terms are as defined previously.

The velocity profile within circular free jets can be calculated from

$$v / v_{max,x} = \exp[-0.693(r / r_{0.5})^2], \qquad (8\text{-}14)$$

where $r_{0.5}$ is the radius from the centerline of the jet to where the local air velocity is one-half that at the centerline. The entrainment ratio is

$$\dot{V}_x / \dot{V}_{inlet} = 2(v_{inlet} / v_{max,x}) \qquad (8\text{-}15)$$

$$= 0.4(x / D_e) \text{ for } 2.5 \leq v_{inlet} \leq 5 \text{ m / s} \qquad (8\text{-}16a)$$

$$= 0.32(x / D_e) \text{ for } 10 \leq v_{inlet} \leq 50 \text{ m / s} \qquad (8\text{-}16b)$$

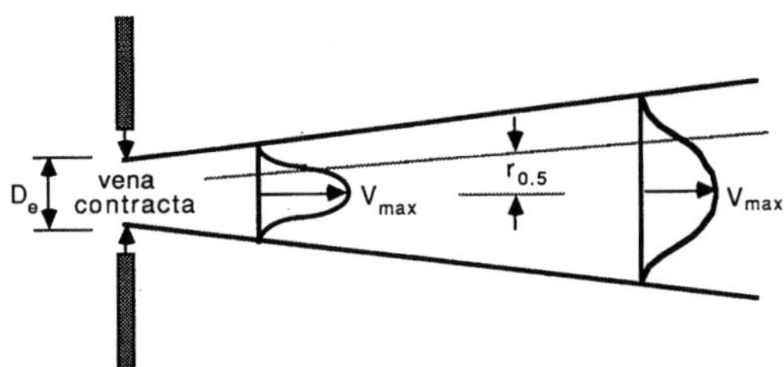

Figure 8-5. Development of velocity profile in a circular free jet.

Example 8-5

Problem: A perforated polyethylene tube has been attached to the discharge side of a fan to distribute fresh air in a greenhouse. Static pressure inside the tube is 50 Pa. Holes in the tube are round, with 50 mm diameters, and characterized by discharge coefficients of 0.6. Fresh air within the duct is at - 5 C and 60% relative humidity and the greenhouse is located at an elevation of approximately 500 m.

Determine how far from the tube the air jets penetrate before air velocity decays to 0.5 m/s.

Solution: First, air velocity at the vena contracta must be determined. Static

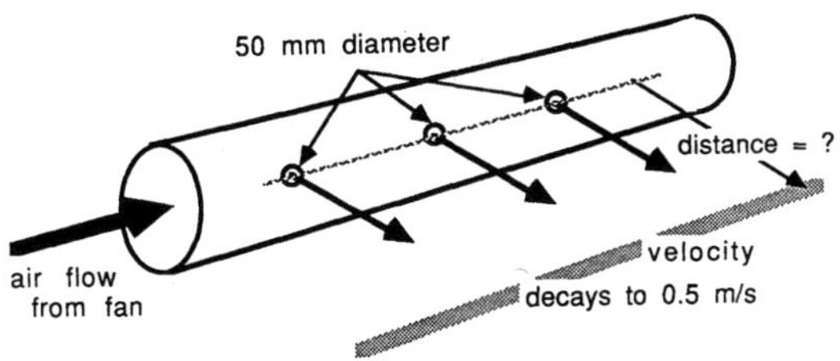

pressure within the polyethylene duct is 50 Pa and Bernoulli's equation can be used to determine the exit velocity. Air density is 1.24 kg/m^3.

$$v_{inlet} = \sqrt{2(\Delta P / \rho)},$$
$$= \sqrt{(2(50\ Pa) / 1.24\ kg / m^3)} = 8.98\ m / s.$$

Note: In this example, air pressure was assumed to be 95.511 kPa. The 50 Pa within the duct were added to the standard atmospheric pressure at 500 m. This small adjustment is not necessary for accuracy because of the small pressure increase within the duct, but was done for completeness.

Equation 8-13 can be used to calculate when the jet velocity will decay to 0.5 m/s. First, the diameter at the vena contracta must be determined. From the definition of the coefficient of contraction, and assuming the coefficient of velocity is equal to unity,

$$Area_{effective} = C_d(Area_{actual}),$$

thus for round inlets,

$$D_e = D \sqrt{C_d}$$

The actual diameter is 0.05 m and

$$D_e = (0.05\ m)\ \sqrt{0.6} = 0.0387\ m.$$

When Equation 8-13 is rearranged,

$$x = Kv_{inlet}D_e / v_{max},$$

and   K = 5.96 (by interpolation for 8.98 m / s exit air velocity), and

$$x = 5.96\ (8.98\ m/s)(0.0387\ m)/(0.5\ m/s) = 4.1\ m.$$

The jet penetrates 4.1 m into the still air before its centerline velocity decays to 0.5 m/s.

8-3.4. Nonisothermal Wall Jets, Temperature Profiles, and Heat Transfer Coefficients. Most ventilation applications in animal housing involve introducing fresh air at one temperature into an airspace which is at a different temperature. The temperature difference between the jet and still air means the jet is "nonisothermal" although, as it entrains still air, its temperature approaches that of the still air. The temperature difference at the inlet can be great during cold weather and a goal of all ventilation is not to chill the animals within the space. Tempering of the fresh air is important. This is especially critical when young animals are housed, for they are more susceptible to cold stress than are mature animals.

Nonisothermal conditions have little effect on velocity profiles within wall jets unless the temperature difference between the jet and the still air is so great as to result in unstable flow. This topic will be left for now, and returned to later.

Jets have temperature profiles, just as they have velocity profiles. In Figure 8-6 is sketched a nonisothermal wall jet and its temperature profile.

For convenience, temperatures can be expressed in nondimensional form as follows:

$$\theta = (t_r - t) / (t_r - t_{inlet}), \text{ and} \tag{8-17}$$

$$\theta_{max} = (t_r - t_{max}) / (t_r - t_{inlet}), \tag{8-18}$$

where t is the temperature at any point within the temperature profile, t_r is room air temperature, and t_{inlet} is air temperature at the inlet. The subscript "max" implies the extreme temperature within the profile, which may be the highest or lowest temperature.

Within the wall jet at any distance x, $0 \leq \theta < 1.0$, and at any distance x along the jet's axis, $0 \leq \theta/\theta_{max} \leq 1.0$.

The maximum jet temperature (θ_{max}) can be calculated using

$$\theta_{max} = 0.587 Re^{0.224}(x / W)^{-0.6}, \tag{8-19}$$

where Re is the jet's Reynolds number at the inlet,

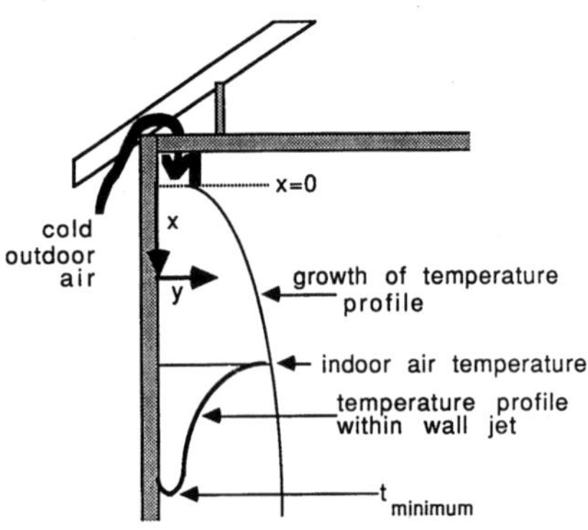

Figure 8-6. Temperature profile in a nonisothermal wall jet.

$$Re = \rho W v_{inlet} / \mu, \qquad\qquad (8\text{-}20)$$

where ρ is air density at the inlet, and μ is the absolute viscosity. The inlet Reynolds number is based on the width of the inlet, not the vena contracta.

The temperature profile thickness increases by the following:

$$Y_{0.5} / W = 28.6 Gr^{-0.282} (x / W)^{0.4}, \qquad\qquad (8\text{-}21)$$

where Gr is the inlet Grashof number,

$$Gr = g\rho^2 \beta W^3 (t_{inlet} - t_r) / \mu^2, \qquad\qquad (8\text{-}22)$$

and W is the width of the inlet, not the vena contracta.

The temperature profile maximum grows as a function of the inlet Grashof number, the distance downstream, and the width of the inlet, as follows:

$$Y_{0.5} = 28.6 W Gr^{-0.282} (x / W)^{0.4}. \qquad\qquad (8\text{-}23)$$

Within the jet's temperature profile,

$$\theta / \theta_{max} = \exp[-0.8(Y / Y_{0.5} - 0.10)^{1.4}]. \qquad\qquad (8\text{-}24)$$

The temperature profile extends farther from the wall than does the velocity profile. This is analogous to boundary layers, where temperature boundary layers are thicker than momentum boundary layers.

Example 8-6

<u>Problem:</u> A continuous slotted inlet such as shown in Figure 8-6 is used to ventilate a swine nursery. The side walls are 2.7 m high. Outdoor conditions are - 15 C and 70% relative humidity, and the elevation is near sea level. Indoor air temperature is 26 C.

Ventilation is by an exhaust system and the pressure difference from inside the barn to the outside is 10 Pa. The coefficient of discharge at the inlet is approximately 0.8.

If the inlet is 10 mm wide, what will be the temperature of the centerline of the air jet when it reaches the floor, how far from the wall will $Y_{0.5}$ be, and what will be the variation of temperature in the dimension perpendicular to the wall?

<u>Solution:</u> As a first step, the Bernoulli equation can be used to find the inlet air velocity, where air density is 1.37 kg/m^3.

$$v_{inlet} = \sqrt{2\Delta P / \rho}$$

$$= \sqrt{2(10\ Pa) / (1.37\ kg / m^3)} = 3.82\ m / s.$$

Air at the inlet can be assumed approximately at outside conditions, and the dynamic viscosity will be approximately 1.567E - 5 kg/ms (see Table 3-1). The inlet Reynolds number is

$$Re = \rho v_{inlet} W / \mu$$
$$= (1.37 \text{ kg} / \text{m}^3)(3.82 \text{ m} / \text{s})(0.010 \text{ m}) / (1.567E - 5 \text{ kg} / \text{ms})$$
$$= 3340.$$

The dimensionless extreme temperature can be determined from Equation 8-19

$$\theta_{max} = 0.587 (3340)^{0.224} (2.7 \text{ m} / 0.010 \text{ m})^{-0.6}$$
$$= 0.135.$$

By the definition of θ_{max},

$$t_{max} = 26 \text{ C} - 0.135(26 \text{ C} - (- 15 \text{ C})) = 20.5 \text{ C}.$$

The action of the wall jet has been to temper the outdoor air until its temperature is close to that indoors. This is an advantage of using the continuous inlet near the ceiling with a wall jet directed down the wall.

The growth of the temperature profile maximum can be determined using Equation 8-23. First, the Grashof number must be calculated:

$$Gr = g\rho^2\beta W^3 (t_r - t_{inlet}) / \mu^2,$$

where, β (the coefficient of thermal expansion) can be approximated by the inverse of the absolute temperature, $1/258.15$ K.

$$Gr = \frac{(9.8 \text{ m} / \text{s}^2)(1.37 \text{ kg} / \text{m}^3)^2(1 / 258.15 \text{ K})(0.010 \text{ m})^3(26 \text{ C} - (-15 \text{ C}))}{(1.567E - 5)^2}$$
$$= 1.19E4$$

The magnitude of $Y_{0.5}$ is,

$$Y_{0.5} = 28.6W(Gr^{-0.282}) (x / W)^{0.4},$$
$$= 28.6 (0.010 \text{ m}) (1.19E4)^{-0.282}(2.7 \text{ m} / 0.010 \text{m})^{0.4}$$
$$= 0.19 \text{ m}.$$

The temperature profile can be predicted using Equation 8-24,

$$\theta = \theta_{max}\exp[- 0.8 (Y / Y_{0.5} - 0.10)^{1.4}]$$
$$= 0.135\exp[- 0.8(Y / 0.19 \text{ m} - 0.10)^{1.4}].$$

The conversion from θ to t is:

$$t = t_r - \theta(t_r - t_{inlet}) = 26 \text{ C} - \theta(41 \text{ K})$$

240

and data for t as a function of Y are:

Y	θ	t
0.02 m	0.135	20.47 C
0.05	0.127	20.80
0.10	0.106	21.66
0.15	0.084	22.56
0.20	0.064	23.38
0.25	0.047	24.07
0.30	0.034	24.61
0.35	0.024	25.03
0.40	0.016	25.34
0.45	0.011	25.55
0.50	0.007	25.71
0.60	0.003	25.88
0.70	0.001	25.95
0.80	0.0004	25.98
1.00	~0	26.00

The temperature profile is graphed in Figure 8-7.

One effect of wall jets during cold weather is to bring cold air along the outside walls, thereby possibly reducing the loss of heat from the building and saving heat. Heat transfer between a wall jet at the airspeeds used for ventilation, and a smooth wall, can be determined from the following equation for the Nusselt number:

$$Nu = 1.11\ Re^{0.57}(x / W)^{0.35}, \qquad (8\text{-}25)$$

where Nu is the local Nusselt number, hx/k, h is the local coefficient of convective heat transfer, W/m^2K, and k is the thermal conductivity of air at conditions within the jet at x. Convective heat exchange with the wall is calculated by

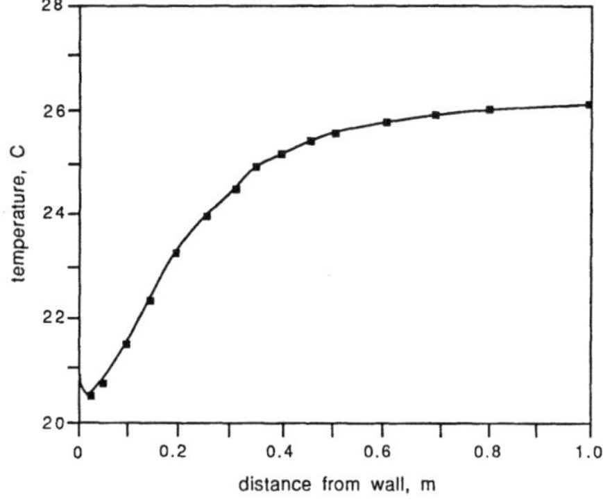

Figure 8-7. Wall jet temperature profile at the floor, Example 8-6.

$$q'' = h(t_{max} - t_{wall}). \qquad\qquad (8\text{-}26)$$

8-4. Slotted Air Inlets

8-4.1. Slotted Inlet Operation. Previous discussion has referred to the type of inlet most commonly used with exhaust ventilation systems, the slotted inlet. The sketch in Figure 8-6 is one type, in Figure 8-8 are sketched several common slotted inlet designs. The baffles may be locally fabricated when the building is constructed, or commercial units may be purchased and installed. Typically such inlets are installed along both long walls of a barn, but not on end walls, and not directly over exhaust fan banks. Being continuous, slotted inlets provide a uniform distribution of fresh air along the length of a building.

The baffled, slotted inlet was invented 40 years ago (Millier, 1950) and has several advantages when used with exhaust ventilation systems. It is an inlet which is relatively easy to construct and install. The baffle along an entire wall can operate as one long unit and can be adjusted as a single unit, simplifying control.

Inlet width can be adjusted to correspond to changes in the required ventilation rate. A recommended practice is to operate the baffles based on the pressure difference between inside the barn and out. Baffles are opened or closed to keep the static pressure difference within a certain range (for example, between 10 and 15 Pa). When the pressure difference rises above the desired range (for example, when additional fans operate as the outdoor air temperature rises) the

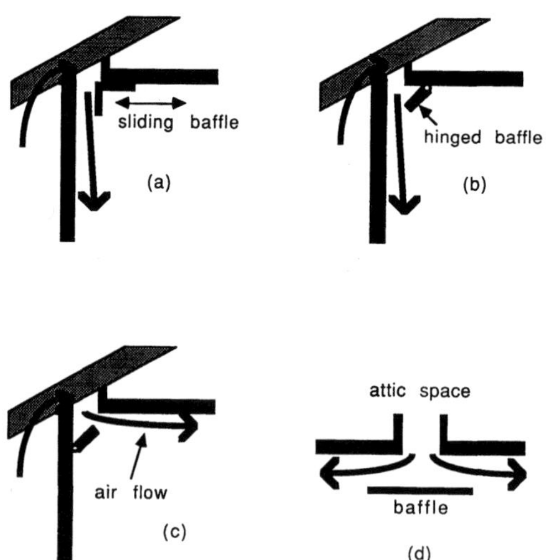

Figure 8-8. Examples of slotted inlet construction: (a) sliding baffle, air directed down the wall; (b) hinged baffle, air directed down the wall; (c) hinged baffle, air directed along the ceiling; and (d) a center-ceiling baffle with air drawn through the attic and directed along the ceiling.

control system opens the baffles until the static pressure difference is brought within the desired range. The reverse happens when fans are deactivated and the static pressure difference falls below the desired range.

It is important to keep the static pressure difference above very low values for two reasons. First, when the pressure difference is low, inlet air velocity is low and air mixing is ineffective. One advantage of wall jets is the way they mix fresh air with air already in the building and temper cold air before it can reach the animals. If air velocity at the inlet is low, mixing will be inadequate.

A second reason to avoid conditions of low static pressure difference is wind effects on ventilation. When wind swirls around a building, regions of partial vacuum are created (for example, leeward of the building). If suction created by the wind is greater than the partial vacuum created within the building by the ventilation system, air will leave the barn through inlets and inlets will become outlets. This greatly distorts ventilation air patterns in the building and can significantly increase the ventilation rate above what is desired, perhaps even causing freezing at some places within the building during cold weather. A static pressure partial vacuum within a barn of at least 10 Pa is generally required to limit wind effects only to times of extreme windiness.

The upper limit of the desired static pressure difference range is a practical consideration related to the fans and their ability to work against a static pressure difference. Propeller fans are designed to move air at low static pressure differences, and begin to lose performance ability to a significant degree when the static pressure difference is above approximately 30 Pa. Also, static pressures much above 30 Pa do little additionally to enhance air mixing within a ventilated space.

Another advantage of slotted inlets, when air is directed to flow down outside walls, is that condensation on the inside surfaces of outside walls is virtually eliminated during cold weather. The fresh air is relatively dry, which keeps the wall surface dry.

Other advantages, such as an ability to temper incoming air quickly and create mixing of fresh air and air within the building, have already been discussed.

Before proceeding, it is important to emphasize one principle related to slotted inlets and exhaust ventilation systems. When the static pressure difference from inside a building to out is maintained within reasonable limits (e.g., between 10 and 30 Pa), the ventilation rate is dictated by fans and air mixing is provided by the inlets. For example, if a period of extreme heat causes stress to animals in a barn ventilated by a slotted inlet exhaust system, and operation is within the static pressure difference range of approximately 10 to 30 Pa, opening the inlets wide to "get more air" may defeat the ability of the ventilation system to create good air mixing and bring fresh air to all the animals. Wider inlets will not cause significantly more air to enter – the airflow rate is dictated by the fans –

and the resulting lower inlet air velocity will reduce air mixing within the barn. It is often tempting, during hot weather, to open all the doors, windows, and inlets in a barn to bring more fresh air to the animals. However, this is exactly what should not be done. To do so (unless there is a strong wind) does not increase the ventilation rate, but instead prevents good ventilation.

8-4.2. Airflow Through Hinged-Baffle and Center Ceiling Slotted Inlets. When the coefficient of discharge is known, airflow through inlets can be calculated if the static pressure difference and inlet areas are known. However, with slotted inlets, the coefficient of discharge is a function of the geometry of the inlet.

The following equations can be used to calculate volumetric airflow rates through slotted inlets:

a. Slotted inlet, baffle attached to the ceiling, airflow down the wall, cases (a) and (b) in Figure 8-8:

$$\dot{V} = 0.0012W^{0.98}\Delta P^{0.49} \tag{8-27}$$

b. Slotted inlet, baffle attached to the wall, airflow across the ceiling, case (c) in Figure 8-8:

$$\dot{V} = 0.00071W^{0.98}\Delta P^{0.49} \tag{8-28}$$

In Equations 8-27 and 8-28, $\dot{V}$ is in m^3/s per meter of inlet length (or m^2/s), W is the actual inlet width in mm, and ΔP is the static pressure difference across the inlet, in pascals. The two equations are based on experiments, thus the exponents of W and ΔP are not 1.0 and 0.5 as would be expected from basic principles. However, the differences are slight.

There is a major difference, however, in the magnitudes of airflows from the two inlets for the same inlet widths and static pressure difference. The coefficient of Equation 8-28 is only 59% of the coefficient of Equation 8-27. This is a significantly smaller coefficient of discharge and is a result of the way airflow within the inlet is restricted by the change of direction to flow along the ceiling. A less restrictive inlet could be designed by having fresh air come straight through the wall instead of over the top plate.

c. Center-ceiling slotted inlet, case (d) in Figure 8-8:

$$\dot{V} = 0.0013W^{0.98}\Delta P^{0.49}(D / T)^{0.08}\exp(- 0.867W / T) \tag{8-29}$$

where, in this case, $\dot{V}$ includes airflow from both sides of the baffle, in m^3/s per meter of baffle length. Variables W and ΔP were previously defined, D is the width of the baffle board and T is the width of the slot through the ceiling.

It must be stressed that Equations 8-27 and 8-28 apply only when there is no airflow restriction of significance other than at the inlet formed by the baffle.

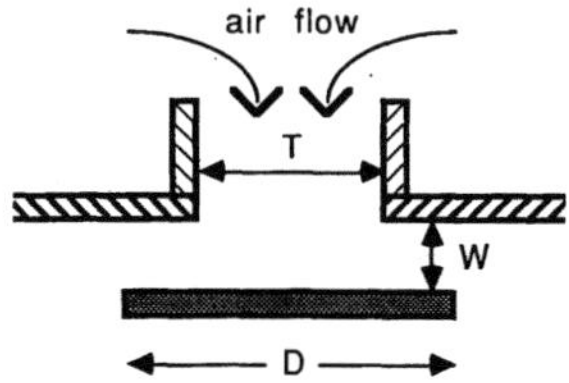

When air flows over the top of the outside wall (over the top plate), there must be no restriction at that point, for example. A rough rule is that upstream restrictions can be neglected if they are more than two and a half times as wide as the largest inlet width. For example, if a baffle is to operate to permit an inlet width as wide as 50 mm, the minimum upstream restriction must be 125 mm.

As livestock buildings become larger, animal stocking densities increase, more ventilation is required, and wider inlets are specified. Common construction techniques may inadvertently cause significant upstream restrictions and the ventilation system will never work as designed. It is the responsibility of the design engineer in such cases to ensure the design is specified to avoid such problems.

There are some limits in applying the center-ceiling inlet equation. When this style inlet is used, air enters first into the attic. A building design must ensure airflow into the attic will be unrestricted. In addition, Equation 8-29 was developed for restrictions

$$2 < D/T < 6, \text{ and}$$
$$0.109 < W/T < 1.46.$$

These ranges encompass typical design practices.

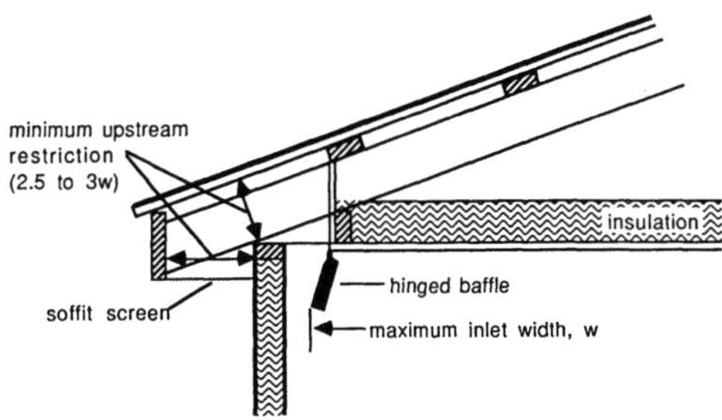

8-5. Air Infiltration

No building is airtight. Small cracks around windows and doors, cracks through the siding, joints between the walls and floor, and entry points for electrical and water service are examples of the many paths by which air can enter a building through inadvertent inlets. When an exhaust ventilation system is used in a barn, the static pressure difference from inside the barn to out induces air to flow through the cracks.

The only agricultural data currently available to predict infiltration as a function of static pressure difference was obtained from several dairy stables more than 30 years ago. Whether it applies to today's construction techniques is questionable, but until better data is available, there is nothing better to use. The *ASHRAE Handbook of Fundamentals* details methods to determine air infiltration around windows and doors (the "crack" method), but the method has not been proven to apply accurately to agricultural buildings.

In Figure 8-9 is sketched the relationship between air infiltration and static pressure difference. The data were obtained from dairy barns ventilated using a negative pressure system, and were presented normalized to a dairy cow animal unit. To apply the data to other animal housing requires imagination and knowledge of how many dairy cows might be housed in the building being designed.

The two curves sketched in Figure 8-9 can be expressed as follows:

$$(a)\ \text{Tight construction: } \dot{V} = 0.017\Delta P^{0.67}, \text{ and} \qquad (8\text{-}30)$$

$$(b)\ \text{Very tight construction: } \dot{V} = 0.006\Delta P^{0.67}. \qquad (8\text{-}31)$$

where $\dot{V}$ is m³/s per dairy cow unit and ΔP is the static pressure difference, pascals.

It is believed current construction is represented more closely by Equation 8-31 than Equation 8-30. If a barn is made especially tight (e.g., with concrete or

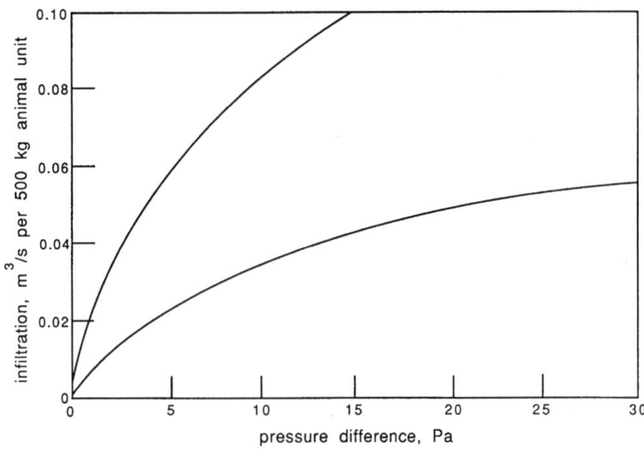

Figure 8-9. Infiltration rates as a function of pressure difference for two Pennsylvania dairy barns.

block walls, or vapor barriers in woodframed construction) a fraction of the infiltration predicted by Equation 8-31 will be experienced. However, what that fraction will be is only a guess unless the building is pressure tested. Applications of the ASHRAE "crack" method have suggested the fraction could be between one-quarter and one-third.

Pressure testing can be accomplished using a "blower door", a calibrated fan which can be mounted in a door (for example), and the airflow/pressure characteristics of the building determined from airflow and pressure data based on the fan's calibration curve.

8-6. System Characteristic Technique

8-6.1. The Technique. Sections 8-4 and 8-5 detail information to calculate ventilation when inlet geometry and the static pressure difference are known. However, the static pressure difference will not be known unless pressure/airflow characteristics of the fans used for ventilation are known. When that data is available, interactions between fans and inlets may be quantified. The method to relate fan and inlet characteristics is known as the "system characteristic technique".

Propeller fans are usually chosen to ventilate agricultural buildings. This type of fan provides the most efficient air moving capability at low static pressure differences. Data are generally available from fan manufacturers to characterize their fans. The general shape of a propeller fan curve is in Figure 8-10. The rate of air delivery is relatively constant when the pressure difference is small. The cutoff point depends on the fan, but is usually at least 100 Pa for fans typically used to ventilate barns and greenhouses.

Data for airflow through inlets has the general form shown in Figure 8-11. When the pressure difference is zero, airflow is zero. Airflow increases according to one of the slotted inlet equations already presented, or can be estimated for other inlets using the Bernoulli equation and a suitable coefficient of discharge.

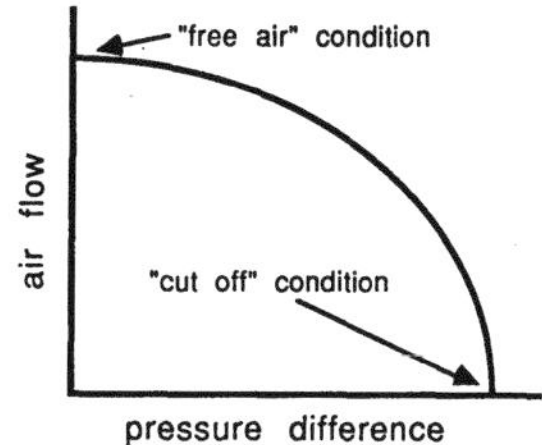

Figure 8-10. Representative curve showing how airflow provided by a (propeller) ventilating fan varies as a function of pressure difference across the fan.

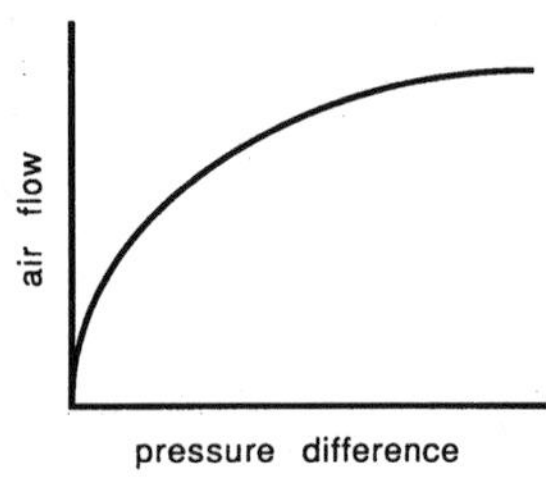

Figure 8-11. Representative curve showing how airflow rate through an inlet varies as a function of the pressure difference imposed across the inlet.

When fans draw air from an airspace and inlets provide fresh air, the airflow/pressure difference curves are as shown in Figure 8-12. When the inlet curve represents all inlets for the airspace and the fan curve represents all fans, the intersection of the two curves determines the operating (or balance) point of the inlet/fan system. The intersection must be the operating point, for it is the only point where the airflow and pressure difference are the same for both the fan and inlets and the two must be equal to satisfy airflow and air pressure continuity.

The system characteristic technique requires fan data be known for a candidate group of fans to be used for ventilation. Several possible inlet widths and the inlet type must be known. Application of the technique is best demonstrated using an example. Example 8-7 applies the design technique to a simple problem. It is a long example, but illustrates how a slotted inlet system might be designed.

Example 8-7

<u>Problem</u>: A barn is designed to house 250 dairy cows. Ventilation is to be an exhaust system, with inlets designed as shown as case (b) in Figure 8-8. A sketch of the barn is shown in Figure 8-13.

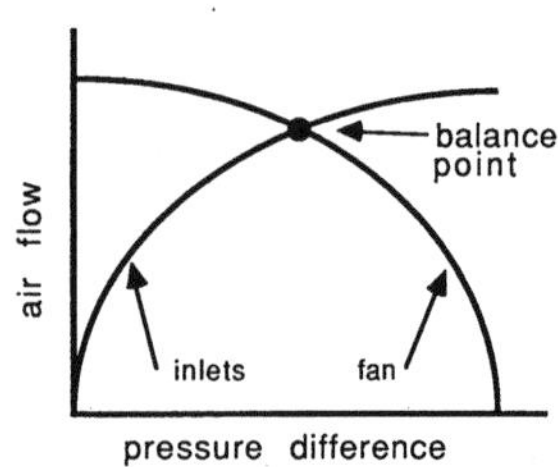

Figure 8-12. Representative curves of fans and inlets; the intersection point is where the combination of fans and inlets will operate in a steady flow state.

A compilation of fan data has been obtained:

Fan Model	Airflow Rate at the Listed Static Pressure Difference			
	0 Pa	10 Pa	20 Pa	30 Pa
A	1.28 m^3/s	1.22 m^3/s	1.13 m^3/s	0.99 m^3/s
B	2.19	2.08	1.93	1.80
C	2.72	2.58	2.39	1.93
D	3.04	2.89	2.68	2.41
E	4.01	3.81	3.53	2.96
F	4.52	4.29	3.98	3.68
G	5.19	4.93	4.57	4.22

Previous analysis has determined the required maximum ventilation rate to be 0.15 m^3/s per cow and the minimum rate to be 0.018 m^3/s per cow. Special care will be taken during construction to ensure air infiltration will be suppressed. As an assumption, the barn will be twice as tight as the "very tight" construction represented by Equation 8-31.

(a) Develop a graph to show the expected operation of the fans and slotted inlet system, at inlet widths of 0, 10, 20, 30, 40, and 50 mm. Use the pressure difference range of 0 to 30 Pa.

(b) Interpret any problems you may notice in the expected operation of the ventilation system.

<u>Solution:</u> As a first step the inlets will be analyzed, including infiltration (the unwanted inlets).

For slotted inlets with hinged baffles on the ceiling and airflow down the walls, Equation 8-26 applies,

$$\dot{V} = 0.0012W^{0.98}\Delta P^{0.49},$$

per meter of inlet. From the plan view of the barn in Figure 8-13, 170 m of the wall will be available on one side of the barn and 160 m on the other. Inlets will not be placed on end walls. However, the full 160 m on the wall with the milking center is not truly available. There are four fan banks on the wall and inlets directly over the fans would lead to some short-circuiting of fresh air.

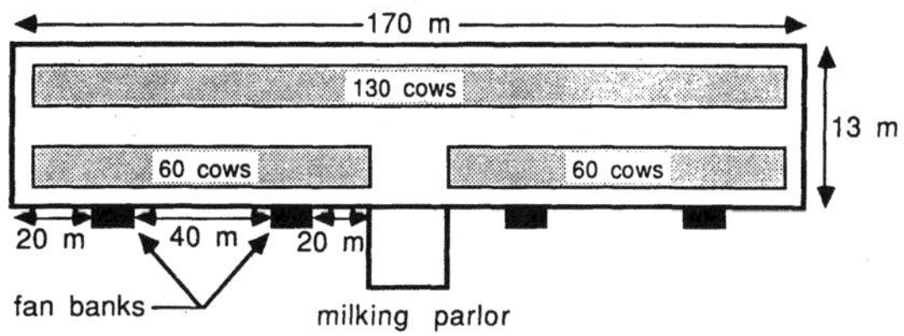

Figure 8-13. Plan view of dairy barn considered in Example 8-7.

Assume a 5 m section over each fan bank will have no inlets, thus, only 140 m of inlet can be on that wall, for a total of 310 m on both walls.

Total airflow through the slotted inlet system will be 310 times the flow per meter or

$$\dot{V}_{\text{slotted inlets}} = 0.372W^{0.98}\Delta P^{0.49}.$$

Infiltration will be at a rate half that indicated by Equation 8-31 or

$$\dot{V}/\text{cow} = 0.003\ \Delta P^{0.67}.$$

There will be space for 250 cows in the barn, so total infiltration will be

$$\dot{V}_{\text{infiltration}} = 0.75\ \Delta P^{0.67}.$$

When airflow through the slotted inlets and airflow from infiltration are added, the following describes all the inlets for the barn:

$$\dot{V} = 0.372W^{0.98}\Delta P^{0.49} + 0.75\Delta P^{0.67}.$$

To develop the system characteristic graph, total airflow as a function of static pressure difference must be calculated in the range from 0 to 30 Pa and for inlet widths of 0, 10, 20, 30, 40, and 50 mm. The equation above for total airflow yields data in the following table.

Static Pressure Difference	Airflow Rate, m^3/s, for Inlet Width, mm, as Shown					
	0	10	20	30	40	50
0 Pa	0.00	0.00	0.00	0.00	0.00	0.00
1	0.75	4.30	7.76	11.18	14.57	17.95
2.5	1.39	6.95	12.36	17.72	23.04	28.33
5	2.20	10.02	17.62	25.15	32.62	40.05
7.5	2.89	12.43	21.70	30.88	39.99	49.06
10	3.51	14.49	25.16	35.73	46.22	
15	4.60	17.99	31.02	43.90		
20	5.58	21.00	35.99	50.83		
25	6.48	23.68	40.41			
30	7.32	26.13	44.42			

Next accumulate airflow data for the fans. The maximum ventilation rate for the barn will be

$$\dot{V}_{\text{max}} = (0.15\ m^3/\text{s-cow})(250\ \text{cows}) = 37.5\ m^3/\text{s}.$$

The minimum rate will be

$$\dot{V}_{\text{min}} = (0.018\ m^3/\text{s-cow})(250\ \text{cows}) = 4.5\ m^3/\text{s}.$$

Table 6-1 contains data to be used as a guide for selecting stages between maximum and minimum ventilation rates. This barn is large, so for now assume a five-stage ventilation system will be used. In a real design, four, five, and six stages could be examined in detail and the best chosen.

With a five-stage system, 10%, 14%, 19%, 35%, and 100% are the idealized stages, but with the data for this example, the minimum stage is 12% of the maximum, not 10%. The staging schedule must be modified and one candidate is: 12%, 17%, 22%, 40%, and 100%.

Fans will be selected to provide this staging, approximately. The five stages are:

Stage	m^3/s
1	4.50
2	6.38
3	8.25
4	15.00
5	37.50

The fan data can be used to obtain a candidate set of fans to install in each bank. The static pressure difference during operation is not yet known, but likely will not be 0 Pa. Neither is it likely to be 30 Pa. For now assume the difference will be approximately 10 Pa and select fans accordingly. One additional criterion in selecting fans will be to make the four fan banks as similar as possible so air distribution will be as uniform as possible.

The following is presented as only one of many possible fan schedules. Symmetry in fan distribution among fan banks is a goal, but stage 4 violates that symmetry somewhat to achieve sufficient ventilation for the stage. A wider selection of fans would make the choice perhaps easier, but the selection which is shown was chosen to illustrate some of the difficulties in finding fan stages which closely match desired ventilation stages.

	Total Air Fow	Fan Bank Number			
Stage		1	2	3	4
1	4.88 m^3/s	A	A	A	A
2	6.60	B	A	A	B
3	8.32	B	B	B	B
4	15.28	A,B	A,B,B	A,B	A,B
5	38.68	A,B,B,F	A,B,B,F	A,B,B,F	A,B,B,F

In the table, letters refer to the fan or fans to operate in each bank in each stage. Each fan bank contains 4 fans, 1 of model A, 2 of model B, and 1 of model F. Control would be implemented so only the fans shown for each bank would operate at the appropriate stages. Control could be by thermostats, or by a computer.

After the fans have been selected, total airflow at each stage and at 0, 10, 20, and 30 Pa can be determined from fan data.

	Airflow Rate, m^3/s, at Static Pressure Differences of			
Stage	0 Pa	10 Pa	20 Pa	30 Pa
1	5.12	4.88	4.52	3.96
2	6.94	6.60	6.12	5.58
3	8.76	8.32	7.72	7.20
4	16.07	15.28	14.17	12.96
5	40.72	38.68	35.88	33.08

As an example calculation, for stage 2 at 20 Pa there will be 2 of model A and 2 of model B operating. At 20 Pa, fan model A delivers 1.13 m^3/s and fan model B delivers 1.93 m^3/s. Two of each provide a total of 2(1.13 m^3/s + 1.93 m^3/s) = 6.12 m^3/s.

The next step is to graph inlet airflow data and fan data on the same axes. Figure 8-14 shows the results. Recall that any intersection of a fan stage curve and an inlet airflow curve is a possible point of operation of the total system. Some would obviously be better than others.

A system characteristic graph such as in Figure 8-14 would not be used for control. Rather, it provides insight into operation of the ventilation system and can be used by a designer to avoid designing a problem ventilation system – one which never seems to work quite right.

One potential problem area when ventilation systems are designed for cold climates is the minimum ventilation stage. The required minimum ventilation rate for this example is 4.5 m^3/s. On Figure 8-14, stage 1 of the fans provides approximately 4.5 m^3/s when the inlet width is 0 mm and approximately 5 m^3/s when the inlet width is 10 mm. However, the pressure difference when the inlets are open 10 mm will be less than 2 Pa, a pressure with no capability to resist wind effects. A pressure difference of at least 10 Pa is preferred.

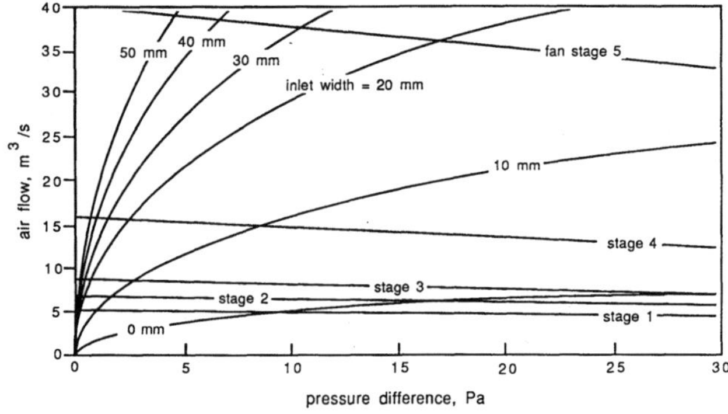

Figure 8-14. System characteristic graph for the dairy barn considered in Example 8-7.

The problem arises due to infiltration effects – infiltration provides all the ventilation needed for the minimum ventilation rate, but there is no guarantee fresh air distribution will be uniform. Likely it will not be. The problem of having low ventilation stages dominated by infiltration is common and indicates the importance of designing buildings to limit infiltration and specifying construction practices to seal all potential points of air infiltration. Also, research is needed to characterize current building designs and construction with regard to their infiltration characteristics.

If the design of the barn in this example were to be continued, an inlet width intermediate between 0 and 10 mm should be considered for adequacy. Practical construction considerations limit the minimum inlet width to no less than 5 mm (it is difficult to build a long inlet which can maintain only a slight opening). If 5 mm were adequate, extra precautions to limit infiltration would not be needed. However, it is likely the 5 mm inlet curve would intersect stage 1 ventilation at a pressure difference significantly less than 10 Pa. It would be a useful exercise to develop the data for an inlet 5 mm wide and add it to Figure 8-14.

The maximum required ventilation rate is 37.5 m^3/s, which can be obtained using an inlet width of only 30 mm. The operating static pressure difference at 30 mm will be approximately 10 Pa. If a greater pressure difference is desired to assure more vigorous air mixing and greater animal comfort during hot weather, a slightly narrower inlet and a different selection of fans could be found to provide such conditions.

A useful design factor which arises from knowing the maximum inlet width is knowledge of the minimum width of restriction upstream of the inlet. In this example, if 30 mm is chosen as the maximum inlet width and the factor of 2.5 times the maximum inlet width is used (as previously discussed), the building design should be specified so no section upstream of the slotted inlet would have a width narrower than 75 mm.

8-6.2. Program SYSCHAR1. The system characteristic design technique is used as an analysis tool to design slotted inlet ventilation systems (or it can be used for other inlet styles or fan and duct ventilation systems). However, it is a long and detailed procedure as Example 8-7 has shown. A computer program to complete the details of design would be useful to permit a designer to explore various options of fan staging and inlet design. Program SYSCHAR1 is provided for such exploration. It is not intended as a general purpose design tool for design engineers, but is adequate for limited applications and is useful as a means to learn to apply the technique and gain understanding of the dynamics of a slotted inlet, negative pressure, ventilation system.

The program is written to permit any of the three types of slotted inlets discussed above. Up to six inlet widths are permitted on one graph, and up to six fan stages. Default fan data are provided but can be changed if specific manufacturer's data is to be used in a real design. The static pressure range of 0 to 25 Pa is used to conform to fan data ranges available from many fan manufacturers.

Infiltration is considered in terms of the dairy barn size equivalent of the building being designed. For example, if a poultry house is considered, the user must determine how many dairy cows would be housed in the same space. A poultry house 100 m long and 14 m wide would be wide enough for two rows of dairy cows. Each cow stall would be approximately 1.2 to 1.3 m wide (typical). If space is left for aisles, etc., one might estimate approximately 140 dairy cows could be housed in the same space.

After the dairy cow equivalent is determined, a scaling factor must be chosen to represent how airtight the building is compared to the "very tight" barn described in Section 8-5. The factor would likely be less than unity and could be less than 0.5. There are no guidelines at present to select the infiltration factor. It will depend on construction materials and techniques, the number of windows and doors, and other openings into the building. This approach is not intellectually satisfying, but is all that is available within agricultural engineering literature at present. To analyze a barn using the ASHRAE crack method would be useful perhaps as a comparison, and resulting data fitted with the form of Equation 8-31 (same exponent) to estimate a suitable infiltration scaling factor. Some evidence exists that the scaling factor should be from one-quarter to one-third for current, typical construction.

Example 8-8

Problem: Use Program SYSCHAR1 to develop a system characteristic graph for the dairy barn described in Example 8-7. Use inlet widths of 0, 5, 10, 20, 30, and 40 mm, and other data as presented in Example 8-7. Limit the choice of fans to those listed as example fans in the program.

Solution: This solution will be understood best if the program is followed on a computer as the solution is developed.

Data required by the program, in the order of entry, are:

    number of cows: 250
    infiltration factor: 0.5 (twice as tight as the standard)
    inlet type: b
    inlet length: 310 m
    number of inlet widths: 6
    inlet widths: 0, 5, 10, 20, 30, and 40 mm
    example selection of fans for stages:

stage 1: 1 model e and 1 model f for 4.48 m³/s at 12.5 Pa
stage 2: add model b (4 each), for 6.76 m³/s at 12.5 Pa
stage 3: add model d (2 each), for 9.12 m³/s at 12.5 Pa
stage 4: add model g (2 each), for 14.66 m³/s at 12.5 Pa
stage 5: add model h (4 each) and model i (2 each), for 38.00 m³/s at 12.5 Pa

The fan selections do not exactly match the required ventilation rates at each stage, but are close. There are 4 fan banks (numbered 1 to 4 from left to right). One way to arrange the fans would be to install

Fan Bank #1	Fan Bank #2	Fan Bank #3	Fan Bank #4
1-model b	1-model b	1-model b	1-model b
1-model e	1-model d	1-model d	1-model f
1-model g	1-model i	1-model i	1-model g
1-model h	1-model h	1-model h	1-model h

At fan stage 1, fan models e and f operate at fan banks 1 and 4 (they operate continuously).

At fan stage 2, model b fans are activated, one at each fan bank.
At fan stage 3, add model d fans at fan banks 2 and 3.
At fan stage 4, add fan models g at fan banks 1 and 4,
At fan stage 5, add 4 fan models h and 2 fan models i.

This is a candidate fan staging sequence, chosen for symmetry and to approximate the need for ventilation at each stage. Many other candidate sequences are possible using the default fan data.

The results of SYSCHAR1 are in Figure 8-15, which is the video screen obtained using the PrtSc function (with GRAPHICS called before SYSCHAR1 to implement graphics printing).

Observations from Figure 8-15 parallel those made in Example 8-7, for the examples are essentially the same. Even if the inlet is only 5 mm wide, at the lowest fan stage the static pressure difference from inside the barn to out will be less than 5 Pa and wind effects could be serious. During periods of extreme cold and wind, closing the inlets entirely might be recommended in spite of the resulting potential nonuniformity of ventilation. The maximum ventilation rate can be achieved when the inlet is 30 mm wide. Even stage 4 ventilation will be adequate with an inlet width of 10 mm. Wider inlets will not increase airflow significantly, and will reduce inlet air velocity and air mixing within the barn.

8-7 Air Outlets

When air enters a space through an opening because of a static pressure difference, airflow is in the form of a jet. Momentum and vorticity within the jet cause the behavior as described in Section 8-3. However when air flows from an airspace toward an outlet, the flow is termed "potential flow". With potential flow, air movement to an outlet is uniformly from all directions.

If one can visualize an imaginary hemisphere surrounding an outlet, air enters the hemisphere uniformly over its surface and the area of the surface is much larger than the area of the outlet unless the hemisphere is small (which means the air is close to the outlet).

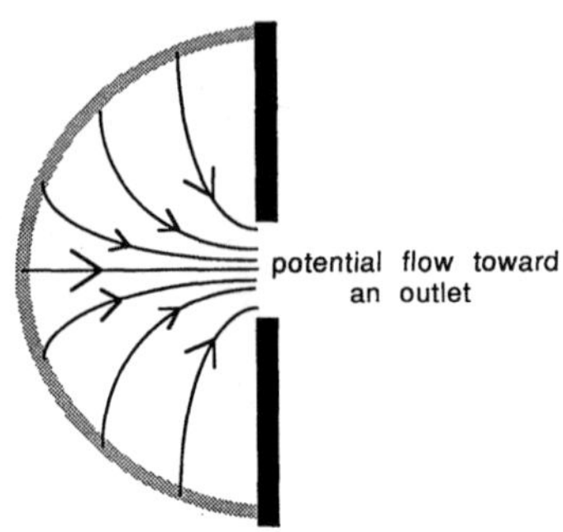

There are two important consequences of potential flow toward an outlet. One is that air velocity will be small except very near the outlet. For example, if an outlet is a circle and the imaginary hemisphere is centered on the outlet and has the same diameter as the outlet, air velocity through the surface of the imaginary hemisphere will be only 50% that of the average air velocity through the outlet. Velocity is inversely proportional to the cross section area of flow. If the hemisphere is twice the diameter of the outlet, air velocity at two outlet

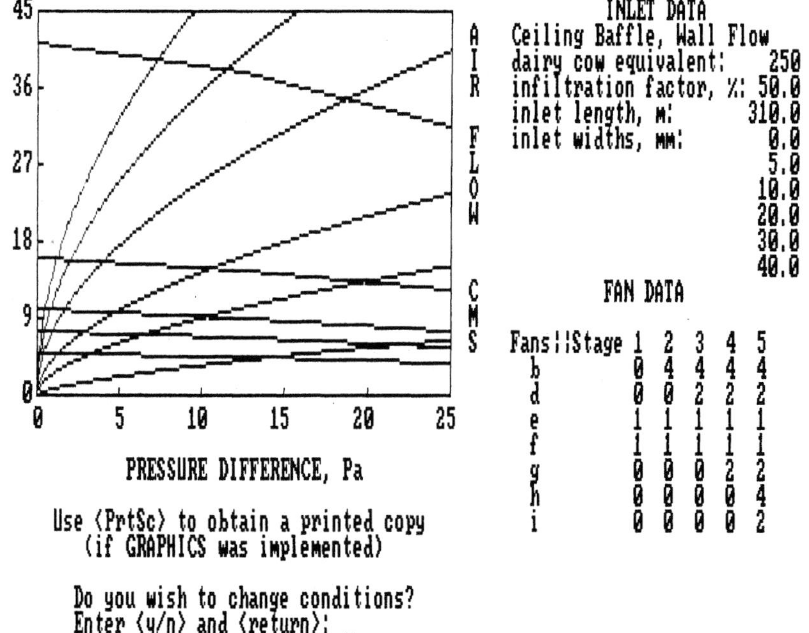

Figure 8-15. Results screen of SYCHAR1 for Example 8-8.

diameters away from the outlet will be only 12% that at the outlet. The decrease of air velocity as a function of distance from an outlet is rapid.

The second consequence of potential flow is that outlets have little effect on airflow patterns within a ventilated space. Airflow patterns are dictated by air flowing through the inlets. Little short circuiting of air from inlets to outlets will be seen and air inlets can be placed close to fans. For example, with slotted inlet ventilation systems, inlets can be placed to within 2 m of fan banks with no discernible effect on airflow patterns. The fresh air does not suddenly curve toward the fans; the fans have no ability to affect the jet's momentum.

SYMBOLS

A	area, m^2
C	coefficient of contraction, discharge or velocity, dimensionless
D	baffle width, center-ceiling slotted inlet, m
D_e	equivalent diameter, m
g	gravitational constant, m/s^2
g_c	units conversion factor, see Equation 8-6
Gr	Grashof number
h	convective heat transfer coefficient, W/m^2K
K	coefficient, see Equation 8-13
m	mass flow rate, kg/s
Nu	Nusselt number
P	pressure, Pa
q''	heat flux, W/m^2
r	radial coordinate in circular free jet, m
Re	Reynolds number
t	temperature, C
T	ceiling slot width in center-ceiling slotted inlet, m
v	velocity, m/s
$\dot{V}$	volumetric flow rate, m^3/s
W	width of slotted inlet, m
x	distance, m
Y	distance from surface into wall jet profile, m
z	elevation above a datum, m
β	coefficient of thermal expansion, K^{-1}
θ	dimensionless temperature
μ	absolute viscosity, kg/ms
ρ	density, kg/m^3

EXERCISES

1. A slotted inlet ventilation system is desired for a barn having negative pressure ventilation. Fresh air will be discharged along a (smooth) ceiling.

The coefficient of discharge at the inlet is expected to be approximately 0.7 and the desired ventilation rate is 0.10 m³/s per m of inlet length. The barn in which the inlet system is to be installed is at an elevation of 2000 m. At outside air conditions of 5 C and 50% relative humidity, calculate and graph, as a function of inlet width

(a) the downstream distance at which the maximum velocity is 1.0 m/s,

(b) the width of the jet when v_{max} = 1.0 m/s, and

(c) the expected static pressure difference from inside the barn to out.

Use the inlet width range of 20 to 50 mm as the x-axis of the graph(s).

2. Air flows within a perforated polyethylene duct. Just upstream of the first exit hole the static pressure is 55 Pa and air velocity is 6 m/s. The duct diameter is 0.4 m and the hole diameter is 50 mm. Standard air conditions may be assumed.

(a) Calculate the airflow (m³/s) expected to exit from the first hole.

(b) Estimate the angle at which the jet exits the air duct.

(c) Estimate the distance from the duct where the maximum velocity within the jet drops to 0.5 m/s.

(d) Calculate the air velocity and static pressure immediately downstream of the hole (within the duct).

3. A 100 mm diameter round hole in a barn wall causes a draft during cold weather. The ventilation system for the barn uses negative pressure. For a pressure difference from inside the barn to outside of 12 Pa, determine how far from the hole an animal must be before the centerline velocity of the resulting jet has decayed to 0.5 m/s.

4. For the ventilation system described in Example 8-3, develop a graph to show decay of the jet's centerline velocity and growth of the half width of the jet as functions of the distance downstream from the inlet, and the inlet width. The x-axis should be the downstream distance, between 0 and 2 m, and three inlet widths should be considered: 5, 10, and 20 mm. Three curves, corresponding to the three inlet widths, should be drawn for centerline velocity, and one for half width ($Y_{0.5}$).

5. For the jets produced from the polyethylene ventilating tube described in Example 8-5, graph the centerline velocity and entrainment ratio as functions of downstream distance for a distance between 0 and 4 m.

6. Use the conditions described in Example 8-6 to graph the temperature profile of the resulting wall jet if the inlet is (5 mm or 20 mm) wide. Compare your results to the results of the example.

7. For the ventilating system and wall jet described in Example 8-6, calculate the dimensionless maximum velocity as a function of the downstream distance. Develop a graph of dimensionless velocity and dimensionless temperature as functions of distance from the wall. Calculate dimensionless velocity using Equation 8-10 and dimensionless temperature using Equation 8-24.

8. Consider the dairy barn of Example 8-7. The inlets have been measured to be 11 mm wide. Stage 4 ventilation is operating and the pressure difference is measured to be 10 Pa. From these data, develop the equation which describes actual infiltration in the barn (using the form of Equations 8-30 and 8-31).

9. A room is ventilated using one of fan model F in Example 8-7, with the fan installed to force air into the room as a positive pressure ventilation system. The air exit is a single opening with an area of 1.5 m^2 and a coefficient of discharge of 0.6. Determine the air pressure which will exist in the room. Assume air density is 1.18 kg/m^3.

REFERENCES

Albright, L. D. 1974. The low-speed non-isothermal wall jet. Journal of Agricultural Engineering Research 19:25-34.

Albright, L. D. 1976. Airflow through hinged baffle slotted inlets. Transactions of the ASAE 19(4):728-732, 735. American Society of Agricultural Engineers, St. Joseph, MI.

Albright, L. D. 1978. Airflow through baffled, center-ceiling slotted inlets. Transactions of the ASAE 21(5):944-947,952. American Society of Agricultural Engineers, St. Joseph, MI.

Albright, L.D. 1979. Designing slotted inlet ventilation by the system characteristic technique. Transactions of the ASAE 22(1):158-161. American Society of Agricultural Engineers, St. Joseph, MI.

ASHRAE. 1989. Handbook of Fundamentals. American Society of Heating, Refrigerating, and Air Conditioning Engineers, Atlanta, GA.

Millier, W. F. 1950. The ventilation of dairy stables with electric fans. Unpublished Ph. D. thesis. Cornell University Libraries, Ithaca, NY.

Walker, J.N. 1977. Review of the theoretical relationships of isothermal ventilating air jets. Transactions of the ASAE 20(3): 517-522. American Society of Agricultural Engineers, St. Joseph, MI.

Walton, H.V. and D.C. Sprague. 1951. Airflow through inlets used in animal shelter ventilation. Agricultural Engineering 32(4):203-205. American Society of Agricultural Engineers, St. Joseph, MI.

Wilson, J.D., M.L. Esmay and S. Persson. 1970. Wall-jet velocity and temperature profiles resulting from a ventilation inlet. Transactions of the ASAE 13(1):77-81. American Society of Agricultural Engineers, St. Joseph, MI.

CHAPTER 9
AIR DISTRIBUTION

9-1. Introduction

Ventilation systems must satisfy several important criteria if they are to be successful. The proper amount of fresh air must be mixed with air already in the building. A function of the fans and their controls is to bring in the correct amount of air. Design techniques to determine the proper amount of ventilation and determine staging have been discussed.

Fresh air should be distributed uniformly throughout a ventilated agricultural building. Slotted inlets and air jets have been found useful for this need, and have been discussed. Air mixing has been found to be enhanced if slotted inlets are on both long walls of a ventilated barn rather than only on the wall opposite the fans.

Another function of ventilation is to create uniform mixing within the ventilated airspace, with the goal of always providing reasonably fresh air at each animal's nose, or fresh air uniformly distributed within the canopy of a greenhouse crop. Whether this goal is reached depends on how well air is directed after leaving the inlets, and whether factors such as thermal buoyancy disrupt distribution.

People who work with ventilation systems soon discover many others in production agriculture who consider themselves to be well informed about air mixing patterns and ventilation in general. Air cannot be seen; it is easy to develop hypotheses about what happens to it within a ventilated space and difficult to refute hypotheses if they are incorrect. Many fanciful ideas have been promulgated and believed.

It is very important to be aware of factors important in determining air mixing within a ventilated airspace.

The most important factor which determines air mixing patterns in a given airspace is placement of inlets. As has been shown, air forms a jet after leaving an inlet, and has momentum. Momentum is removed from a jet by friction with the adjoining wall (if a wall jet). Momentum exchange with surrounding air is thought to be conservative – momentum diffuses into the still air by entrainment but is not lost. The jet simply enlarges by entrainment and slows as the momentum is distributed over a greater mass of air.

If slotted inlets are used for ventilation, inlet placement and the initial direction of the entering wall jet dictate the form of air circulation within the airspace. The action of jets is to create large zones with rotational circulation within the space. Three examples of circulations which have been observed are shown in Figure 9-1. Each system's ability to deliver fresh air to any specific place within an airspace can be seen to differ greatly from the others. However, each system creates a pair of counter-rotating vortices within the airspace – if inlets are symmetrically placed and controlled – and if there are not large obstructions to airflow. If inlets are on only one side of the building and the space is not too wide, one circulation zone is established.

Momentum at an inlet is sufficient to keep the jet moving in a single direction unless:

(a) the jet strikes an obstruction large enough to divert its path,

(b) thermal effects (buoyancy) divert its path,

(c) the jet slows sufficiently that eddies in the room air and thermal effects cause the jet to break up, and lose identity, or

(d) inlet air velocity is sufficiently slow that there is too little momentum to establish stable air mixing patterns within the airspace.

A side effect of momentum has been demonstrated when slotted inlet ventilation is used in a barn having a slotted floor over a manure pit for waste collection. Momentum can carry some air through the floor, into the manure pit,

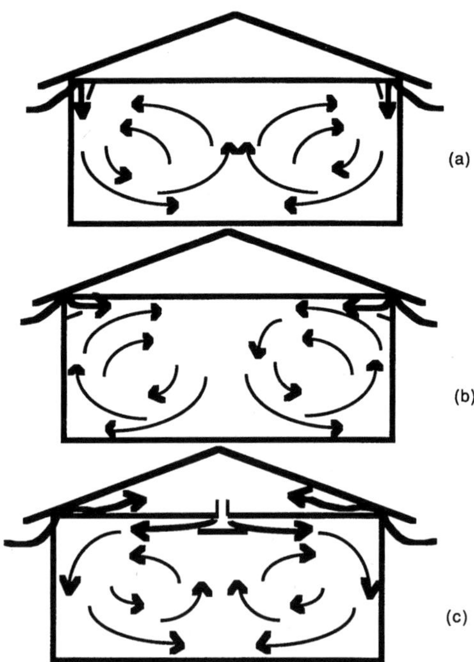

Figure 9-1. Air circulation patterns induced by common slotted inlet systems: (a) with baffles hinged on the ceiling and airflow down the wall; (b) with baffles hinged on the outside wall and airflow across the ceiling; and (c) with center-ceiling slotted inlets.

and back up into the ventilated space. Noxious and toxic gases may be carried in this manner to the animals, and cause loss of performance, morbidity, and perhaps even death. A separate exhaust ventilation system for such manure pits usually is required.

Air infiltration is important in determining air distribution and mixing. In one way the effect is direct. If one section of a building contributes considerably more infiltration than other sections, the goal of having fresh air evenly distributed around a building is defeated. Pressure within a single airspace is everywhere uniform, thus the amount of fresh air brought to each section of a building depends on the relative amount of inlet area within that section. Infiltration openings act as additional inlets. For example, one door left open can completely defeat the ability of a negative pressure ventilation system to provide adequate air distribution and mixing.

Infiltration also indirectly affects air distribution uniformity, and the effect is important. If, as was shown in the discussion of system characteristic graphs in Chapter 8, infiltration provides much or most of the fresh air required at the lowest ventilation stage (or stages), and at a relatively low static pressure difference, the building is not sufficiently airtight to be buffered against wind effects. When static pressure is low, wind can draw air out of inlets, and cause an excess of air to enter other inlets – on the upwind side of the building. The effect is great nonuniformity of air distribution, and an inability to maintain desired air temperatures during the coldest weather.

A second effect of the low static pressure difference is to prevent air entering through planned inlets from achieving sufficient velocity to create air mixing. The patterns shown in Figure 9-1 are never established.

Wind pressure coefficients can be used to estimate how seriously wind will affect airflow through inlets. Wind pressure coefficients are the same as used to estimate wind loading when designing the structure of a building. A wind pressure coefficient is defined as the ratio between the over or under pressure created at a point on a building's surface, divided by the stagnation pressure of the free wind. If v_w is wind velocity, ρ is air density, P_g is the gage pressure on the building surface, and C_p is the pressure coefficient,

$$C_p = P_g / (\rho v_w^2 / 2) \qquad (9\text{-}1)$$

Table 9-1. Example wind pressure coefficients around buildings

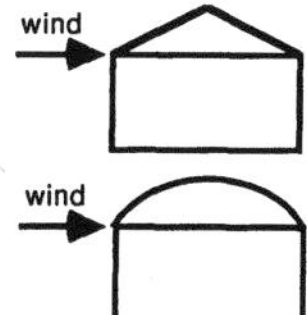

gable roof building, windward wall: +0.7
gable roof building, leeward wall: -0.5

arched roof building, windward wall: +0.8
arched roof building, leeward wall: -0.5

Wind pressure coefficients will be described in more detail when natural ventilation is discussed, but for now, some typical values are as listed in Table 9-1. The shapes are frequently used for both animal buildings and greenhouses.

Example 9-1

<u>Problem:</u> Wind is in a direction perpendicular to the long side of a gable-roofed dairy barn. An exhaust ventilation system has been designed to maintain a partial vacuum inside the building of 10 Pa. Determine the wind speed at which air will be drawn out inlets on the leeward side of the building.

<u>Solution:</u> From Table 9-1, the wind pressure coefficient for the leeward wall of a gable roof building is -0.5. Equation 9-1 can be rearranged to determine wind speed:

$$v_w = (2P_g/\rho C_p)^{1/2}$$

Air will begin to flow out the inlets when gage pressure outside the building is lower than inside, or $P_g \sim -10$ Pa. Air density can be assumed to be based on standard air conditions, $\rho \sim 1.2$ kg/m^3.

The wind speed at which $P_g \sim -10$ Pa is

$$v_w = [(2)(-10 \text{ Pa}) / (1.2 \text{ kg/m}^3)(-0.5)]^{1/2}$$
$$= 5.8 \text{ m/s} = 21 \text{ km/hr.}$$

This wind speed is commonly reached during winter in most regions of the world, indicating again the importance of maintaining static pressure differences of at least 10 Pa from inside to outside a barn ventilated using an exhaust, or negative pressure, system.

Exhaust fan placement is relatively less important in determining air mixing than is inlet placement. The preferred location for exhaust fans is on the side of a building which is leeward of winter winds. Otherwise, wind effects can seriously reduce the ability of a fan to provide sufficient air movement. The positive pressure caused by wind on the windward side can even, in extreme cases, totally block airflow through a fan facing into it.

Fans are usually grouped into banks to simplify electrical wiring. The distance between banks is usually limited to approximately 40 m, or no more than twice the width of the building.

As has been discussed in Chapter 8, short circuiting of air from inlets to fans is not likely because of the limited region of influence of air outlets (and fans on their upstream side). However, if the distance between inlets and a fan is less than approximately 2 m, some short circuiting may be observed.

Air mixing can be enhanced by using circulation fans, or horizontal airflow as the technique has been termed for greenhouse applications. Horizontal airflow can be achieved by installing circulation fans in two rows, facing in opposite directions with fans approximately 12 m apart, along the two long walls of the airspace. The fans (approximately 0.4 to 0.6 m diameter) direct air in a race-track fashion around the building and mix air continuously to help alleviate thermal stratification during times of limited ventilation when there is little air mixing induced by airflow through inlets.

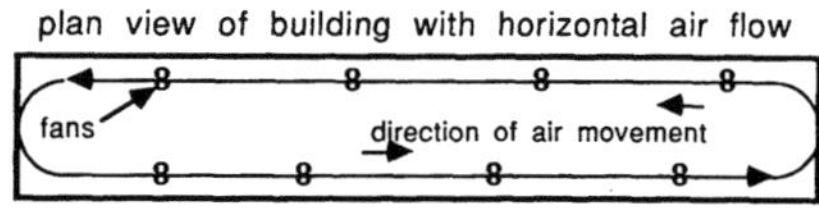

Analysis of air mixing patterns is still primarily in a research phase, with little yet available in practical terms except general guidelines related to slotted inlet ventilation systems. However, some research results can be used to avoid certain problems related to inlet jet instability and poor airflow patterns. The remainder of this chapter will discuss some of those results and applications. Individuals working with ventilation, however, should keep current with research literature as more refined design techniques are developed.

9-2. Inlet Air Jet Stability, the Effect of Thermal Buoyancy

Thermal buoyancy is a major effect which causes inlet air jets to be unstable. The problem of instability generally arises during cold weather when an inlet jet is directed across the ceiling of a ventilated building, with inlet types as shown in Figures 8-8(c) and 8-8(d). Figure 9-2 shows typical airflow patterns for stable and unstable jet flow.

During warm weather, or when the ventilation rate is sufficiently great to insure a robust air jet at the inlet, the jet flow is stable and remains as shown in Figure 9-2(a). During cold weather and times of limited ventilation, inlet air velocity will typically be low, which suppresses the Coanda effect to some extent. The temperature difference between the jet and the air inside the building is large during cold weather and a low inlet velocity does not promote rapid temperature change within the jet. Thermal buoyancy forces can be sufficiently large that the jet detaches from the ceiling, falls to the floor, and may cause significant local chilling for the animals at that point, as shown in Figure 9-2(b) and 9-2(c). If buoyancy forces are great enough, the jet falls almost immediately and airflow patterns are stable, although undesirable.

The criterion of a threshold value of an Archimedes number has been proposed (Randall, 1979) as a means to assess whether jet detachment or instability will occur. The Archimedes number, Ar, can be viewed as the ratio between buoyancy and dynamic pressures and, for a horizontal jet in an airspace, is calculated as

$$Ar = gD\,(T_w - T_o)\,/\,Tu^2, \qquad (9\text{-}2)$$

where g is the gravitational constant, D is a characteristic dimension of the airspace, T_w and T_o represent, in some unspecified form, indoor and outdoor temperatures, T is a form of average temperature, and u is a characteristic air velocity.

The Archimedes number has been shown useful to predict airflow pattern stability in some instances, however, research has not yet proven the Archimedes number as a useful criterion for airflow pattern stability in animal housing ventilated by slotted inlets.

9-3. Inlet Air Jet Stability, the Effect of Obstructions

Obstructions to wall jet flow are common in agricultural buildings, especially if the wall jet flows across a ceiling. Rows of lights are common obstacles,

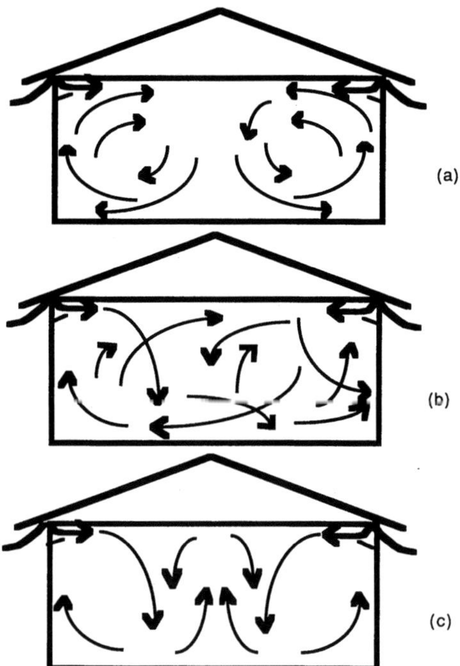

Figure 9-2. Airflow patterns in a ventilated space, showing: (a) stable attachment of a horizontal jet to the ceiling; (b) unstable jet flow when thermal buoyancy effects are strong but do not dominate the airflow; and (c) stable detachment and flow toward the floor when thermal buoyancy is strong.

although framing members and supports for poultry cages, feed distribution systems, and water pipes, as examples, may be found on ceilings and walls. If an obstacle is large relative to a wall jet, the jet will detach at an angle of 90° from the wall (ceiling). If an obstacle is intermediate sized, detachment will be at an angle less than 90°, and reattachment may occur. If the obstacle is small relative to the jet, there will be no detachment, although the velocity profile will be disrupted for some distance downstream from the obstacle.

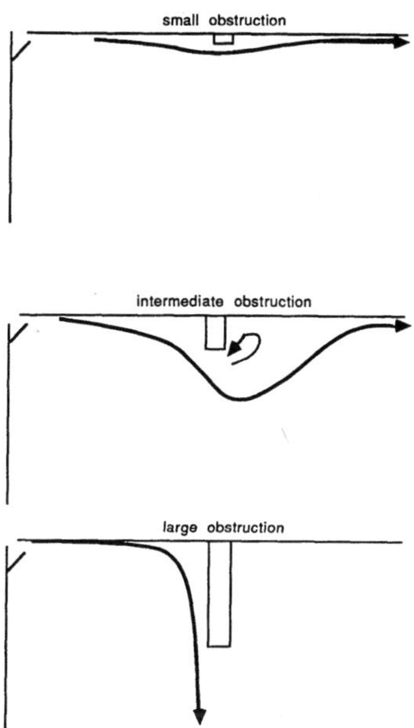

One effect of an intermediate sized obstacle when there is reattachment is to slow the jet and increase the rate of air entrainment more rapidly than would otherwise be expected. If the obstacle is approximately as large as the jet's velocity profile, temporary jet detachment will occur (at an angle greater than 90° from the direction of the original path), with rapid slowing of the jet in the vicinity of the obstacle. The jet will later attach itself to the wall (or ceiling) downstream from the obstacle. The Coanda effect is a powerful mechanism causing reattachment.

A criterion of whether an obstacle is sufficiently large to cause permanent detachment is the ratio of the height of the obstacle relative to the width of the jet where the velocity is one-half the velocity maximum within the profile, or $Y_{0.5}$. If the obstruction height is greater than $2Y_{0.5}$, King (1970) found detachment to be at 90° from the wall, and permanent. The growth of $Y_{0.5}$ has been described in Chapter 8, therefore, this criterion provides some guidance for the design engineer to avoid creating a jet detachment problem when slotted inlet ventilation systems are used.

Example 9-2

<u>Problem:</u> A slotted inlet system of ventilation is to be used in a swine barn, where airflow will be directed to flow across the ceiling as shown in Figure 8-8(c). A row of light fixtures is to be attached to the ceiling, parallel with the long axis of the building, and perpendicular to the direction of the air jet resulting from the slotted inlet ventilation. The fixtures will extend down from the ceiling a distance of 0.2 m and be solid barriers to airflow. What is the minimum distance from the inlet the lights can be and not cause permanent detachment of the ventilating jet?

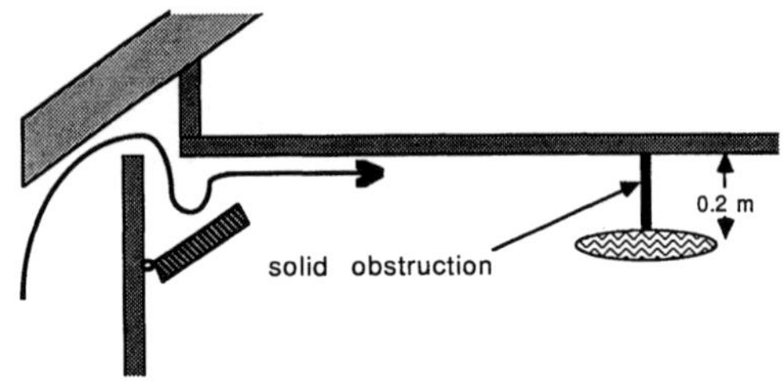

Hinged baffle slotted inlet with air flow across the ceiling

<u>Solution:</u> As described in Chapter 8, $Y_{0.5}$ in a two-dimensional wall jet grows at an angle of approximately 4°. The criterion for detachment is that the obstacle must be no thicker than $2Y_{0.5}$, or $Y_{0.5} = 0.1$ m in this example. If $Y_{0.5}$ is assumed to be zero at the inlet (the conservative assumption), then x, the downstream distance at which $Y_{0.5}$ equals 0.1 m is

$$x \; = \; 0.1\,\text{m} / \tan 4° = 1.43\,\text{m}.$$

If this distance is too far from the wall for light placement (based on criteria of lighting design), then either the light fixtures must be mounted so as not to extend so far from the ceiling, or a different style must be chosen to hang from the ceiling, permitting air to flow smoothly along the ceiling.

9-4. Air Mixing Pattern Stability

Some possible actions of an inlet jet in driving airflow patterns have been discussed, and idealized airflow patterns were shown in Figure 9-1. Jet momentum was proposed by Kaul (1975) to determine whether an inlet jet contains sufficient momentum to keep the circulation zones shown in Figure 9-1 stable. Barber et al. (1982) developed the momentum concept further into an inlet jet momentum number as a criterion for determining when air mixing patterns are stable and able to resist the disruptive effects of thermal buoyancy.

$$J = \dot{V}v_i / gV, \qquad\qquad (9\text{-}3)$$

where $\dot{V}$ is the total ventilation rate for the airspace, m³/s, v_i is the averaged inlet air velocity, m/s (averaged over the inlet area); g is the gravitational constant; and V is the volume of the ventilated space, m³. The criterion to determine whether the overall mixing pattern is stable (not just whether the inlet jet is stable) is: stability occurs when $J > J_{min}$ (7.5E - 04 at - 20 C outdoors). It is frequently difficult to achieve a stable value of J using an exhaust ventilation system during the coldest weather. It should be emphasized the jet momentum number criterion is a proposal; the concept has not been verified in a wide variety of applications for agricultural building ventilation.

Example 9-3

Problem: A poultry house with a slotted inlet ventilation system will have a minimum winter ventilation rate of six air changes per hour (winter minimum air exchange rates for most animal housing buildings are generally four to six per hour, based on moisture control). The building has a volume of 1000 m³. What must be the inlet air velocity during times of minimum ventilation to maintain stable air mixing patterns within the ventilated space at J_{min}= 7.5E - 4?

Solution: The inlet jet momentum number provides the criterion for stability. Air exchange is six building volumes per hour, and the building's volume is 1000 m³, thus, the ventilation rate is

$\dot{V}$ = ((6 volumes / hr) (1000 m³ / volume) / 3600 s / hr)
  = 1.67 m³ / s.

Inlet air velocity needed for a jet momentum number of 7.5E - 4 is

v_i = gJV / $\dot{V}$
  = (9.8 m / s²) (7.5E - 4) (1000 m³) / (1.67 m³ / s)
  = 4.4 m / s.

It should be noted that although this is the solution to the design question, the solution may not be simple to achieve. As was shown in Chapter 8, minimum winter ventilation rates can be dominated by infiltration, which prevents achieving a sufficient static pressure difference. A significant static pressure difference is needed to cause an inlet air velocity of 4.4 m/s. Recall this inlet velocity is averaged over the inlet area; it is not the velocity at the vena contracta. Air velocity at the vena contracta will equal 4.4 m/s divided by the coefficient of discharge of the inlet. A coefficient of 0.8 means a pressure difference of approximately 20 Pa will be needed. Such a pressure difference is likely to bring in much more air by infiltration than is needed in total. This adds one more reason why construction practices which limit air infiltration should be specified as part of a design, and recirculation may be needed for cold climates.

9-4.1. Air Recirculation to Promote Mixing Pattern Stability

We have seen how, during the coldest weather and lowest ventilation rate, slotted inlets may not provide sufficient pressure difference to resist wind effects. Ventilation during such times may also not provide sufficient momentum in the inlet jet to promote air mixing stability, or mixing pattern suitability. This situation is especially serious in cold climates where ventilation at low rates may be required for many hours during a year. Air recirculation systems integrated into fresh air distribution have been found to help alleviate this problem and promote good air mixing. Two such methods are illustrated in Figure 9-3 (Hodgkinson and Barber, 1986).

The first method shown in Figure 9-3, a polyethylene air recirculation duct suspended beneath a center-ceiling baffle, is a quick way for retrofit in a barn already having a center-ceiling inlet. The recirculated air intercepts the fresh, cold air entering through the inlet, accelerates it, and provides the momentum to maintain air mixing pattern stability. In addition, the immediate mixing prevents thermal buoyancy from causing the ventilating jet to fall from the ceiling, and also tempers the fresh air so draftiness is a less serious concern.

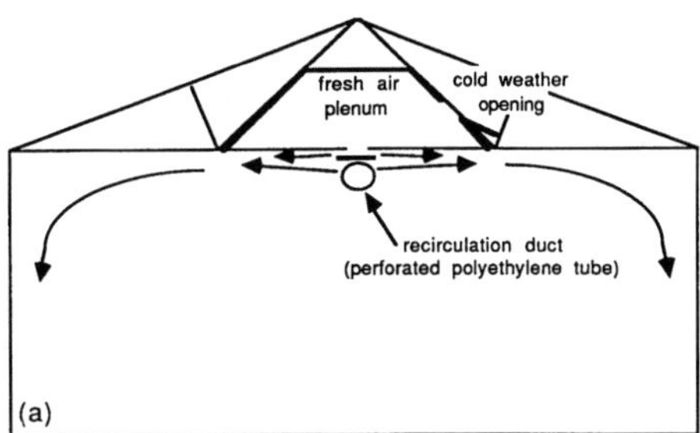

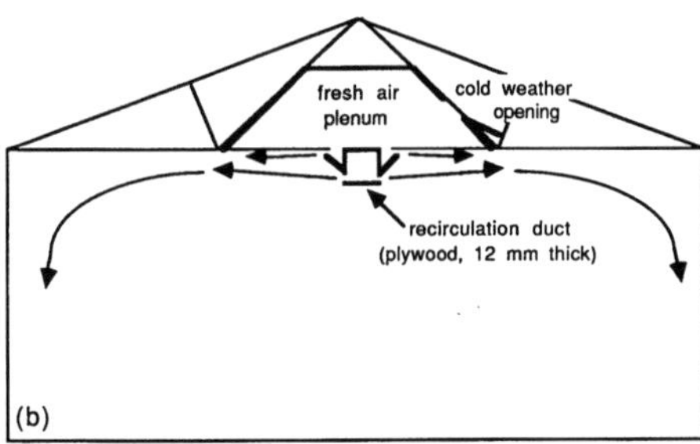

Figure 9-3. Two air recirculation methods to aid stability of inlet jets during times of minimum ventilation, and cold weather.

The second method shown in Figure 9-3 requires somewhat more construction, but does not limit overhead space as severely and is easier to clean (dust can accumulate rapidly in such air ducts). The slotted inlet in the recirculation duct can be either continuous, or discontinuous to increase the speed of air exiting the duct (increase inlet jet momentum). (Thin spacers can be used along the edge of the fresh air inlet baffle to make the inlet discontinuous and increase inlet air velocity.) A ratio of recirculated to fresh air of 3:1 has been found satisfactory with either system.

It should be noted there may be times when it is undesirable for a center-ceiling inlet to take air directly from the attic space. During summer, with sun shining on the roof, attic air may be 5 K or more above outdoor air temperature. To mitigate this problem, a large plenum space in the attic can be constructed, using the center sections of the trusses as a frame for the plenum. In this method, air enters from either or both gable ends of the attic directly into the plenum. Of course, the additional heating of ventilation air by the sun, and by heat loss through the ceiling, may be desirable during winter, and the plenum could have a removable or hinged section openable for cold weather ventilation.

9-5. Inlet Jet Momentum Numbers as a Criterion to Control Air Inlet Baffles

The concept of using the inlet jet momentum number as a means to characterize air mixing pattern stability has already been stated to be still a matter primarily for research, with field applications yet to come. However, recent research results have provided incentive to speculate how the concept might be used. This section shows how imposing inlet jet momentum number limits can lead to criteria which may be useful in environment control systems.

We have seen how, in negative pressure ventilation systems, the volumetric airflow rate is determined primarily by the fans, while air distribution and mixing within the ventilated space are determined primarily by inlet placement and control. The system characteristic technique was introduced to show how inlets and fans interact to determine possible operating points of a slotted inlet ventilation system.

We also saw how, during times of low ventilation rates and if infiltration is significant, control of a negative pressure ventilation system to avoid wind effects can be difficult or virtually impossible to achieve.

An aspect of ventilation not yet addressed by our discussion of the system characteristic graph concerns the potential problem of having too much air mixing, with resulting draftiness near the floor, especially during cold weather. Small and young animals may be subjected to drafts which people, dressed warmly, do not sense. These drafts can arise in any ventilation design, although the primary focus here will be on negative pressure, slot-ventilated systems.

As a general rule, inlets can be controlled so as to limit the partial vacuum within a ventilated space to a range thought to be adequate to avoid most wind effects. The partial vacuum range is typically chosen to be around 10 to 15 Pa to avoid problems with winds up to 30 to 40 km/hr. However, winds of these speeds are not common, therefore the pressure difference and resulting high inlet air velocity may be more than needed for much of the time. Can this be a problem? Might it create draftiness near the floor? We can answer those questions by amplifying our use of the system characteristic graph.

As has been stated, the proposed minimum inlet jet momentum number to provide air mixing stability is approximately 7.5E - 04. This threshold has been observed to apply to slotted inlet ventilation systems, and is adequate for some ventilation systems using perforated polyethylene tubes. There may be justification in thinking the threshold is only a weak function of inlet type. The incoming air is a continuous source of momentum. Large-scale circulation of air within the ventilated space causes a loss of momentum through the boundary layers to the airspace surfaces. Air circulation stabilizes such that momentum removal is balanced by momentum addition. Because animal housing systems are generally similar in geometry, and their geometry is independent of inlet locations, similar intensities of air circulation will induce similar rates of momentum removal. This reasoning leads one to conclude the required rate of momentum addition to insure stable air mixing patterns may also be somewhat independent of inlet type.

Evidence exists that inlet jet momentum numbers which are too high may be undesirable. This is physically reasonable. The higher the inlet jet momentum number, the higher will be air recirculation velocities within the ventilated space. If air velocities are too high, draftiness may occur in some locations depending on inlet placement and the initial direction of the incoming fresh air. For example, using perforated polyethylene tubes in swine housing, Ogilvie et al. (1988) found that inlet jet momentum numbers greater than 15E - 04 induced excessive air velocity near the floor and potential stress on the pigs. If this upper limit can be generalized, ventilation systems must work within a relatively narrow range of inlet jet momentum numbers for control to be correct.

Of course, air temperature must also affect the suitable value of air velocity near the floor to prevent chilling. However, although air temperature and floor airspeed will dictate a suitable value of inlet jet momentum number, the momentum number will still be the criterion for inlet baffle control.

Because system characteristic graphs have been shown to be useful in visualizing how ventilation system control can function, the band of acceptable inlet jet momentum numbers will be added to the graph.

The inlet jet momentum number can be expressed as (see Section 9-4):

$$J = \dot{V} v_1 / gV. \tag{9-4}$$

We will focus for now on slotted inlets. In Section 8-4.2 it was shown how airflow rates through slotted inlets can be expressed as

$$\dot{V} = KLW^{0.98}\Delta P^{0.49} \tag{9-5}$$

where L is the total length of the inlet and K is a coefficient depending on the type of inlet.

From Equation 9-3

$$\dot{V} = gJVv_i^{-1} \tag{9-6}$$

and from continuity considerations

$$\dot{V} = c_d v_i WL, \text{ or} \tag{9-7}$$

$$v_i = \dot{V}c_d^{-1}W^{-1}L^{-1}. \tag{9-8}$$

Thus,

$$\dot{V} = gJVc_d\dot{V}^{-1}WL, \text{ or} \tag{9-9}$$

$$(\dot{V})^2 = gJVc_dWL. \tag{9-10}$$

From the Bernoulli relationship for fluid flow,

$$v_i = (2/\rho)^{0.5}\Delta P^{0.5}. \tag{9-11}$$

Combining Equations 9-5, 9-7, and 9-11 (or by substituting Equation 9-11 directly into 9-4, but then how the method applies to slotted inlets may not be clear):

$$c_d = (\rho/2)^{0.5}KW^{-0.02}\Delta P^{-0.01}, \text{ and} \tag{9-12}$$

$$(\dot{V})^2 = (\rho/2)^{0.5}gJVLW^{0.98}\Delta P^{-0.01} \tag{9-13}$$

Dividing Equation 9-13 by Equation 9-5 leads to

$$\dot{V} = (\rho/2)^{0.5}gJV\Delta P^{-0.5} \tag{9-14}$$

Equation 9-14 provides a means to superimpose a line of constant inlet jet momentum number onto a system characteristic graph. Airspace volume and air density may be treated as constant for a given building. This leaves only the jet momentum number as a parameter when adding the equation to the system characteristic graph.

9-5.1. Program SYSCHAR2. Program SYSCHAR2 is in executable form, and is a modification of Program SYSCHAR1. The only change in its use, compared to SYSCHAR1, is to determine beforehand the desired lower and upper limits of the inlet jet momentum number. The limits 7.5E-04 and 15E-04 are suggested values, but obviously are not mandatory. Of course, limits greatly different from these two may not be sensible. Output of the program is a system characteristic graph, with traces of the specified acceptable inlet jet momentum numbers superimposed.

9-5.2. Interpreting Results of Program SYSCHAR2. Before discussing the output of the program, we will begin with an example. The barn for analysis, a single-story, sloping, tie-stall dairy barn, is designed to house 62 milking cows and is detailed in Plan Number 855, available through the Northeast Dairy Practices Council, Riley-Robb Hall, Cornell University, and other Cooperative Extension sources. The plan is typical of those available through numerous state cooperative extension offices.

Example 9-4

<u>Problem:</u> Determine how inlet jet momentum number limits of 7.5E - 04 and 15E - 04 for the ventilation system of Plan 855 will dictate how the inlets must be controlled, and what other ventilation decisions will be affected. Assume the barn is fully stocked and the minimum and maximum ventilation rates have been determined to be 1.2 and 8.8 m^3/s, respectively. A five-stage ventilation system is in place, with stages of 14%, 20%, 30%, 50%, and 100% of full capacity.

Air inlets with hinged baffles direct air to flow down the walls, and total inlet length is 80 m. Infiltration is expected to be equivalent to one-quarter of the "very tight" data of ASAE Engineering Practice EP270.5. Inlet widths range from 5 to 50 mm. The barn's floor area is 508 m^2 and wall height is 2.7 m.

<u>Solution:</u> It is suggested to follow this solution on a computer using SYSCHAR2. Data required by the program, in order of entry, are:

    number of cows: 62
    infiltration factor: 0.25
    inlet type: b
    inlet length: 80
    number of inlet widths: 6
    inlet widths: 5, 10, 20, 30, 40, 50
    fan data changes?: no

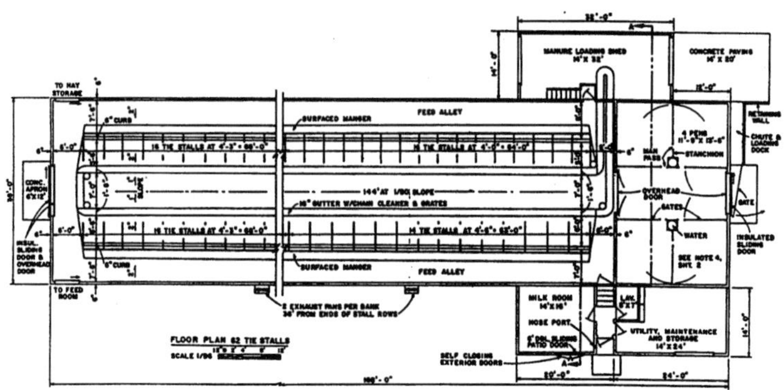

Figure 9-4. Northeast Dairy Practices Plan Number 855.

example selection of fans for five stages:

    stage 1: 4 of model a (1.16 m³/s at 12.5 Pa)
    stage 2: 6 of model a (1.74 m³/s at 12.5 Pa)
    stage 3: add 2 of model b (2.88 m³/s at 12.5 Pa)
    stage 4: add 2 of model c (4.24 m³/s at 12.5 Pa)
    stage 5: add 4 of model d (8.96 m³/s at 12.5 Pa)

As in Example 8-8, assume 4 fan banks and a suitable arrangement of the fans.

A copy of the results screen of SYSCHAR2 for this example is below. The two lines of constant inlet jet momentum number are indicated. The band of acceptable conditions defined by the two numbers shows several problems of control may be expected with this ventilation system.

First, at fan stages 1 and 2, it will not be possible to achieve adequate air mixing within the space by using fresh air alone. There will be insufficient momentum in the fresh air to drive the desired air mixing patterns at a satisfactory speed. What can be done? One possibility would be to install recirculation fans.

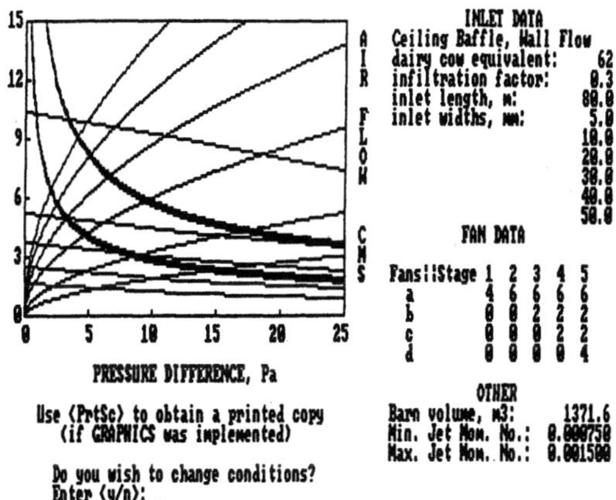

The inlet jet momentum number could be a guide to determine how much recirculation would be needed. One criterion would be to have sufficient recirculation fan capacity so that the total momentum added to all air within the airspace (fresh plus recirculated) brings the jet momentum number up to at least the lower limit of 0.00075. For example, suppose ventilation is at the lowest stage. The system characteristic graph can be used to estimate the expected airflow rate, and the corresponding pressure difference. The pressure difference determines the expected inlet air velocity, and these factors plus other known data (e.g., building volume) can be used to calculate actual momentum in the fresh air. Then, if the velocity of air leaving the recirculation fans is known, the volumetric flow rate required to bring the total jet momentum number up to 0.00075 could be determined. Control would be instituted so the recirculation fans would be activated whenever the system operates at the lowest two stages.

Ventilation stage 3 intersects the 5 mm inlet width line within the acceptable jet momentum number band. Control of the system would be such that, at stage 3, recirculation fans would be deactivated and inlet width would remain at 5 mm (or the pressure difference would be approximately 18 Pa).

At stage 4, inlet width could be between 10 and 30 mm. The inlet width could be either fixed at some value within this range whenever ventilation would be at stage 4, or inlet width could be varied depending on some other condition such as ambient windspeed.

Problems may exist for ventilation stage 5. Fortunately, this stage will be activated only during warm weather, when draftiness and chilling may not be a concern and the upper jet momentum number of 15E - 04 is no longer applicable. However, to maintain the spirit of staying within our jet momentum number band, one solution would be to increase the inlet width, perhaps to 60 mm, to bring the intersection of operating conditions within the desired band. If windiness were an occasional problem, the inlet width could be decreased to bring the pressure difference to a desired level (such as 10 or 15, or more) and the inlet reopened later when wind effects decrease.

Numerous decisions are required to implement control based on the inlet jet momentum number. One might conclude that control using an environmental control computer may be best. Such controllers are being developed, and may soon be widely used.

9-6. Characterizing Incomplete Mixing

At numerous points in previous chapters, the assumption of complete air mixing has been made. For example, the sensible energy balance used to calculate required ventilation rates was used based on complete mixing. Research has shown air mixing in fact is seldom complete. The following phenomena act to prevent complete mixing:

1. short circuiting
2. secondary recirculation zones
3. air mixing zones cascaded in series

9-6.1. Short Circuiting. The concept of short circuiting is shown in the sketch of Figure 9-5. This type of short circuiting differs from the idea of fresh air entering through an inlet and being diverted directly to the outlet (fan).

As shown in Figure 9-5, fresh air enters the space and entrainment causes mixing with air already in the space. Near the outlet, air from the enhanced jet passes out of the airspace. Some air which leaves through the outlet is air which was part of the initial air jet, and which did not return to the general circulation

276

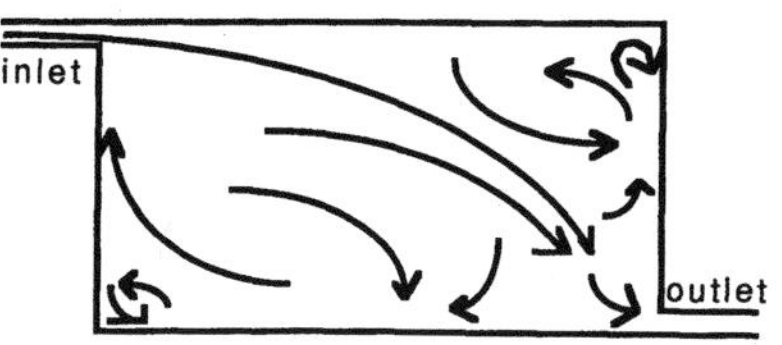

Figure 9-5. Airflow pattern in a simple airspace ventilated by a slotted inlet, showing short-circuited air flow from inlet to outlet.

area in the middle of the airspace and accumulate heat, moisture, and contaminants. This is short circuiting.

It has been estimated that in slot-ventilated enclosures, as much as a third of the ventilating air leaves this way and never contributes to air freshness within the airspace. The fraction involved in such short circuits is a function of numerous geometric and airflow factors and is not fixed. However, the fraction is always greater than zero. Conservative design practice would add this factor when calculating design airflows based, for example, on sensible energy and moisture balances.

9-6.2. Secondary Recirculation Zones. Two secondary recirculation zones are shown in Figure 9-5, one at the upper right and one at the lower left. Such zones always exist in corners. Fresh air cannot enter the secondary zone directly, but instead must diffuse across the mixing layer between the primary circulation zone and the secondary zone. Within the secondary zone, the humidity ratio and concentration of other contaminants will be greater than in the primary zone of air circulation, if such sources are present within the zone.

Secondary recirculation zones can become large in certain situations. For example, consider an animal housing facility as sketched in Figure 9-6, where stock are kept in small pens. The action of the ventilating jet causes a primary circulation zone above the pens, but fresh air enters the pens only after diffusing through the mixing layer between the primary and secondary zones.

The secondary zones in Figure 9-5 may be insignificant for the stock in the airspace, but in Figure 9-6, all animals would be subjected to contaminant and

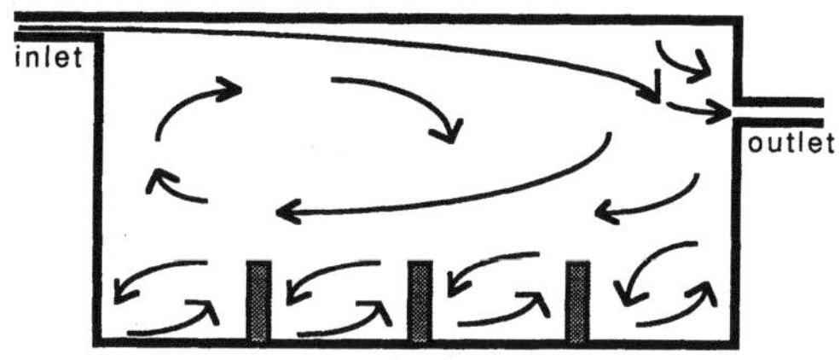

Figure 9-6. Air mixing patterns in a slot-ventilated enclosure where stock is kept in small, solid-sided pens.

humidity levels greater than design values. Good design practice would avoid creating such potential dead air zones, or if unavoidable, the ventilation system should be altered to bring fresh air to each pen. For example, in the barn sketched in Figure 9-6, a slotted inlet system may not be best. Instead, perhaps ducted airflow, or perforated polyethylene tubes for air distribution may be preferred.

9-6.3. Air Mixing Zones in Series. When an airspace ventilated by a slotted inlet is long in the direction of airflow, relative to its height, the inlet jet may not have sufficient momentum to induce one large recirculation zone within the space. Such a situation is sketched in Figure 9-7. Each of the three zones, in turn, exhibits an increasing air temperature, humidity, and contaminant level, but only the final of the three zones has environmental parameters at design values. The ratio of length to height is termed the "aspect ratio". Some data have shown when the aspect ratio is 3 or greater, mixing zones in series will exist. In practical terms, airspaces ventilated using a continuous slotted inlet on only one side of the space may have effective aspect ratios greater than 3.

The humidity ratio of the third zone will be at the design value, but the second and first zones will have humidity ratios below the design value. The first and second zones will be, in effect, overventilated. Thus, a concern might arise as to possible chilling in all but the final zone in series. Location of the control sensor(s) would also be a highly significant concern.

A redesign with inlets on both sides and the outlet in the center would help alleviate the problem of potentially having nonuniform conditions. This is one situation where location of the exhaust significantly affects air mixing patterns.

9-7. Airflow Visualization

Numerous experimental means are available to visualize streamline patterns in a ventilated space. Some are suited only for laboratory applications, others can be used in production agricultural buildings.

Smoke is a commonly used means to visualize flow in an airspace. Smoke may be from a "smoke bomb", although these often produce too much smoke and fill the air so completely that airflow patterns are obscured. Small smoke generators using titanium tetrachloride can be used for local visualization. A common beehive smoker is equally useful to see local effects, and is inexpensive.

Figure 9-7. Airspace ventilated by a continuous slotted inlet, demonstrating air mixing zones in series.

Small, neutrally buoyant soap bubbles have been used both in laboratory scale, and production scale. Neutral buoyancy is achieved by filling small soap bubbles with helium. If the proper soap solution is used, the bubbles are long lasting.

Other techniques to visualize flow, such as using silk threads tied along walls, using helium-filled balloons, and using aerosols other than smoke have all been tried and can be useful in limited applications.

If a system has been designed and constructed, and ventilation is not satisfactory, it is helpful to "see" air move, for a visual impression can often help quickly trace a problem – poor air mixing, an unanticipated air leak, or other problems.

9-8. Air Mixing Models

A great deal of engineering research and design information exists related to mixing theory. However, applications have generally arisen in chemical engineering, food processing, and other industrial and manufacturing situations. Only limited research has been completed which applies to air mixing and ventilation systems in agricultural buildings.

Today the mixing factor concept is perhaps the most popular for quantifying air mixing, and will be described. However, before that can be done, some basic concepts are required.

9-8.1. Idealized Air Mixing Models. The air mixing problem can be idealized as shown in Figure 9-8. A single airspace is ventilated with a single inlet and single outlet. The fresh ventilation air contains a concentration, c_i, of some contaminant. Within the airspace, additional contamination is produced (at a rate of c_s), and the exhaust air contains a contaminant concentration of c_e. Within the airspace, the average contaminant concentration is c. The fresh air ventilation mass flow rate is m_i, the exhaust ventilation rate is m_e, and within the airspace the mass of air is M. The contaminant might be a gas, aerosol, or perhaps excess heat content.

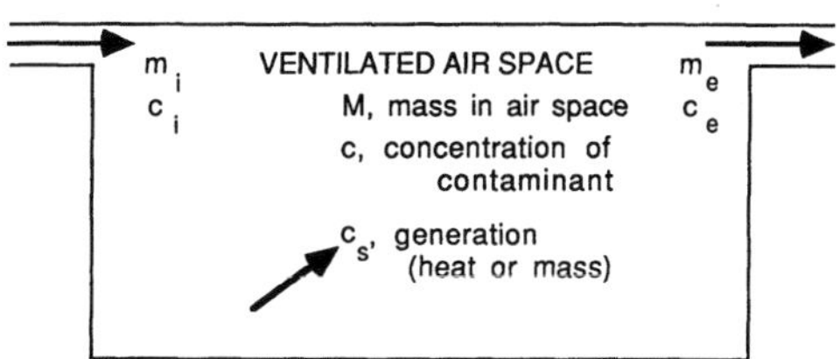

Figure 9-8. Air mixing within a ventilated airspace, with ventilation providing dilution of a contaminant.

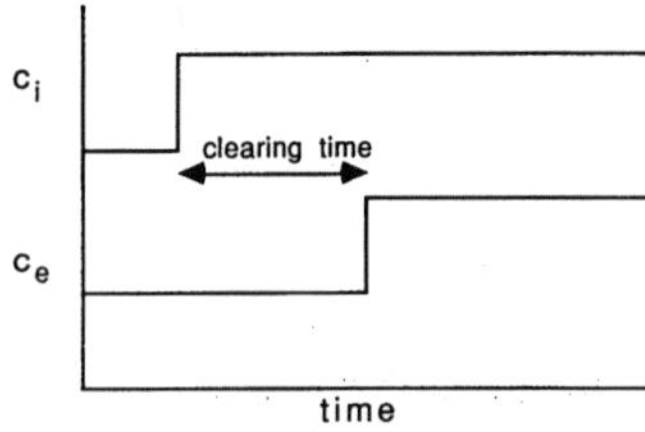

Figure 9-9. Contaminant concentration in incoming and exhaust ventilation air with plug flow.

9-8.2. Plug Flow, Perfect Non-Mixing. An extreme mixing situation is to have no mixing at all, and, for example, a step change in contaminant concentration in the incoming ventilation air. The changes of contaminant concentration in the fresh air, and exhaust air, which characterize this situation are shown in Figure 9-9. Clearing time is a function of distance from the inlet, being short near the inlet and longer away from the inlet. This is not a type of mixing which could characterize ventilation.

9-8.3. Perfect Mixing. In perfect mixing, air at the inlet is instantaneously and uniformly dispersed (not mixed) throughout the airspace. The changes of contaminant concentration for a step change of inlet air concentration are shown in Figure 9-10. This too is physically unrealistic as a model for ventilation. The model could be approached only if the airspace is very small and the ventilation airflow rate very large.

9-8.4. Dispersed Plug Flow. This type of mixing is similar to plug flow mixing, however, there is back mixing at interfaces between regions having different contaminant concentrations. Contaminant concentrations of inlet and exhaust air are shown in Figure 9-11 for a step change of inlet concentration. This model can have application in ventilation, perhaps in laminar airflow rooms, but is not generally applicable to ventilation. In laminar airflow rooms, one wall is essentially all inlet, and the opposite wall essentially all outlet. Air flows slowly (in laminar flow) from inlet to outlet and mixing (except by diffusion as shown in Figure 9-11) is minimized. Of course, movement of objects and thermal buoyancy within the airspace are two of many factors within the airspace which can cause mixing and deviation from the ideal.

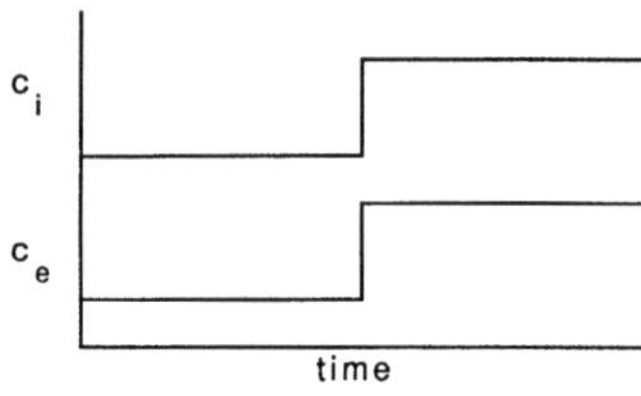

Figure 9-10. Contaminant concentration in incoming and exhaust air with perfect mixing.

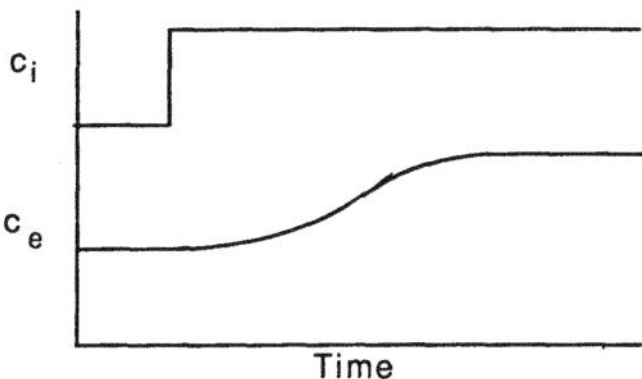

Figure 9-11. Contaminant concentration in incoming and exhaust air with dispersed plug flow.

9-8.5. Mixed Flow. Mixed flow is a general term, and the mixing may be complete, or not. Within any ventilated airspace there is a network of airflow regions connected in series and parallel, as indicated previously in this chapter. Contaminant concentrations in inlet and exhaust air for a step change of inlet concentration and mixed flow are shown in Figure 9-12. If mixing is <u>complete</u>, contaminant concentration at the exhaust equals the concentration within the airspace, and within the airspace the concentration is uniform. If mixing is <u>incomplete</u>, these two criteria are not met.

If there is sufficiently vigorous internal mixing within an airspace, the complete mixing model can approximate what occurs in a ventilated space. In fact, complete mixing is frequently the assumed condition regardless of actual mixing. For example, in previous chapters complete mixing and uniform temperature everywhere within an airspace have been assumed, and in practice this assumption is adequate for many situations.

If mixing is complete, c is uniform and equal to c_e and

$$\frac{d(cM)}{dt} = c_i m_i - cm_e + c_s, \qquad (9\text{-}15)$$

where t is time, and other terms are as defined above. If the derivative is expanded, the substitution $\rho V = M$ is made, $m_i = m_e$ is assumed, and the result is rearranged, we have

$$\frac{dc}{dt} + \frac{cm_i}{\rho V} = \frac{1}{\rho V}(c_i m_i + c_s) \qquad (9\text{-}16)$$

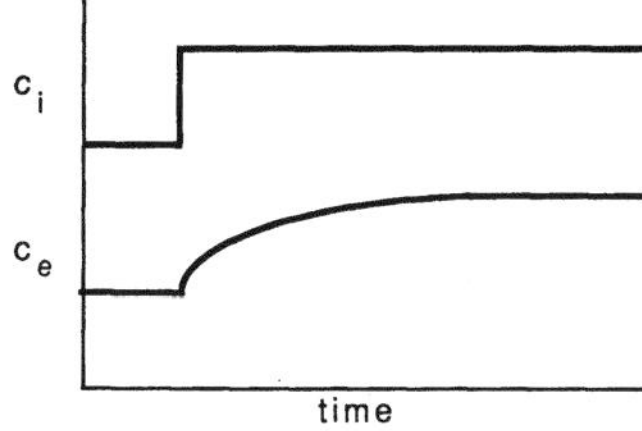

Figure 9-12. Contaminant concentrations in incoming and exhaust air characterizing mixed airflow.

If c_s, c_i, ρ and m_i are constant, Equation 9-16 is a linear, first order, differential equation. If $c = c_0$ at $t = 0$,

$$c = (1/z)(c_i z + \frac{c_s}{\rho V}) + c_0 \exp(-zt) - (1/z)(c_i z + \frac{c_s}{\rho V})\exp(-zt) \quad (9\text{-}17)$$

where $z = m_i/\rho V$.

A further simplification of Equation 9-17 is the case of a step change of contaminant concentration in the inlet air to zero after a steady state at a non-zero level (c_0) has been achieved, there is no production of contamination within the airspace, and the concentration is expressed as an excess above the level in the incoming ventilation air.

This simple situation is shown in Figure 9-13. The contaminant concentration within the airspace can be determined using

$$c = c_0 \exp(-\dot{V}t / V) \quad (9\text{-}18)$$

where $\dot{V}$ is the volumetric rate of ventilation, V is the volume of the airspace, and t is time. The value of t at which $c = c_0 / e$ is the time constant of the mixing process. Equation 9-18 also applies to temperature histories when there is a step change of inlet air temperature (except c_0 is replaced by the initial difference of temperature between the inlet air after the step change and air within the space), and c is the temperature difference between air within the space and the inlet air. The temperature difference decays to zero as ventilation proceeds.

9-8.6. Mixing Factor Concept. Incomplete mixing has been modeled in numerous ways. A simple approach is to define a mixing factor, K, where

$$K = \frac{\text{the effective air exchange rate}}{\text{the actual air exchange rate}} \quad (9\text{-}19)$$

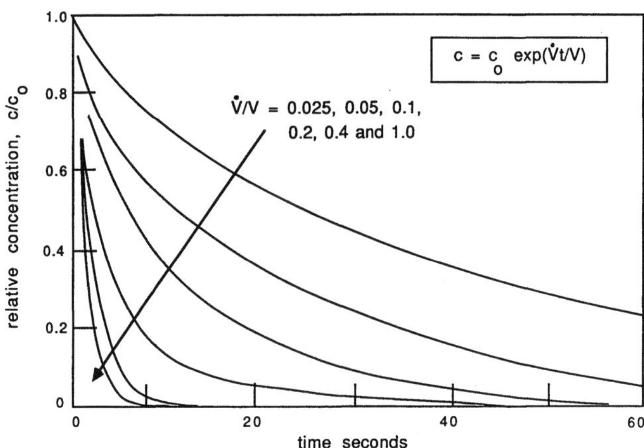

Figure 9-13. Complete mixing in ventilation, with a step change of inlet conditions, showing concentration changes with time.

Each location within an airspace could be characterized by its own value of mixing factor, but in practice a single value of the factor is chosen to represent the airspace. This is adequate if values of K are relatively insensitive to location, which is often true for much of the major recirculation zone of a ventilated airspace. (The major zone shown, for example, in Figure 9-5.)

When the simple mixing model embodied in Equation 9-18 is assumed, the mixing factor appears as follows:

$$c = c_0 \exp(-K\dot{V}t / V), \qquad (9\text{-}20)$$

and K can be determined experimentally by measuring the decaying concentration of a tracer gas, or temperature, and graphing the data on a semi-logarithmic plot. If $\dot{V}$ and V are known, K can be calculated from the experimentally determined value of the slope of the graph.

Complete mixing leads to a value of K of 1.0. Values of K less than 1.0 indicate short circuiting, and values greater than 1.0 indicate stagnant zones. However, in a real airspace both occur, and in some regions of the airspace the two effects may cancel so the resulting data yields a value of K very close to 1.0. Values of K between 0.3 and 0.7 have been recommended to apply to agricultural buildings and single airspaces ventilated by jets. The larger value is thought to be more appropriate for larger airspaces such as in typical barns and greenhouses.

9-8.7. Other Models. Research has indicated the mixing factor approach is not a good representation of air mixing in ventilated agricultural buildings when air quality is analyzed in fine detail. However, it is a useful concept to apply in design. One result is that a designer may find it to be more desirable to increase ventilation rates for summer conditions by 30% to 40% to compensate for incomplete mixing and short circuiting of the ventilating air jet. A discussion of more sophisticated models can be found in Barber (1981), for example.

SYMBOLS

Ar	Archimedes number, see Equation 9-2
c	constituent concentration
c_s	constituent production rate
c_d	coefficient of discharge
C_p	wind pressure coefficient
D	characteristic dimension, see Equation 9-2
g	gravitational constant, m/s^2
J	inlet jet momentum number
K	coefficient, see Equation 9-5
K	mixing factor, see Equation 9-19
L	inlet length, m
m	mass flow rate, kg/s

M   mass, kg
P   pressure, Pa
t   time, s
T   temperature, K
u   characteristic air velocity, see Equation 9-2, m/s
v   velocity, m/s
V   volume, m^3
$\dot{V}$ volumetric flow rate, m^3/s
W   inlet width, mm
x   distance from inlet, m
Y   distance from surface into wall jet profile, m
ρ density, kg/m^3

EXERCISES

1. A dairy barn is 14 m wide, 70 m long, and has sidewalls 3 m high. By
    ventilation graph analysis techniques, it has been determined the
    minimum ventilation rate for moisture control during winter is 5 air
    changes per hour. The barn is ventilated using a slotted inlet system with
    airflow down the sidewall, and inlets along the entire length of each side
    of the barn. The coefficient of discharge of the inlet system is estimated to
    be approximately 0.8.

    For times of minimum ventilation, determine the inlet width necessary to
    maintain stability of the air mixing patterns within the barn. Calculate the
    static pressure difference required at this condition, assuming there is no
    infiltration. Speculate on the reasonableness of the required static pressure
    if typical infiltration were to exist.

2. A poultry house having a total volume of 3000 m^3 is ventilated at a rate of
    500 m^3/s. Carbon dioxide was added to the air to build its concentration to
    3000 ppm everywhere in the airspace. The carbon dioxide addition was
    stopped and its concentration was monitored using a small, nondispersive
    infrared gas analyzer. The history of the concentration was as follows:

Time, Seconds	Concentration, ppm
0	3000
5	2135
10	1644
15	1365
20	1207
25	1118
30	1067
50	1007
100	1000
200	1000
1000	1000

The final concentration of 1000 ppm resulted from respiration of the poultry. From these data, determine the <u>effective</u> volumetric rate of ventilation.

REFERENCES

Barber, E.M. 1981. Quantitative prediction of incomplete mixing in the large ventilated airspace. Unpublished Ph.D. dissertation, University of Guelph, Guelph, Ontario.

Barber, E.M. and J.R. Ogilvie. 1982. Incomplete mixing in ventilated airspaces. Part I. Theoretical considerations. Canadian Agricultural Engineering 24 (1): 25-29.

Barber, E.M., S. Sokhansanj, W.P. Lampman and J.R. Ogilvie. 1982. Stability of airflow patterns in ventilated airspaces. Paper Number 82-4551. American Society of Agricultural Engineers, St. Joseph, MI.

Barber, E.M. and J.R. Ogilvie. 1984. Incomplete mixing in ventilated airspaces. Part II. Scale model study. Canadian Agricultural Engineering 26 (2): 189-196.

Baturin, V.V. 1972. Fundamentals of Industrial Ventilation. (O.M. Blunn, translator). Pergamon Press of Canada, Toronto.

Choi, H.L., L.D. Albright, M.B. Timmons and Z. Warhaft. 1988. An application of the k-epsilon turbulence model to predict air distribution in a slot-ventilated enclosure. Transactions of the ASAE 31(6): 1804-1814.

Hodgkinson, D.G. and E.M. Barber. 1986. Integrated air inlet and recirculation systems for livestock buildings. Paper Number 86-110. Canadian Society of Agricultural Engineering, Ottawa, Canada. 11 pp.

Kaul, P., W. Maltry, H.J. Muller and V. Winter. 1975. Scientific technical principles for the control of the environment in livestock houses and stores. Translation 430, British Society for Research in Agricultural Engineering. National Institute for Agricultural Engineering Research, Silsoe, Bedford, Great Britain (unpublished).

King, F.C., Jr., G.M. White and J.N. Walker. 1970. The effect of surface obstructions on air wall jets. Paper Number 70-403. American Society of Agricultural Engineers, St. Joseph, MI. 32 pp.

Ogilvie, J.R., E.M. Barber, N. Clarke and V. Pavlicik. 1988. Design of recirculation air ducts for swine barn ventilation systems. Proceedings, III International Livestock Environment Symposium. American Society of Agricultural Engineers, St. Joseph, MI. pp 51-58.

Randall, J.M. 1975. The prediction of airflow patterns in livestock buildings. Journal of Agricultural Engineering Research 20: 199-215.

Randall, J.M. and V.A. Battams. 1979. Stability criteria for airflow patterns in livestock buildings. Journal of Agricultural Engineering Research 24: 361-374.

Timmons, M.B., L.D. Albright, K.E. Torrance and R.B. Furry. 1980. Experimental and numerical study of air movement in slot-ventilated enclosures. Transactions ASHRAE 86(1): 221-239.

CHAPTER 10
VENTILATION CONTROL
AND QUANTIFICATION OF PERFORMANCE

10-1. Introduction

Negative pressure ventilation systems with continuous slotted inlets have been a focus of this text to this point. This chapter will also emphasize slotted inlet ventilation, although the concepts of control and fan system duty factors apply regardless of the type of ventilation.

In feedback control, a signal (i.e., a temperature measurement) is obtained which is information for the control system to use to determine what control action should be taken. The information is compared to a predetermined reference input (i.e., the thermostat setpoint) and if the difference between the two is sufficient to cause some action, a control element (the heater or fans) is activated in a manner to reduce the difference between the reference input (setpoint) and the controlled variable (air temperature) .

A simple feedback control system is that which activates a heater or fan to control air temperature in a building.

A building has an air temperature, A, in response to outside air temperature, B, and heat sources within the building, including the heater, H. A thermostat, T, measures A and controls the operation of H. If A is above the thermostat setpoint, nothing happens. If A is below the thermostat setpoint, H is activated until A is brought back to the desired temperature.

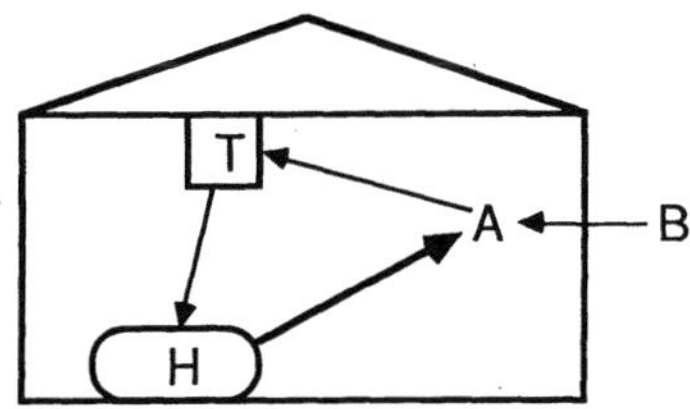

In Figure 10-1, a generalized strategy to control air temperature in a building, either by venting when needed or heating, is sketched. Outside air temperature and other heating loads influence inside air temperature. Inside air temperature, in conjunction with thermostat setpoints, controls vents, fans or heaters. The influence of vents, fans, or heaters on inside air temperature is the negative feedback mechanism. The feedback is negative because the greater the action of the control element, the less need for more action. (For example, the more heat that was delivered during the recent past, the less will be needed during the immediate future.)

Control engineers have developed a classification system for different types of control. Common types are: on/off control, proportional control, integral control, derivative control, and combinations of one or more (for example, proportional plus integral control). These are all types of feedback control.

<u>On/Off control.</u> The first type of control, on/off, is the simplest and can be activated either manually (e.g., a light switch) or with an automatic system (e.g., a thermostat). On/off control is typically applied to fans in agricultural buildings. Single speed fans are used, and they are either on or off. On/off control is also frequently used for heating systems in both animal housing and greenhouses. The advantage is simplicity of control; the disadvantage is less accurate control because of an inability to modulate the response according to intensities of other conditions.

One way in which on/off control is less accurate is the manner in which it can lead to overshooting the controlled variable. For example, consider a steam heating system for a greenhouse during very cold weather. After the system has operated for some time, air temperature reaches the temperature setpoint and the call for heating ceases. The heating pipes still contain both leftover steam and a significant amount of heat within the thermal mass of the pipes. However, the rate of heat loss from the greenhouse during cold weather is great, so the extra heat does not create a significant degree of overshoot of greenhouse temperature. The same heating system, however, is also activated during mild weather, and in that situation, extra heat within the heating pipes can cause air temperature within the greenhouse to overshoot beyond the setpoint by several degrees. Fortunately, most biological systems are forgiving of this type of inadequate temperature control, although energy is wasted and control is not optimum.

<u>Proportional control.</u> In proportional control, the rate at which the controlled parameter (e.g., heat, ventilation) is delivered is proportional to the difference between the reference input and the controlled output (e.g., air temperature).

Proportional control is commonly used with variable speed fans and to control heating systems. In ventilation systems, fan speed and, thus, airflow rate are

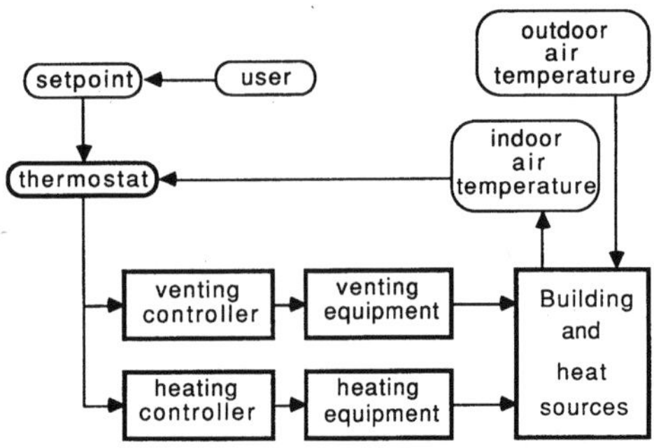

Figure 10-1. Generalized scheme for controlling building air temperature using venting and heating.

functions of the amount by which airspace temperature deviates from the desired temperature. As the temperature error increases (air temperature falls farther below the setpoint), fan speed of variable speed fans decreases in an attempt to limit heat losses and maintain the desired temperature.

Proportional control may be applied to heating systems. If it is used, for example, to control greenhouse heating in the system described above, control is more stable and results in less overshooting of temperature setpoints. As indoor air temperature approaches the setpoint, the rate at which steam is delivered slows and less heat is left within the heating pipes at the end of the cycle. Thus, there is less overshoot during mild weather.

However, proportional control requires a more sophisticated control system. Whenever possible for environmental control in agricultural buildings, on/off control has traditionally been used because of its simplicity. Proportional control has found more use in environmental control of commercial and industrial buildings. However, as computerized control systems are developed, sophisticated control strategies can be implemented through software development. One development can be applied in many buildings, significantly lowering the cost of application.

Integral control. Integral control limits small long-term offsets from setpoints and is not frequently used for environmental control in agriculture. Integral control acts to integrate error over time and bring the average error to zero. Its response is slow, but its long-term accuracy in obtaining an average is good.

One possible application is for temperature control. If the air temperature remains slightly below the thermostat setpoint for an extended period of time, integral control acts to integrate the error (difference between actual and desired temperatures) and eventually forces a correction.

Derivative control. Derivative control is used when a rapid response is desired. The time derivative of error determines the magnitude of the response.

Derivative control is useful to prevent sudden large departures of the controlled variable from the setpoint, but also has inherent instability and can cause responses which overshoot the desired setpoint. Derivative control is not frequently used for environmental control (although it can be part of a larger control strategy, especially as computerized control systems become more common).

Other control methods such as pseudo-derivative-feedback control (Phelan, 1977) and adaptive control are currently being explored by researchers for applications to environmental control in agriculture.

As long as items of environmental control equipment (fans, heaters, etc.) are in good repair, their capacity to deliver the desired flux of fresh air, heat, etc., can be considered constant over the long term. Although the static pressure differences from inside a building to outside influences the airflow rate through a fan, operation is usually within a sufficiently narrow range that fan capacity can also be considered constant in the short term.

The initial focus of this chapter will be on aspects of ventilation control and how control influences the duty factor of a ventilating system and thereby the cost of ventilation.

10-2. Duty Factors, Staged Fan Systems Using Single Speed Fans

A fan system duty factor is defined as the average ventilation rate for a year (m^3/s) divided by the maximum ventilating capacity of the fan system. The average ventilating rate is composed of operation during all times of the year — times when operation is at the lowest stage, the highest stage, and all stages between. If the duty factor and efficiencies of the fans are known, the yearly cost of ventilation can be calculated. The yearly cost is sensitive to ventilation stages and thermostat setpoints, as well as the inherent efficiencies of fans, and thermal parameters of the building being ventilated.

Ventilation staging was discussed in Chapter 6. Methods to choose temperature setpoints were described and a procedure was outlined to choose ventilation stages between the maximum and minimum ventilation rates.

Staged control can be viewed in several ways. Consider the hypothetical situation of sensible heat production within a barn which remains constant while the outdoor air temperature varies. The ventilation rate as a function of indoor air temperature would be as shown in Figure 10-2. The sketch demonstrates a system with four fan stages, $\dot{V}_1$, $\dot{V}_2$, $\dot{V}_3$, and $\dot{V}_4$. Three thermostat setpoints are required for control, one setpoint between each adjoining pair of stages. A fourth setpoint, for freeze prevention, might be recommended but would not be a factor in determining the fan system duty factor unless environmental control was poorly designed and ventilation was deactivated to prevent freezing for extended periods of a typical year.

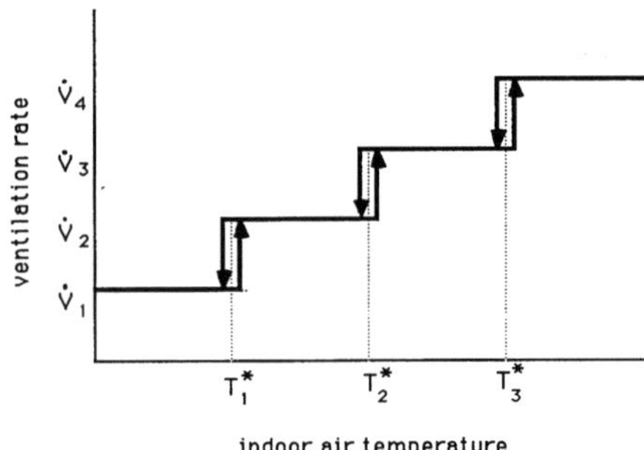

Figure 10-2. An example of ventilation control using single-speed, staged fans, as a function of indoor air temperature. Starred values are thermostat setpoints.

Suppose an animal housing facility is being ventilated but outdoor air temperature is very low. Ventilation will be at stage 1, the lowest stage, with indoor air temperature below the lowest thermostat setpoint. Then the outdoor air temperature rises. The first effect is to permit indoor air temperature to rise to the lowest thermostat setpoint. Thermostats have hysteresis, which is a lag in response caused by, among other factors, thermal mass. Indoor air temperature rises slightly above the first thermostat setpoint before the second stage of ventilation is activated.

After the second stage of ventilation is activated, the increased ventilation rate causes barn temperature to begin to fall – outdoor air temperature is not sufficiently high to permit the barn to maintain air temperature at the first thermostat setpoint when the ventilation rate is as high as stage 2. After passing the hysteresis region, the thermostat deactivates stage 2 ventilation. However, conditions which caused activation of stage 2 still exist and the barn temperature rises again through the hysteresis region.

As long as outside air temperature remains within some narrow range, fan control will alternate between stages 1 and 2 and indoor air temperature will remain within the narrow region defined by the hysteresis of the thermostat. When outdoor air temperature is in the lower parts of the narrow (and still undefined) temperature range, ventilation is primarily in stage 1, with limited excursions to stage 2. As outdoor air temperature rises through this narrow range, ventilation gradually changes from mostly stage 1 to mostly stage 2.

An outdoor air temperature is eventually reached where indoor air temperature can be sustained at or above the first thermostat setpoint with constant stage 2 ventilation. As outdoor air temperature continues to rise, stage 2 ventilation is maintained and indoor air temperature rises toward the second thermostat setpoint. When the second setpoint is reached, the same oscillation between stages 2 and 3 occurs as has been described between stages 1 and 2. If the outdoor air temperature rise continues, the same scenario holds to eventual continuous operation at stage 4.

A different way to describe ventilation staging is in Figure 10-3. The <u>average</u> ventilation rate is a smooth function of outdoor air temperature even though regions of oscillation between stages exist. The average ventilation rate can be calculated as a function of outdoor and indoor temperatures and thermal parameters of the building.

A third means to view ventilation staging is shown in Table 10-1. The table contains values of outdoor air temperature which correspond to steady-state ventilation at combinations of ventilation stages and thermostat setpoints. For example, at an outdoor air temperature of $(t_o)_{1,\,2}$ and a ventilation rate of $\dot{V}_2$, indoor air temperature will remain steady and equal to the first setpoint, SP_1. The indoor air temperature will also be steady and equal to SP_1 when the ventilation rate is $\dot{V}_1$ but the outdoor air temperature is $(t_o)_{1,\,1}$. The ventilation

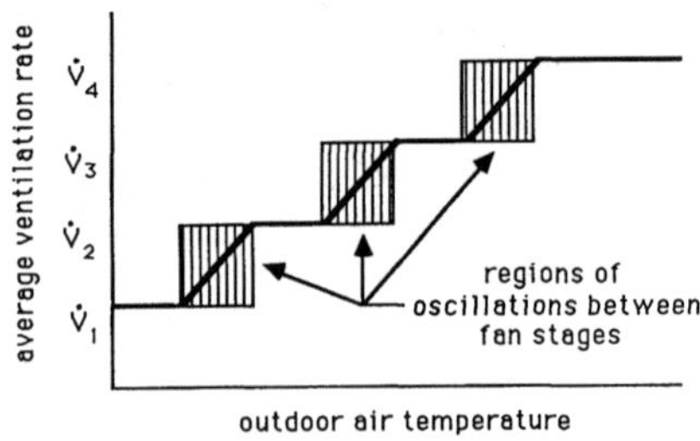

Figure 10-3. An example of ventilation control using single-speed, staged fans, as a function of outdoor air temperature.

rate $\dot{V}_1$ is less than $\dot{V}_2$, thus, $(t_o)_{1,1}$ must be less than $(t_o)_{1,2}$. Similarly, SP_2 is higher than SP_1, thus, $(t_o)_{2,2}$ is higher than $(t_o)_{1,2}$.

Values of outdoor air temperature in Table 10-1 are calculable using a steady-state thermal energy balance,

$$t_o = [- q_{prod} + (\Sigma UA + FP + mc_p)t_i] / (\Sigma UA + FP + mc_p), \quad (10\text{-}1)$$

where q_{prod} is the sensible heat production within the airspace, $\Sigma UA + FP$ is the structural heat loss factor(HLF), and m and c_p are the mass flow rate of ventilation air and specific heat of air, respectively.

Equation 10-1 is more convenient to use when the sensible heat production of animals is treated as a function of indoor air temperature (expressed for the entire herd or barn),

$$q_{prod} = a + bt_1 + ct_1^{\;2}. \qquad (10\text{-}2)$$

Sensible heat production can be expressed as a second order polynomial using program POLYNOM, for example. When sensible heat production is expressed in this form, indoor air temperature may be calculated by

$$t_i = \frac{- (b - HLF - mc_p) - \sqrt{(b - HLF - mc_p)^2 - 4c(a + HLFt_o + mc_pt_o)}}{2c} \qquad (10\text{-}3)$$

Table 10-1. Outdoor air temperatures corresponding to n thermostat setpoints and n + 1 ventilation stages

Thermostat Setpoint	Ventilation Rate			
	$\dot{V}_1$	$\dot{V}_2$	$\dot{V}_3 \ldots \dot{V}_n$	$\dot{V}_{n+1}$
SP_1	$(t_o)_{1,1}$	$(t_o)_{1,2}$		
SP_2		$(t_o)_{2,2}$	$(t_o)_{2,3}$	
SP_3			$(t_o)_{3,3}$	
$\vdots$				
SP_n			$(t_o)_{n,n}$	$(t_o)_{n,n+1}$

where HLF is the building's Heat Loss Factor ($\Sigma UA + FP$) and outdoor air temperature may be calculated using

$$t_o = \frac{- ct_1^{\,2} - (b - HLF - mc_p)t_i - a}{HLF + mc_p}. \tag{10-4}$$

Note: In Equations 10-3 and 10-4, the constant term "a" may include sensible heat produced by sources such as lights and motors. Heat from such sources is not produced as a function of indoor air temperature.

The outdoor air temperature ranges in Table 10-1 have known probabilities of occurrence for any location where weather records are available. For example, weather bin data such as in Appendix 6-2 can be used to estimate the number of hours during a year when outdoor air temperature is within a certain range, and the probability is that number of hours divided by the total number of hours in a year (8760).

Each temperature range also has a known average ventilation rate associated with it. For example, the range $(t_o)_{1,\,2}$ to $(t_o)_{2,\,2}$ defines when the average ventilation rate is $\dot{V}_2$. Temperature ranges when ventilation oscillates between stages require an assumption. It will be assumed when outdoor air temperature is between $(t_o)_{1,\,1}$ and $(t_o)_{1,\,2}$, for example, the average ventilation rate is $(\dot{V}_1 + \dot{V}_2) / 2$. That is, the probability of t_o within the range $(t_o)_{1,\,1}$ to $(t_o)_{1,\,2}$ is uniformly spread over the temperature range and the range is small enough that sensible heat production is effectively constant.

The probability of a temperature range is denoted by P. For example, $P[(t_o)_{1,\,1} < t_o < (t_o)_{1,\,2}]$ is the probability outdoor air temperature will be within the indicated temperature range during a year. It equals the number of hours during a year the outdoor air temperature is within the range $(t_o)_{1,\,1}$ to $(t_o)_{1,\,2}$ divided by the number of hours in a year. The sum, $\Sigma P[\text{all ranges}]$, equals unity.

The average ventilation rate for a year is the weighted average of ventilation rates within each temperature range possible for the year. The average is weighted by the probabilities of occurrence of the ranges,

$$\dot{V}_{ave} = \Sigma(\dot{V}_{range}P[\text{range}]) \tag{10-5}$$

For data in Table 10-1,

$$
\begin{aligned}
\dot{V}_{ave} = {} & \dot{V}_1 P[-\infty < t_o < (t_o)_{1,\,1}] \\[6pt]
& + \tfrac{1}{2}(\dot{V}_1 + \dot{V}_2)P[(t_o)_{1,\,1} < t_o < (t_o)_{1,\,2}] \\[6pt]
& + \dot{V}_2 P[(t_o)_{1,\,2} < t_o < (t_o)_{2,\,2}] \\[6pt]
& + \tfrac{1}{2}(\dot{V}_2 + \dot{V}_3)P[(t_o)_{2,\,2} < t_o < (t_o)_{2,\,3}] \\[6pt]
& + \dot{V}_3 P[(t_o)_{2,\,3} < t_o < (t_o)_{3,\,3}] +
\end{aligned}
\tag{10-6}
$$

The duty factor is,

$$\text{duty factor} = \dot{V}_{ave} / \dot{V}_{n+1}. \tag{10-7}$$

10-3. Ventilating Efficiency Ratios and the Cost of Ventilation

Fan efficiency can be expressed in several ways, but the way most informative to agricultural producers relates the useful output (m³/s of air delivery) to input which must be paid for (kWh of electricity). The ratio of these two parameters will be termed the Ventilating Efficiency Ratio (VER),

$$\text{VER} = (\text{m}^3/\text{s of air delivery}) / (\text{kW of electricity}) \tag{10-8}$$

A recent trend has been to rate agricultural ventilation fans in terms of their VER (termed the cfm/watt ratio in IP units). Data, when available, usually is presented for several static pressure differences. Data for the difference which represents expected operating conditions should be chosen.

A further caution is that only data which represents operation of the fans as installed should be used. That is, safety guards, louvers, and other attachments should have been in place during tests. If not, significant differences from operation with all attachments can result.

Three factors are needed to calculate the cost to operate a ventilating system for a year: the duty factor (or average ventilating rate for the year), the cost of electricity, and the VER of the fan system. The cost is calculated using

$$\text{cost} = \frac{(8760 \text{ hrs / yr})(\text{max capacity})(\text{duty factor})(\text{unit elect. cost})}{\text{Ventilating Efficiency Ratio}} \tag{10-9}$$

Example 10-1

Problem: Determine the cost to operate a fan (5.5 m³/s capacity) for a year if the VER is 8 m³/s-kW, the cost of electricity is expected to be $0.11/kWh, and the expected duty factor is 0.6.

Solution: Equation 10-9 provides the solution directly,

$$\text{cost} = \frac{(8760 \text{ hrs / yr})(5.5 \text{ m}^3 / \text{s})(0.6)(\$0.11 / \text{kWh})}{8 \text{ m}^3 / \text{s-kW}}$$

$$= \$397.49 / \text{yr}$$

Note: Although it is not frequently realized, the cost to operate a ventilating fan during its lifetime is usually far greater than its purchase cost. It is usually recommended to invest slightly more at the time of installation to purchase energy efficient fans as the VER of seemingly identical fans may differ by 50% or more.

10-4. The Cost of Ventilating Animal Housing, an Example

Procedures have been presented at various places in this text to:

(a) determine heat loss characteristics of a building,

(b) determine ventilation requirements and methods to stage ventilation between the maximum and minimum ventilation rates,

(c) determine appropriate thermostat setpoints,

(d) estimate ventilating system duty factors, and

(e) calculate how much ventilation systems cost to operate.

The following example is presented to demonstrate an integration of the procedures.

Example 10-2

Problem: A mechanically-ventilated, tie-stall dairy barn is being designed for construction near Denver, Colorado. The barn is to house 150 milking cows, averaging 550 kg body weight. A design has been chosen which results in a structural heat loss factor (walls, ceiling, perimeter, etc.) of 900 W/K. General lighting, designed at approximately 10 W/m^2, is expected to operate around the clock. The floor area of the barn will be 1500 m^2.

Minimum and maximum ventilation rates will be 2.9 and 21 m^3/s, respectively. Fans are being considered which have a VER of 9 m^3/s-kW and electricity is expected to cost $0.10/kWh.

Choose fan stages and thermostat setpoints and estimate the yearly cost to ventilate the barn.

Solution: First, fan stages and thermostat setpoints will be determined. A five-stage fan system is assumed for now, with thermostat setpoints at 9, 13, 17, and 21 C. (These are example setpoints and not proposed as optimum.)

The lowest fan stage is 2.9/21 or 13.8% of the highest. To follow the concept of fan staging presented in Table 6-1, the following stages are chosen:

Stage	Percentage of Maximum	m^3/s
1	13.8	2.9
2	19	4.0
3	25	5.3
4	45	9.5
5	100	21.0

A next step is to determine sensible heat production within the barn. Lighting will produce a constant $(10 \text{ W/m}^2)(1500 \text{ m}^2) = 15,000$ W. Appendix 5-1 contains data to estimate sensible heat production from animals. For 500 kg dairy cows,

Air Temp,C	Sensible Heat Prod., W/kg
-1	1.9
10	1.5
15	1.2
21	1.1
27	0.6

These data were examined in Example 6-4, and polynomials fitted to the sensible heat data using program POLYNOM. The second order polynomial yielded:

$$q'_{prod} = 1.8603 - 0.03074t - 5.268E - 04t^2,$$

expressed in W/kg for a 500 kg dairy cow.

Transforming q''_{prod} to the entire herd,

$$q_{prod} = (150 \text{ cows})(500 \text{ kg})(550/500)^{0.734}(q'_{prod})$$
$$= 149,600 - 2473t - 42.4t^2,$$

where t is indoor air temperature, C.

Sensible heat production within the airspace includes animal heat and heat from lights. Heat from lights is not a function of indoor air temperature, thus total heat production within the barn is

$$q_{barn} = 164,600 - 2473\,t - 42.4t^2.$$

The elevation of the barn is not specified in the example, but being near Denver it is likely to be approximately 1500 m. Atmospheric pressure is estimated to approximate 85 kPa. The density of air is required to determine outdoor air temperatures as in Table 10-1. A changing air temperature but constant relative humidity within the barn of 60% will be assumed. Air densities are:

296

Indoor Air Temperature, C	Air Density, kg/m^3
9	1.05
13	1.03
17	1.02
21	1.00

The change of air density over the range of thermostat setpoints is small, an average value of $1.02 \ kg/m^3$ is sufficiently accurate to be used as a constant, but the data were checked to be sure.

The point has been reached where the table of outdoor air temperatures as a function of thermostat setpoints and ventilation rates can be calculated. Equation 10-4 applies.

As an example calculation, consider 13 C setpoint and the third stage of ventilation, $5.3 \ m^3/s$. The mass flow rate is

$$m = (5.3 \ m^3/s)(1.02 \ kg/m^3) = 5.41 \ kg/s.$$

The outside air temperature corresponding to this combination of indoor air temperature and ventilation rate is

$$t_o = \frac{-(-42.4)(13)^2 - [-2473 - 900 - (5.41)(1006)](13) - 164,600}{900 + (5.41)(1006)} = -6.75$$

Similar calculations yield the following data for outdoor air temperature, C:

| Thermostat | Ventilation Rate, kg/s | | | | |
Setpoint, C	2.90	4.08	5.41	9.69	21.42
9	- 27.4	- 18.8 C			
13		- 12.0	- 6.75 C		
17			- 0.39	6.63 C	
21				12.2	16.8 C

There are now nine temperature ranges for which the probability of occurrence during a typical year must be determined. Using the data for Denver, the probability of the first temperature range is

$$
\begin{aligned}
P[-\infty < t_o < -27.4 \ C] = \ & P[-\infty < t_o < -34.4] + P[-34.4 < t_o < -28.9] \\
& + P[-28.9 < t_o < -27.4] \\
= \ & [0 \ hrs + 1 \ hr + (8 \ hrs)(-28.9 + 27.4) \\
& / (28.9 + 23.3)] / 8760 \\
= \ & [0 \ hrs + 1 \ hr + 2 \ hrs] / 8760 \ hrs \\
= \ & 0.0003
\end{aligned}
$$

Probabilities of the other temperature ranges are obtained in the same manner, leading to the following data:

Temperature Range	Hours Per Year in the Range (a)	Probability of the Range
to -27.4	3	0.0003
-27.4 to -18.8	35	0.0040
-18.8 to -12.0	157	0.0179
-12.0 to -6.75	366	0.0418
-6.75 to -0.39	1132	0.1292
-0.39 to 6.63	1833	0.2092
6.63 to 12.2	1485	0.1696
12.2 to 16.8	1227	0.1401
16.8 up	2522	0.2879

(a) rounded to the nearest hour

The average ventilation rate for the year can now be calculated using Equation 10-6. Note that volumetric airflow rates can be used again, mass flow rates were necessary only for the sensible energy balance.

$$
\begin{aligned}
\dot{V}_{ave} &= (2.9\ \mathrm{m^3/s})(0.0003) + 1/2(2.9\ \mathrm{m^3/s} + 4.0\mathrm{m^3/s})(0.0040) \\
&\quad + (4.0\ \mathrm{m^3/s})(0.0179) + 1/2(4.0\ \mathrm{m^3/s} + 5.3\ \mathrm{m^3/s})(0.0418) \\
&\quad + (5.3\ \mathrm{m^3/s})(0.1292) + 1/2(5.3\ \mathrm{m^3/s} + 9.5\ \mathrm{m^3/s})(0.2092) \\
&\quad + (9.5\ \mathrm{m^3/s})(0.1696) + 1/2(9.5\ \mathrm{m^3/s} + 21\ \mathrm{m^3/s})(0.1401) \\
&\quad + (21\ \mathrm{m^3/s})(0.2879), \\
&= 0.00087 + 0.01380 + 0.07160 + 0.19437 + 0.68476 + 1.54808 \\
&\quad + 1.61120 + 2.13653 + 6.04590 \\
&= 12.30711,\ \text{or } 12.3\ \mathrm{m^3/s}.
\end{aligned}
$$

The duty factor is thus $(12.3\ \mathrm{m^3/s})/(21\ \mathrm{m^3/s})$, or 0.59. Typical duty factors for fan systems in ventilated animal housing are in the range 0.5 to 0.7 in moderate climates.

The yearly cost of ventilation can now be determined, using Equation 10-9.

$$
\text{cost} = \frac{(8760\ \mathrm{hrs/yr})(21\ \mathrm{m^3/s})(0.59)(\$0.10)}{9\mathrm{m^3/s\text{-}kW}}
$$

$$
= \$1206,\ \text{or approximately } \$8\ \text{per cow-yr.}
$$

It should be noted that an overall VER of 9 $\mathrm{m^3/s\text{-}kW}$ is relatively high if numerous small fans (e.g., less than one meter diameter) are used in the system. A lower VER value raises the cost of ventilation.

10-5. Program DUTYFACT

The work required to complete Example 10-2 is substantial, and the effort involved is a disincentive to repeat the calculations for different fan staging schedules, thermostat setpoints, and building thermal parameters. Also,

Example 10-2 was greatly simplified from a computational standpoint by assuming the VER is uniform among all the fans. In fact, each fan model has its own VER, ranging from approximately 3 for small fans (and large fans which are not efficient) to 10 and larger for efficient, large diameter ventilating fans (e.g., 1 m diameter and larger).

A computer program which includes the more general case of differing VER values, and permits the user to vary parameters easily, would be a design tool useful for an engineer to demonstrate the effect of ventilation control on the yearly cost of ventilation. Program DUTYFACT is provided as an example of a program designed for such simulations. It is intended for demonstration purposes, not as an everyday design tool.

Several assumptions are included in DUTYFACT. A simple model of atmospheric pressure as a function of elevation is included to calculate air density,

$$\text{air pressure} = 101.325 \exp(-0.00011943\ z) \\ -6.799E-06z - 6.976E-08z^2, \tag{10-10}$$

where z is elevation in m and air pressure is expressed in kPa.

In addition, because the actual operation static pressure difference is not known, the VER of each fan (a function of static pressure difference) is assumed equal to the VER at a pressure difference of 12.5 Pa. This is an intermediate value and represents a desirable pressure difference in a well constructed and managed barn. Note: DUTYFACT assumes once a fan or group of fans is activated, it remains active for all higher stages. Fans can be turned off or changed, however, at higher stages by specifying them by letter and setting their number to zero (or some other number if appropriate).

Example 10-3

Problem: Reconsider the barn described in Example 10-2. Use program DUTY-FACT to calculate the cost of ventilating the barn when the fan staging schedule described in Example 10-2 is used, and recalculate the cost when the range of ventilation is segregated into equal increments. Use the default fan data in DUTYFACT to select fans for staging and maintain the setpoints used in Example 10-2.

Solution: Program DUTYFACT is used and data is entered as follows:

1. elevation above sea level = 1500 m,

2. ΣUA value = 900 W /K,

3. FP value = 0 W/K (UA and FP are not separated in the statement of the

problem. For convenience, the Heat Loss Factor is consolidated into UA. In the sensible energy balance, the two are added so the consolidation has no effect),

4. (a), (b), and (c) are as calculated: 164,600 W, - 2473 W/C, and 42.2 W/C^2 (Note: for the entire barn),

5. the cost of electricity is $0.10 per kWh (somewhat higher than current agricultural rates, if available, but not unusually so), and

6. the default fan data are not changed.

A. The cost of ventilation with fan staging as done in Example 10-2.

Fan Stage	Ventilation Rate, m^3/s
1	2.9
2	4.0
3	5.3
4	9.5
5	21.0

An example fan staging schedule to approximate these ventilation rates is:

Fan Stage	Fans to Operate	m^3/s at 12.5 Pa
1	5 - model b	2.85
2	7 - model b	3.99
3	7 - model b	
	2 - model c	5.35
4	7 - model b	
	2 - model c	
	2 - model e	9.35
5	7 - model b	
	2 - model c	
	2 - model e	
	2 - model d	
	2 - model j	21.21

Note: This number of fan models would likely not be used in an actual design. A more compatible selection, with real data, would be used to approximate the proper staging.

The setpoints entered in DUTYFACT are 9, 13, 17, and 21 C. The city to be chosen is Denver and the weather bin data need not be altered. When all the data are entered, the results of the program are as follows.

Thermostat Setpoint, C	Ventilation Rate, m^3/s				
	2.85	3.99	5.35	9.35	21.21
9	- 27.46	- 18.93 C			
13		- 12.19	- 6.69 C		
17			- 0.34	6.44 C	
21				12.00	16.83 C

The control table of outdoor air temperatures differs little from the one in Example 10-2, and would not be expected to differ. There are only slight changes of the ventilation rates and other data are identical. The probabilities of outdoor temperature ranges are provided by the program and need not be listed here; they differ little from the data calculated in Example 10-2.

The average ventilation rate at 12.5 Pa is calculated by the program as 12.38 m^3/s, with a duty factor of 0.58, again not very different than before. However, the cost of ventilation is greatly different. The average VER is 4.83 and the yearly cost to ventilate is $2244.84. In Example 10-2, the VER was stated to be 9, and such a value might represent very efficient, large diameter fans. However, for much of the year smaller and less efficient fans operate, thus, the cost is not solely a function of the efficiency of the largest fans. Also, the default fan data do not list fans with VER values as high as 9, but the default data are more realistic.

The probabilities of the outdoor air temperature ranges should be examined with some care, for they provide insight into strategies to lower the cost of mechanical ventilation. In this example,

Temperature Range	Probability of the Range	Fan Operation
to - 27.46	0.000349	stage 1 only
- 27.46 to - 18.93	0.003854	stages 1 and 2
- 18.93 to - 12.19	0.016545	stage 2 only
- 12.19 to - 6.69	0.043454	stages 2 and 3
- 6.69 to - 0.34	0.130651	stage 3 only
- 0.34 to 6.44	0.201918	stages 3 and 4
6.44 to 12.00	0.169281	stage 4 only
12.00 to 16.83	0.147024	stages 4 and 5
16.83 up	0.286924	stage 5 only

The VER values for the temperature ranges are

Fan Stage	VER, m^3/s-kW
1	3.660
2	3.660
3	3.631
4	4.075
5	5.642

As the stages increase, so too does the VER because larger and more efficient fans are activated.

The probability data show a third of the year involves the outdoor air temperature range from - 6.69 C to 6.44 C. This range involves either stage 3 by itself or alterations between stages 3 and 4. Yet, the VER of stage 3 is relatively low. A strategy to lower the cost of ventilation would thus be to obtain more efficient fans for stage 3 as a first priority. This, of course, involves also selecting efficient fans for stages 1 and 2 because of the assumption the stages are additive.

Fan Stage	Ventilation Rate, m^3/s
1	2.90
2	7.43
3	11.95
4	16.48
5	21.00

With the default fan data in DUTYFACT, a candidate fan schedule is:

Fan Stage	Fans to Operate	m^3/s at 12.5 Pa
1	5 - model b	2.85
2	5 - model b	
	4 - model d	7.57
3	5 - model b	
	8 - model d	12.29
4	5 - model b	
	8 - model d	
	2 - model e	16.29
5	5 - model b	
	8 - model d	
	2 - model e	
	2 - model f	21.25

The same type of data as described in A above is presented by DUTYFACT, but need not be listed here. The example asks for the cost of ventilation, which in this second fan staging approach, is $2892.51 per year. This is significantly greater than the cost in part A. However, the cost difference cannot be generalized, for it depends on the probability distribution of weather and the fans selected for stagings and their VER values. However, it is not unusual to find that staging the fans with small initial steps and large final steps results in a lower overall cost of ventilation than the cost for equal increments of stages. This is an additional advantage of avoiding large steps between the lowest fan stages.

Example 10-4

Problem: Continue Examples 10-2 and 10-3, but in this example use the fan staging of Example 10-2 and examine the effects of changing various setpoints.

Solution: Program DUTYFACT can be used. For a variety of setpoint schedules, the resulting costs of ventilation are:

Setpoints	Yearly Ventilation Cost (Electricity)
9, 13, 17, 21	$2244.84
9, 13, 19, 21	$2180.09
9, 15, 19, 21	$2169.21
9, 13, 21, 25	$1916.93
9, 15, 21, 25	$1905.41
9, 15, 21, 22	$2052.22
5, 13, 17, 21	$2245.37
9, 15, 17, 21	$2234.09

The setpoint schedules do not vary greatly from the original because the primary purpose of the schedule is to promote cow comfort, not lower the cost of ventilation. However, some effects can be seen. The magnitudes of the effects depend on the fans chosen for the fan schedule and the expected weather bin data, and so should not be seen as general rules for choosing setpoints.

The effect of changing the lowest setpoint is not great in this example. Lowering just the lowest setpoint from 9 to 5 C raised the yearly cost of ventilation by only approximately $10. This is an outgrowth of the weather data. With Denver weather, there are few hours of extreme cold where only stage 1 ventilation would be needed when the first setpoint is 9 C. Lowering the setpoint has little effect, but might be of some benefit to the cows by avoiding a few hours with conditions of minimum ventilation and high relative humidity. This observation may not be true for a region with a colder climate.

The effects of the middle setpoints are greater. Weather data for Denver show a peak in the bin data for outside temperatures between 10.0 and 15.6 C. Raising the middle setpoints so the control table of outdoor air temperatures shows more temperatures in the upper part of this range will yield more significant electricity savings. For example, raising the setpoint schedule to 9, 13, 21, and 25 C is projected to save more than $300, but a further change to 9, 15, 21, and 25 C leads to only $10 additional savings.

It must be emphasized that electricity cost is not the only criterion for choosing setpoints; it is not even the major criterion. Animal comfort and productivity as a function of ambient temperature must be the primary concern. However, within the range of thermoneutrality, the effects of various setpoint schedules can be seen using the duty factor approach, and some savings achieved. In the future, work to optimize setpoint schedules could include animal productivity models, ventilation cost models, heating cost models, etc., to achieve an optimum program of environmental control.

10-6. Quantifying Environment Control Effectiveness

Quantifying the interactions of ventilation control, building design, animal population, and weather can be useful for more than determining the cost of mechanical ventilation.

Environment control in confinement animal housing involves a relatively limited set of environment modification alternatives. Ventilation is the primary and frequently the only means available to modify environment in an animal barn. As a consequence, there will be times during a year when the aerial environment in a typical barn is less than desired, and may even be such as to stress the animals. A means to assist a building designer to estimate the fraction of a typical year that a building will provide conditions within desirable limits would be useful.

An Environment-Control Effectiveness Index (EEI) can be defined as the fraction or percentage of a year when conditions within a housing space are acceptable for the animals within the space. Values of the index will range from 0.0 (never suitable) to 1.0 (always suitable).

The concept of "Acceptable Weather Space" (AWS) has been proposed (Cole, 1982) as a means to visualize weather conditions under which mechanical ventilation can be expected to provide suitable indoor air temperatures and relative humidities. The AWS concept can be understood by beginning with two more fundamental concepts, the Climate Space (CS) and the Production Space (PS).

An animal's CS is defined as the range of weather (environmental) conditions within which the animal can survive. Each type of animal has its own CS, and many animals (but obviously not all) tolerate a CS which is quite wide in terms of air temperature and relative humidity. However, in confinement animal agriculture, simple survival is not enough, animals must also be productive.

An animal's PS is defined as the range of conditions (typically air temperature and relative humidity) in which the animal will remain productive up to or nearly up to its genetic potential, assuming feeding is proper. That is, milk production will not be suppressed due to heat stress, feed conversion and growth will be near optimum, etc. In terms of temperature, the PS approximates the thermoneutral region for the animal in question. The PS is obviously more limited than the CS, but is a subset of it. For highest productivity, the ideal is to keep conditions always within the PS, although highest profitability may result if some deviation out of the PS is permitted during extreme weather.

It is unlikely a conditioned space can be held within a desired PS at all times unless sophisticated and expensive environment conditioning equipment is installed, to include: air conditioning, heating, ventilating, humidifying, and dehumidifying equipment. Environment design must be a compromise between the cost of total environment control and the cost of straying outside the PS occasionally. Practical experience has shown the usual best compromise in confinement animal housing is to have ventilation, evaporative cooling in warm and dry climates, and supplemental heat for small or sick animals, but seldom more. Thus, a barn with few options for environment modification will provide conditions within the PS for only part of the time, which will be quantified here through the Environment-Control Effectiveness Index (EEI).

Conditions inside a barn can be related to outdoor conditions through use of conventional energy and mass balances. For simplicity, at this point only ventilation will be considered, although the concept to be developed applies equally well when more extensive environment modification equipment is available, and to both confinement animal housing systems (barns) and controlled environment agriculture (greenhouses). The details, of course, become more complicated.

Simple, steady-state energy and mass balances for a ventilated airspace are:

$$q_s = (\Sigma UA + FP + mc_p)(t_i - t_o) \qquad (10\text{-}11)$$

$$\dot{m}_p = m(W_i - W_o) \qquad (10\text{-}12)$$

Equations 10-11 and 10-12 connect outdoor conditions to the resulting indoor conditions. The line in Figure 10-4(a) from point o (outdoors) to point i (indoors) can be termed a "conditioning line". For specific conditions at point i, fixed building thermal characteristics, unchanging animal population heat, and moisture production parameters (which is to say, a steady-state), only one outdoor combination of temperature and relative humidity will map to point i. The location of point o along the conditioning line is determined by the only parameter in Equations 10-11 and 10-12 which can change, the ventilation rate. If ventilation is high, point o is close to point i. If ventilation is limited, point o is not close to point i.

Each point i within a conditioned space can be uniquely mapped to an outdoor condition if the ventilation rate at point i is known. Of course, it is possible to find indoor conditions where mapping will take point o outside the realm of possible conditions. For example, moisture production in Figure 10-4(b) is zero and the sensible heat production is unchanged, thus point o must fall above the saturation line. The same is true if point i is at the dry bulb temperature shown, but near saturation (Figure 10-4(c)). Point i cannot be reached from any possible point o by using only ventilation (as long as the sensible heat production is not changed). Ventilation and humidification are required, with the amount of humidification determined by outdoor conditions.

Another assumption in this development is that the PS of the animal group being housed can be described by limits of relative humidity and dry bulb temperature. Assume (constant) lower and upper limits of both relative humidity and temperature. For example, a PS could be defined as between 15 and 25 C, and 30 and 80% relative humidity. Such a PS can be outlined on a psychrometric chart, as shown in Figure 10-5.

Staged ventilation will be assumed within the PS in Figure 10-5. Thermostat setpoints control the ventilation stages and can be superimposed onto the PS. A three-stage ventilation system with two thermostat setpoints, sp_1 and sp_2, is shown in Figure 10-6.

Individual points within the PS, as well as entire regions of the PS, map to uniquely defined outdoor conditions when the ventilation rate is fixed. For example, region 1 of the PS in Figure 10-6 maps as shown in Figure 10-7. One conditioning line, from c to c', is shown. The entire subregion could be mapped point by point or only the corner points could be mapped as a reasonable approximation. The width of the subregion of the PS is usually sufficiently narrow that only corner points are needed and relative humidity contours can be approximated as straight lines.

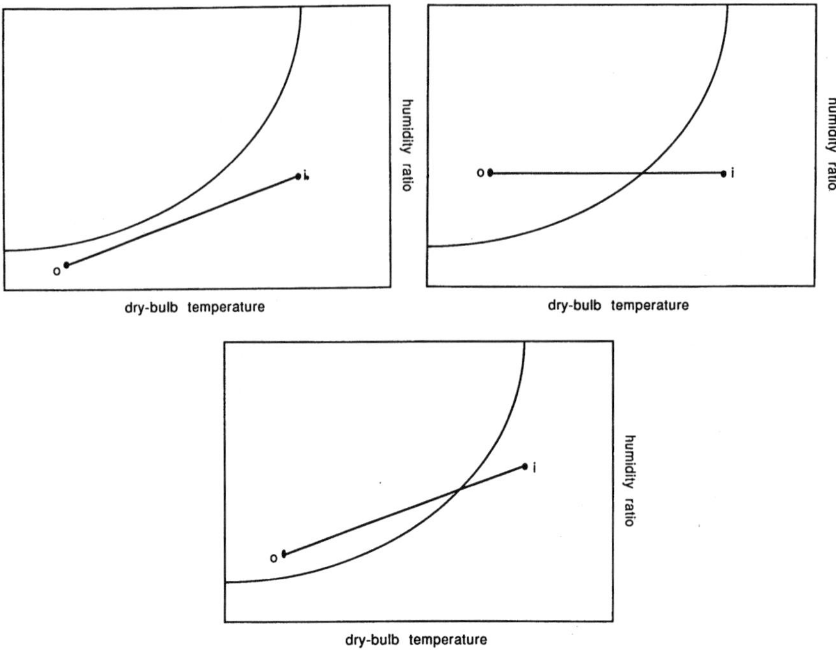

Figure 10-4. Sketch showing conditioning line between outdoor conditions of dry-bulb temperature and relative humidity (point o) and specified indoor conditions (point i). A possible situation is shown in (a). Impossible cases are in (b) and (c) where the outdoor conditions must be above the saturation line on the psychrometric chart to produce point i inside a building. In (b) there is no moisture production within the space, but there is sensible heat production. In (c) the desired indoor conditions are at too high a relative humidity, with insufficient moisture production, to achieve that state using only indoor air for ventilation. Thus, (b) and (c) demonstrate two causes of the same result, the PS can not be achieved by ventilation alone, for moisture production is too low.

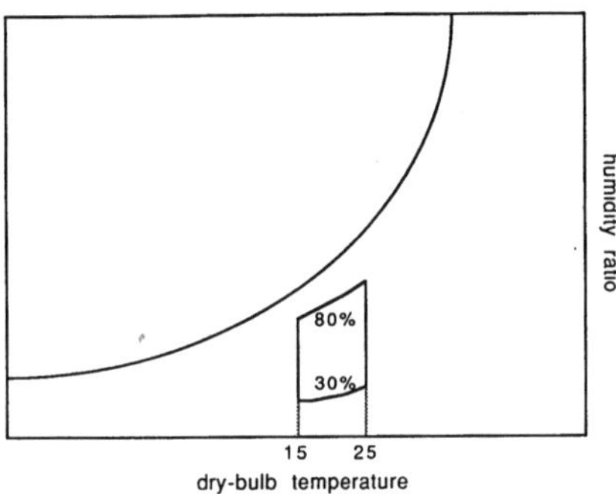

Figure 10-5. Sketch of Production Space defined by upper and lower limits of dry-bulb air temperature and relative humidity.

306

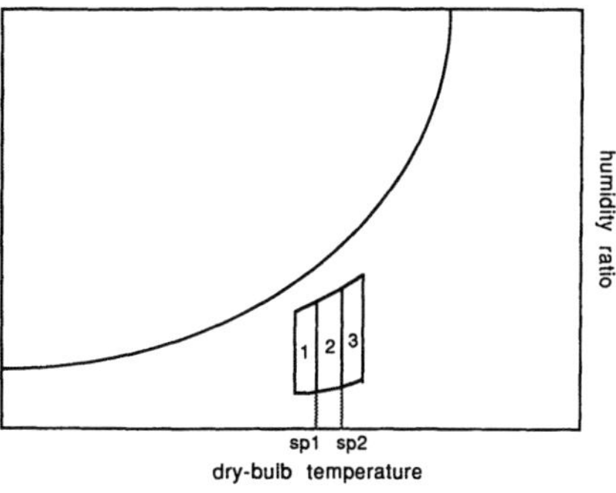

Figure 10-6. Sketch of Production Space demarcated by three ventilation stages and two thermostat setpoints.

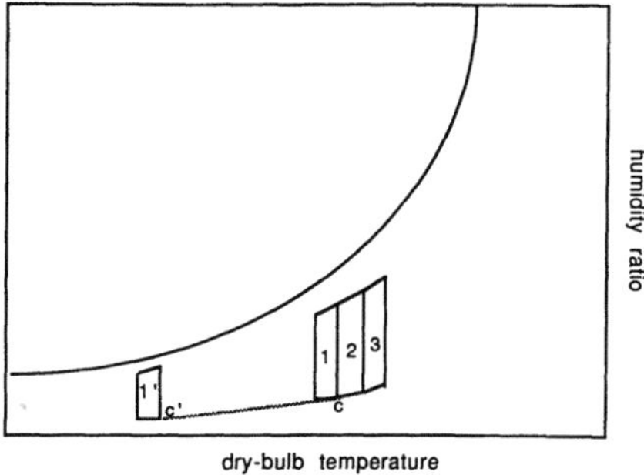

Figure 10-7. Sketch of how a single subregion of a Production Space maps to a region of outdoor conditions - a Weather Space.

The mapped region (1') is almost but not quite identical to region 1 of the PS. Animal heat and moisture production vary from point to point in subregion 1, thus conditioning lines which define the corner points of subregion 1' are not quite parallel and are not the same length.

The entire PS maps as shown in Figure 10-8 because each subregion of the PS is characterized by its own ventilation rate. In actual practice, it is likely the subregion of the PS having the highest ventilation rate will overlap the lower temperature region of the PS, and others may also. However, for clarity the mapped subregions in Figure 10-8 are separated from the PS.

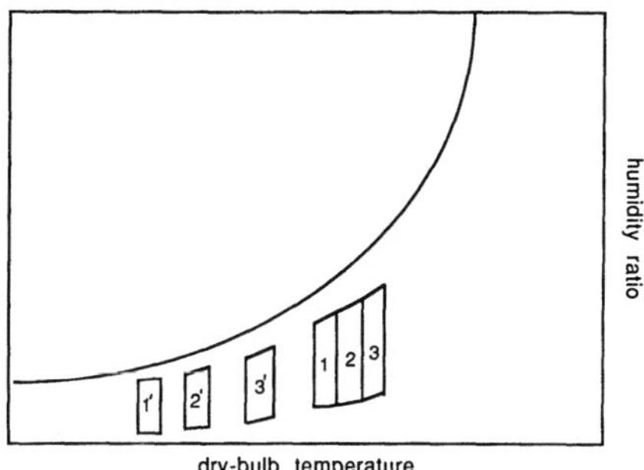

Figure 10-8. Illustration of mapping all subregions of a Production Space with staged ventilation.

A staged ventilation system has been assumed. At each thermostat setpoint a limit cycle is established and control is as shown in Figures 10-2 and 10-3. The result of cycling is to connect the mapped subregions shown in Figure 10-8 (1' to 3'). The right boundary of mapped subregion 1' is the same as the left boundary of mapped subregion 2', which is also true of the right boundary of 2' and the left boundary of 3'. Thus, the mapped subregions connect at their corner points by straight lines, as shown in Figure 10-9 (Cole, 1982).

In Figure 10-9, regions a and b are identified with times of limit cycling, while regions 1', 2', and 3' are identified with times of steady ventilation rates. What is important, however, is that regions 1', 2', 3', a and b form the outline of all possible outdoor conditions which map (permit) the indoor environment to be within the Production Space. This is an important point to realize. The implication is, whenever weather is outside the boundary of the mapped sub-regions, ventilation alone cannot provide conditions within the building which are inside the PS and production may be suppressed, with the degree of suppression being a function of how far actual conditions in the barn are from the PS. All points within the mapped subregions permit ventilation and operation within the PS. These subregions form what is termed the "Acceptable Weather Space" (AWS). Whenever weather falls within the AWS, indoor conditions are within the PS.

Next consider weather and its variation. Each combination of dry bulb temperature and relative humidity on the psychrometric chart can be identified with a probability of occurrence. Extreme conditions have a low probability of occurrence, while moderate conditions (e.g., 20 C and 50% relative humidity) have a relatively high probability in most temperate climates.

When a two-dimensional probability function, such as characterizes weather summarized as a probability density function on a psychrometric chart, is integrated over a region of its domain, the result is the probability of being within that region. If the probability function of weather is known, integrating over the mapped subregions in Figure 10-9 leads to what was defined previously as the Environment-Control Effectiveness Index (EEI).

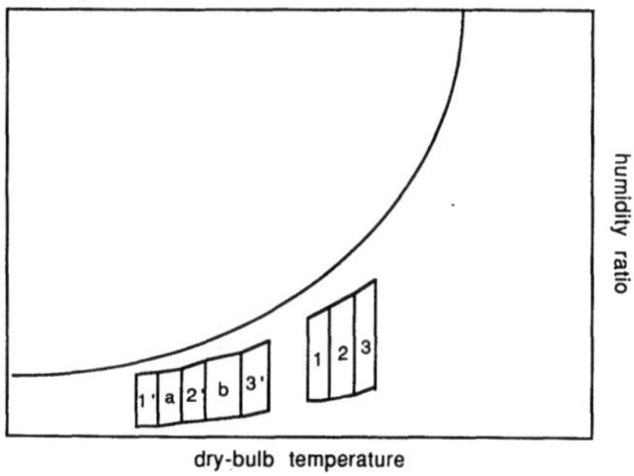

Figure 10-9. An illustration of how mapped subregions of a Production Space connect to form an Acceptable Weather Space.

308

A probability function to describe local climates cannot be readily found in analytical form, but can be obtained from weather records in discrete form – the probability of being at each of the combinations of dry bulb temperature and relative humidity which characterize the climate of the region of interest. As a practical matter, most climate zones fall within the ranges from - 30 to 40 C and 0 to 100% relative humidity. The probability function frequently will not be smooth and monotonically changing. Localized peaks and valleys may result from local weather peculiarities and other factors such as proximity to a large body of water. There frequently will be a peak near freezing at high humidity, a condition which characterizes times of snow and cold rain and nights during the spring and autumn in cool and moderate climates.

Hourly weather data today is sometimes available on a magnetic storage medium and, thus, can be summarized with some degree of ease into probability data. Few years of such data may be available, but as time passes the data will become more extensive. At least five years of data are suggested to approach a usable estimate of the long-term average, and 10 years is significantly better. It must be noted, however, that a designer wishing to obtain a probability function for a region of interest will likely find it necessary to generate one from local weather data. A probability function is not a format in which weather records are typically kept. Weather stations at land-grant universities and state experiment stations are possible sources of data.

10-6.1. Program WEATHER

Program WEATHER is provided in executable form to accomplish calculation of an EEI if probability data of dry bulb temperature and relative humidity is available. An example of such data is provided as data file WEATHER.DAT. The data characterizes a relatively humid climate having cold winters and moderate summers. The data are based on four years of hourly weather records from Ithaca, New York, and are only a rough estimate of the true data.

Program WEATHER is intended to illustrate EEI calculations; it is not a general purpose building design and evaluation program. The program calculates the EEI by following the sequence described in the previous section. The PS is defined by the user. Conditioning lines are calculated to define how subregions of the PS map to subregions of the AWS. Corners of the AWS subregions are connected using straight lines. The weather data file is read and the probability of being within the AWS is calculated.

The file WEATHER.DAT contains probability data for each combination of dry bulb temperature and relative humidity between - 30 and 35 C, and 0 to 100% relative humidity, as follows. If another data file is generated, it must conform to this format.

-30	0	0.0000000000
-30	1	0.0000000000
-30	2	0.0000000000
—	—	—
-30	100	0.0000000000
-29	0	0.0000000000
-29	1	0.0000000000
—	—	—
-20	58	0.0000293204
—	—	—
0	50	0.0000586407
0	51	0.0001466018
—	—	—
9	97	0.0011141735
—	—	—
35	100	0.0000000000

Each data line contains three items: dry bulb temperature (-30 to 35 C); relative humidity (0 to 100%), and the probability of that combination occurring (expressed as a decimal).

10-6.2. Example Use of WEATHER

This section presents an example to illustrate how WEATHER is to be used, and the results one can expect from it. It is best to follow the example on a computer, using WEATHER as we go along.

Example 10-5

<u>Problem:</u> Consider the barn described in Example 9-4, to be located near Ithaca, NY, and to house 62 milking cows averaging 550 kg body mass. A five-stage ventilation system is planned, with the ventilation stages described in the example (1.16, 1.74, 2.88, 4.24, and 8.96 m^3/s). The barn will be located at an elevation 300 m above sea level, and have ΣUA and FP values of 640 and 105 W/K, respectively. The Production Space is between 5 and 25 C, and 30 and 80% relative humidity. Thermostat setpoints of 8, 12, 16, and 20 C will be specified. Determine how well this design will provide the cows with conditions within the PS.

<u>Solution:</u> Program WEATHER and data files WEATHER.DAT and WEASPACE.DAT are needed to solve this problem. The first data file contains the weather probability data, while the second contains other input required by the program; for this example, data consists of default values in the file but can be changed if desired. The program permits this change part way through its sequence of menus.

Output from Program WEATHER can be obtained in several ways. One is as a printed summary, a copy of which is in Figure 10-10. Another is as a graph, a copy of which is in Figure 10-11. The graph is presented with axes of dry bulb temperature and relative humidity rather than the conventional axes of the psychrometric chart. To span the dry bulb temperature range from the lowest temperature of the AWS to the highest temperature of the PS, the relative humidity lines in the low temperature area of the graph would be compressed into such a small region as to be indecipherable.

A preliminary interpretation of the output might be that some error has occurred. The lowest subregion of the PS has been collapsed into a subregion of the AWS of zero height between temperatures of - 21.02 and - 16.40 C. This is, in fact, not an error. It instead indicates the environment control, as specified, is inappropriate at the lowest stage – the moisture removal capacity of a ventilation rate of 1.16 m³/s is never sufficient in the lowest stage to keep the barn's relative humidity less than 80%. The other subregions of the AWS extend over a wide range of relative humidity and thus contribute to ventilation sufficiency. The net result is that the EEI is reduced somewhat by the ineffectiveness of the lowest ventilation stage. To summarize the control, the EEI (0.7363) shows the barn will provide conditions within the PS for slightly less than three-quarters of the hours in a typical year.

```
THE SUITABILITY INDEX IS:    0.7363  <<<<=========

ANIMALS:                              dairy
                          Number:       62
                     Weight, kg:       550.0

BUILDING:                elevation, m:   300.0
                         UA value, W/K:  640.0
                         FP value, W/K:  105.0

PRODUCTION SPACE:        Upper Temp, C:    25.0
                         Lower Temp, C:     5.0
                          Upper RH, %:     80.0
                          Lower RH, %:     30.0

VENTILATION:  5 Stages

     Stages, m3/s:        1.16      1.74      2.88      4.24      8.96
     Set Points, C:            8.00      12.00     16.00     20.00

WEATHER DATA FILE:  WEATHER.DAT
OTHER DATA FILE:    WEASPACE.DAT
CHANGED DATA FILE:  WEASPACE.DAT

ACCEPTABLE WEATHER SPACE SUB REGION COORDINATES
-----------------------------------------------------
                             Temp.   Rel. Hum.

     Stage 1       upper left:   -21.02     0.00
                   upper right:  -16.49     0.00
                   lower right:  -16.40     0.00
                   lower left:   -20.95     0.00

     Stage 2       upper left:   -10.46    81.69
                   upper right:   -4.74    88.96
                   lower right:   -4.65     0.00
                   lower left:   -10.39     0.00

     Stage 3       upper left:     0.70   100.00
                   upper right:    6.01    98.85
                   lower right:    6.09     1.17
                   lower left:     0.77     0.00

     Stage 4       upper left:     8.80    96.76
                   upper right:   13.86    91.74
                   lower right:   13.93    17.54
                   lower left:     8.86    16.10

     Stage 5       upper left:    16.88    87.05
                   upper right:   22.65    83.55
                   lower right:   22.68    25.74
                   lower left:    16.91    26.06
```

Figure 10-10. Printed summary output of Program WEATHER for Example 10-5.

For comparison, the EEI was recomputed using slightly different ventilation stages: 1.8, 2.5, 3.5, 5.3, and 8.8 m^3/s, but with all other data the same. The resulting EEI is 0.7731 and the output graph is in Figure 10-12. With this change, even though the barn cannot hold temperature within the PS when it is colder than - 14 C outdoors, (compared to - 21 C in the base case) the EEI is approximately four percentage points better. This consists of a yearly equivalent of 322 hrs (= (0.7731 - 0.7363)8760 hrs/year) additional time in the PS, because the relative humidity criteria are satisfied at lower outdoor temperatures.

10-7. Ventilation Control in Greenhouses

In principle, the cost of ventilating a greenhouse could be determined using a sensible heat balance and determining times when fans would be needed. However, the procedure would be complicated by the interaction of solar gain and outside air temperature. Weather bin data which combine both solar insolation and air temperature are available from ASHRAE.

Greenhouses are not ventilated continuously during cold weather, thus in moderate climates the fan system duty factor should be less than the duty factor for animal housing in a similar climate. Also, temperature needs of greenhouse crops are sufficiently narrow that little opportunity exists to modify thermostat setpoints to save electricity. The primary opportunity to save electricity when

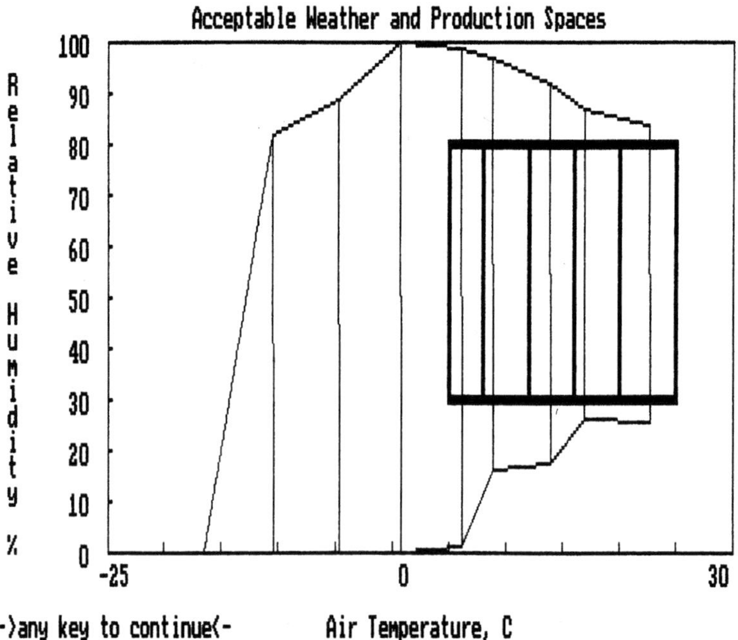

Figure 10-11. Copy of output graphics screen of Program WEATHER for Example 10-5.

ventilating greenhouses is to choose ventilating fans which are inherently energy efficient, and then maintain the fans to prevent a significant degradation of energy efficiency. That is, fan belts should be kept in adjustment, bearings should be oiled, louvers should be properly adjusted, etc.

A common recommendation for greenhouse fan systems is to have a three-stage system. The lowest stage is for winter ventilation and is typically only about 15% of the maximum installed ventilation capacity. Winter ventilation needs are small and too much air movement may cause severe localized chilling of plants. The second stage is typically approximately half the maximum ventilation rate.

Care is required to prevent simultaneous operation of both heating and ventilation in greenhouses. A setpoint for venting at least 3 K above the heating setpoint is recommended. Fan operation then begins at least 3 K above the heating setpoint. (In greenhouses, natural ventilation or "venting" may provide sufficient air movement during times of moderate need, such as during cool weather or warm and cloudy days with wind.) When computerized control systems are installed, where air temperature is always measured from the same sensor or sensors, the problem of simultaneous heating and venting may be avoided.

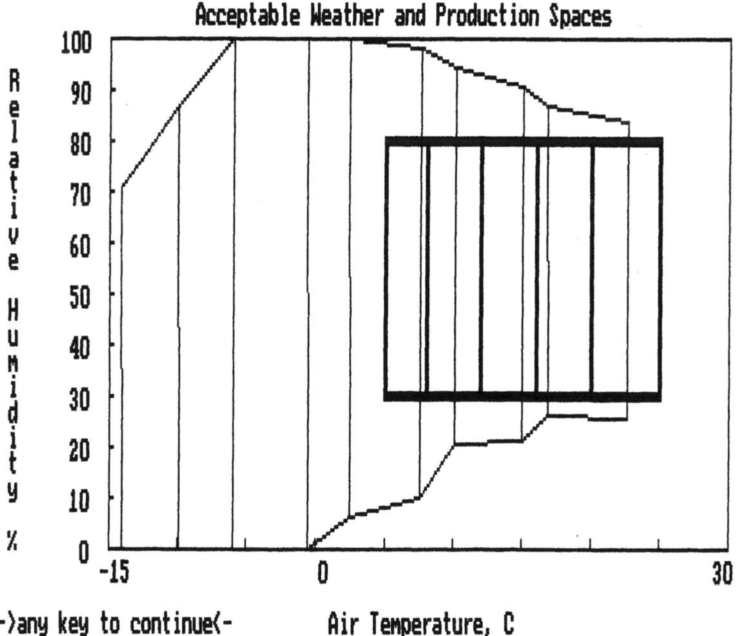

Figure 10-12. Copy of output graphics screen of Program WEATHER for Example 10-5, with ventilation stages changed.

SYMBOLS

a,b,c	coefficients for metabolic (and other) heat generation
A	area, m^2
c_p	specific heat, J/kg K
F	perimeter heat loss factor, W/m K
HLF	building heat loss factor, $\Sigma UA + FP$, W/K
m	mass flow rate, kg/s
$\dot{m}$	water vapor production, kg/s
P	perimeter, m
P[]	probability
q	heat generation or flow, W
t	temperature, C
U	unit area thermal conductance, W/m^2 K
$\dot{V}$	volumetric flow rate, m^3/s
W	humidity ratio, kg/kg
z	elevation above sea level

EXERCISES

1. A plant growth chamber is be ventilated with a two-stage ventilation system. When the temperature in the chamber is below 20 C, a small fan (0.5 m^3/s) is used. When the chamber's temperature is above 20 C, the smaller fan is inactivated and a larger fan (5 m^3/s) is used. The Ventilating Efficiency Ratio (VER) of the small fan is 2 m^3/s-kW, and the VER of the large fan is 5 m^3/s-kW. The heat source in the chamber is 2 kW of fluorescent lights (2 kW, including ballasts), and the UA value for the chamber is 60 W/K. Seventy-five percent of the heat load from the lights appears as sensible heat. The chamber is located outdoors near Fort Knox, Kentucky.

   From this data, estimate the yearly cost of electricity to ventilate this chamber if the unit cost of electricity is $0.11/kWh.

2. A simple ventilating system for an animal housing facility has a single thermostat setpoint and two ventilation stages, 2 and 20 m^2/s. Each stage has a Ventilating Efficiency Ratio of 6 m^3/s-kW and electricity costs $0.18 per kWh. Your calculations have shown the following duty factors

Setpoint, C	Duty Factor	Setpoint, C	Duty Factor
5	0.68	13	0.51
6	0.67	14	0.49
7	0.66	15	0.47
8	0.65	16	0.45
9	0.63	17	0.43
10	0.61	18	0.41
11	0.58	19	0.39
12	0.54	20	0.38

for the ventilation system as a function of the thermostat setpoint (to switch between the two stages):

The production of the animals is also a function of indoor air temperature (the setpoint). Data have shown net income (excluding the cost of ventilation) as a function of air temperature is as follows:

Setpoint, C	Net Income, $	Setpoint, C	Net Income, $
5	43,000	13	49,950
6	47,500	14	49,800
7	48,500	15	49,650
8	49,000	16	49,400
9	49,600	17	49,100
10	49,700	18	48,700
11	49,900	19	48,200
12	50,000	20	47,600

Based on these data, what is the thermostat setpoint for highest profitability?

3. A greenhouse operator in Eugene, Oregon has a separate small building for germinating seeds. Flats of soil are seeded, placed in this building under lights for 10 days, and then taken from the building where the seedlings are transplanted into pots and moved to the greenhouse.

The germination room is well insulated ($\Sigma UA + FP = 65$ W/K) and is ventilated using fans in four stages: 0.5, 1.0, 2.0, and 5.0 m^3/s (total for each stage). The heat load in the room is from lights; total wattage is 50 kW, and it is estimated 70% of the total heat is converted into sensible heat. The rest appears as latent heat and is removed from the room as humidity in the air. There are no other heat sources.

What will be the average ventilation rate (m^3/s) for the room if thermostat setpoints are 22, 24, and 27 C? What is the highest temperature the grower can expect to occur in the germination room during an average year? What fraction (or percentage) of the year can she not expect to maintain the 22 to 27 C temperature range?

4. This problem involves use of the program DUTYFACT. Consider an animal housing facility having a UA value of 850 W/K and an FP factor of 180 W/K. There are 1300 animal units in the barn, each producing sensible heat as a function of indoor air temperature as follows:

$$q(\text{watts}) = 58 - 1.33t - 0.0052t^2$$

where t is air temperature, C. The barn is located near South Bend, Indiana, at an elevation of approximately 200 m.

The calculated minimum ventilation rate is to be 0.01 m^3/s per animal unit, and the maximum will be 0.085 m^3/s. The fans available for use have the following performance data. Note: The airflow data are in cfm and the VER data are in cfm/watt. See the note at the end of the problem statement for conversion factors.

Model	cfm@0"	VER@0"	cfm@.05"	VER@.05"	cfm@.1"	VER@.1"
a	4192	16.4	3870	14.9	3461	13.3
b	4637	13.4	4372	12.5	4030	11.4
c	5769	21.0	5181	18.5	4305	15.1
d	6436	18.1	5926	16.5	5317	14.3
e	8496	22.7	7617	19.8	6481	16.4
f	9572	21.0	8796	18.7	8004	16.6

This problem is to estimate the yearly cost of ventilation for several possible control strategies. The base case is for six ventilation stages, with the stages incremented according to equal intervals of outdoor air temperature. The estimated unit cost of electricity will be $0.09/kWh.

Another possibility is to stage the fans according to even increments of ventilation rate using a six-stage system. Determine the yearly cost of ventilation for this possibility.

For both designs, use setpoints of 13, 16, 19, 22, and 25 C.

Think about why the difference in duty factors and costs should arise.

For selecting fans, assume there will be six fan banks. This should not dictate fan selection, but the expected number of fan banks should be kept in mind when selecting fans so the end results are approximately symmetrical among the fan banks.

Note: This problem and DUTYFACT are based on the SI system. Fan data is available today in the US in IP units. One way to approach the disparity is to convert all fan data to SI units. Conversion factors are:

$$2119 \text{ cfm} = 1.0 \text{ m}^3/\text{s}; \quad 2.119 \text{ cfm} / \text{W} = 1.0 \text{ m}^3/\text{s-kW}$$

REFERENCES

Albright, L.D. and G.W. Cole. 1980. A procedure to estimate fan system duty factors. Transactions of the ASAE 23 (3): 661-670, 675. American Society of Agricultural Engineers, St. Joseph, MI.

Albright, L.D., A.N. Rousseau and B.L. Brockett. 1988. A method to quantify the suitability of mechanical ventilation to provide desired indoor conditions. Proceedings, Livestock Environment III, pp. 195-202. American Society of Agricultural Engineers, St. Joseph, MI.

ASHRAE. 1985. Bin and degree hour weather data for simplified energy calculations (5 computer disks). American Society of Heating, Refrigerating, and Air Conditioning Engineers, Atlanta, GA.

AMCA. Standard 210: Laboratory methods of testing fans for rating purposes. Air Movement and Control Association, Arlington Heights, IL.

Cole, G.W. 1982. A method to predict the performance of mechanically ventilated systems. Transactions of the ASAE 25(2):419-424. American Society of Agricultural Engineers, St. Joseph, MI.

Phelan, R.M. 1977. Automatic Control Systems. Cornell University Press, Ithaca, NY. 280 pp.

CHAPTER 11
NATURAL VENTILATION

11-1. Introduction

Natural ventilation is ventilation induced by so-called natural means: wind forces and thermal buoyancy. The focus of this chapter will be on naturally ventilated buildings for relatively cold climates where ventilation must be controlled or limited for at least part of the year. The only housing that may be required, where weather is seldom or never cold, is a naturally ventilated sun shade. This ventilation is not designed, it just happens.

As electricity prices have risen relative to the general inflation rate, there has been a revival of interest in using natural ventilation for certain types of animal housing. Great interest has been expressed relative to housing dairy cows and beef animals. Cows tolerate low air temperatures with little effect on milk production. They eat more feed to compensate for extra heat loss during cold weather, but there are few other effects.

When milking herd size is greater than approximately 80 cows, current dairy science recommendations are to use free-stall rather than tie-stall housing to decrease construction costs per cow and increase labor efficiency. Free-stall barns are usually naturally ventilated. Naturally ventilated dairy barns are also

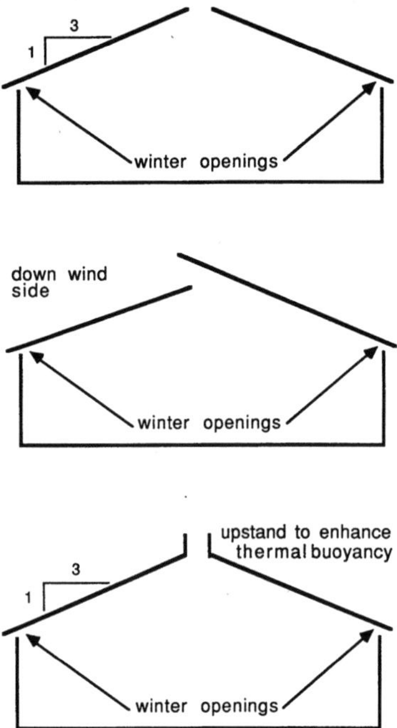

Figure 11-1. Typical designs of naturally ventilated barns.

known as "cold housing" because airflow rates are sufficient during the winter to keep indoor air temperature to within a few degrees of outdoor air temperature. Typical designs for naturally ventilated barns are sketched in Figure 11-1.

Considerable interest also exists to use naturally ventilated housing for swine and broilers, but there is only limited interest to apply natural ventilation to laying-hen housing where temperature control is a critical matter and animal stocking density is very high. Although natural ventilation can provide a great deal of air exchange, air velocity through inlets is not sufficient to achieve good air mixing within the ventilated space if there are many obstructions to airflow as there are in laying-hen poultry houses.

Natural ventilation is frequently part of the ventilation schedule for greenhouses and is an intermediate step before fans are activated. In moderate climates, natural ventilation may be capable of providing the majority of ventilation during a year. Where weather is seldom hot and a crop is grown which tolerates occasional high temperatures, some commercial greenhouse ranges are still ventilated only by natural means.

Natural ventilation can be very vigorous and may be designed to provide as much or more airflow than can mechanical ventilation designed using conventional techniques, given a proper design and suitable weather conditions. Much natural ventilation system design to date has been based on rules of thumb, but analysis tools are available to develop designs on a sounder theoretical basis.

The rules of thumb to design naturally ventilated barns are relatively simple. It is recommended a naturally ventilated building have a roof slope no less than 1:4 (14°). This is recommended to promote ventilation by thermal buoyancy during cold weather when there is little wind. Thermal buoyancy can be viewed as a mirror image of water flowing to a floor drain. A drain in a flat floor causes problems, water does not flow toward the drain. If a roof slope is too shallow, warmed air has little inducement to rise toward the exit, which should be located at the highest point in the airspace. In addition, the underside of the roof should be smooth.

Another rule of thumb for natural ventilation is that sidewall openings should be large during warm weather (1 to 1.5 m, at least) to promote ventilation by the wind. With openings this wide, even slight breezes provide many air exchanges per hour within the building, and wind is almost never totally still. Sidewall openings must be closed during cold weather to prevent cold winds from causing stress on the animals. Vents only approximately 0.1 to 0.2 m wide under the eaves are left open to permit entry of fresh air, due either to wind effects or thermal buoyancy.

Simple rules have been developed to determine the size of the ridge vent (air exit). A ridge vent approximately 0.15 m wide has been found to work well when a barn is no more than 12 m wide. For every additional 3 m of width (above 12 m), the ridge vent should be an additional 0.05 m wide. For example, a barn 15 m wide should have an unobstructed ridge vent 0.2 m wide.

When naturally ventilated barns are used in cold climates, care must be taken to prevent excessive snow entry through the roof vent. For example, in the design sketched in Figure 11-1a, the ridge vent is a simple slot along the ridge of the roof. Some persons object to a hole in the roof, for rain can enter. Unless the climate is very rainy, the amount of water from rain is insignificant compared to the water added to the floor by the animals, but rain may still be undesirable in certain situations.

An initially attractive solution, but ultimately a naive one, is to place a simple raised roof section over the ridge vent. The raised roof stops rain but acts as a snow fence whenever there is wind-driven snow. Wind passing over a building accelerates as the streamlines converge and high-speed wind can carry more snow than can low-speed wind. As wind passes under the raised roof section, streamlines diverge and the wind slows. Snow carrying capacity is reduced and the effect is similar to the action of a snow fence. In some instances, such designs have led to 2 m deep snowdrifts down the centers of barns, a problem somewhat greater than a little rain.

There is some evidence that baffles located along a roof, parallel to a raised roof section, may help prevent the snow fence effect described above. The baffle deflects air to flow over the top of the raised roof section, creating a suction at the ridge vent rather than an overpressure. It has been suggested the baffle be designed so $H = 2W$.

Use of such baffles is coming into general practice although their effectiveness under all conditions has not been demonstrated. As an alternative, commercially available ridge vent covers may be installed which have been designed to enhance wind suction at the ridge vent and increase ventilation, while preventing snow entry. Care is needed if such a unit is specified to allow for sufficient air movement through the ridge vent during times of no wind.

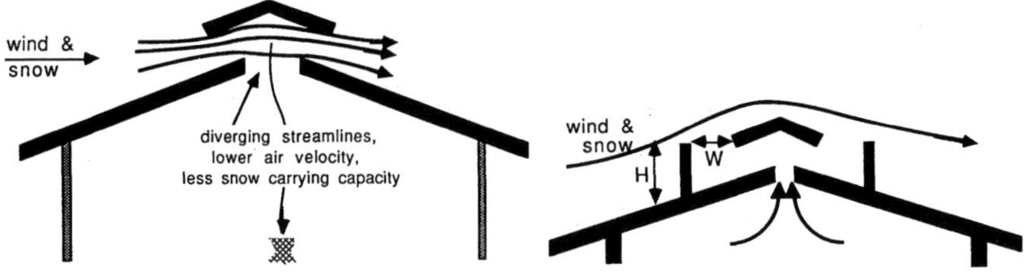

11.2. Natural Ventilation Due to Thermal Buoyancy, An Empirical Approach

A simple model exists to predict thermally-induced natural ventilation where there is one inlet and one outlet, but its use should be limited to making initial or field estimates. If the areas of inlet and outlet are equal, there is no wind and the coefficient of discharge of each opening is 0.65, airflow can be estimated by

$$\dot{V} = 2A[g\Delta h(T_i - T_o) / T_i]^{1/2}, \qquad (11\text{-}1)$$

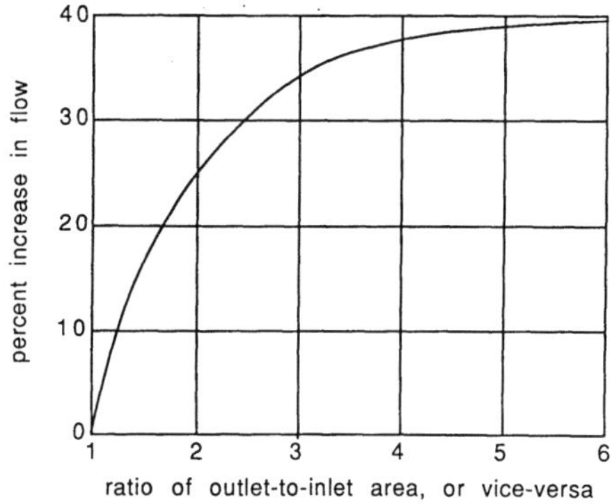

where $\dot{V}$ is in m³/s, g is the gravitational constant, A is the area of one of the openings (either opening, they are equal), Δh is the distance, m, between the two openings, and T_i and T_o are indoor and outdoor air temperatures, K, respectively. If the coefficient of discharge is not 0.65, Equation 11-1 must be multiplied by the expected coefficient, divided by 0.65. Coefficients of discharge will be assumed in this chapter to be independent of the flow rates through the openings. Evidence exists that this may not be true for airflow through small openings (infiltration) but is thought to be a reasonable assumption for larger, ventilation openings.

When the two areas are not equal, the smaller of the two is used in Equation 11-1 and $\dot{V}$ is adjusted using the graph in Figure 11-2.

Example 11-1

Problem: Estimate the ventilation rate due to thermal buoyancy in a dairy barn 75 m long having a continuous ridge vent 0.2 m wide, and continuous vents under each eave 0.25 m wide. Outdoor air temperature is - 20 C and indoor air temperature is - 10 C. The building is 12 m wide and the roof slope is 1:3.

Figure 11-2. Increase in ventilation rate through two openings due to thermal buoyancy alone when one opening is larger than the other.

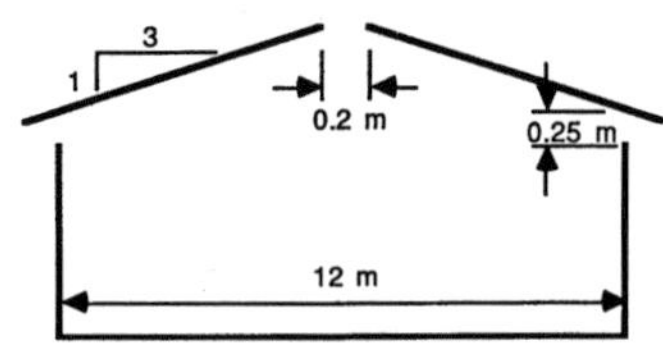

<u>Solution:</u> Equation 11-1 applies, but is based on the smaller of the two openings. Note in this example that although there are two eave vents, they are at the same elevation and can be combined into one vent for computational purposes, with a combined area of 37.5 m^2. This assumes the coefficients of discharge of the two openings are equal. The smaller area is (75 m)(0.2 m) = 15 m^2.

The elevation difference between the two openings is the total rise of the roof. A roof slope of 1:3 over a distance of 6 m leads to a rise of 2 m between the inlet and outlet, plus the distance from the top of the eave vent to its center, less the distance from the edge of the roof vent to the peak. (These corrections could be ignored as small, but are 0.125 - 0.03 = 0.095 m.) The ridge vent lies in a horizontal plane. Air temperatures are: T_i = 263.15 K and T_o = 253.15 K. Were the inlet and outlet areas equal, the ventilation rate would be

$$\dot{V} = 2(15 \text{ m}^2) \, [(9.8 \text{ m/s}^2) \, (2.095 \text{ m}) \, (263.15 \text{ K} - 253.15 \text{ K}) \, / \, (263.15 \text{ K})]^{1/2}$$
$$= 26.5 \text{ m}^3/\text{s}.$$

The inlet and outlet areas are not equal, thus the ventilation rate will be greater than 26.5 m^3/s by the factor shown in Figure 11-2. The ratio of areas is 37.5 m^2/ 15 m^2 = 2.5, and the factor is approximately 1.32. The predicted ventilation rate is, thus,

$$\dot{V} = 1.32 \, (26.5 \text{ m}^3/\text{s}) = 35 \text{ m}^3/\text{s}.$$

Note that a barn this size would house approximately 100 cows, therefore, the ventilation rate is equal to 0.35 m^3/s per cow. This ventilation rate is more than double the design maximum summer ventilation rate for dairy cows in the northern areas of the United States. Natural ventilation even by thermal buoyancy alone can be vigorous.

Note a weakness of this example. The indoor air temperature was assumed to be 10 K above the outdoor air temperature. With vigorous ventilation, such a large temperature difference would not be likely. The method of this example does not directly include a sensible energy balance and, thus, is awkward to use for design purposes unless the indoor air is heated to a specified temperature.

11-3. Thermal Buoyancy, the Stack Effect

Thermal buoyancy-induced ventilation through two openings at different elevations in an otherwise closed airspace is a classical problem of fluid mechanics, the so-called chimney problem. Consider a building as shown in Figure 11-3. There is no wind; the only forces acting to create ventilation are those of thermal buoyancy. The lower of the two ventilation openings can be single or can be a combination of two openings, one on each side of the building, each at the same elevation and with the same coefficient of discharge.

The two openings are separated by a difference of elevation, $\Delta h = h_1 - h_2$. The upper opening lies in a horizontal plane and the height of the lower opening is assumed small compared to the elevation difference, Δh. Air outside the building has a density of ρ_o and ρ_i is the density of the indoor air. Point 1 is located outside the building at the elevation of the upper opening. Points 2 and 3 are outside and inside the lower opening, at the midpoint of that opening. Point 4 is just inside the upper opening.

If one begins at any point around the building and traces pressure changes along any path, the pressure changes should sum to zero upon return to the starting point. If the path traverses the points 1, 2, 3, 4, and back to 1, some of the pressure differences will be due to fluid statics and some due to fluid dynamics. The differences are:

$$
\begin{aligned}
\text{1 to 2: } & P_1 - P_2 = -\rho_o g \Delta h && \text{(a fluid statics difference),}\\
\text{2 to 3: } & P_2 - P_3 = \rho_o v_2^2 / 2 && \text{(a fluid dynamics difference),}\\
\text{3 to 4: } & P_3 - P_4 = \rho_i g \Delta h && \text{(a fluid statics difference),}\\
\text{and \quad 4 to 1: } & P_4 - P_1 = \rho_i v_4^2 / 2 && \text{(a fluid dynamics difference).}
\end{aligned}
$$

If the differences are summed,

$$
2 g \Delta h (\rho_o - \rho_i) = \rho_o v_2^{\,2} + \rho_i v_4^{\,2}. \tag{11-2}
$$

For now assume ρ_o and ρ_i are known. Equation 11-2 contains two unknowns, the two velocities. However the velocities can be related through the mass flow continuity relationship:

$$
\rho_o C_{d2} A_2 v_2 = \rho_i C_{d4} A_4 v_4. \tag{11-3}
$$

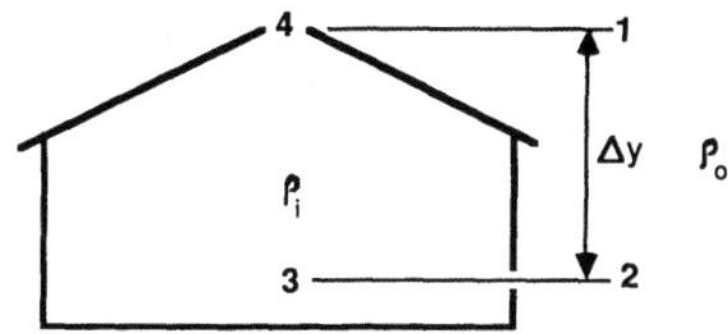

Figure 11-3. Cross-section of a simple building ventilated by thermal buoyancy.

Equations 11-2 and 11-3 can be solved to yield one equation in one unknown,

$$v_4 = \left(\frac{2g\Delta h(1 - \rho_i / \rho_o)}{(\rho_i / \rho_o) + (\rho_i / \rho_o)^2 (C_{d4}A_4 / C_{d2}A_2)^2} \right)^{1/2} \qquad (11\text{-}4)$$

If air is assumed to act as a perfect gas,

$$\rho_i / \rho_o = T_o / T_i, \qquad (11\text{-}5)$$

where T_i and T_o are indoor and outdoor air temperatures (K), respectively.

Using the perfect gas relationship, Equation 11-4 can be restated in terms of temperature, a more useful form,

$$v_4 = \left(\frac{2g\Delta h(T_i - T_o)}{T_o + (T_o^2 / T_i)(C_{d4}A_4 / C_{d2}A_2)^2} \right)^{1/2} \qquad (11\text{-}6)$$

This form is still not as useful for design as it might be, for the temperature difference is not known a priori. However, if q_{prod} is the sensible heat produced within the building, a simple sensible heat balance is

$$q_{prod} = (\Sigma UA + FP)(T_i - T_o) + c_p \rho_i C_{d4} A_4 v_4 (T_i - T_o) \qquad (11\text{-}7)$$

where c_p is the specific heat of air, 1006 J/kgK.

Equation 11-7 can be solved for $(T_i - T_o)$ and Equation 11-6 modified as follows:

$$v_4 = \left(\frac{2g\Delta h q_{prod}}{T_o[1 + (T_o / T_i)(C_{d4}A_4 / C_{d2}A_2)^2][\Sigma UA + FP + c_p \rho_i C_{d4} A_4 v_4]} \right)^{1/2}$$
$$(11\text{-}8)$$

This completes the sequence of equations needed to determine the rate of ventilation due to thermal buoyancy. However, Equation 11-8 cannot be solved directly. It is implicit. Indoor air temperature is not known independently. An iterative solution is possible as follows:

1. Begin with an initial estimate of v_4 (for example, 1 m/s).

2. Solve for ρ_i using a rearrangement of Equations 11-2 and 11-3,

$$\rho_i^2 [(C_{d4}A_4 / C_{d2}A_2)^2 (v_4^2 / \rho_o)]$$
$$+ \rho_i(v_4^2 + 2g\Delta h) - 2g\Delta h \rho_o = 0 \qquad (11\text{-}9)$$

    which is a quadratic equation in ρ_i. When a value of v_4 is available, all terms in the equation except ρ_i are known.

3. Use the perfect gas relationship to estimate the value of T_4,

$$T_4 = \rho_o T_o / \rho_i. \qquad (11\text{-}10)$$

4. Substitute the new values of ρ_i and T_4 into Equation 11-8 to obtain a new estimate of v_4.

5. Return to step 1 and continue to iterate until a stable solution for v_4 is obtained.

Note how this approach differs from the method of Section 11-2. Inlet and outlet areas may differ without need for a correction. Sensible heat production within the building is directly incorporated into the approach, obviating the need to assume an indoor air temperature. However, the assumption must still be made that only two ventilation areas are involved (one or both may be combined areas if the combination is of areas at the same elevation, having the same coefficient of discharge). Also, note this method assumes air velocity is uniform across each of the inlets/outlets. In reality this is only an approximation.

The Neutral Pressure Level concept must be imposed to account for air velocity variations across an opening (see Section 11-4). If the openings are small compared to the separation between them, the approximation of uniform velocity used in this section should be adequate.

Example 11-2

<u>Problem:</u> The recommended design of greenhouses for natural ventilation (ref: Ventilation of Agricultural Structures, edited by Hellickson and Walker) is:

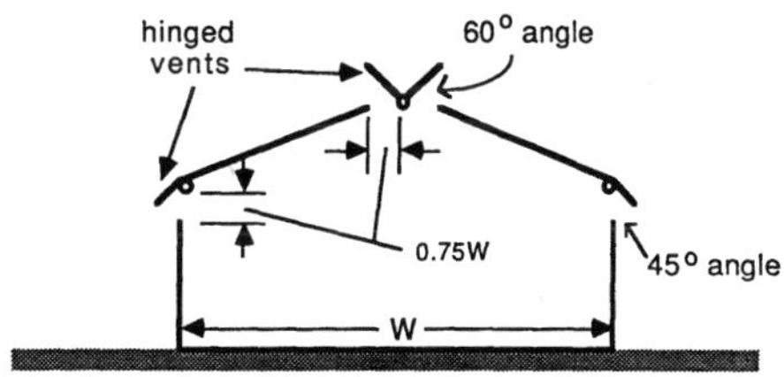

You are designing a greenhouse according to the above plan for a location at 1000 m elevation. The greenhouse is to be 10 m wide and 120 m long. The height to the eaves is to be 2.5 m and the roof slope is to be 1:3. The total heat loss factor (ΣUA + FP) has been calculated to be 8500 W/K. During the middle of the day, solar insolation (outside, on a horizontal surface) is expected to be as high as 700 W/m^2. The overall transmissivity of the greenhouse is expected to be 0.68. Of the solar insolation which passes into the greenhouse, 35% appears as sensible heat in the airspace. The rest is either transformed to latent heat, reflected back to the outside, or used in photosynthesis.

The inlets and outlets will each be 0.75 m wide and will extend the full length of the greenhouse. If the outside air temperature is 20 C and there is no wind, what will be the ventilation rate of the greenhouse (m^3/s and air exchanges per hour)?

Solution: The expected air density outside the greenhouse can be found if a relative humidity is assumed. It was not specified, so assume 50%; air density will be 1.18 kg/m^3.

The total sensible heat load can also be calculated directly. It is:

$$q_{prod} = (700 \text{ W/m}^2)(0.68)(0.35)(1200 \text{ m}^2) = 200,000 \text{ W.}$$

The areas A_2 and A_4 are each 2(0.75 m)(120 m), or 180 m^2. The distance from ridge to eave is (1/3)(5 m), or 1.67 m. The centers of the ridge vents are not exactly at the ridge and the centers of the wall vents are not exactly at eave height. The small differences could be omitted or could be included. There would be little difference in the result. If included, the center-to-center distance from the eave-to-ridge vents is

$$\Delta h = 1.67 \text{ m} +1/2(0.75 \text{ m}) - 1/2(1/3)(0.75 \text{ m}) = 1.92 \text{ m.}$$

A further adjustment could be made because the actual wall opening is between the edge of the hinged vent section and the wall. It will be ignored as a small effect. The coefficient of discharge at each opening is not specified. Each opening is approximately represented as a sharp edged orifice – no special transition sections are included to facilitate airflow through the opening, but neither are there additional flow restrictions. Thus, coefficients of discharge will be assumed to be 0.6 and will cancel. This is a conservative estimate of C_d.

Data are now available to begin the iterative solution procedure outlined above. The procedure begins with an assumption of air velocity at the ridge vent, v_4. Assume v_4 is 1.0 m/s.

The next step is to solve for ρ_i using Equation 11-9,

$$\rho_i^2[(180 \text{ m}^2 / 180 \text{ m}^2)^2 (1.0 \text{ m/s})^2 / 1.18 \text{ kg/m}^3]$$

$$+ \rho_i[(1.0 \text{ m/s})^2 + 2(9.8 \text{ m/s}^2) (1.92 \text{ m})]$$

$$- 2(9.8 \text{ m/s}^2)(1.92 \text{ m})(1.18 \text{ kg/m}^3) = 0, \text{ or}$$

$$0.8475\rho_i^2 + 38.632\rho_i - 44.406 = 0.$$

This is in the form

$$a\rho_i^2 + b\rho_i + c = 0.$$

Although the quadratic has two possible solutions, air density must be a positive number and the only solution which permits that is

$$\rho_i = [-b + \sqrt{(b^2 - 4ac)}]/2a, \text{ or}$$

$$\rho_i = 1.122 \text{ kg}/m^3$$

This corresponds to an indoor air temperature of

$$T_4 = (1.18 \text{ kg/m}^3)(293.15 \text{ K}) / (1.122 \text{ kg/m}^3)$$
$$= 308.34 \text{ K}.$$

Sufficient data are now available to calculate the next estimate of v_4 (Equation 11-8):

$$v_4 = \left(\frac{2(9.8 m/s^2)(1.92 m)(200,000 W)}{293.15 K[1+(293.15 K/308.34 K)(180 m^2/180 m^2)^2][8500 W/K+\beta]} \right)^{1/2}$$

where $\beta = (0.6)(1006 \text{ J/kgK})(1.122 \text{ kg/m}^3)(180 \text{ m}^2)(1.0 \text{ m/s})$
$= 121,900 \text{ W/K}.$

The solution for v_4 is

$$v_4 = 0.318 \text{ m/s}.$$

This estimate of v_4 is used for the next iteration, and iterations continue until convergence. The sequence of iterations leads to the following data:

Iteration	T_i	ρ_i	v_4
1	308.34	1.122	0.318 m/s (as calculated above)
2	294.72	1.174	0.513
3	297.21	1.164	0.420
4	295.88	1.169	0.457
5	296.38	1.167	0.441
6	296.16	1.168	0.448
7	296.25	1.168	0.445
8	296.21	1.168	0.446
9	296.23	etc.	0.445
10	296.22		0.446
11	296.23		0.446
12	296.22		0.446
13	296.23		etc.
14	296.23		
15	296.23		
	etc.		

Convergence is to $v_4 = 0.446$ m/s at an inside air temperature of 23.08 C, or approximately 23 C. The outlet area is 180 m^2, thus, the ventilation rate is

$$\dot{V} = C_{d4}A_4v_4 = 0.6(180 \text{ m}^2)(0.446 \text{ m/s}) = 48.2 \text{ m}^3/\text{s}.$$

The volume of the greenhouse is approximately 4000 m^3, thus, the ventilation provides one air change every 83 s. The air exchange rate is 0.72 air exchanges per minute or the approximate exchange rate one would calculate for a maximum required summer ventilation rate.

The solution showed only a 3 K temperature rise of air temperature above the outdoors, yet the solar insolation is high. This is an indication that natural ventilation can provide significant cooling even under severe conditions. The need for forced ventilation in greenhouses may be dictated more by a need for fans to move air through evaporative coolers and to compensate for inadequate inlet and outlet areas and poor air distribution, than a need for simple ventilation.

Finally, note that although q_{prod} was treated as a constant in this example, it can just as easily be permitted to vary as a function of indoor air temperature, T_4.

11-4. Thermal Buoyancy, the Concept of a Neutral Pressure Plane

In Section 11-3, a pressure loop was used to obtain a solution to the chimney, or stack, problem. Now consider pressure variation within a building in more detail.

Begin by visualizing a closed building which is heated and warmer inside than out, and indoor air temperature is everywhere the same. If a small opening is made at the top of the building, and a corresponding opening made at the bottom of the building, warm air will flow out of the top opening and be replaced by cold air flowing into the bottom opening. The direction of airflow indicates air pressure within the building is higher than outside in the upper regions of the building and is lower in the lower regions. On the average, air pressure within the building equals atmospheric pressure but is not uniform in the vertical direction. It is uniform in the horizontal plane.

Air pressure within the building varies as sketched in Figure 11-4. The sketch is for a building with no inlets or outlets. In this case, the pressure difference shows linear variation from top to bottom with the positive pressure at the top equal in magnitude to the negative pressure at the bottom. The pressure difference is zero at the midpoint of the building. This is the neutral pressure plane – the level where indoor and outdoor air pressures are the same. The variation of pressure difference is a linear function of elevation and can be calculated as a fluid statics problem.

Consider next the effect of inlets and outlets on the pressure difference and the level of the neutral pressure plane. If the areas and coefficients of discharge of the inlet and outlet are identical, the level of the neutral pressure plane does not change and the pressure distribution is represented again by the sketch in Figure 11-4. If the area of the top opening is larger than the area of the bottom opening, the level of the neutral plane changes to be nearer the larger of the openings,

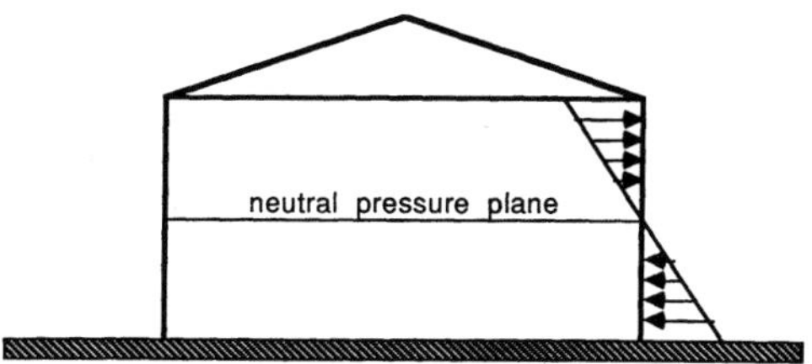

Figure 11-4. The difference between inside air pressure and outside air pressure in a heated building with no air inlets or outlets.

which is at the top. The pressure distribution changes to that shown in Figure 11-5 and the neutral pressure level moves as shown. However the variation of pressure difference over elevation remains linear and, if air densities are not changed, will have the same slope.

If the converse occurs and the largest opening is at the bottom of the building, the neutral pressure plane moves to near the lower opening (as shown in Figure 11-6). By anticipating the effect of the pressure differences, one could reason that the neutral pressure plane must move toward the larger of the two openings, for airflow will be a function of the pressure difference, and for the two airflows to balance, the opening with the largest area must be the one with the smallest pressure difference.

Figure 11-6 shows an additional possibility in the relationship between the neutral pressure plane and a vent opening. The neutral pressure level lies within the vent opening. This is always possible when there are vents on walls, especially if the wall vent is much larger than the top vent (or if there is no top vent). The elevation of the plane must be between the highest and lowest point of the airspace. The horizontal ceiling in Figure 11-6 is a limit of how high the neutral pressure plane can be – the limit when the top vent is very large and the bottom vent area is nearly zero. The neutral pressure plane can be no higher than the top of the highest vent, nor lower than the bottom of the lowest vent.

There has been no stated limitation on the number of vent openings. Only two have been shown in the figures for simplicity, but any number may be present and not change the general picture. As an approximation, if only two vents are present, the analytical method of Section 11-3 may suffice, and the concept of

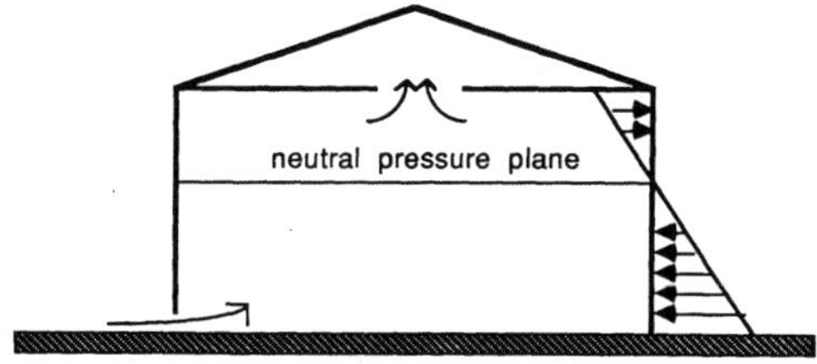

Figure 11-5. The difference between inside and outside air pressures in a heated building with a large vent opening at the ceiling and a small vent opening near the floor.

the neutral pressure plane is not needed. This approximation is valid if each vent is in a horizontal plane; or if in a vertical plane, the separation between the vents is much greater than their widths.

The pressure difference from inside to outside in the buildings (stack effect), shown in Figures 11-4 to 11-6, can be calculated from

$$P_s = (\rho_o - \rho_i)g(h_p - h),$$
(11-11)

where P_s is the stack pressure (negative above the neutral pressure plane), ρ_o and ρ_i are air densities outside and inside the building, g is the gravitational constant, and h is the distance above a reference datum plane (h_p is the distance above the datum to the neutral pressure plane). A suggested datum plane is the building's floor.

If there is a vent opening, airflow velocity, v, through the opening can be determined from the Bernoulli equation,

$$v = \sqrt{2P_s / \rho},$$
(11-12)

where air density, ρ, represents either the outside or inside air, depending on the direction of airflow.

Equations 11-11 and 11-12 can be combined in the following way.

$$v = \frac{\left|2g(\Delta\rho / \rho)(h_p - h)\right|^{3/2}}{2g(\Delta\rho / \rho)(h_p - h)}$$
(11-13)

The form of Equation 11-13 is chosen to retain the vector identity of v. If $h_p >$ h, v is positive. If $h_p <$ h, v is negative and flow is out of the building at that point. Positive and negative flow directions are arbitrary but must be consistent. For now, consider flow out of the building to be defined as negative airflow. Note, to be precise, the value chosen for ρ in Equation 11-13 depends on the direction of airflow and equals ρ_o if flow is into the building and ρ_i if flow is out. The term $\Delta\rho$ is always positive for $\rho_o > \rho_i$.

The air velocity defined in Equation 11-13 is a function of distance from the neutral pressure plane and thus will not be constant across any vent opening

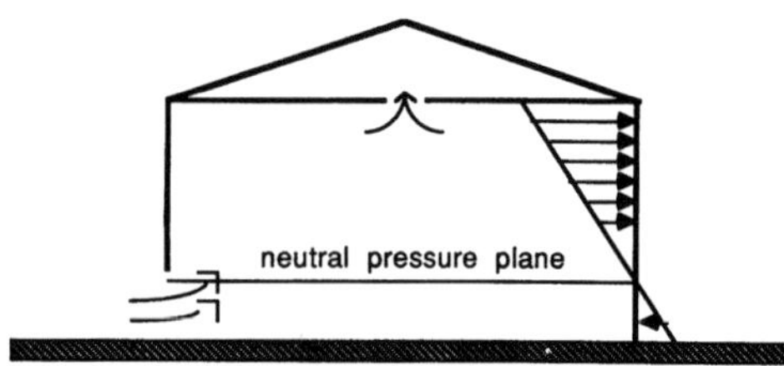

Figure 11-6. The difference between inside and outside air pressure in a heated building with a small vent opening at the ceiling and a large vent opening near the floor.

except one which is horizontal, as in a flat ceiling. The volumetric airflow rate is the integral of velocity over area, or

$$\dot{V} = C_d \int_{area} v\,dA \qquad \text{(for all openings)} \qquad (11\text{-}14a)$$

$$= C_d v A \qquad \text{(for horizontal openings)} \qquad (11\text{-}14b)$$

where C_d is the coefficient of discharge of the opening.

Mass flow continuity must hold, $\Sigma \rho \dot{V} = 0$, assuming no mass addition or removal within the space. The sum of mass flows over all openings of the airspace must be zero,

$$\sum_n C_{dn}\rho_n \int_{area_n} \frac{\left|2g(\Delta\rho/\rho)(h_p - h)\right|^{3/2}}{2g(\Delta\rho/\rho)(h_p - h)}\,dA$$
$$+ \sum_m C_{dm}\rho_m \frac{\left|2g(\Delta\rho/\rho)(h_p - h)\right|^{3/2}}{2g(\Delta\rho/\rho)(h_p - h)} W_m L_m = 0. \qquad (11\text{-}15)$$

where the index m sums air mass flow over all horizontal openings (e.g., ridge vents) and n sums over all others (e.g., sidewall vents). The width of horizontal vents is W and the length is L, and the vents are assumed to be rectangular.

Each integral in Equation 11-15 is in the form

$$\int_{area} \frac{\left|\alpha(h_p - h)\right|^{3/2}}{\alpha(h_p - h)}\,dA \qquad (11\text{-}16)$$

where $\alpha = 2g\Delta\rho/\rho$.

If each non-horizontal vent opening is rectangular and of length L,

$$dA = L\,dh, \qquad (11\text{-}17)$$

where L is the length of the opening, and h is the width dimension. The terms of the continuity equation related to non-horizontal openings are now

$$\Sigma C_{dn}\rho_n L_n (\alpha_n)^{1/2} \int_{bottom}^{top} \frac{\left|(h_p - h)\right|^{3/2}}{(h_p - h)}\,dh. \qquad (11\text{-}18)$$

The integral

$$\int_{bottom}^{top} (h_p - h)^{-1} \left|(h_p - h)\right|^{3/2} dh \qquad (11\text{-}19)$$

has an analytical solution

$$-(2/3)\left|(h_p - h)\right|^{5/2}/(h_p - h)\Big|_{bottom}^{top} \qquad (11\text{-}20)$$

thus the final form of the continuity equation is

$$(-2/3)\sum_n C_{dn}\rho_n L_n \alpha_n^{1/2}\left|(h_p - h)\right|^{5/2}(h_p - h)^{-1}\Big|_{bottom}^{top}$$

$$+ \Sigma C_{dm}\rho_m L_m W_m \alpha_m^{1/2}\left|(h_p - h)_m\right|^{3/2}(h_p - h)^{-1} = 0. \qquad (11\text{-}21)$$

Although it may not be obvious initially, Equation 11-21 is one equation with a single unknown, h_p. If h_p can be determined, airflow through each vent can be calculated and the total ventilation rate determined. Even instances where the neutral pressure plane intersects an opening can be included by treating such openings as two, one with positive and one with negative airflow.

The solution procedure for h_p is numerical and most easily accomplished using iteration. One approach is to begin with $h_p = 0$ (at the floor), solve the continuity equation, and note the sign of Σ(flows). The sum should be negative, all flow is out of the airspace because the neutral pressure plane is below every opening. Increment h_p in the positive direction (for example, add 0.5 m) repeatedly until the sign of the net flow is positive, an indication the current trial value of h_p is above the actual level of the neutral pressure plane. Increment h_p back in smaller steps until the sign of the net flow changes again to negative. Continue this alternate marching toward the solution using increasingly smaller steps until an acceptable level of accuracy has been obtained, such as a value of h_p accurate to within 0.005 m. Then calculate the airflow through each vent and from them determine the net ventilation rate. Another approach is to use the traditional numerical solution technique of bisection.

Use of the neutral plane concept relaxes one assumption of the thermal buoyancy analysis in Section 11-3, the need for there to be only two vents or vents at only two elevations in the building. Integration can be over as many vents as desired.

The development so far in this section has been based on the assumption that air densities inside and outside the building are known. Outdoor air properties will be known as design conditions but indoor air properties depend on a sensible energy balance for the airspace which is being ventilated. Thus the energy balance and the continuity equation are mutually dependent, as was the case described in Section 11-2.

To incorporate sensible heat production within the building into the solution for the ventilation rate is somewhat less convenient than was the situation with only two vents described in Section 11-3. Computer implementation is required, for the solution is iterative. One way to approach a solution is through the following steps:

1. Assume an indoor air temperature and air density.

2. Solve Equation 11-21 for h_p.

3. Determine the net ventilation rate by calculating individual airflows in Equation 11-21 and summing either the positive or negative airflows (the two must balance when Equation 11-21 is solved).

4. Solve a sensible energy balance for the airspace, using the calculated ventilation rate to determine a new estimate of the indoor air temperature and density.

5. Return to step 2 and continue to iterate until a stable solution is obtained. Note that animal heat production may still be a function of indoor air temperature and be changed during each iteration.

11-5. Wind Pressures on Buildings

The concept that wind causes pressures on a building to differ from atmospheric pressure in the free stream has already been introduced. Generally, positive pressures (relative to atmospheric pressure in the undisturbed flow) arise wherever wind must decelerate and negative pressures (suction) arise wherever air must accelerate or flow separates. The constantly changing nature of wind (gustiness) causes an ever changing pressure field to exist around a building. However, for analysis purposes, constant and steady wind pressures are usually assumed. There is some indication that gustiness affects natural ventilation to a significant degree, but the subject is still a matter of research more than design application.

Regions of acceleration and deceleration can be predicted in a broad sense, but no analytical model currently exists which can accurately predict pressures due to wind effects. As examples of general regions of flow separation, acceleration and deceleration, consider the simple building cross section sketched in Figure 11-7. Wind is at right angles to the long dimension of the building.

On the windward wall, region A, streamlines are constricted and flow slows. A stagnation point may or may not exist on the wall (in general, at least one will be somewhere on the wall) but airspeed will be below that of the free stream. This is a region of positive pressure due to airflow deceleration.

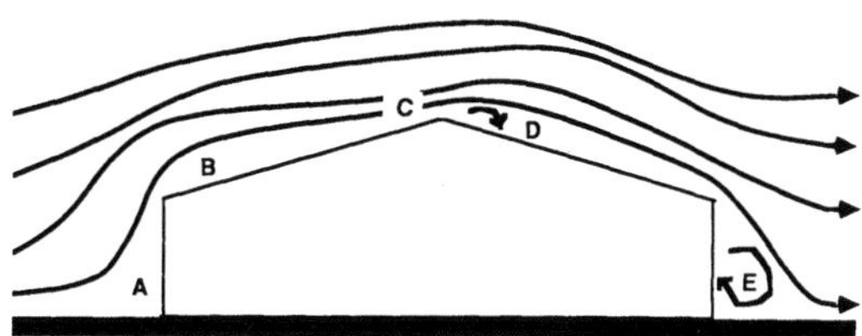

Figure 11-7. Streamlines of wind flowing over a long building, showing areas of flow deceleration (A,C) and acceleration (B,D,E).

In region B the air must change direction rapidly, accelerating to turn the corner at the eave. There will likely be a point of flow separation at the eave and a region of strong turbulence on the lower part of the windward roof. Separation and rapid acceleration to turn the corner causes region B to be one of negative pressure unless the roof slope is steep. If the roof is steep, deceleration may predominate to cause a net positive pressure in region B, but for most agricultural buildings, region B has a negative gage pressure.

In region C, the area near the ridge on the windward side, streamlines are forced above the building, causing deceleration in the forward motion direction. This produces a region of net positive pressure. For buildings used in agriculture, their shapes generally lead to region C being larger than region B, and the windward roof is predominantly characterized by a positive pressure.

However, at the peak and beyond (region D), air must rapidly change direction, accelerating over the ridge and probably separating at the peak. A region of very turbulent flow usually exists on the leeward side of the roof, and the pressure is one of uplift. At the ridge the effect is similar to that of a venturi, thus ventilation openings at the peak are strongly affected by the wind, and wind promotes natural ventilation.

On the leeward wall, flow separation at the leeward eave causes a region of negative pressure, shown as region E. The leeward side of a building is one of negative gage pressure.

A similar picture can be drawn in plan view for wind flowing around a building. An example is in Figure 11-8. When the wind is exactly perpendicular to the long axis of the building, the pressure field should be symmetrical. The windward side of the building, region A, is a region of net positive pressure due to flow deceleration. At the two ends, region B, acceleration as streamlines separate from the building at the windward corners can cause negative pressure zones. The wake of the building, region C, causes the leeward side of the building to be a zone of negative pressure (although not necessarily a uniform negative pressure).

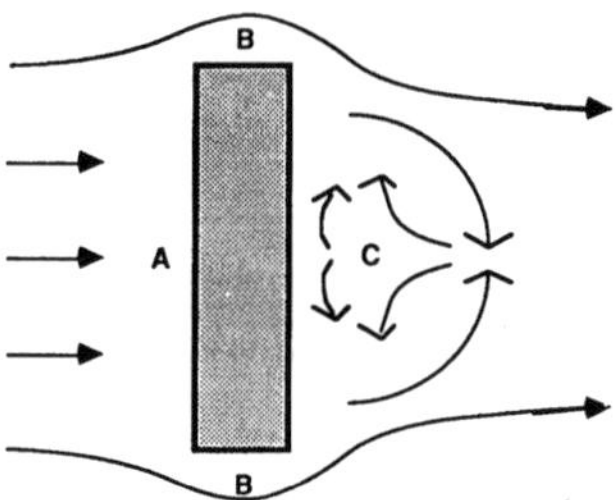

Figure 11-8. Streamlines of wind flowing around a building (plan view), showing an area of flow deceleration and net positive pressure (A) and areas of flow acceleration and net negative pressure (B,C).

The pressure field is very sensitive to the direction of wind. For example, if the wind direction shown in Figure 11-8 changes as shown in Figure 11-9, the pressure field changes greatly. The windward side and end walls are regions of flow deceleration and net positive pressure. Regions B start at the points of flow separation at the corners of the building and are characterized by strong negative pressures. Region C will also likely have a negative pressure because of flow over the roof and separation at the leeward eave, but the gage pressure will be less than in region B on the leeward wall.

Much speculation is possible regarding the distribution of wind pressures around buildings, but only very general rules appear to be possible. Airflow is always three-dimensional, which complicates the picture considerably. Upwind terrain affects the shape of the local planetary boundary layer, which also complicates the situation. The presence of other buildings in the vicinity will make the situation unique and not subject to general rules. These complications make the design of natural ventilation systems as much art as science, and have worked against codification of a standard design procedure.

In spite of uncertainty, it is possible to apply analysis techniques to the design and possibly even the control of natural ventilation systems, as long as the uncertainty of wind effects is always borne in mind. One approach a design engineer can take is to evaluate a potential design using a variety of possible wind loadings. Note this differs from structural design techniques where generally only the worst case is interesting. In naturally ventilated buildings, the varying ventilation rate is likely to be more important than the maximum ventilation rate, which may occur for only a short time.

Generalized data exist to characterize the effects of wind around buildings. In Appendix 11-1 is a collection of wind pressure coefficient data for several common building shapes. The wind pressure coefficient is defined as the ratio of the actual gage pressure at a point on a building, divided by the stagnation pressure of the wind, where the stagnation pressure is determined using the Bernoulli equation,

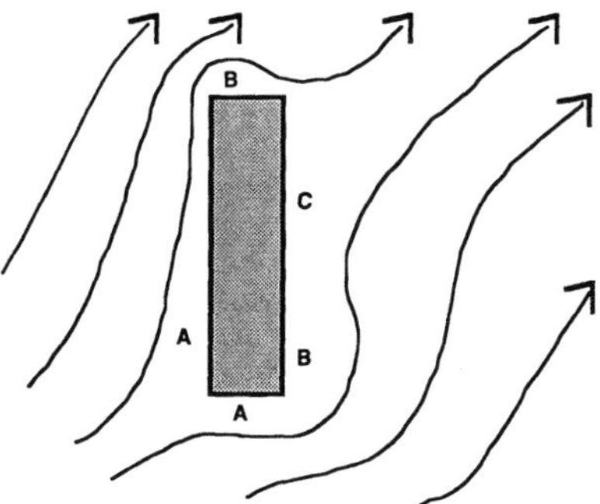

Figure 11-9. Streamlines of wind oblique to the long axis of a building (plan view) showing areas of deceleration and positive pressure (A,C) and areas of acceleration and negative pressure (B,C).

$$P_{stagnation} = 1/2\rho v_{wind}^2, \text{ and} \qquad\qquad (11\text{-}22)$$

$$P_{actual} = C_p P_{stagnation} \qquad\qquad (11\text{-}23)$$

where C_p is the wind pressure coefficient, and may be either positive or negative.

Finally, pressure coefficients for wind effects are generally determined based on the wind velocity at eave height. Weather data frequently exist only for wind at a standard height above the ground, for example, 10 m. The planetary boundary layer causes windspeed at the standard height not to equal the windspeed at eave height. To correct from one to the other, use

$$v_x = v_o(h_x / h_o)^a \qquad\qquad (11\text{-}24)$$

where v_x is the windspeed at height (h_x) above the ground, v_o is windspeed at the standard height (h_o), and a is an exponent whose value depends on the upwind terrain.

Values of the exponent, a, can range from 0.14 for very smooth terrain (such as the surface of a large lake), to 0.20 for smooth terrain such as agricultural fields on flat land, to 0.28 when there are scattered trees and small buildings upwind, and finally to 0.40 for rough terrain with tall trees and buildings. For most agricultural applications, values between 0.20 and 0.28 are appropriate.

Example 11-3

<u>Problem:</u> Estimate the eave-height windspeed corresponding to a 10 m windspeed of 25 m/s, when the eave height is 3 m and upwind terrain is characterized by hedge rows, fields, and several farm buildings.

<u>Solution:</u> The value of the exponent, a, in Equation 11-24 which appears most appropriate is 0.28. All other data are known, thus,

$$
\begin{aligned}
v_{3m} &= v_{10m}(h_{3m} / h_{10m})^{0.28} \\
&= (25 \text{ m/s})(3 \text{ m} / 10 \text{ m})^{0.28} = 17.8 \text{ m/s.}
\end{aligned}
$$

Note the wind stagnation pressure is proportional to the square of velocity, thus, the wind pressure decreases by approximately 50%, while the windspeed decreases by approximately 30%.

11-6. Natural Ventilation Due to the Wind, an Empirical Approach

Airflow through a building caused by wind is a function of windspeed and direction, wind pressure coefficients, and the size of the ventilation openings.

However, it is unlikely the speed of airflow through an opening will equal the windspeed itself. An opening effectiveness applies as follows:

$$\dot{V} = EAv_w \qquad (11\text{-}25)$$

where $\dot{V}$ is airflow, m³/s, A is the free opening area, m², v is windspeed, m/s, and E is effectiveness which is dimensionless. Recommended values for E range from 0.5 to 0.6 when the wind direction is perpendicular to the opening, to 0.25 to 0.35 for diagonal winds.

Example 11-4

Problem: Consider the greenhouse described in Example 11-2, however with only the two side vents open. Meteorological data suggest an average windspeed (10 m height) in the location will be 7.4 m/s during the summer, and the surrounding terrain is such that the exponent, a, in Equation 11-24 will be 0.28. Estimate the average ventilation rate due to wind.

Solution: First, windspeed at the height of the midpoint of the sidewall vent can be estimated using Equation 11-24.

$$v_w = (7.4 \text{ m/s})(2.125 \text{ m} / 10 \text{ m})^{0.28}$$
$$= 4.8 \text{ m/s}.$$

Wind is seldom perpendicular to the long wall of a building. Experience has shown an opening effectiveness of approximately 0.35 applies to agricultural buildings. The area of each vent is 90 m², thus,

$$V = (0.35)(90 \text{ m}^2)(4.8 \text{ m/s})$$
$$= 151 \text{ m}^3/\text{s},$$

which is one air exchange approximately every 26 s.

Note there are several shortcomings to this approach. In the example, each vent was open the same amount, but the technique does not require that be true. One would expect large differences in the ventilation rate depending on how far open the second vent might be. Also, wind pressure forces with diagonal winds cause very uneven pressure distributions around the building. This effect is acknowledged by the empirical effectiveness coefficient, but the same value is used in all conditions regardless of actual pressure distributions.

11-7. Natural Ventilation Due to the Wind, the Wind Pressure Coefficient Method

Consider the building cross section sketched in Figure 11-10. Wind causes the exterior pressures P_{e1}, P_{e2}, and P_{e3} external to vents 1, 2, and 3. There is some internal air pressure which is not necessarily equal to the ambient, free stream air pressure. No heat is added to the building (this is wind-induced ventilation only), thus, outdoor air density applies everywhere. External pressures can be expressed in the form (with C_p defined based on eave height wind)

$$P_{en} = C_{pen}(1/2\rho_o v_w^2),\qquad(11\text{-}26)$$

where v_w is the windspeed at eave height. The indoor air pressure can also be expressed in the same form,

$$P_i = C_{pi}(1/2\rho_o v_w^2).\qquad(11\text{-}27)$$

The indoor wind pressure coefficient, C_{pi}, is defined in a manner parallel to C_{pe}, it is the ratio between the existing indoor gage pressure and the wind stagnation pressure. The indoor air pressure differs from free stream atmospheric pressure and is affected by the external pressure coefficients and the sizes of the openings. External pressure coefficients are thought not to be affected by the sizes of vents unless the vents are large (for example, greater than 15% of the wall area). The value of the indoor pressure coefficient must be bounded by the range defined by the highest and lowest external pressure coefficients.

At each vent,

$$\Delta P_n = (C_{pen} - C_{pi})(1/2\rho_o v_w^2),\qquad(11\text{-}28)$$

based on the definitions of pressure coefficients, where v_w is windspeed at eave height, and

$$\Delta P_n = 1/2\rho_o v_n^2\qquad(11\text{-}29)$$

based on the Bernoulli equation, where v_n is the velocity of air moving through the inlet.

The two pressure differences are the same, thus,

$$v_n = v_w \frac{(C_{pen} - C_{pi})}{\sqrt{|(C_{pen} - C_{pi})|}}.\qquad(11\text{-}30)$$

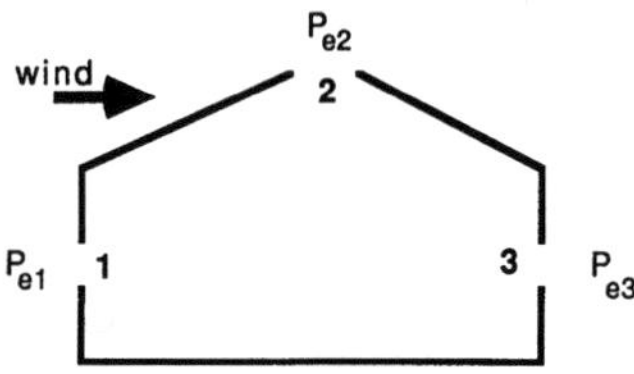

Figure 11-10. Wind pressures caused by wind perpendicular to the long axis of a single air-space building. Openings 1, 2 and 3 have exterior pressures of P_{e1}, P_{e2}, and P_{e3}; indoor pressure is P_i. Each is gage pressure.

Equation 11-30 is in a form which retains the vector identity of velocity. A positive velocity is defined as flow into the building, negative flow is out. The sign of airflow is retained in this general way because it is not always obvious whether a vent will act as an inlet or outlet. For example, in Figure 11-10, vent 1 can reasonably be expected to be an inlet. However, vents 2 and 3 can be shown to act either as two outlets, or one inlet and one outlet depending on the magnitudes of the pressure coefficients and areas of the openings.

Mass flow continuity also must apply,

$$\sum_n C_{dn}\rho_n A_n v_n = 0, \qquad\qquad (11\text{-}31)$$

where C_{dn} is the coefficient of discharge of vent n.

Between Equations 11-30 and 11-31, there are three parameters with known values (C_{pen}, v_w ,and A_n), and two are unknown (C_{pi} and v_n).

With two equations and two unknowns, a solution is attainable. However, the solution is not in closed form. Iteration must be applied. The strategy for solution is to search for a value of C_{pi} such that flow continuity is achieved. If a very low value of C_{pi} is used (such as - 2.0) all flow is into the building. If a high value of C_{pi} is used, all flow is out.

A simple procedure to search for the actual value of C_{pi} is to begin with a value larger or smaller than the actual value can ever be (such as - 2 or + 2). Inlet air velocities are determined using Equation 11-30, and the net flow and its sign are calculated using Equation 11-31. The value of C_{pi} is incremented (for example, by + 0.2 if the beginning value is - 2.0), and the net flow again calculated. When the sign of the net flow changes, the increment is reversed in smaller steps. Alternate searching around the solution continues until an acceptable level of accuracy is achieved. Also, as with the solution of wind-induced ventilation, the bisection technique also can work well, beginning with C_{pi} values which must bracket the solution, such as - 2.0 and + 2.0.

Example 11-5

Problem: A dairy barn is to be naturally ventilated. The ridge vent is to be 0.2 m wide, and for cold weather there will be only two sidewall vents, one on each side of the barn just under the eaves and 0.4 m wide. The barn is 150 m long, with vents the full length of the barn. The external pressure coefficients are expected to be: + 0.7 on the windward wall, - 0.7 at the ridge vent, and - 0.5 at the leeward wall vent. Coefficients of discharge are estimated to be 0.6 at the sidewall vents, and 0.7 at the ridge vent.

Estimate the volumetric rate of ventilation when (eave height) windspeed is 5 m/s and air density is 1.2 kg/m^3.

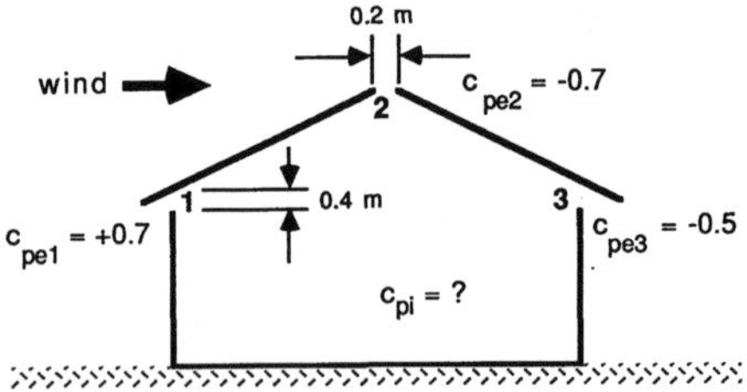

<u>Solution:</u> Values of external pressure coefficients bound the possible value of C_{pi}. Thus, the first estimate of C_{pi} can be 1.0, with the decrement being - 0.5. For purposes of nomenclature, the windward vent will be #1, the ridge vent #2, and the leeward wall vent #3.

At an internal pressure coefficient of 1.0, air velocities through the three vents are:

$$v_1 = (5 \text{ m/s}) \frac{(+0.7 - 1.0)}{\sqrt{|(+0.7 - 1.0)|}}$$

$$= -2.74 \text{ m/s, (out of the building)}$$

$$v_2 = (5 \text{ m/s}) \frac{(-0.7 - 1.0)}{\sqrt{|(-0.7 - 1.0)|}}$$

$$= -6.52 \text{ m/s, and}$$

$$v_3 = (5 \text{ m/s}) \frac{(-0.5 - 1.0)}{\sqrt{|(-0.5 - 1.0)|}}$$

$$= -6.12 \text{ m/s.}$$

The net calculated volumetric ventilation rate is (with the assumption air density is 1.2 kg/m^3 for all openings and thus volumetric airflow continuity may be imposed).

$$\dot{V} = \Sigma C_{dn} A_n v_n$$

$$= (0.6)(60 \text{ m}^2)(- 2.74 \text{ m/s}) + (0.7)(30 \text{ m}^2)(- 6.52 \text{ m/s})$$

$$+ (0.6)(60 \text{ m}^2)(- 6.12 \text{ m/s})$$

$$= -455.9 \text{ m}^3/\text{s.}$$

This obviously does not satisfy flow continuity. For the next iteration, try $C_{pi} = 0.5$, and continue incrementing until the net flow is satisfactorily close to zero. A sample iteration sequence is as follows.

Iteration	C_{pi}	v_1	v_2	v_3	$\dot{V}_1$	$\dot{V}_2$	$\dot{V}_3$	$\Sigma\dot{V}$
1	1.0	- 2.74	- 6.52	- 6.12	- 98.6	- 136.9	- 220.5	- 455.9
2	0.5	2.24	- 5.48	- 5.00	80.5	- 115.0	- 180.0	- 214.5
3	0.0	4.18	- 4.18	- 3.54	150.6	- 87.8	- 127.3	- 64.5
4	- 0.2	4.74	- 3.54	- 2.74	170.8	- 74.2	- 98.6	- 2.1
5	- 0.21	4.77	- 3.50	- 2.69	171.7	- 73.5	- 96.9	1.3
6	- 0.206	4.76	- 3.51	- 2.71	171.3	- 73.8	- 97.6	- 0.1
7	- 0.2062	4.76	- 3.51	- 2.71	171.4	- 73.8	- 97.6	0.0

As an approximation, C_{pi} = -0.2, giving a net flow of 2.1 m^3/s which is negligible compared to the total ventilation rate. Coefficients of discharge and pressure coefficients are not known with degrees of surety that C_{pi} can be stated to many significant digits.

The net ventilation rate is approximately 170 m^3/s. Both the ridge and leeward wall vent are outlets; only the windward wall vent is an inlet for the airspace. Note: For a typical barn 150 m long, a ventilation rate of 170 m^3/s is equivalent to an air exchange rate of approximately once per minute. Wind can produce rapid ventilation (also note the ventilation rate is linearly proportional to windspeed).

Previously it was stated that flow from the ridge and leeward wall vents in the building sketched in Figure 11-10 could not be specified as to direction. Consider the case of the windward wall and leeward wall vents changed to areas of only 6 m^2 each (one-tenth their original size). With that change, C_{pi} is reduced to - 0.632, and flow is into the windward and leeward wall vents, and out the ridge vent. In spite of the negative wind pressure coefficient at the leeward wall vent, the vent still acts as an inlet for ventilation. It would be a useful exercise to complete the analysis for this new situation to verify the results and then explain why, physically, the leeward wall vent becomes an inlet.

11-8. Combined Thermal Buoyancy and Wind Ventilation, Quadrature

When thermal buoyancy and wind cause natural ventilation, the net rate of ventilation is not the sum of the two effects acting independently. Airflow through vents is not linearly proportional to pressure difference. Pressure differences add, not flow rates.

Field experience and comparison with relatively elaborate natural ventilation computer models have shown wind-induced and thermal buoyancy-induced ventilation rates approximately add through quadrature,

$$\dot{V}_{total} = \sqrt{(\dot{V}_{wind}^2 + \dot{V}_{thermal}^2)}. \tag{11-32}$$

Evidence exists that addition by quadrature is in error for certain conditions, but in general can be used with sufficient accuracy for rough design purposes. The error can be as much as 25% when the two contributions are approximately equal or the neutral pressure level falls within a vent opening and the vent acts as both an inlet and an outlet. As already stressed, data to represent the various coefficients are not known with a high degree of accuracy.

The error arising from quadrature should not be a concern for most design purposes when indoor air temperature is not affected by wind-induced ventilation (e.g., when heating to a thermostat setpoint), but therein lies the major problem in applying quadrature to agricultural buildings. When wind-induced ventilation affects indoor air temperature, and thereby thermally-induced ventilation, quadrature is inappropriate as an analysis tool. Such interaction is typical in animal barns where the only source of heat is the animal population, and added ventilation from the wind changes the indoor air temperature, which in turn affects the stack effect.

As has already been suggested, the procedure thought to be correct for superposing wind and thermal buoyancy effects is to add the pressure differences due to wind and thermal buoyancy at each ventilation opening, describe the resulting air velocity by using the Bernoulli equation, and integrate airflows over the areas of the openings. This, however, is still a matter more for research than engineering design. In Appendix 11-2 is the development of one method to superimpose the effects.

SYMBOLS

a	exponent, see Equation 11-24
A	area, m^2
C_d	coefficient of discharge, dimensionless
C_p	wind pressure coefficient, dimensionless
E	ventilation opening effectiveness
F	perimeter heat loss factor, W/m K
g	gravitational constant, m/s^2
h	elevation above a datum, m
h_p	elevation of neutral pressure level, m
L	slotted inlet length, m
P	pressure, Pa
P	slab-on-grade perimeter, m
q	heat production, W
T	temperature, K
U	unit area thermal conductance, W/m^2 K
v	velocity, m/s
$\dot{V}$	volumetric flow rate, m^3/s
W	slotted inlet width, m
α	see Equation 11-16
ρ	density, kg/m^3

EXERCISES

1. The method to determine the rate of ventilation caused by wind, which was developed in this chapter, relied on iteration to reach a solution. However, if there are only two ventilation openings in the building, the solution can be in closed form. Derive this closed-form solution.

2. Develop a computer program to solve the general problem of determining

the rate of ventilation through an agricultural building induced by the wind alone.

3. Develop a computer program to solve the general problem of determining the rate of ventilation through an agricultural building induced by thermal buoyancy alone.

4. Develop a combined computer program to include the programs from Exercises 2 and 3 above, and calculate the combined ventilation rate (using quadrature).

5. Rederive Equation 11-3 assuming the coefficients of discharge of the two ventilation openings are not equal.

6. Recompute a solution to Example 11-2 assuming the corrections for elevation differences to the centers of the vents are omitted. How large is the effect of the omission?

7. An open window in a building has an area of 3 m^2 and a wind-induced pressure across it (from outside to inside) of $+ 8$ Pa. Estimate the volumetric rate of airflow through the window at standard air density and at an air density of 0.9 kg/m^3.

8. Windspeed has been measured as 8 m/s at an elevation of 5 m above the ground and the upwind terrain is relatively smooth with some hedge rows and scattered trees. Calculate the pressure caused by the wind 2 m above the ground if the wind pressure coefficient is $+ 0.65$.

9. A barn ventilated only by the wind has two vent openings. The windward wall vent has a total area of 25 m^2 and a wind pressure coefficient of $+ 0.7$. The leeward wall vent has an area of 50 m^2 and the wind pressure coefficient is $- 0.5$.

   Calculate the wind pressure coefficient which would be necessary at the ridge vent (if one were to be installed) in order that the barn ventilation rate not change. (Hint: must equal inside pressure coefficient.)

10. You have built a "solar sponge" solar collector to be attached to houses for heating air by solar energy. It is a box, insulated on all but one side. The uninsulated side is glass and faces south. Air from the house enters at the bottom of the box, is heated within the box, and flows by thermal buoyancy to the top of the box and back into the house. The overall heat loss factor (ΣUA) is 15 W/K. The distance from the top of the box to the bottom is 2 m.

    You have built the box and attachments so resistance to airflow through the box will be negligible – the only resistances are at the top and bottom openings, which are rectangular and open directly into the house. Each opening is 1.3 m wide and 0.15 m high. The coefficient of discharge at each opening is expected to be 0.5. You want to estimate the temperature of air entering the house (C) and the airflow rate (m^3/s) when the solar energy absorbed within the box is 1000 W.

For this problem, set up the equations needed to estimate the temperature and airflow rates, determine values of all required parameters, and provide calculations showing the first two iterations in the chain of iterations needed to reach a solution. Assume air temperature in the house is 20 C and air density is 1.2 kg/m^3.

As an alternative problem, develop a computer program to solve this problem, where each parameter can be entered by the user rather than fixed within the computer code. Explore the effects of each parameter on the expected airflow rate and air temperature.

REFERENCES

ASHRAE. 1989. Handbook of Fundamentals. American Society of Heating, Refrigerating, and Air Conditioning Engineers, Atlanta, GA.

Brockett, B.L. 1986. A control model for natural ventilation systems in agricultural buildings. M.S. thesis, Cornell University Libraries, Ithaca, NY.

Brockett, B.L. and L.D. Albright. 1987. Natural ventilation in single airspace buildings. Journal of Agricultural Engineering Research 37:141-154.

Bruce, J.M. 1978. Natural convection through openings and its application to cattle building ventilation. Journal of Agricultural Engineering Research 23:151-167.

Bruce, J.M. 1982. Ventilation of a model livestock building by thermal buoyancy. Transactions of the ASAE 25(6):1724-1726.

Chastain, J.P. 1987. Pressure gradients and the location of the neutral pressure axis for low-rise structures under pure stack conditions. M.S. thesis, University of Kentucky Libraries, Lexington, KY.

Hellickson, M.L. and J.N. Walker. 1982. Ventilation of agricultural structures (Ch. 5). American Society of Agricultural Engineers, St. Joseph, MI.

Timmons, M.B. and G.R. Baughman. 1981. Similitude analysis of ventilation by the stack effect from an open ridge livestock structure. Transactions of the ASAE 24(4):1030-1034.

CHAPTER 12
AIRFLOW IN DUCTS

12-1. General

Some applications of ventilation in agricultural buildings require air distribution through ducts. As examples, air distribution in refrigerated storages for horticultural products may be through ducts. Heated air may by distributed through ducts in growth rooms, laboratories, and growth chambers. Fresh air may be distributed using perforated polyethylene tubes in barns and greenhouses. Grain may be dried using heated air distributed through ducts. Although the majority of ventilation situations in agricultural buildings do not require knowledge of air duct design, the topic is sufficiently important to require a knowledge of analysis and design techniques.

Air duct design in commercial, office, and industrial buildings is a complex process, and extremely laborious if done by hand. Such duct systems are extensive and have many branches. During the past several years, computer programs have become commercially available which greatly reduce the labor required for such designs.

Applications in agricultural buildings are usually simpler than in commercial and industrial buildings. However, although such designs are simpler, poor designs still lead to improper air distribution and poor ventilation, and the economic consequences can be large. For example, it is not unusual to store at least a million dollars worth of a horticultural product in one storage building. Poor air distribution could cause rapid spoilage of the stored product and enormous consequences for the designer.

This chapter describes analysis procedures to determine frictional and dynamic energy losses for air flowing through air ducts, and then describes a method useful to design simple air duct systems. Perforated polyethylene ducts will then be characterized and an analysis and design technique detailed.

12.2 The Bernoulli Equation

As has already been seen, airflow can be analyzed using the Bernoulli equation. The equation can be cast into several forms, the form most useful for airflow is expressed in terms of pressure. Previous uses of the Bernoulli equation (e.g., in Section 8-2) have been based on constancy of total pressure along a streamline.

When air flows through ducts, its total pressure is not constant. <u>Fans</u> within a duct add total pressure to the air. <u>Friction losses</u> due to air drag along duct walls and <u>dynamic losses</u> caused by turbulence in duct fittings (for example) act to remove energy from the air reducing total pressure. (Recall, the Bernoulli equation expresses the conservation of energy along a streamline.) The Bernoulli equation can be restated to incorporate losses as follows:

$$\frac{\rho v_1^2}{2} + P_1 = \frac{\rho v_2^2}{2} + P_2 + (\Delta P_t)_{1\text{-}2} \qquad (12\text{-}1)$$

where $(\Delta P_t)_{t\text{-}2}$ is the total pressure loss, Pa, between stations 1 and 2 in the system. Total pressure at point 1 in Equation 12-1 has two components, dynamic or velocity pressure and static pressure, P, each with units of pascals. Note elevation differences between points 1 and 2 are ignored; air density is sufficiently small in practical applications that potential energy changes are negligibly small compared to static pressure and velocity pressure changes. Also note air density is assumed constant between stations 1 and 2.

Several basic concepts are critically important if one is to understand analysis of pressure changes through air ducts. First, the total pressure of air prior to entry into the duct (i.e., prior to being drawn into a fan) equals atmospheric pressure within the space (which could be the outdoors) from which the air is drawn. Within this space, air velocity is assumed to be negligible, thus, total pressure is comprised entirely of static pressure. The total pressure of air exiting a duct system equals the total pressure of ambient air within the space of discharge. Although air may exit the duct system having velocity pressure (as a jet, for example) velocity pressure is entirely lost as the air slows and mixes with ambient air.

A typical situation is: atmospheric pressure within the space from which air is drawn and within the space to which air is discharged, are the same. For example, air could be drawn from outdoors, distributed throughout a building through an air duct system, and discharged into rooms within the building, rooms which are all at atmospheric pressure. Air starts at atmospheric pressure, gains and loses total pressure as it passes through the duct system, and returns to a state of atmospheric pressure. The net change of total pressure of the air in such a system is zero. This is not a universal situation, but is most typical. <u>The net change is zero.</u>

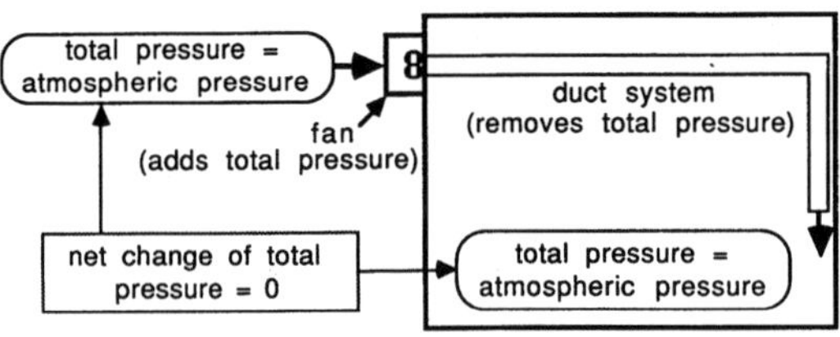

air flow balances so total losses = total gains

The observation that losses and gains sum to zero provides one basis for air duct design. Airflow rates adjust themselves so air velocities are great enough to cause sufficient losses. Energy gains from fans are entirely dissipated by friction and dynamic losses. Air obviously cannot plan ahead to balance energy gains and losses deliberately, but the laws of physics operate so a balance is achieved.

(Note: For the purposes of this text, assume the only mechanisms adding energy to air are fans. There will not be sufficient stack effect, wind effect, etc., to cause air movement through ducts without the motive power of fans.)

Another assumption central to the analysis and design procedures to follow is that air is incompressible. Air density along the path of flow will be assumed equal to atmospheric conditions. It would be a useful exercise to check the adequacy of this assumption. Typical maximum static pressures throughout a modest-sized air duct system are generally less than 1000 Pa and air velocities are generally held to less than 10 m/s.

Another basic concept of air duct design relates to interactions of total, static, and velocity pressures. Stated briefly, the airflow rate through a duct system is self-adjusted so friction and dynamic losses just balance pressure gains from fans. Duct geometry (cross-section areas of the ducts) determines velocities at all points within ducts (except at junctions), and velocities determine velocity pressures. Velocity pressure and static pressure sum to comprise total pressure, thus, if there is a total pressure change, velocity pressure adjusts (if necessary) to maintain flow continuity and static pressure adjusts to maintain equality between total pressure and the sum of velocity and static pressures.

12-2.1. Pressure Change Examples. Adjustments of total, velocity, and static pressures can be seen in the sketches of a simple duct system in Figure 12-1.

In Figure 12-1(a), the fan is at the entry and total pressure is greater than atmospheric pressure everywhere except at the entrance of the fan. At the entrance, pressures are somewhat ill-defined. Air approaching the fan gains velocity and thereby velocity pressure. Static pressure is reduced to compensate and there is little total pressure change until the air reaches the fan. Flow of air approaching the fan can be treated as inviscid with no losses, a good approximation. However, the approach is from all directions and the static pressure cannot be easily represented in a two-dimensional sketch.

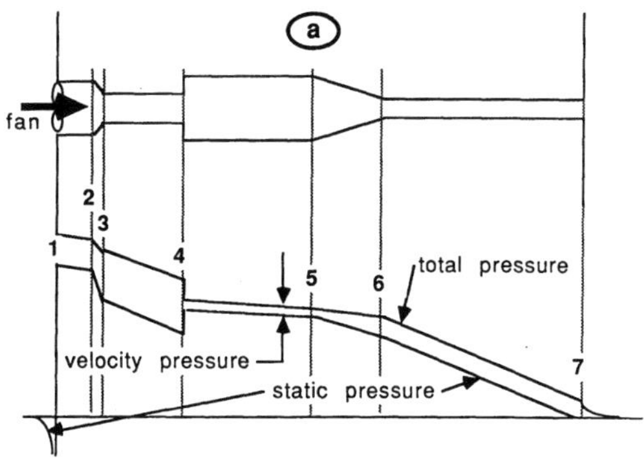

Figure 12-1. Pressure changes as air flows through a simple duct system, (a) with the fan at the entry.

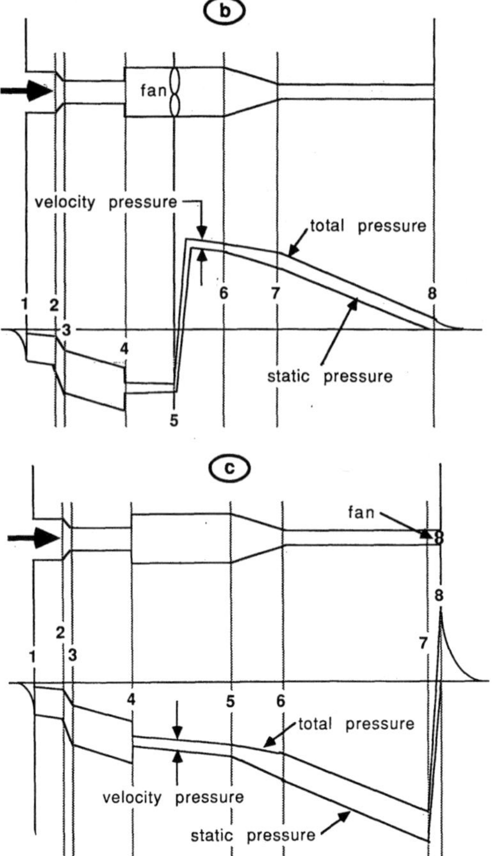

Figure 12-1. Continued. (b) with the fan in the central part of the duct system, and (c) with the fan at the exit.

At the fan, section 1, air gains total pressure, velocity pressure, and static pressure. Velocity pressure is dictated by flow continuity. Static pressure adjusts so the sum of static and velocity pressures equals total pressure. From sections 1 to 2, airflow is through a smooth duct of constant cross section. Energy is lost by friction, but flow continuity demands the velocity pressure remain constant. Static pressure thus drops at the same rate as does total pressure.

Between sections 2 and 3 there is a rapid contraction to a duct of small diameter. The contraction introduces some turbulence and there is a total pressure decrease. However, the velocity pressure becomes larger due to higher air velocity and, as a consequence, static pressure decreases rapidly. A significant fraction of the static pressure change is converted to velocity pressure, the rest is total pressure loss.

Between sections 3 and 4, the duct is of constant diameter and velocity pressure remains the same. However, air velocity is higher than in section 1-2 and there is more friction loss per unit length of duct, thus, the rate of total pressure loss (and thereby static pressure loss) is greater than in section 1-2.

At section 4, a sudden expansion introduces a great deal of turbulence with a resulting sudden loss of a significant amount of total pressure. (Whenever streamlines diverge, dynamic losses can be expected to be significant and much greater than where streamlines converge.)

After section 4, air velocity is slower than before. The velocity pressure decreases to such an extent that static pressure actually increases. This is termed <u>static pressure recovery.</u>

Between sections 4 and 5, air velocity is low and friction losses are small. The result is to have only a slight decrease of total pressure due to friction, and a corresponding slight decrease of static pressure.

Between sections 5 and 6, total pressure decreases due to two effects. The transition section is long so there is some friction loss. There is also turbulence, thus, some dynamic loss. The two appear as a lower total pressure. Static pressure decreases more rapidly, however, because air accelerates and the velocity pressure increases.

Between sections 6 and 7, the duct is again constant and friction losses cause total pressure and, therefore, static pressure to decrease.

At the exit, point 7, static pressure and velocity pressure become somewhat ill-defined again. At some distance from the exit, all pressure is lost and the air returns to atmospheric conditions. There is velocity pressure at the exit if a jet is formed, but whatever excess (above atmospheric) pressure energy exists at the exit is dissipated rapidly and not linearly.

The pressure lines in Figures 12-2(b) and 12-2(c) can be described in similar fashion. The only difference is the location of the fan. As can be seen, total pressure can be negative in part or all of a duct system depending on the location of the fan. However, the fan adds sufficient total pressure so losses are balanced by the gain.

12-2.2. System Characteristics. The air duct system and fan interact and a balance of airflow is achieved. The system characteristic concepts developed in Section 8-6 apply equally well to a system comprised of air ducts as to a combination of slotted inlets and infiltration. A conceptual system characteristic graph is sketched in Figure 12-2.

The operating or balance point is where the fan and system curves intersect. The fan curve must be obtained from laboratory data (probably from the manufacturer of the fan). The system curve must be developed by a design engineer. Usually the system curve is developed first and various fan curves assessed to determine which fan provides adequate airflow with that particular system. In a complete design, an engineer might develop several system curves to determine the most economical combination of fan and system. Analysis techniques to develop a system graph follow.

12-3. Friction Losses.

12-3.1. Darcy-Weisbach and Colebrook Equations.

Air is a viscous fluid. When it flows in contact with a solid surface, momentum is transferred to the surface because of viscous drag. Calculation of pressure drop, ΔP, due to friction for fluid flow in conduits has long been accomplished using the Darcy-Weisbach equation,

$$\Delta P = f(L / D)(\rho v^2 / 2), \qquad (12\text{-}2)$$

expressed in terms of pressure loss, not energy loss. The length of duct is L, the diameter is D, and f is the so-called <u>friction factor.</u>

Air velocity, v, is treated as constant because only ducts of constant cross-section are considered for practical applications.

The friction factor depends on the relative roughness of the duct wall and the state of turbulence – the Reynolds number. The relative roughness is ε/D, where ε is the absolute roughness, and D is the duct diameter. The Reynolds number is as computed before, $Re = \rho v D/\mu$. Air properties can be found in Table 3-1.

When the flow is completely laminar, the friction factor is a function of only the Reynolds number. When the flow is completely turbulent, the friction factor is a function of only the relative roughness. For the transition region between, the friction factor can be estimated using the Colebrook equation,

$$f = (1.14 + 2\mathrm{Log}(D / \varepsilon) - 2\mathrm{Log}\{1 + 9.3 / [Re(\varepsilon / D \sqrt{f}]\})^{-2} \qquad (12\text{-}3)$$

where the Log is to base 10.

Equation 12-3 is implicit in f, so the solution of the equation must be by trial and error. A value of f can be assumed and substituted in the right-hand side to calculate an improved value of f. The improved value is then substituted into

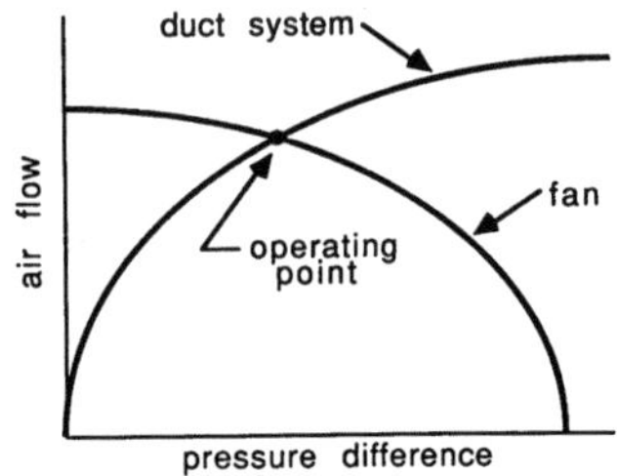

Figure 12-2. Typical fan and system characteristic curves for an air duct.

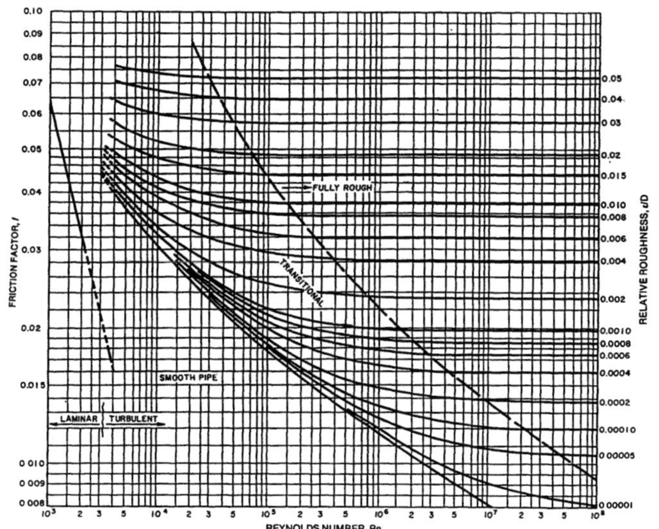

Figure 12-3. Moody diagram for estimating the friction factor to be used in the Darcy-Weisbach equation.

the right-hand side, and iteration continues until the calculated value of f stabilizes.

An alternative method to determine f is to use a graphical representation of Equation 12-3, the Moody chart, shown in Figure 12-3.

To use either the equation to calculate f, or the Moody diagram, a value of the Reynolds number must be calculated and the relative roughness must be known. The absolute roughness of some common materials is in Table 12-1.

Example 12-1

<u>Problem:</u> Air at standard atmospheric pressure and 300 K flows through a round sheet metal duct (galvanized steel with 1 slip joint every 0.75 m) with a

Table 12-1. Absolute roughness of some materials commonly used for air ducts.

Material	Absolute Roughness, mm
PVC plastic pipe	0.01 to 0.05
Aluminum	0.04 to 0.06
Galvanized steel, continuously rolled, longitudinal seams	0.06 to 0.12
Galvanized steel, 1 slip joint every 0.75 m	0.15 (averaged)
Fibrous glass duct, rigid	0.90
Concrete	0.3 to 3.0
Flexible metal duct, fully extended	1.2 to 2.1
Flexible fabric and wire ducts, all types, fully extended	1.0 to 4.6

diameter of 0.5 m. The duct is 50 m long. What is the expected pressure loss due to friction from one end of the duct to the other when the airflow rate is 1 m^3/s?

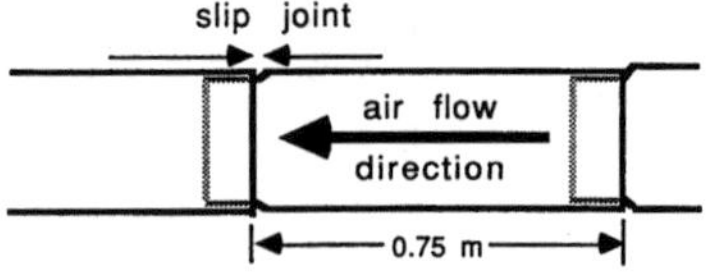

<u>Solution:</u> The Moody diagram will be used to estimate the friction factor and the Colebrook equation will then be used to check the accuracy of the estimate. The relative roughness, Reynolds number, and friction factor will be used in the Darcy-Weisbach equation to calculate the pressure loss due to friction.

The absolute roughness of the duct material is found from Table 12-1 to be 0.15 mm. The duct diameter is 500 mm, thus, the relative roughness is

$$\varepsilon/D = 0.15 / 500 = 0.0003.$$

From Table 3-1, air properties needed to calculate the Reynolds number are: ρ = 1.1774 kg/m^3, and μ = 1.983E - 5 kg/m-s (at 300 K). Air velocity in the duct is found by dividing the cross-sectional area into the volumetric flow rate, or

$$v = (1\ m^3/s)/[(\pi)(0.25\ m)^2] = 5.09\ m/s.$$

The Reynolds number is

$$Re = \rho v D/\mu$$
$$= (1.1774\ kg/m^3)(5.09\ m/s)(0.5\ m)/(1.983E - 5\ kg/m\text{-}s)$$
$$= 1.51E + 5$$

From the Moody diagram, for this combination of relative roughness and Reynolds number, the friction factor is approximately 0.018. Checking this value using the Colebrook equation,

$$f = \{1.14 + 2Log(500/0.15)$$
$$- 2Log[1 + 9.3/((151{,}000)(0.0003)(\sqrt{0.018}))]\}^{-2}$$
$$= 0.0184$$

which is close to the estimate of 0.018. A second iteration using the Colebrook equation yields an improved estimate of f = 0.0183 and further iterations show this to be a stable result.

The Darcy-Weisbach equation can now be used to estimate the pressure drop due to friction,

$$\Delta P = (0.0183)(50\ m / 0.5\ m)(1.204\ kg/m^3)(5.09\ m/s)^2/2,$$
$$= 28.5\ Pa.$$

354

12-3.2. Friction Charts. Many air duct systems are built using galvanized sheet steel, where the duct is made in short sections which are joined together using slip joints. There is usually 1 slip joint every 0.75 m. Because this is a standard method (although obviously not the only one), a standardized graph has been developed for duct design. It is in Figure 12-4.

Figure 12-4 can be used if two of the following four parameters are known: duct diameter, air velocity (averaged over the cross section of the duct), friction loss in Pa/m, or airflow rate. When two are known, the other two can be determined using the graph. For analysis purposes, friction loss is usually a parameter to be calculated, but for design purposes the friction loss may be known (specified) and a duct diameter needed to achieve that pressure loss.

The friction loss graph is useful without correction for duct materials with absolute roughnesses of approximately 0.15 mm, air temperatures within 20 K of 20 C, air which is relatively dry, elevations up to 500 m, and air pressures within 5 kPa of ambient atmospheric pressure.

Correction factors are available when these standard conditions are not approximated. Each factor is multiplicative.

To correct for air density and viscosity, where the subscript s refers to standard

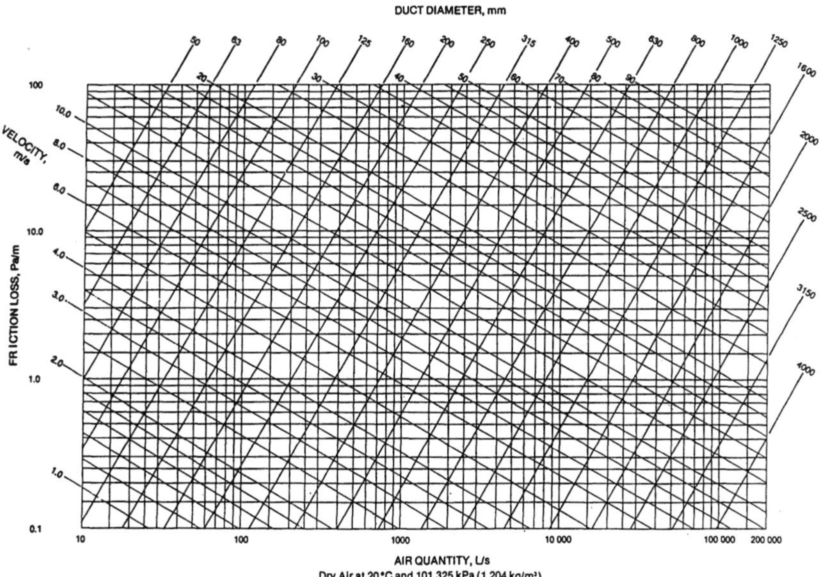

Figure 12-4. Pressure loss in straight sheet metal ducts for air at 20 C, 50% relative humidity, and 101.325 kPa, and an absolute duct roughness of 0.15 mm. Taken by permission from the *ASHRAE Handbook of Fundamentals*.

conditions and a to actual, corrections for temperature and barometric pressure are required.

$$C_{temperature} = (293.15 / T_a)^{0.825},$$ (12-4)

where T_a is air temperature in Kelvin.

$$C_{elevation} = (1 - 2.2557E - 5z)^{4.733},$$ (12-5)

where z is elevation in m.

To correct for humidity,

$$C_{humidity} = [1.0 - 0.378(P_s / P_{atm})]^{0.9}$$ (12-6)

where P_s is the saturation water vapor pressure at the actual conditions, and P_{atm} is atmospheric pressure.

These corrections apply whether the friction drop graph or pressure drop equations are used.

Example 12-2

Problem: Reconsider Example 12-1, except for an application in a building at an elevation of 2000 m, where heated air at 60 C is to be humidified to 50% relative humidity and distributed using the duct.

Solution: The pressure drop for standard air conditions may be obtained using the friction drop graph, Figure 12-4. The duct diameter is 0.5 m and 1.0 m³/s of air flows through it. For this combination of conditions, the friction drop graph shows the pressure drop per m of length will be approximately 0.58 Pa/m. The duct length is 50 m, thus, the total pressure drop is estimated to be 29 Pa.

The correction for temperature is

$$C_{temperature} = (293.15 / 333.15)^{0.825}$$
$$= 0.900.$$

The correction for elevation is

$$C_{elevation} = [1.0 - (2.2557E - 5)(2000 \text{ m})]^{4.733}$$
$$= 0.804.$$

The correction for relative humidity requires a value for saturation partial pressure of water vapor at a dew point temperature of 60 C. At 2000 m, the standard atmospheric pressure is 79.495 kPa. Program PLUS may be used to estimate the saturation pressure, 10.10 kPa. (Note: The saturation pressure is at the dew point of the actual conditions, which corresponds to the actual vapor partial pressure at 60 C dry bulb temperature and 50% relative humidity.) The correction factor is

$$C_{humidity} = [1.0 - 0.378 \ (10.10 \ kPa \ / \ 79.495 \ kPa)]^{0.9}$$
$$= 0.957.$$

The three correction factors combine with the pressure loss for standard conditions by multiplication, thus, the actual pressure difference from one end of the duct to the other will be

$$\Delta P_a \ = (29 \ Pa) \ (0.900) \ (0.804) \ (0.957)$$
$$= 20 \ Pa.$$

12-3.3. Rectangular Ducts. When ducts are not circular (many rectangular ducts are used in the HVAC industry), the hydraulic diameter is used in place of the diameter for a round duct.

The hydraulic diameter for analysis of pressure drops in air ducts can be calculated in one of two ways. The first is the common definition of the hydraulic diameter,

$$D_{equivalent} = 4(\text{cross-sectional area / perimeter}). \qquad (12\text{-}7)$$

This is the equivalent diameter to use in the Darcy-Weisbach equation and the friction loss graph (Figure 12-4) <u>if the graph is entered using the velocity and diameter</u>. The friction loss will be correct, but the flow rate as obtained from the graph will be incorrect. The flow rate must be calculated independently of the graph.

The alternative, <u>if one wishes to enter the friction loss graph using the airflow rate and duct diameter as the known parameters</u>, is the following form of equivalent diameter:

$$D_{equivalent} = 1.30(ab)^{0.625}(a + b)^{-0.25}, \qquad (12\text{-}8)$$

where a and b are the two sides of the rectangle.

The value of equivalent diameter obtained from Equation 12-8 cannot be used in the Darcy-Wiesbach equation. If the equivalent diameter obtained using Equation 12-8 is used to enter the friction loss graph, along with a value for the flow rate, the friction loss will be correct. However, the velocity obtained from the graph will be incorrect and must be calculated independently.

Example 12-3

<u>Problem:</u> Reconsider Example 12-1 if the duct is redesigned to be rectangular with a cross-section of 0.2 m by 1.0 m.

Solution: As in Example 12-1, the same approach will be used, with the same value of relative roughness and an air velocity of

$$v = 1 \text{ m}^3\text{/s} / (0.2 \text{ m}^2) = 5.0 \text{ m/s}.$$

The hydraulic diameter (in the form of Equation 12-7) is

$$D = 4(\text{area}) / (\text{perimeter}) = 4(0.2 \text{ m}^2) / (1.4 \text{ m}) = 0.57 \text{ m}.$$

The Reynolds number is

$$\text{Re} = (1.1774 \text{ kg/m}^3)(5.0 \text{ m/s})(0.57 \text{ m}) / (1.983\text{E} - 5 \text{ kg/m-s})$$
$$= 169,000.$$

If the friction factor value of 0.0183 from Example 12-1 is assumed for an initial value (which also agrees with the Moody diagram for this example), the Colebrook equation can be used with iterations to estimate a friction factor for this example of 0.0181. (It would be a useful exercise to use the Colebrook equation and verify this result.)

The pressure drop due to friction is now

$$\Delta P = (0.0181)(50 \text{ m} / 0.57 \text{ m})(1.1774 \text{ kg/m}^3)(5.0 \text{ m/s})^2 / 2,$$
$$= 23.3 \text{ Pa}.$$

The velocity is slightly less and the effective diameter is slightly greater, therefore, the pressure drop is less.

12-3.4. Computational Form for Velocity Pressure and Reynolds Number.

For simplicity, when air is close to standard conditions, the velocity pressure, $\rho v^2/2$, can be condensed into a simpler, computational form. For air velocity in m/s and velocity pressure in Pa,

$$\rho v^2/2 \sim (v / 1.29)^2. \qquad (12\text{-}9)$$

The Reynolds number can be calculated for standard air conditions as follows:

$$\text{Re} \sim 66,300Dv \qquad (12\text{-}10)$$

where D is duct diameter in meters and v is air velocity in m/s.

12-3.5. Program FRICTION.

FRICTION, which substitutes for the friction loss graph and correction procedures, is within the collection of computer programs which accompany this text. The program first computes friction loss for dry air at standard conditions and then corrects for elevation, temperature, and relative humidity.

Example 12-4

<u>Problem:</u> Use program FRICTION to recalculate the pressure loss for example 12-2.

<u>Solution:</u> In the order entered, the data needed for FRICTION is:

    absolute roughness: 0.15 mm
    duct diameter: 0.5 m
    duct length: 50 m
    elevation: 2000 m
    air temperature: 60 C
    humidity calculation basis: relative humidity
    relative humidity: 50%
    airflow basis: flow rate
    airflow rate: 1.0 m^3/s

The results of FRICTION: the pressure drop is 19.55 Pa, air velocity is 5.09 m/s, and the friction factor is 0.0125.

Note: This value of the friction factor is corrected for temperature, elevation, and relative humidity, it is not the value obtained directly from the friction factor chart, Figure 12-4.

The results from FRICTION and the results of Example 12-2 differ slightly. In Example 2, air viscosity was approximated by the value at 300 K. This introduces a slight error in the calculation of the Reynolds number. In FRICTION, the Reynolds number is calculated according to Equation 12-10. Also, the atmospheric pressure in FRICTION is approximated as a function of elevation by

$$\text{air pressure} = 101.325 \exp(-0.00011943\,z) - 6.799\text{E}-06z - 6.976\text{E}-08z^2$$

where z is the elevation above sea level. The error in this approximation is only a few tenths of a kPa over the range of elevations where buildings are constructed, but it does introduce a slight difference in the elevation correction factor (Equation 12-5).

12-4. Dynamic Losses

Dynamic losses arise from disturbances in airflow through ducts. Disturbances cause increased turbulence. Through turbulence, energy loss appears as a pressure drop. Disturbances can be caused by entrances, flow transition sections, exits, junctions, elbows, in-line heaters, and other changes of the duct cross-section and direction of airflow.

Dynamic losses are separate from friction losses and are assigned to the element (i.e., an elbow) giving rise to the flow disturbance. They are calculated as a fraction of the velocity pressure existing at the element.

$$\Delta P_{dynamic} = C_{loss}(\rho v^2 / 2) \qquad (12\text{-}11)$$

Experimental data are available for local loss coefficients, C_{loss}, for common fittings used in air ducts. Each loss coefficient is referenced to a velocity pressure, $\rho v^2/2$, either upstream or downstream of the fitting. It is important to ensure the proper velocity is used to determine the velocity pressure.

Common nomenclature and sections for the reference velocity for fittings and other elements of duct systems are in Figure 12-5. Local loss coefficient data for some common air duct fittings are in Appendix 12-1. A larger selection of data is available in the *ASHRAE Handbook of Fundamentals*.

Example 12-5

<u>Problem:</u> A round, 0.6 m diameter, main air duct carrying 1.5 m³/s of air at standard conditions is joined at a 45° angle by a round, 0.3 m diameter, lateral

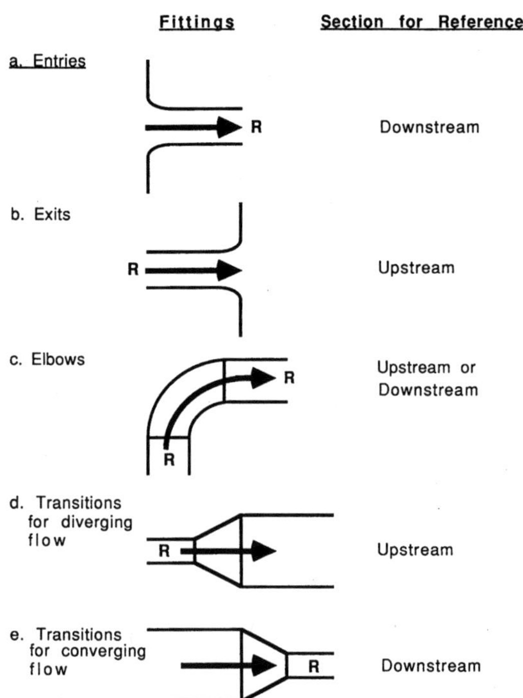

Figure 12-5. Nomenclature and reference sections for velocity pressure and dynamic loss coefficients for common air duct fittings. R marks reference section velocity used to calculate dynamic losses.

<u>Fittings</u>   <u>Section for Reference</u>

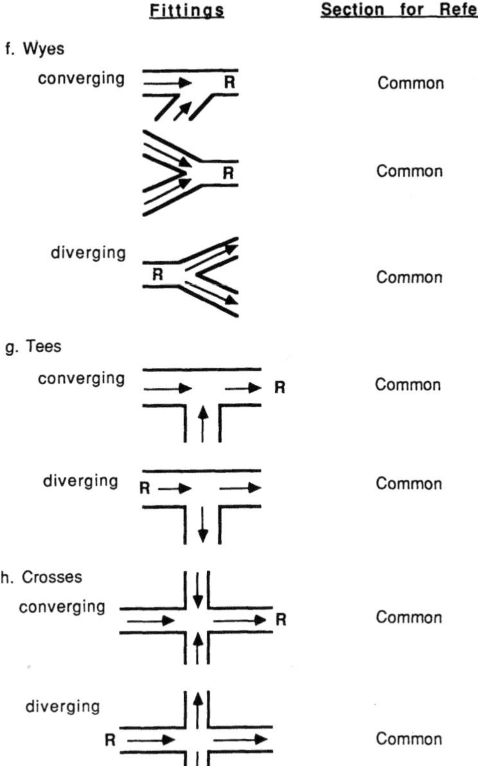

Figure 12-5. Continued. Nomenclature and reference sections for velocity pressure and dynamic loss coefficients.

duct carrying 0.5 m³/s of air also at standard conditions. What is the pressure change for air flowing through the main air duct? What is the pressure change for air flowing from the lateral into the main air duct?

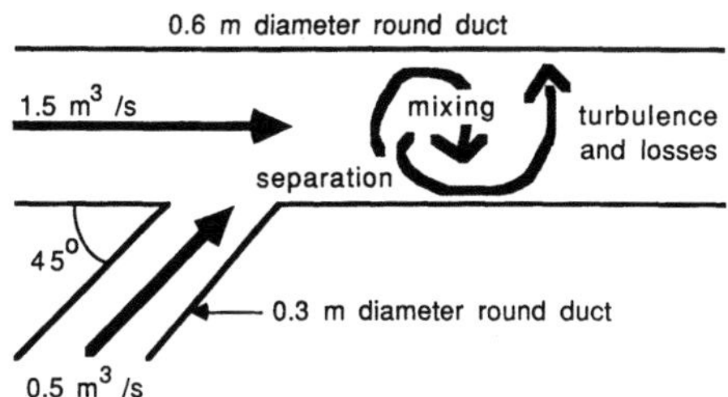

<u>Solution:</u> The fitting is a converging 45° wye and is in Appendix 12-1.

Entering the wye from the main duct,

$$\dot{V} = 1.5 \text{ m}^3/\text{s},$$
$$A = 0.283 \text{ m}^2, \text{ and}$$
$$v = 5.31 \text{ m/s}.$$

Entering the wye from the lateral duct,
$\dot{V}$ = 0.5 m³/s,
A = 0.0707 m², and
v = 7.07 m/s.

Leaving the wye,
$\dot{V}$ = 2.0 m³/s,
A = 0.283 m², and
v = 7.07 m/s.

For this wye, in Appendix 12-1, the dynamic loss coefficient for the branch duct is found from:

$$v_b/v_c \quad = 7.07 \text{ m/s} / 7.07 \text{ m/s} = 1.0,$$
$$A_b/A_c \quad = 0.0707 \text{ m}^2/0.283 \text{ m}^2 = 0.25,$$

and $C_{c,b}$ can be found from interpolation of data in the table,

$$C_{c,b} \quad = 0.31 + (0.05/0.10)(0.38 - 0.31) = 0.345.$$

The pressure drop is referenced to the velocity pressure in the common section of the wye, where V = 7.07 m/s.

$$\Delta P_{dynamic} \quad = C_{c,b}(v_c / 1.29)^2,$$
$$= 0.345(7.07 / 1.29)^2 = 10.4 \text{ Pa}.$$

The dynamic loss coefficient for the main duct is found from:

$$v_s/v_c \quad = 5.31 \text{ m/s} / 7.07 \text{ m/s} = 0.75,$$
$$A_b/A_c \quad = 0.25 \text{ as before,}$$

and $C_{c,s}$ can also be found by interpolation:

$$C_{c,s} = 1/2 \, [\, 1/2(- 0.18 + 0.01) + 1/2(- 0.03 + 0.07)] = - 0.03.$$

The dynamic pressure drop is also referenced to the velocity pressure in the common section, where V = 7.07 m/s.

$$\Delta P_{dynamic} \quad = C_{c,s} (v_c / 1.29)^2,$$
$$= -0.03(7.07 / 1.29)^2 = - 0.9 \text{ Pa}.$$

The result for the lateral air duct appears reasonable. There is energy loss in the air flowing from the lateral duct due to mixing and turbulence when the two airstreams meet.

However, the negative dynamic pressure drop for air flowing through the main duct may not be intuitively obvious. The result says the energy content actually increases – the loss is negative. The explanation is that air in the main duct accelerates through the wye. Acceleration is a gain of kinetic energy, and because the gain by acceleration is larger than the dynamic loss, a net gain is observed. This is not to say energy is created. If there were no losses, the pressure change would equal the difference between the velocity pressures before and after the wye,

$$P_{before} = (v_s / 1.29)^2 = (5.31 / 1.29)^2 = 16.9 \text{ Pa, and}$$
$$P_{after} = (v_c / 1.29)^2 = (7.07 / 1.29)^2 = 30.0 \text{ Pa.}$$

The difference is a gain of 13.1 Pa. In fact, a gain of only 0.9 Pa was calculated. The difference, 12.2 Pa, is lost as a dynamic loss and is expressed as a decrease of static pressure. However, air flowing through the main duct experiences a net gain of 0.9 Pa in its energy content and is expressed as a total pressure gain.

12-5. Combined Friction and Dynamic Losses, an Example

System characteristic curves are useful for design. When the system characteristic curve is known, fan characteristic curves can be graphed on the same axes, and the combination of fan and system is chosen so that operation provides the proper airflow rate.

To graph a system curve, not every point along the curve must be completely recalculated unless one or more of the dynamic loss coefficients is a function of the Reynolds number. The analysis procedure for dynamic losses was based on the velocity pressure, $\rho v^2/2$. If, for example, air velocity doubles, the velocity pressure increases by a factor of 4, <u>and so do dynamic losses</u>. The same is not true of the friction loss except at very high airflow rates where the Reynolds number is sufficiently high that only relative roughness affects the friction loss gradient – the extreme right side of the Moody diagram, Figure 12-3.

Most air duct applications do not encompass this part of the Moody diagram for all sections of the duct system, thus, friction losses must be determined for a series of airflow rates before the system characteristic graph can be drawn. Program FRICTION can be used to reduce the effort required to obtain such a set of data.

The following example is presented to demonstrate the technique of developing data for a system characteristic graph.

Example 12-6

<u>Problem:</u> Consider an air duct system similar to the one shown in Figure 12-1 (a). The duct system is to be constructed as follows:

Section	Description
duct section 1 to 2	0.6 m diameter round galvanized duct
transition from 2 to 3	60° contraction, round
duct section 3 to 4	0.3 m diameter, round
transition at 4	180° conical diffuser
duct section 4 to 5	1.1 m diameter, round
transition from 5 to 6	10° contraction, round
duct section 6 to 7	0.4 m diameter, round
exit at 7	30° conical, round exit, with a diameter of 0.8 m at the exit.

Duct section lengths are listed below.

Develop a graph for the system characteristic curve for this air duct system if it is located at 1000 m elevation, and the air to flow through it will be at 0 C and 90% relative humidity (as perhaps to a refrigerated storage). Develop the system characteristic graph in the flow rate range from 0 to 2 m^3/s.

<u>Solution:</u> Friction losses will be calculated first. Four duct sections contribute friction loss.

Section	Duct Dia.	Section Length
1 to 2	0.6 m	20 m
3 to 4	0.3	30
4 to 5	1.1	40
6 to 7	0.4	60

Note: The small duct diameter in section 3 to 4 is not realistic for a design. In a real air duct system, such constricted flow would usually not be permitted. Its effect will be seen in the high pressure loss it contributes.

A data table can be developed using either program FRICTION, or the friction drop graph, corrected for elevation, temperature, and relative humidity. FRICTION provides the following.

	Pressure Loss Due to Friction, Pa, for Airflow Rates, m^3/s							
Section	0.25	0.50	0.75	1.00	1.25	1.50	1.75	2.00
1 to 2	0.35	1.22	2.56	4.35	6.59	9.27	12.4	16.0
3 to 4	15.3	55.8	120	209	322	458	619	803
4 to 5	0.04	0.13	0.27	0.45	0.68	0.94	1.25	1.60
6 to 7	7.45	26.7	56.9	98.0	150	212	285	369
Total:	23.1	83.9	180	312	479	680	918	1190

The pressure drop due to dynamic losses will be calculated for one airflow rate, and data for other flow rates scaled from that point. To minimize possible error, it is best to select the calculated point somewhere in the middle of the characteristic curve. Air duct systems usually are designed for air velocities between 5 and 10 m/s. The midpoint, 7.5 m/s, will be assumed in duct section 6 to 7, the middle-sized duct section. The duct diameter from section 6 to 7 is 0.4 m, thus, the airflow rate for the analysis will be

$$\dot{V} = Av = \pi(0.2 \text{ m})^2 (7.5 \text{ m /s}) = 0.942 \text{ m}^3 \text{/s}.$$

This value of airflow rate is approximately in the middle of the desired range (0 to 2.0 m^3/s) of airflow rates. As shall be seen, a value of $\dot{V} = 1.0$ m^3/s could just as well have been chosen and would have made the rest of the calculations easier.

Another assumption is required before beginning. The duct system contains one fan, located at the entrance. Assume construction is such that there is not a separate entrance – air enters directly into the fan. Dynamic pressure loss due to air entering the fan is included in the fan data.

The dynamic losses can now be calculated. There are four fittings where dynamic losses occur: the converging section at 2; the sudden expansion at 4; the converging section at 5; and the exit at 6. Each fitting is listed in Appendix 12-1 and dynamic loss coefficients may be obtained from there.

The transition at 2 is a 60° converging section, termed in Appendix 12-1 "contraction, round". The ratio of upstream to downstream areas, A_1/A_0, is the ratio of the square of the diameters, or (0.6 m/0.3 m)^2= 4.0. For the 60° convergence, the local loss coefficient is 0.07. The dynamic pressure loss is referenced to the downstream flow (the loss coefficient is designated C_0, and the subscript corresponds to the section wherein the velocity pressure is used, A_0).

The downstream velocity is 13.33 m/s, thus,

$$\Delta P_{\text{dynamic},2} = 0.07(13.33 / 1.29)^2 = 7.5 \text{ Pa}.$$

The next dynamic loss is at the sudden expansion at 4. In Appendix 12-1 this is termed a "conical diffuser, round", with a 180° angle. The ratio of areas is A_1/A_0 = (1.1/0.3)^2 = 13.4. Interpolating within the local loss coefficient table, C_0 = 0.86. The reference velocity pressure in this case is upstream, so

$$\Delta P_{\text{dynamic, 4}} = 0.86(13.33 / 1.29)^2 = 91.8 \text{ Pa}.$$

The next dynamic loss is at 5, another converging section, except this time with a convergence angle of 10°. The ratio A_1/A_0 = (1.1/0.4)^2 = 7.56. The local loss coefficient is 0.05, and

$$\Delta P_{\text{dynamic},5} = 0.05(7.50 / 1.29)^2 = 1.7 \text{ Pa}.$$

The final dynamic loss is at the exit, which is a 30° conical round transition section. The area ratio, A_1 / A_0 = (0.8 / 0.4)^2 = 4. The local loss coefficient is

0.70 and the dynamic loss is

$$\Delta P_{dynamic,7} = 0.70(7.50 / 1.29)^2 = 23.7 \text{ Pa.}$$

The sum of dynamic pressure losses is 7.5 + 91.8 + 1.7 + 23.7, or 124.7 Pa, at an airflow rate of 0.942 m³/s.

Dynamic pressure losses, ΔP, and airflow rates, $\dot{V}$, scale by the following where the subscript o refers to original data and s to scaled data:

$$\Delta P_s = \Delta P_o(\dot{V}_s / \dot{V}_o)^2$$

From this equation, a data set for a system characteristic graph follows. Friction loss is added to obtain total pressure loss.

Airflow Rate	Dynamic	Friction	Total
	Pressure Losses		
0.00 m³/s	0.00 Pa	0.00 Pa	0.00 Pa
0.25	8.78	23.1	31.9
0.50	35.13	83.9	119.
0.75	79.05	180.	259.
1.00	140.38	312.	452.
1.25	219.58	479.	699.
1.50	316.19	680.	996.
1.75	430.37	918.	1348.
2.00	562.11	1190.	1752.

The system characteristic graph is in Figure 12-6. The form of the system characteristic graph in Figure 12-6 differs from the form used for slotted inlet systems, the axes are reversed. The form of Figure 12-6 is most commonly used for air duct systems, and is the form most easily graphed from the data and the manner in which data are accumulated.

12-6. System Effect Factors

Most fans available to ventilate agricultural buildings are sold as packaged units and are not intended to be attached to ducts. Performance data for such fans generally apply to the fans as installed and can be used directly.

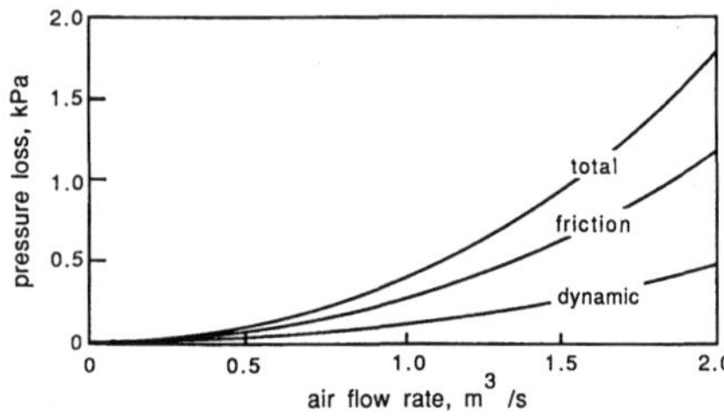

Figure 12-6. System characteristic graph for Example 12-6, showing dynamic, friction and total pressure loss effects.

However, when fans are used to distribute air through air ducts, fan attachment to the ducts influences the pressure difference imposed on the fan and the resulting airflow rate. Such fans are tested while attached to wind tunnels in manners seldom reproduced exactly in field installations, and field installation procedures are not sufficiently standardized that fixed factors can be imposed to estimate the effect of attachment.

In Figure 12-7 representative system characteristic curves are sketched, one for an air duct system by itself and one representing attachment of a fan to the duct system. The difference between the system characteristic curve with and without the fan attachment is termed the system effect factor. As drawn, the factor results in decreased airflow and a higher operating pressure for the fan. An effect in this direction is not mandatory; it is possible for attachment to be such that airflow can increase and the pressure difference decrease, although the reverse is most common.

The Air Movement and Control Association (AMCA) is a voluntary association of fan manufacturers charged with developing standards for testing fans, louvers, etc., among other responsibilities. AMCA has developed design data for numerous ways in which fan and duct systems can be coupled. Such data is published, for example, in AMCA Publication 201: Fans and Systems. For example, system effect factors can be reduced by following some general installation rules:

1. Do not place elbows at the fan inlet. An elbow induces rotational motion in the air entering the fan and reduces fan performance.

2. Be sure the fan outlet area and attached duct area are the same size and shape. If not, the sudden contraction or expansion will decrease system performance.

3. Do not place elbows near the fan outlet. Sufficient straight duct length (at least 2.5 duct diameters) should be installed to permit the air velocity profile within the duct to develop and stabilize before the airflow is disturbed again.

4. Do not place obstructions such as dampers, heaters, or pipes in the plane of

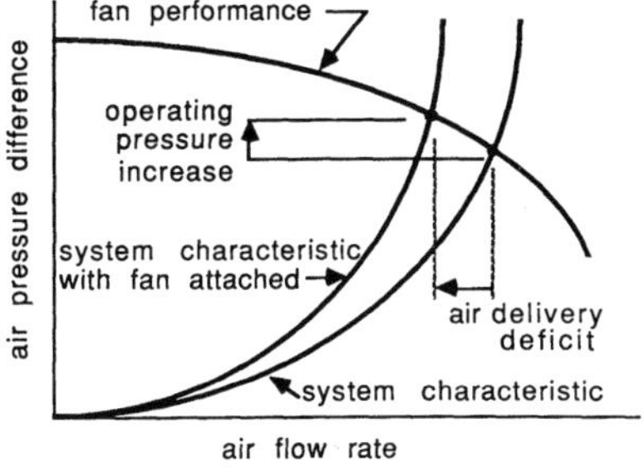

Figure 12-7. System performance with and without a system effect factor.

the fan inlet or outlet.

5. Design the inlet to the fan such that air enters the fan evenly, uniformly, and without spin.

The topic of system effect factors should be studied in detail by any engineer designing large fan and duct systems. AMCA Publication 201 is the reference for more detail than can be presented in this text.

12-7. Design of Air Duct Systems

Several methods to design air ducts are commonly used in the HVAC industry. The common methods are: 1) the equal friction method, 2) the constant velocity method, 3) the velocity reduction method, and 4) the static regain method. Briefly, the four methods are based on the following criteria for sizing ductwork:

1. <u>Equal Friction</u>. Duct sizes are chosen to provide a uniform pressure drop gradient due to friction everywhere in the system.

2. <u>Constant Velocity</u>. Duct sizes are chosen so air velocity is everywhere the same in the system.

3. <u>Velocity Reduction</u>. Duct sizes are chosen so air velocity is reduced after each junction by a predetermined amount.

4. <u>Static Regain</u>. Duct sizes are chosen so static pressure regain (due to lower air velocity) after each junction offsets the pressure loss in the following duct section.

A duct design method, suitable for simple air duct systems such as might be installed in an agricultural building, will be outlined in this text. (This design is not necessarily suitable for large and complex air duct systems.) The method will be termed the <u>critical path</u> method, and grows from methods used to design water piping systems. For more details about the four standard duct design methods, consult the *ASHRAE Handbook of Fundamentals*.

The critical path method is derived from the equal friction method, but differs from it in that a uniform gradient of friction loss is maintained in only one section of the duct system - the critical path. The critical path is defined as the path from the entrance of the system (normally where the fan is placed) to the farthest outlet. Gradients of friction loss in laterals leaving the critical path differ from that in the main path so as to dissipate all pressure available at the beginning of each lateral without the need for dampers at the ends of the lateral sections.

For example, consider an air duct system where there is one long, main duct and one short lateral diverting air from near the fan.

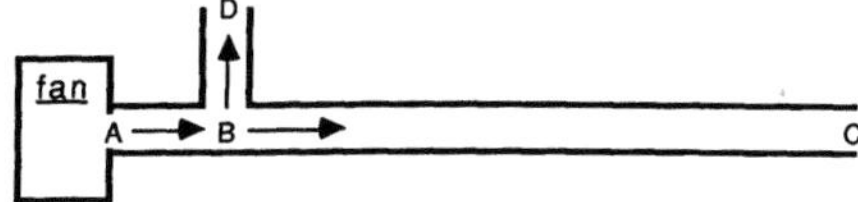

Suppose the airflow rate desired at D is the same as at C. If a constant friction loss gradient is imposed (by having ducts BD and BC the same size) the only way the system could balance would be if a damper were installed at D, in effect to dissipate the remaining pressure at D – the pressure not dissipated by friction. However, if the duct size from B to D were smaller than from B to C, and chosen properly, there might be no need for a damper; the increased gradient of friction pressure drop would dissipate all the pressure available at B while maintaining the desired airflow rate.

Complex air duct systems, such as in large office buildings, cannot be designed using such a simple approach for there are laterals arising from laterals, etc., etc., etc. Airflow balance in such systems is a combination of art and science, and is accomplished by installing dampers at the ends of all duct runs and tuning the systems after installation. The procedure is sufficiently intricate that subcontracting firms have arisen which specialize in balancing air delivery systems.

When an air duct system is relatively simple, such fine tuning may not be necessary. Example 12-7 is an example of such a situation. The duct system can be designed to provide sufficiently acceptable performance. If the duct system cannot be designed to match the desired airflow rates exactly, an evaluation is necessary to assess the acceptable level of deviations. In some cases, small deviations (i.e., 15% or 20%) may be accepted as insignificant in field practice. Of course, post-construction balancing may be required if the system's performance does not match its design specifications, for whatever reasons.

Example 12-7

<u>Problem:</u> Consider a refrigerated storage building for apples. There are six cold storage rooms, each supplied with cold air from a central refrigeration system. Each room is to receive 1 m^3/s of conditioned air and the air duct system is to be configured as sketched in the plan view in Figure 12-8. At each outlet (E, F, G, H, I, and J) there is to be a 90° elbow (not shown) and an abrupt exit with a screen.

Design the duct system (choose duct sizes), specify all connections, and determine the pressure which must be supplied by a fan at point A. Standard atmospheric conditions may be assumed sufficiently close to operating conditions that corrections are not necessary and Figure 12-4 may be used directly.

<u>Solution:</u> To begin, the critical path must be chosen. In this example, the choice is clear, either path ABCDG or path ABCDJ is critical, for the duct system is symmetrical.

The critical path design procedure as used in this example relies on the equal friction method to design the critical path, but before the critical path method may be used, a design criterion must be chosen. The criterion is either: (a) maximum air velocity in section AB, or (b) a design value for the friction loss gradient along the critical path.

If a maximum air velocity is chosen, a value of 10 m/s is typical. High airspeed systems with air velocities up to 25 m/s may be used if duct size restrictions exist. The maximum value is chosen to limit both the noise generated by air flowing through the duct (a result of turbulence), and the friction loss. In agricultural applications where noise may be less a concern than in office buildings, for example, and the air duct system is sufficiently short that friction and dynamic losses are relatively small, a higher value for maximum air velocity may be chosen.

In commercial applications with long duct lengths, the design value for friction loss is frequently approximately 1 Pa/m. However, when duct lengths are short, dynamic losses may outweigh friction losses. In such situations, the criterion for design may be to limit dynamic losses (and thereby total losses) to some predetermined value. Dynamic losses can be reduced by lowering air velocities through the use of larger duct sizes, and choosing smooth transition sections, with no immediate concern for friction loss – which is small.

In this example, the criterion will be chosen to be a maximum air velocity of 10 m/s in section AB. This assumption will establish the desired value of friction loss per unit length throughout the critical path.

<u>Friction losses.</u> The airflow rate in section AB is 6 m^3/s, with a desired air velocity of 10 m/s. If this criterion results in a pressure drop from entrance to exit which exceeds that obtainable from a reasonable fan for the application, the design will be changed. Similarly, if the pressure drop is low and can be increased without incurring excessive operating cost, the design also may be changed.

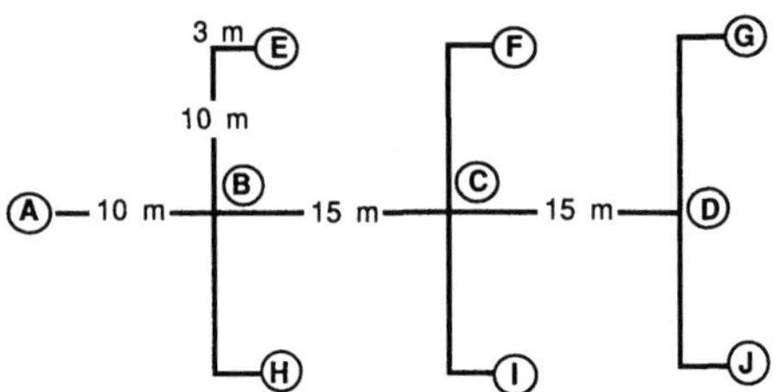

Figure 12-8. Layout of air duct system for Example 12-7.

The friction loss graph, Figure 12-4, shows for the given flow rate and assumed air velocity, the most appropriate duct sizes are 800 and 1000 mm. The closest to the intersection of the conditions is the 800 mm duct, so that size will be chosen. Note: Duct sizes listed on the friction drop graph will be assumed for now to represent industry standards of available sizes. That is, while it is possible to specify a duct diameter of 832 mm, for example, such a size would necessitate a special order and be significantly more expensive. The 800 mm size is standard and readily available.

For an airflow of 6 m^3/s and a duct diameter of 800 mm, air velocity will be 12.0 m/s and the friction loss will be approximately 1.7 Pa/m from Figure 12-4 (standard atmospheric conditions were specified and corrections are not required). (Note: The scale is logarithmic and not easily interpolated by eye.) If program FRICTION is used, the conditions result in a calculated air velocity of 11.94 m/s and a friction loss of 1.61 Pa/m.

The remainder of this example will be solved using Figure 12-4, thus, values read from it will be assumed to be correct; they usually would be sufficiently precise to be within the precision of data related to the air duct system. However, if corrections to the data were necessary, FRICTION, in conjunction with Figure 12-4, could provide corrected data readily. Figure 12-4 would be used to choose duct sizes closest to desired conditions.

The critical path will be designed using equal friction and the friction loss of 1.7 Pa/m will be the standard, based on section AB.

The fitting at point B can be either two sequential wye sections or a 90° cross. For this example, 90° cross fittings will be specified for points B, C, and D. In an actual design, both wyes and crosses might be investigated (unless other factors demand one or the other) as to cost and pressure drop, and the best specified. As will be seen, there are no pressure loss differences.

After point B, airflow is 4 m^3/s. A friction loss of 1.7 Pa/m is desired and Figure 12-4 shows a duct diameter of 630 mm is closest to providing the combination. With a diameter of 630 mm, air velocity is approximately 13.0 m/s and friction loss is approximately 2.6 Pa/m. Note this violates the criterion of equal friction, but unless custom-fabricated ducts are specified, complete uniformity of the pressure loss cannot be achieved. Equal friction is a desired but not required design criterion.

In section CD, airflow is 2 m^3/s. The duct size which most closely provides a friction loss of 1.7 Pa/m is 500 mm and results in an air velocity of 10.0 m/s and a friction loss of approximately 2.1 Pa/m.

In section DG, the airflow rate is 1 m^3/s. The duct diameter which provides a friction loss closest to 1.7 Pa/m is 400 mm and results in an air velocity of 8.0 m/s and a friction loss of approximately 1.8 Pa/m.

The basic data for duct sizes in the critical path are now complete,

Section	Length	Diameter	ΔP, Pa/m	ΔP, Pa	Air Velocity
AB	10 m	800 m	1.7	17	12.0 m/s
BC	15	630	2.6	39	13.0
CD	15	500	2.1	32	10.0
DG	13	400	1.8	23	8.0
			Total:	111	

The differing duct diameters require additional transition pieces. For example, either before or after the cross at B, there must be a converging section from 800 mm diameter to 630 mm. To reduce dynamic losses, the transition likely would be after the 90° cross, but both options could be investigated in a design and the least cost option chosen.

The next step is to determine dynamic losses, and calculate the total pressure change from one end of the duct system to the other at the required airflow rates from friction and dynamic losses.

<u>Dynamic loss at B</u>. At B, the first dynamic loss is associated with the 90° diverging cross. For this fitting, (data in Appendix 12-1 for the main section and diverging flow) data for a diverging wye (main) apply. For a diverging wye, and a downstream-to-upstream flow ratio of (4 m³/s)/(6 m³/s) = 0.67, the fitting dynamic loss coefficient can be interpolated to be 0.05. Note: For this design, it will be assumed the converging section is after the cross and the area of flow does not change within the cross. In that situation, the downstream-to-upstream velocity ratio equals the downstream-to-upstream airflow ratio. The dynamic loss is based on the upstream air velocity and is

$$\Delta P_{B1} = C(v / 1.29)^2 = 0.05(12.0 / 1.29)^2 = 4.3 \text{ Pa.}$$

The next fitting is the converging section. For the design, a 30° angle of convergence will be specified. Obviously, if further analysis shows a different angle to be preferred either because of cost or pressure loss, this can be adjusted. The upstream-to-downstream area ratio is (800 mm/630 mm)^2 = 1.6. This falls slightly outside the table in Appendix 12-1 for a contraction. It appears safe to assume a fitting loss coefficient of 0.05 applies, and dynamic loss is based on the downstream air velocity. The downstream velocity is 13.0 m/s, that in section BC. The dynamic friction loss is

$$\Delta P_{B2} = C(v / 1.29)^2 = (0.05)(13.0 / 1.29)^2 = 5.1 \text{ Pa.}$$

The total dynamic loss at B is the sum of that at the cross and at the converging section, or

$$\Delta P_B = 4.3 \text{ Pa} + 5.1 \text{ Pa} = 9.4 \text{ Pa.}$$

<u>Dynamic loss at C.</u> At C, the downstream-to-upstream air velocity ratio equals

$(2 \text{ m}^3\text{s})/(4 \text{ m}^3/\text{s}) = 0.5$. Data in Appendix 12-1 for a diverging wye (main) show the fitting loss coefficient to be approximately 0.09. The dynamic loss is based on upstream air velocity, 13.0 m/s. The pressure drop due to the cross will be

$$\Delta P_{Cl} = C(v \, / \, 1.29)^2 = 0.09(13.0 \, / \, 1.29)^2 = 9.1 \text{ Pa}.$$

The contraction is from a duct diameter of 630 mm to a diameter of 500 mm. The area ratio is $(630 \text{ mm} \, / \, 500 \text{ mm})^2 = 1.6$, and a fitting loss coefficient of 0.05 is appropriate, as before. Dynamic loss is based on downstream air velocity and is

$$\Delta P_{C2} = C(v \, / \, 1.29)^2 = (0.05) \, (10.0 \, / \, 1.29)^2 = 3.0 \text{ Pa}.$$

Total dynamic pressure loss at C is the sum of ΔP_{Cl} and ΔP_{C2} or 12.1 Pa.

<u>Dynamic loss at D</u>. The fitting at D is a round, 180° diverging wye with the area reduction built into the wye. The downstream-to-upstream velocity ratio is $(8.0 \text{ m/s})/(10.0 \text{ m/s}) = 0.8$. The (half) divergence angle is 90°, thus the fitting dynamic loss coefficient is 1.0, the dynamic loss is based on downstream air velocity (8.0 m/s) . The dynamic loss is

$$\Delta P_D = C(v \, / \, 1.29)^2 = (1.0)(8.0 \, / \, 1.29)^2 = 38.5 \text{ Pa}.$$

<u>Dynamic loss between D and G</u>. Between points D and G there are two 90° elbows. Many elbow designs are available for use. The smooth radius, round elbow with a radius-to-diameter ratio of 1.0 will be specified for this example. The fitting dynamic loss coefficient is 0.22 and the air velocity is 8.0 m/s. Thus, the dynamic pressure loss is

$$\Delta P_{D\text{-}G} = 2C(v \, / \, 1.29)^2 = 2(0.22) \, (8.0 \, / \, 1.29)^2 = 16.9 \text{ Pa}.$$

<u>Dynamic loss at G</u>. The exit at G with a screen is abrupt. For this design, a screen with a 50% free area will be chosen. The dynamic loss coefficient is 1.0 plus that due to the screen. From Appendix 12-1, for a screen free area of 0.5, the screen effect is an additional loss coefficient of 1.7. The total dynamic loss coefficient is 2.7 and the appropriate air velocity is 8.0 m/s. The dynamic pressure loss is

$$\Delta P_G = C(v \, / \, 1.29)^2 = (2.7)(8.0 \, / \, 1.29)^2 = 103.8 \text{ Pa}.$$

The total pressure loss from A to G can now be determined. It is the sum of friction and dynamic losses along the critical path. The friction loss is 111 Pa and the sum of the dynamic losses is

$$\begin{aligned}
\Delta P_{\text{dynamic}} &= \Delta P_B + \Delta P_C + \Delta P_D + \Delta P_G \\
&= 9.4 \text{ Pa} + 12.1 \text{ Pa} + 38.5 \text{ Pa} + 16.9 + 103.8 \text{ Pa} \\
&= 180.7 \text{ Pa},
\end{aligned}$$

or approximately 181 Pa. The sum of friction and dynamic losses is 292 Pa.

The type of fan to be used at A could be a tubeaxial fan. For such a fan type, a pressure of 292 Pa (plus whatever system effect factor might apply) is not a high pressure. (It is enough pressure to raise water in a column approximately 30 mm.) A centrifugal fan (designed for greater pressure differences) could be used but would not be required.

<u>Pressure profile along the critical path</u>. Total pressures at the points along the critical path relative to atmospheric pressure are:

Point	Pressure
A	292 Pa
B	275 (that at A, less friction loss from A to B)
C	227 (that at B, less friction loss from B to C and dynamic loss at B)
D	183 (that at C, less friction loss from C to D and dynamic loss at C)
G (J)	0 (back to atmospheric pressure, or that at D, less friction loss from D to G and, dynamic loss at D, dynamic losses from D to G, and the dynamic loss at G)

Pressures at B and C are those just before the fittings, or the pressures imposed on air exiting to the laterals at those points.

An important point is that the pressure change from A to G must be the same pressure change as from A to all other exits.

<u>Design of laterals</u>. The laterals now can be designed. The laterals BE and BH must be designed to dissipate the pressure existing at B, 275 Pa, for an airflow rate of 1.0 m^3/s. The laterals CF and CI must be designed to dissipate the pressure existing at C, 227 Pa, at an airflow rate of 1.0 m^3/s.

Looking ahead, most of the pressure loss occurs due to the exit. One way to adjust the pressure drop at the exit is to choose a proper screen (a suitable free area). Thus, for this example, design of laterals is simplified because the size of lateral ducts is less important – sizes need not be selected to dissipate most of the available pressure. Fittings, as well as duct diameters, can be adjusted to balance airflows.

Airflow is 1 m^3/s in each lateral. Section DG was designed to have a diameter of 400 mm, and an air velocity of 8 m/s. A design is likely to be less expensive if fewer sizes of ducts and fittings are specified, so all laterals will be chosen to be 400 mm diameter. Because airflows are alike, the friction loss in each lateral will equal that in DG, or 23 Pa. Also, dynamic losses through the elbows will be identical to those of DG, or 16.9 Pa. The dynamic loss through the crosses and at the exits must still be determined, and the exits chosen to balance flow.

Lateral BE. Data for branches from diverging 90° crosses are in Appendix 12-1. For diverging flow from a cross, the coefficients are the same as for a diverging wye, which is also in the appendix. For a divergence angle of 90°, a branch-to-upstream area ratio (A_b/A_c) of (400 mm/800 mm)^2 = 0.25, and a branch-to-upstream airflow ratio ($\dot{V}_b/\dot{V}_c$) of (1.0 m^3/s)/(6.0 m^3/s) = 0.17, the dynamic loss coefficient ($C_{c,b}$) must be determined by two-way interpolation.

For A_b/A_c = 0.2,
    and for $\dot{V}_b/\dot{V}_c$ = 0.1, $C_{c,b}$ = 1.3;
    and for $\dot{V}_b/\dot{V}_c$ = 0.2, $C_{c,b}$ = 1.9.
    Thus, for $\dot{V}_b/\dot{V}_c$ = 0.17,
$$C_{c,b} = 1.3 + (0.07 / 0.10)(1.9 - 1.3) = 1.7.$$
For A_b/A_c = 0.3,
    and for $\dot{V}_b/\dot{V}_c$ = 0.1, $C_{c,b}$ = 1.1;
    and for $\dot{V}_b/\dot{V}_c$ = 0.2, $C_{c,b}$ = 1.4.
    Thus, for $\dot{V}_b/\dot{V}_c$ = 0.17,
$$C_{c,b} = 1.1 + (0.07 / 0.10)(1.4 - 1.1) = 1.3.$$
For A_b/A_c = 0.25,
$$C_{c,b} = 1.7 + (0.05 / 0.10)(1.3 - 1.7) = 1.5.$$

At the fitting, dynamic pressure loss is based on upstream air velocity, 12 m/s. The dynamic loss is

$$\Delta P_B = C(v / 1.29)^2 = 1.5(12 / 1.29)^2 = 130 \text{ Pa.}$$

At B, 275 Pa are available. The loss at B is 130 Pa, friction loss is 23 Pa, and dynamic loss in the elbows is 16.9 Pa. Therefore, the exit must absorb

275 Pa - 130 Pa - 23 Pa - 16.9 Pa = 105 Pa.

A method to design the exit is to determine the dynamic loss coefficient required. Air velocity is 8 m/s, thus,

$$C = \Delta P/(v / 1.29)^2 = 105/(8 / 1.29)^2 = 2.73.$$

The dynamic loss coefficient for a screened exit is 1.0 plus the screen effect, which in this situation must be an additional loss coefficient of 1.73. The table for screen obstructions in Appendix 12-1 shows a 50% screen and provides almost exactly this value. Thus, an abrupt exit with a 50% free area screen would be specified and the design of laterals BE and BH is complete.

Lateral CF. The procedure to design lateral CF will parallel that for lateral BE. The pressure available at C is 227 Pa. Lateral CF will be chosen as a 400 mm round duct for the same reasons as in the design of lateral BE. As before, divergence is at 90° and the branch-to-upstream area ratio is (400 mm/630 mm)^2 = 0.40. The branch-to-upstream airflow rate is (1.0 m^3/s)/(4.0 m^3/s), or 0.25. From the data table for a 90° diverging wye, for A_b/A_c = 0.4,

for $\dot{V}_b / \dot{V}_C = 0.2$, $C_{c,b} = 1.1$, and
for $\dot{V}_b / \dot{V}_c = 0.3$, $C_{c,b} = 1.3$, thus,
for $\dot{V}_b / \dot{V}_c = 0.25$, $C_{c,b} = 1.2$.

Dynamic loss is based on upstream air velocity, so

$$\Delta P_C = C(v / 1.29)^2 = 1.2(13 / 1.29)^2 = 122 \text{ Pa}.$$

As before, the exit must be designed to use

$$227 \text{ PA} - 122 \text{ Pa} - 23 \text{ Pa} - 16.9 \text{ Pa} = 65 \text{ Pa}.$$

The dynamic loss coefficient to do this is

$$C = \Delta P/(v / 1.29)^2 = 65/(8 / 1.29)^2 = 1.69.$$

The exit itself contributes a dynamic loss coefficient of 1.0, leaving 0.69 as the required contribution of the screen. From the appendix table for screens, a free area ratio of 0.65 provides a loss coefficient of 0.75 and a free area ratio of 0.70 provides a dynamic loss coefficient of 0.58. The free area ratio of 0.65 provides the closer approximation of the desired value. If the small difference were not acceptable, a strategy would be to use different elbows in laterals CF and CI to balance airflow.

In summary, duct sizes are as listed above, and fittings are:

B: a round 90° cross from a duct diameter of 800 mm in the main to a duct diameter of 400 mm in the laterals.

C: a round 90° cross from a duct diameter of 630 mm in the main to a duct diameter of 400 mm in the laterals.

D: a round, 180° diverging wye from a duct diameter of 500 mm to duct diameters of 400 mm.

All elbows: round, 90° smooth radius, with r/D = 1.0.

Exits E, H, G, and J: abrupt exits with screens having a free area ratio of 0.50.

Exits F and I: abrupt exits with screens having a free area ratio of 0.65.

This completes the design, except for choosing the fan and specifying attachment of the fan to the duct system. The fan must provide 6 m^3/s at a pressure of 292 Pa, plus whatever system effect factors might arise.

Although Example 12-7 is long, it illustrates the type of steps required to obtain a useful design. The example develops a design without having to retrace any design steps to assure meeting required airflow rates. Sometimes this is possible, and sometimes not. For example, when the point is reached of designing a lateral, it may be discovered the critical path must be redesigned so design of laterals is more reasonable. For example, more air pressure may be needed at the junction of the lateral to preclude the need for oversized ducts and fittings in the lateral. Designing air ducts by hand can be very time-consuming.

12-8. Computerized Design Procedures

The effort required to design a significant air duct system works against a desire to try many designs to find the optimum in terms of installation and operating cost. Also, the amount of required detail can lead to occasional errors.

Therefore, computerized duct design methods are desirable, have been developed, and are currently marketed. The computerized methods do not optimize design, but reduce the drudgery so many trial designs can be developed and the best chosen. Such programs are currently proprietary and beyond the level which could be included as a teaching program for this text. Although some require mainframe computing capability, several are available for personal computers of 256K and 512K capacity.

12-9. Fan Operation Cost

The operating cost of a fan and duct system arises from the electricity required by the fans. The required amount of electricity is a function of airflow rate, the pressure difference, and the ventilating efficiency ratio of the fan as installed. Fan efficiency depends on fan design and motor characteristics.

The power input to air, W, in watts, equals the product of volumetric flow rate, $\dot{V}$, m^3/s, and total pressure difference, ΔP, Pa.

$$W = \dot{V}\Delta P. \tag{12-12}$$

For example, the system described in Example 12-7 was characterized by an airflow rate of 6 m^3/s through the fan at a pressure difference of 292 Pa. The rate at which power needed to be added to the air was (6 m^3/s)(292 Pa) = 1752 W. The need for electricity is 1752 W, divided by the product of the fan and motor efficiencies at the point of operation.

Motor efficiencies are usually high (0.7 or higher) but the mechanical efficiency of a fan may not be (although fan design today is sufficiently sophisticated that efficiencies can be as high as 90%). However, the cost of operation is directly proportional to the pressure difference, which is a function of how the duct system is designed (and system effect factors). Optimization procedures should incorporate consideration of operating cost. Because system effect factors can be relatively significant (especially in small air duct systems), and fan attachments generally contribute a small part of total system cost, rapid payback can result from expenditures to limit system effect factors.

12-10. Fan Laws

Fan "laws" relate the performance of one fan to a similar fan, and performance of a fan at conditions different from which fan data was obtained. Fan laws are developed from two observations, airflow through fans is proportional to rotational speed, and pressure loss for air flowing through ducts is proportional to dynamic pressure, $\rho v^2/2$.

Fans (here taken to mean the impellor and housing, less the motor and attachments to duct systems) are generally supplied with data which relate airflow rates and pressure differences. The data may be in the form of a graph or tabulated.

However, fan data is not available for a finely graded series of rotational speeds, air densities, or perhaps even all size models of a fan series. Fan "laws" make it possible to extend laboratory data on fan performance to fans operated under conditions differing from the test conditions, and to fans which are dynamically and geometrically similar but which were not tested. The laws can be used to predict the airflow rate, $\dot{V}$; power requirement, W; and pressure difference, ΔP; as functions of rotational speed, RPM; fan diameter, D; and air density, ρ. For subscript 1 referring to known conditions and subscript 2 referring to predicted conditions,

$$\dot{V}_2 / \dot{V}_1 = (RPM_2 / RPM_1)(D_2 / D_1)^3 \qquad (12\text{-}13)$$

$$W_2 / W_1 = (RPM_2 / RPM_1)^3 (D_2 / D_1)^5 (\rho_2 / \rho_1) \qquad (12\text{-}14)$$

$$\Delta P_2 / \Delta P_1 = (RPM_2 / RPM_1)^2 (D_2 / D_1)^2 (\rho_2 / \rho_1). \qquad (12\text{-}15)$$

Limitations are placed on use of the fan laws. The laws are restricted to fans which are geometrically and dynamically similar (homologous). The laws should be used only to predict performance of a larger diameter fan from known performance of a smaller diameter fan. Neither the diameter ratio nor the speed ratio should be greater than 3, and the product of the two should not be greater than 3.

Example 12-8

<u>Problem:</u> A vaneaxial fan has been tested in a laboratory and found to provide the following airflow and pressure characteristics when operated at 500 rpm, under conditions of standard air (air density of 1.204 kg/m^3).

Airflow, m^3/s	Pressure Difference, Pa
6.0	20
5.3	400
4.0	580
3.0	520
2.0	540
1.0	680

The fan is to be used in an application where air density is expected to be approximately 1.1 kg/m^3, and 5 m^3/s of airflow are required. Determine the rotational speed required for satisfactory performance and the required power input to the air at operating conditions.

Solution: The diameter is unchanged, but air density and rotational speed are changed.

Airflow is proportional to rotational speed if the fan diameter is unchanged, thus, by Equation 12-13

$$RPM_2 = RPM_1(V_2 / V_1) = 500(5.0 \ m^3/s)/(5.3 \ m^3/s)$$
$$= 472.$$

Pulley sizes should be chosen to provide this speed, or as close to it as possible. Using Equation 12-15, the pressure difference at 472 rpm is

$$\Delta P_2 = (400 \ Pa)(472 \ rpm / 500 \ rpm)^2(\rho_2 / \rho_1)$$
$$= 326 \ Pa.$$

The air duct system must be designed to load the fan with 326 Pa when 5 m^3/s of airflow.

12-11. Perforated Polyethylene Air Ducts

Air ducts made of polyethylene have been employed in agriculture for many

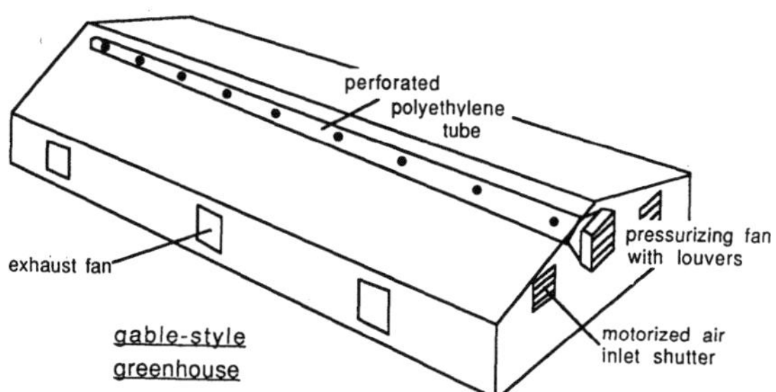

Figure 12-9. Example application of perforated polyethylene tubes to ventilate an agricultural building (from Duncan and Walker, 1973).

years as a means to distribute fresh air to ventilated airspaces. Such ducts are formed as long cylinders of polyethylene sheet. One end is closed and the other may have a fan to force air into the duct, or may open to the outdoors with the ventilated airspace held under a negative pressure by exhaust fans, an action which draws fresh air into the polyethylene ducts. If a fan is attached to the duct, an air straightener after the fan is recommended to reduce the swirling motion of air leaving the fan blades.

Along the length of the cylinder is a line of small holes (e.g., 50 mm diameter). The holes are closely spaced and common practice is to have two rows, one on each side of the duct. Air pressure within the duct forces air out the holes and is an inexpensive means to distribute air throughout an airspace. A common use is to retrofit older buildings in which modern systems such as hinged baffle slotted inlets cannot be readily installed. Another common use is to distribute ventilation air during winter in greenhouses, where temperature uniformity is important for uniform and proper plant growth.

Air issues from the small holes in the tubes at a high velocity, but being air jets of small diameter, the air slows rapidly and entrainment is vigorous. This tempers cold air quickly. All air from the tubes may be fresh, or recirculation may be designed into the ventilation system, with the ratio of fresh to recirculated air ranging from 0% to 100%.

Perforated polyethylene air ducts ("poly-tubes") with uniformly spaced holes may be designed so more air issues from holes near the beginning of the duct, near the end, or air distribution may be relatively uniform. When air recirculation is desired, to have more distributed at the end farthest from the recirculation fan is preferred to avoid a significant degree of short-circuiting back to the fan. Poly-tubes with nonuniformly spaced holes can be tailored to provide air distribution patterns of any reasonable form. When holes are uniformly spaced, airflow per hole at the beginning of a poly-tube is approximately half the flow rate per hole at the end.

Airflow from the holes in poly-tubes conforms to the Bernoulli equation. Air within the tube has both static and velocity pressure and the static pressure at a hole determines the velocity of air flowing from the hole. The coefficient of flow discharge is typically between 0.61 and 0.64.

Conditions can arise (i.e., a very long tube with many holes) where air velocity at the beginning of the poly-tube is so high that static pressure is below atmospheric pressure (as in a venturi). When this occurs, airflow is unstable, and flapping of the plastic occurs. Conversely, poly-tubes which are too short and have too few holes can constrict airflow, overload the fan, and provide poor ventilation. A rule which avoids both problems is to ensure the ratio of total hole area to duct cross sectional area (the aperture ratio) is within the range of 1.6 to 2.0.

Static pressure recovery after each exit hole is a factor which determines the relative distribution of air from one end of a poly-tube to another. If a poly-tube has a uniform diameter (generally the case), flow continuity demands air velocity to decrease after each hole, and the result is static pressure recovery.

Air exits the holes at an angle, having a forward component of velocity everywhere except perhaps at the final hole. The angle of emergence varies from nearly 60° from a perpendicular to the tube at the first hole, to nearly 0° at the final hole. Although not a planned effect, this too acts to limit short-circuiting of ventilation air back to a recirculation fan.

SYMBOLS

A	area, m^2
C	correction factor, dimensionless
D	diameter, m
f	friction factor, dimensionless
L	length, m
P	pressure, Pa
Re	Reynolds number, dimensionless
T	temperature, K
v	velocity, m/s
$\dot{V}$	volumetric flow rate, m^3/s
z	elevation above sea level, m
ε	absolute surface roughness, mm
μ	dynamic viscosity, kg/ms
ρ	density, kg/m^3

EXERCISES

1. Calculate the pressure drop of 30 C air flowing with a mean velocity of 8 m/s in a circular sheet metal duct 300 mm in diameter and 15 m long, using (a) equations and (b) the pressure drop graph.

2. A rectangular duct has dimensions of 0.25 by 1.0 m. Determine the pressure drop per meter length when 1.2 m^3/s of air flows through the duct. Assume standard conditions.

3. A sudden enlargement in a circular duct measures 0.2 m diameter upstream and 0.4 m downstream of the section change. The upstream static pressure is 150 Pa and the downstream static pressure is 200 Pa. What is the volumetric flow rate of 20 C air flowing through the enlargement?

4. Measurements made on a newly installed air handling system were: 1200 rpm fan speed, 3 m³/s flow rate, 350 Pa fan discharge pressure, and 1.8 kW supplied to the fan motor. These measurements were made at an air temperature of 18 C, but the system is to operate eventually with air at 50 C. If fan speed is unchanged, what will be (a) the airflow rate (m³/s), (b) the pressure from the fan (Pa), and (c) the power required by the motor?

5. A round duct has a diameter of 1.5 m and an absolute roughness of 5 mm. If air flows through the duct so the Reynolds number is 1.5E6, determine the friction factor which would be used in the Darcy-Weisbach equation to calculate pressure loss due to friction. Make your determination based on calculations, not on an estimate from the Moody diagram. Then double check your result using the Moody diagram.

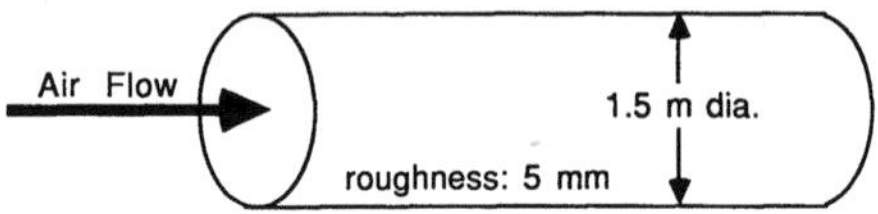

6. You have designed the following air duct system to distribute air in a refrigerated apple storage building. The fan must supply 2 m³/s at what pressure?

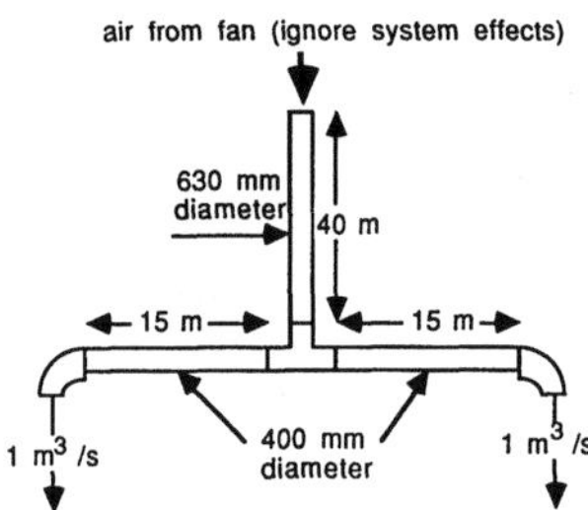

7. An air distribution system made of circular duct is designed as shown. The fitting at the junction of the branch is a 45° diverging conical wye, and the elbow in the branch has a 45° smooth radius with no vanes and r/D of 2.0. Each exit is abrupt with no diffusing section or screen. The 15 and 50 m sections of the main are 400 and 315 mm diameter, respectively. What diameter should the branch be to use up all the available pressure without needing a damper? Assume standard air conditions. Note: The diverging wye will be 400 mm diameter, followed by a converging section to reduce duct diameter to 315 mm.

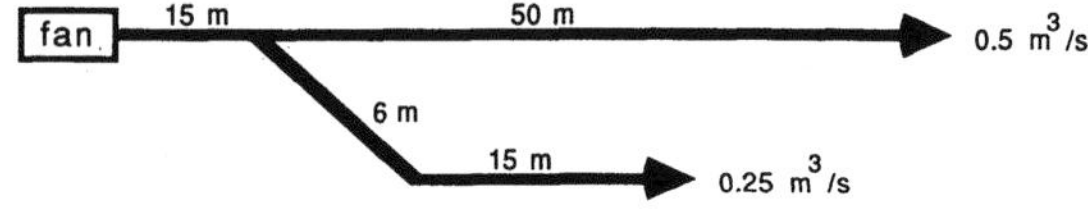

8. Consider the air duct system shown. The fan operates at 300 Pa total pressure difference. You can assume there is no significant system effect,

and the outlet is a bell mouth exit, with no additional outlet coefficient. The duct is made from a material with an absolute roughness factor of 0.15 mm and is round. Air density is 1.1 kg/m^3, the fan rotates at 580 rpm, and the motor on the fan draws 1.3 kW.

(a) Estimate the airflow rate through the duct.

(b) Determine the required fan rotation speed, total pressure, and motor power requirement if the air density changes to 0.9 kg/m^3 and the same volumetric flow rate of air is desired.

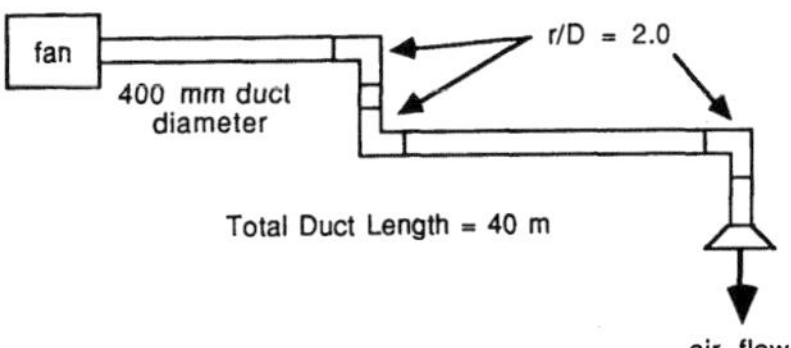

9. Air is carried through a horizontal, galvanized metal duct. Duct diameter is 630 mm and air velocity is 10 m/s. Air density is 1.2 kg/m^3. The duct is 80 m long in the 630 mm diameter section. After a converging section (convergence angle is 120°) the diameter is reduced to 400 mm. After the convergence the duct continues for another 45 m. The fan is located at the beginning of the 630 mm section and supplies 550 Pa total pressure.

Fifteen meters after the convergence there is a smooth, round hole (75 mm diameter). What will be the volumetric airflow rate (m^3/s) from the hole?

REFERENCES

AMCA. 1975. Publication 201: Fans and Systems. Air Movement and Control Association, Inc., Arlington Heights, IL.

ASHRAE. 1989. Handbook of Fundamentals. American Society of Heating, Refrigerating, and Air-Conditioning Engineers, Atlanta, GA.

ASHRAE. 1988. Handbook: Equipment Volume. American Society of Heating, Refrigerating, and Air-Conditioning Engineers, Atlanta, GA.

Bailey, B. 1982. The design of film-plastic ventilation ducts. Report 39 of the National Institute of Agricultural Engineering, Wrest Park, Silsoe, Bedford, Great Britain.

Duncan, G. A. and J. N. Walker. 1973. Poly-tube heating-ventilation systems and equipment. Publication AEN-7, Department of Agricultural Engineering, University of Kentucky, Lexington, KY.

Hellickson, M. A. and J. N. Walker, eds. 1983. Ventilation of Agricultural Structures. American Society of Agricultural Engineers, St. Joseph, MI.

Saunders, D. and L. D. Albright. 1984. Airflow from perforated polyethylene tubes. Transactions of the ASAE 27(4):1144-1149 American Society of Agricultural Engineers, St. Joseph, MI.

Stoecker, W. F. and J. W. Jones. 1982 . Refrigeration and Air Conditioning. McGraw-Hill Book Co., New York, NY.

APPENDIX 3-1
SELECTED VALUES OF THERMAL CONDUCTIVITY[a]

Material	Thermal Conductivity, W/mK
Metals	
Aluminum (alloy 1100)	221
Brass, red (85% Cu, 15% Zn)	150
Brass, yellow (65% Cu, 35% Zn)	120
Copper (electrolytic)	393
Gold	297
Iron, cast	47.7 (327 K)
Iron, wrought	60.4
Lead	34.8
Nickel	59.5
Silver	424
Steel, mild	45.3
Tin	64.9
Zinc, galvanizing	110
Wood	
Ash, white	0.172
Elm, American	0.153
Fir, white	0.12
Mahogony	0.13
Maple, sugar	0.187
Oak, white	0.176
Pine, white	0.11
Spruce	0.11
Other	
Brick, building	0.7
Cardboard	0.07
Cellulose	.057
Charcoal, wood	0.05 (473 K)
Concrete stone	0.93
Cork granulated	0.048 (268 K)
Cotton, fiber	0.042
Earth, dry and packed	0.064
Glass, soda-lime	1.0 (366 K)
Ice, 0 C	2.24
Leather	0.16
Marble	2.6
Paper	0.13
Plaster	0.74 (348 K)
Sand, dry	0.33
Sawdust	0.05
Snow, fresh at 32 C	0.598

Selected from the 1985 ASHRAE Handbook of Fundamentals.

a. values at room temperature unless noted in parentheses.

385

DESIGN HEAT TRANSMISSION COEFFICIENTS
Taken from the 1985 ASHRAE Handbook of Fundamentals, American Society of Heating, Refrigerating and Air Conditioning Engineers, Atlanta GA. (Used by permission.)

Thermal Properties of Typical Building and Insulating Materials—Design Values[a]

Description	Density kg/m³	Conductivity λ W/m·°C	Conductance (C) W/m²·°C	Resistance (R) Per meter thickness (1/λ) m·°C/W	Resistance (R) For thickness listed (1/C) m²·°C/W	Specific Heat kJ/(kg·°C)
BUILDING BOARD						
Boards, Panels, Subflooring, Sheathing						
Woodboard Panel Products						
Asbestos-cement board	1920	0.576	—	1.74	—	1.01
Asbestos-cement board .... 3.18 mm	1920	—	187.4	—	0.005	
Asbestos-cement board .... 6.35 mm	1920	—	93.72	—	0.011	
Gypsum or plaster board .... 9.53 mm	800	—	17.61	—	0.056	1.09
Gypsum or plaster board .... 12.70 mm	800	—	12.61	—	0.079	
Gypsum or plaster board .... 15.88 mm	800	—	10.11	—	0.099	
Plywood (Douglas Fir)	544	0.115	—	8.68	—	1.22
Plywood (Douglas Fir) .... 6.35 mm	544	—	18.18	—	0.055	
Plywood (Douglas Fir) .... 9.53 mm	544	—	12.10	—	0.083	
Plywood (Douglas Fir) .... 12.70 mm	544	—	9.09	—	0.11	
Plywood (Douglas Fir) .... 15.88 mm	544	—	7.33	—	0.14	
Plywood or wood panels .... 19.05 mm	544	—	6.08	—	0.16	1.22
Vegetable Fiber Board						
Sheathing, regular density .... 12.70 mm	288	—	4.32	—	0.23	1.30
.... 19.84 mm	288	—	2.78	—	0.36	
Sheathing intermediate density .... 12.70 mm	352	—	4.66	—	0.21	1.30
Nail-base sheathing .... 12.70 mm	400	—	5.00	—	0.20	1.30
Shingle backer .... 9.53 mm	288	—	6.02	—	0.17	1.30
Shingle backer .... 7.94 mm	288	—	7.27	—	0.14	
Sound deadening board .... 12.70 mm	240	—	4.20	—	0.24	1.26
Tile and lay-in panels, plain or acoustic	288	0.058	—	17.35	—	0.59
.... 12.70 mm	288	—	4.54	—	0.22	
.... 19.05 mm	288	—	3.01	—	0.33	
Laminated paperboard	480	0.072	—	13.88	—	1.38
Homogeneous board from repulped paper	480	0.072	—	13.88	—	1.17
Hardboard						
Medium density	800	0.105	—	9.51	—	1.30
High density, service temp. service underlay	880	0.118	—	8.47	—	1.34
High density, std. tempered	1008	0.144	—	6.94	—	1.34
Particleboard						
Low density	592	0.078	—	12.84	—	1.30
Medium density	800	0.135	—	7.36	—	1.30
High density	1000	0.170	—	5.90	—	1.30
Underlayment .... 15.88 mm	640	—	6.93	—	0.14	1.22
Wood subfloor .... 19.05 mm		—	6.02	—	0.17	1.38
BUILDING MEMBRANE						
Vapor—permeable felt	—	—	94.86	—	0.011	
Vapor—seal, 2 layers of mopped 0.73 kg/m² felt	—	—	47.43	—	0.021	
Vapor—seal, plastic film	—	—	—	—	Negl.	
FINISH FLOORING MATERIALS						
Carpet and fibrous pad	—	—	2.73	—	0.37	1.42
Carpet and rubber pad	—	—	4.60	—	0.22	1.38
Cork tile .... 3.18 mm	—	—	20.45	—	0.049	2.01
Terrazzo .... 25.40 mm	—	—	71.00	—	0.014	0.80
Tile—asphalt, linoleum, vinyl, rubber	—	—	113.6	—	0.009	1.26
vinyl asbestos						1.01
ceramic						0.80
Wood, hardwood finish .... 19.05 mm			8.35		0.12	
INSULATING MATERIALS						
Blanket and Batt[b]						
Mineral Fiber, fibrous form processed from rock, slag, or glass						
approx.[c] 76.2–101.6 mm	4.8–32.0	—	0.52	—	1.94[b]	
approx.[c] 88.9 mm	4.8–32.0	—	0.44	—	2.29[b]	
approx.[c] 139.7–165.1 mm	4.8–32.0	—	0.30	—	3.34[b]	
approx.[c] 152.4–177.8 mm	4.8–32.0		0.26		3.87[b]	
approx.[c] 215.9–228.6 mm	4.8–32.0		0.19		5.28[b]	
approx. 304.8 mm	4.8–32.0	—	0.15		6.69[b]	

Thermal Properties of Typical Building and Insulating Materials—Design Values[a]

Description	Density kg/m³	Conductivity (λ) W/m·°C	Conductance (C) W/m²·°C	Resistance (R) Per meter thickness	Resistance (R) For thickness listed	Specific Heat kJ/(kg·°C)
Board and Slabs						
Cellular glass	136	0.050	—	19.85	—	0.75
Glass fiber, organic bonded	64–144	0.036	—	27.76	—	0.96
Expanded perlite, organic bonded	16.0	0.052	—	19.29	—	1.26
Expanded rubber (rigid)	72.0	0.032	—	31.58	—	1.68
Expanded polystyrene extruded						
Cut cell surface	28.8	0.036	—	27.76	—	1.22
Smooth skin surface	28.8–56.0	0.029	—	34.70	—	1.22
Expanded polystyrene, molded beads	16.0	0.037	—	23.25	—	—
	20.0	0.036	—	27.76	—	—
	24.0	0.035	—	28.94	—	—
	28.0	0.035	—	28.94	—	—
	32.0	0.033	—	30.19	—	—
Cellular polyurethane[c] (R-11 exp.)(unfaced)	24.0	0.023	—	43.38	—	1.59
Foil-faced, glass fiber-reinforced cellular						
Polyisocyanurate (R-11 exp.)[d]	32.0	0.020	—	49.97	—	0.92
Nominal 12.70 mm		—	1.58	—	0.63	
Nominal 25.40 mm		—	0.79	—	1.27	
Nominal 50.80 mm		—	0.39	—	2.53	
Mineral fiber with resin binder	240	0.042	—	23.94	—	0.71
Mineral fiberboard, wet felted						
Core or roof insulation	256–272	0.049	—	20.40	—	
Acoustical tile	288	0.050	—	19.85	—	0.80
Acoustical tile	336	0.053	—	18.74	—	
Mineral fiberboard, wet molded						
Acoustical tile[e]	368	0.060	—	16.52	—	0.59
Wood or cane fiberboard						
Acoustical tile[e] ... 12.70 mm	—	—	4.54	—	0.22	1.30
Acoustical tile[e] ... 19.05 mm	—	—	3.01	—	0.33	
Interior finish (plank, tile)	240	0.050	—	19.85	–	1.34
Cement fiber slabs (shredded wood with Portland cement binder	400–432	0.072–0.070	—	13.88–13.12	–	—
Cement fiber slabs (shredded wood with magnesia oxysulfide binder)	352	0.082	—	12.15	–	1.30
LOOSE FILL						
Cellulosic insulation (milled paper or wood pulp)	36.8–51.2	0.039–0.046	—	25.68–21.72	—	1.38
Sawdust or shavings	128–240	0.065	—	15.41	—	1.38
Wood fiber, softwoods	32.0–56.0	0.043	—	23.11	—	1.38
Perlite, expanded						
	32.0–65.6	0.039–0.045		25.68–22.90		
	65–118	0.045–0.052		22.90–19.43		
	118–176	0.052–0.060		19.43–16.66		
Mineral fiber (rock, slag or glass)						
approx. 95.3–127.0 mm	9.6–32.0	—	—		1.94	0.71
approx. 165.1–222.3 mm	9.6–32.0	—	—		3.34	
approx. 190.5–254.0 mm	9.6–32.0	—	—		3.87	
approx. 260.4–349.3 mm	9.6–32.0	—	—		5.28	
Mineral fiber (rock, slag or glass)						
approx. 83.8 mm (closed sidewall application)	32.0–56.0	—	—	—	2.46	
Vermiculite, exfoliated	112–131	0.068	—	14.78	—	1.34
	64.0–96.0	0.063	—	15.75	—	
FIELD APPLIED						
Polyurethane foam	24.0–40.0	0.023–0.026	—	43.38–36.50	—	
Ureaformaldehyde foam	11.2–25.6	0.032–0.040	—	24.78–31.58	—	
Spray cellulosic fiber base	32.0–96.0	0.035–0.043	—	23.11–28.94	—	
PLASTERING MATERIALS						
Cement plaster, sand aggregate	1865	0.720	—	1.39	—	0.84
Sand aggregate ... 9.53 mm	—	—	75.54	—	0.014	0.84
Sand aggregate ... 19.05 mm	—	—	37.83	—	0.026	0.84
Gypsum plaster:						
Lightweight aggregate ... 12.70 mm	720	—	17.72	—	0.056	
Lightweight aggregate ... 15.88 mm	720	—	15.17	—	0.069	
Lightweight agg. on metal lath ... 19.05 mm.	—	—	12.10	—	0.083	
Perlite aggregate	720	0.216	—	4.65	—	1.34

Description	Density kg/m³	Conductivity (λ) W/m·°C	Conductance (C) W/m²·°C	Resistance (R) Per meter thickness	Resistance (R) For thickness listed	Specific Heat kJ/(kg·°C)
PLASTERING MATERIALS						
Sand aggregate	1680	0.806	—	*1.25*	—	0.84
Sand aggregate............12.70 mm	1680	—	63.05	—	*0.016*	
Sand aggregate............15.88 mm	1680	—	51.69	—	*0.019*	
Sand aggregate on metal lath ....19.05 mm	—	—	43.74	—	*0.023*	
Vermiculite aggregate	720	0.245	—	*4.09*	—	
MASONRY MATERIALS						
Concretes						
Cement mortar	1856	0.720	—	*1.39*	—	
Gypsum-fiber concrete 87.5% gypsum,						
12.5% wood chips	816	0.239	—	*4.16*	—	0.88
Lightweight aggregates including ex-	1920	0.749	—	*1.32*	—	
panded shale, clay or slate; expanded	1600	0.518	—	*1.94*	—	
slags; cinders; pumice; vermiculite;	1280	0.360	—	*2.78*	—	
also cellular concretes	960	0.245	—	*4.09*	—	
	640	0.166	—	*5.97*	—	
	480	0.130	—	*7.70*	—	
	320	0.101		*9.92*		
Perlite, expanded	640	0.134		*7.50*		
	480	0.102		*9.79*		
	320	0.072		*13.88*		1.34
Sand and gravel or stone aggregate						
(oven dried)	2240	1.296	—	*0.76*		0.92
Sand and gravel or stone aggregate						
(not dried)	2240	1.728	—	*0.56*		
Stucco	1856	0.720	—	*1.39*		
MASONRY UNITS						
Brick, common[f]	1920	0.720	—	*1.39*	—	0.80
Brick, face[f]	2080	1.296	—	*0.76*	—	
Clay tile, hollow:						
1 cell deep ....76.2 mm	—	—	7.10	—	*0.14*	0.88
1 cell deep ....101.6 mm	—	—	5.11	—	*0.20*	
2 cells deep...........................152.4 mm	—	—	3.75	—	*0.27*	
2 cells deep...........................203.2 mm	—	—	3.07	—	*0.33*	
2 cells deep...........................254.0 mm	—	—	2.56	—	*0.39*	
3 cells deep...........................304.8 mm	—	—	2.27	—	*0.44*	
Concrete blocks, three oval core:						
Sand and gravel aggregate ....101.6 mm	—	—	7.95	—	*0.12*	0.92
............203.2 mm	—	—	5.11	—	*0.20*	
............304.8 mm	—	—	4.43	—	*0.23*	
Cinder aggregate ....76.2 mm	—	—	6.59	—	*0.15*	0.88
............101.6 mm	—	—	5.11	—	*0.20*	
............203.2 mm	—	—	3.29	—	*0.30*	
............304.8 mm	—	—	3.01	—	*0.33*	
Lightweight aggregate ....76.2 mm	—	—	4.49	—	*0.22*	0.88
expanded shale, clay, slate ....101.6 mm	—	—	3.81	—	*0.26*	
or slag; pumice ....203.2 mm	—	—	2.84	—	*0.35*	
............304.8 mm	—	—	2.50	—	*0.40*	
Concrete blocks, rectangular core.[g]						
Sand and gravel aggregate						
2 core, [h] ....203.2 mm, 16.3 kg	—	—	5.45	—	*0.18*	0.92
Same with filled cores[i]	—	—	2.95	—	*0.34*	0.92
Lightweight aggregate (expanded shale,						
clay, slate or slag, pumice):						
3 core, [h] ....152.4 mm, 8.6 kg	—	—	3.46	—	*0.29*	0.88
Same with filled cores[i]	—	—	1.87	—	*0.53*	
2 core, [h] ....203.2 mm, 10.9 kg	—	—	2.61	—	*0.38*	
Same with filled cores[i]	—	—	1.14	—	*0.89*	
3 core, [h].................304.8 mm, 17.3 kg	—	—	2.27	—	*0.44*	
Same with filled cores[i]	—	—	0.97	—	*1.02*	
Stone, lime or sand	—	1.800	—	*0.56*	—	0.80
Gypsum partition tile:						
76.2 • 304.8 • 762.0 mm, solid	—	—	4.49	—	*0.22*	*0.80*
76.2 • 304.8 • 762.0 mm, 4-cell	—	—	4.20	—	*0.24*	
101.6 • 304.8 • 762.0 mm, 3-cell	—	—	3.41	—	*0.29*	

Thermal Properties of Typical Building and Insulating Materials—Design Values[a]

Description	Density kg/m^3	Conductivity λ W/m·°C	Conductance (C) W/m·°C	Resistance (R)		Specific Heat kJ/ (kg · °C)
				Per meter thickness (1/λ) m·°C/W	For thickness listed (1/C) m²·°C/W	
METALS (See Chapter 39, Table 3)						
ROOFING						
Asbestos-cement shingles	1920	—	27.04	—	0.037	1.01
Asphalt roll roofing	1120	—	36.92	—	0.026	1.51
Asphalt shingles	1120	—	12.89	—	0.077	1.26
Built-up .... 9.53 mm	1120	—	17.04	—	0.058	1.47
Slate .... 12.70 mm	—	—	113.6	—	0.009	1.26
Wood shingles, plain and plastic film faced	—	—	6.02	—	0.17	1.30
SIDING MATERIALS (on flat surface)						
Shingles						
Asbestos-cement	1920	—	26.98	—	0.037	
Wood, 406.4 mm, 190.5 mm exposure	—	—	6.53	—	0.15	1.30
Wood, double, 406.4 mm, 304.8 mm exposure	—	—	4.77	—	0.21	1.17
Wood, plus insul. backer board, 7.94 mm	—	—	4.03	—	0.25	1.30
Siding						
Asbestos-cement, 6.35 mm, lapped	—	—	27.04	—	0.037	1.01
Asphalt roll siding	—	—	36.92	—	0.026	1.47
Asphalt insulating siding (12.70 mm bed.)	—	—	3.92	—	0.26	1.47
Hardboard siding, 11.11 mm	640	0.215	—	4.65		1.17
Wood, drop, 25.4 · 203.2 mm	—	—	7.21	—	0.14	1.17
Wood, bevel, 12.7 · 203.2 mm, lapped	—	—	6.99	—	0.14	1.17
Wood, bevel, 19.1 · 254.0 mm, lapped	—	—	5.40	—	0.18	1.17
Wood, plywood, 9.53 mm, lapped	—	—	9.03	—	0.10	1.22
Aluminum or Steel[j], over sheathing						
Hollow-backed	—	—	9.14	—	0.11	1.22
Insulating-board backed nominal 9.53 mm	—	—	3.12	—	0.32	1.34
Insulating-board backed nominal 9.53 mm, foil backed			1.93		0.52	
Architectural glass	—	—	56.80	—	0.018	0.84
WOODS (12% Moisture Content)[k,l]						
Hardwoods						1.63
Oak	659-749	0.161-0.180	—	6.18-5.55	—	
Birch	682-726	0.167-0.176	—	6.04-5.69	—	
Maple	637-704	0.157-0.171	—	6.52-6.11	—	
Ash	614-670	0.153-0.164	—	6.52-6.11	—	
Softwoods						1.63
Southern Pine	570-659	0.144-0.161	—	6.94-6.18	—	
Douglas Fir-Larch	536-581	0.137-0.145	—	7.36-6.87	—	
Southern Cypress	502-514	0.130-0.132	—	7.70-7.56	—	
Hem-Fir, Spruce-Pine-Fir	392-502	0.107-0.130	—	9.37-7.70	—	
West Coast Woods, Cedars	347-502	0.098-0.130	—	10.27-7.70	—	
California Redwood	392-448	0.107-0.118	—	9.37-8.47	—	

Notes

[a] Except where otherwise noted, all values are for a mean temperature of 23.9°C. Representative values for dry materials, selected by ASHRAE TC 4.4, are intended as design (not specification) values for materials in normal use. Insulation materials in actual service may have thermal values that vary from design values depending on their in-situ properties (e.g., density and moisture content). For properties of a particular product, use the value supplied by the manufacturer or by unbiased tests.

[b] Does not include paper backing and facing, if any. Where insulation forms a boundary (reflective or otherwise) of an air space, see Tables 1, 2A and 2B for the insulating value of an air space with the appropriate effective emittance and temperature conditions of the space.

[c] Values are for aged, unfaced, board stock. For change in conductivity with age of expanded urethane, see Chapter 20, Factors Affecting Thermal Conductivity.

[d] Time-aged values for board stock with gas-barrier quality (0.025 mm thickness or greater) aluminum foil facers on two major surfaces.

[e] Insulating values of acoustical tile vary, depending on density of the board and on type, size and depth of perforations.

[f] Face brick and common brick do not always have these specific densities. When density differs from that shown, there is a change in thermal conductivity.

[g] Data on rectangular core concrete blocks differ from the above data on oval core blocks, due to core configuration, different mean temperatures, and possibly differences in unit weights. Weight data on the oval core blocks tested are not available.

[h] Weights of units approximately 193.7 mm high and 400.1 mm long. These weights are given as a means of describing the blocks tested, but conductance values are all for 0.093m² of area.

[i] Vermiculite, perlite, or mineral wool insulation. Where insulation is used, vapor barriers or other precautions must be considered to keep insulation dry.

[j] Values for metal siding applied over flat surfaces vary widely, depending on amount of ventilation of air space beneath the siding; whether air space is reflective or nonreflective; and on thickness, type, and application of insulating backing-board used. Values given are averages for use as design guides, and were obtained from several guarded hotbox tests (ASTM C236) or calibrated hotbox (BSS 77) on hollow-backed types and types made using backing-boards of wood fiber, foamed plastic, and glass fiber. Departures of±50% or more from the values given may occur.

[k] Forest Products Laboratory Wood Handbook, U.S. Dept. of Agriculture #72, 1974, Tables 3 and 4.

[l] L. Adams: Supporting cryogenic equipment with wood (*Chemical Engineering*, May 17, 1971).

Material	Surface Emittance	Condition
A. Metals		
Aluminum	0.04-0.06	polished, 200-600 C
	0.09	commercial, 95 C
	0.20-0.33	oxidized, 90-540 C
Brass	0.07	polished, 40 C
	0.46-0.56	oxidized, 40-260 C
Chromium	0.08-0.27	polished, 40-540 C
Copper	0.04	polished, 40 C
	0.05	tarnished, 40 C
	0.76	black oxidized, 40 C
Gold	0.02-0.04	polished, 90-590 C
Iron and steel		
mild steel	0.14-0.32	polished, 150-480 C
rolled sheet	0.66	new, 40 C
steel	0.80	oxidized, 40 C
cast iron	0.44	new, 40 C
	0.57-0.66	oxidized, 40-260 C
iron	0.61	red rust, 40 C
	0.85	heavy rust, 40 C
stainless steel	0.07-0.17	new, 40 C
	0.50-0.70	used, 230-900 C
wrought iron	0.35	new, 40 C
	0.94	oxidized, 20-360 C
Nickel	0.05-0.07	polished, 40-260 C
	0.35-0.49	oxidized, 40-260 C
Silver	0.01-0.03	polished, 40-540 C
	0.02-0.04	oxidized, 40-540 C
Tin	0.06	bright, 40 C
Zinc	0.23	galv., bright, 40 C
	0.28	galv., gray, 40 C
	0.21	dull, 40-260 C
B. Nonmetals		
Brick	0.93	red building, 40 C
	0.75	fireclay, 980 C
Carbon, soot	0.95	40 C
Concrete	0.94	rough, 40 C
Glass	0.94	smooth crown, 40 C
Ice	0.97	smooth, 0 C
	0.99	rough crystals, 0 C
	0.99	hoarfrost, -20 C
Paints	0.27-0.62	aluminum, var. ages and compositions
	0.90	black gloss
	0.89-0.97	white paint
	0.92-0.96	various colors
Paper	0.95	white, 40 C
	0.92-0.94	any color, 40 C
	0.91	roofing, 40 C
Plaster	0.92	rough, 40-260 C
Snow	0.82	-10 C
Water	0.96	>0.1 mm thick, 40 C
Wood	0.80-0.90	various, 40 C

I. Configuration 1, parallel rectangular planes,
$$X = a/c; \quad Y = b/c; \quad W^2 = 1+X^2; \quad Z^2 = 1+Y^2$$

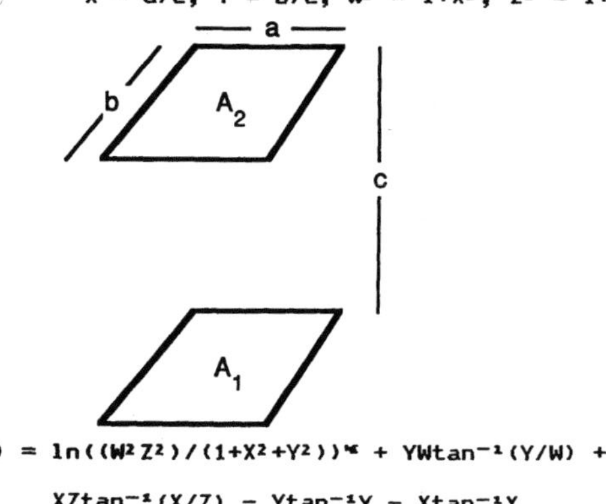

$$F_{1-2}(\pi XY/2) = \ln((W^2 Z^2)/(1+X^2+Y^2))^{\frac{1}{2}} + YW\tan^{-1}(Y/W) +$$

$$XZ\tan^{-1}(X/Z) - Y\tan^{-1}Y - X\tan^{-1}X.$$

II. Configuration 2, rectangular planes intersecting at right
angles, $X = a/b; \quad Y = c/b; \quad Z^2 = X^2+Y^2$

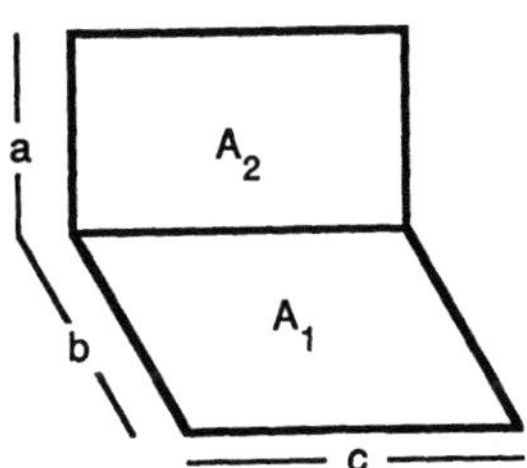

$$F_{1-2}(\pi Y) = \left(\tfrac{1}{4}\right)\ln\left(\frac{(1+X^2)(1+Y^2)}{1+Z^2}\right) + \left(\tfrac{1}{4}\right)Y^2\ln\left(\frac{Y^2(1+Z^2)}{(1+Y^2)Z^2}\right) +$$

$$\left(\tfrac{1}{4}\right)X^2\ln\left(\frac{X^2(1+Z^2)}{Z^2(1+X^2)}\right) + Y\tan^{-1}(1/Y) +$$

$$X\tan^{-1}(1/X) - Z\tan^{-1}(1/Z).$$

III. **Configuration 3, small area facing rectangular plane,**
     $X = a/c$; $Y = b/c$; $W^2 = (1+X^2)$; $Z^2 = (1+Y^2)$

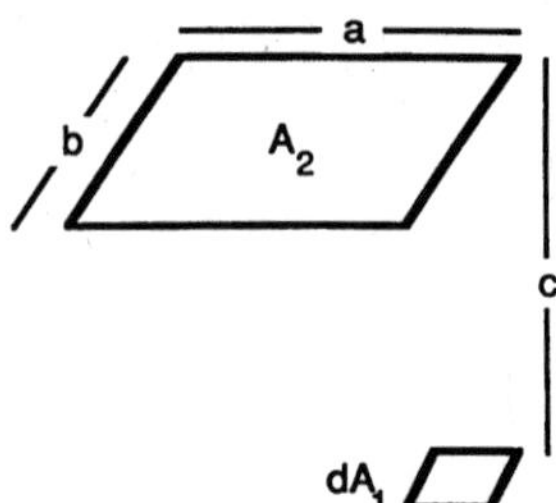

$$F_{1-2}(2\pi) = (X/W)\tan^{-1}(Y/W) + (Y/Z)\tan^{-1}(X/Z)$$

IV. **Configuration 4, small area perpendicular to rectangular plane,** $X = a/b$; $Y = c/b$; $Z = (X^2+Y^2)^{-1/2}$

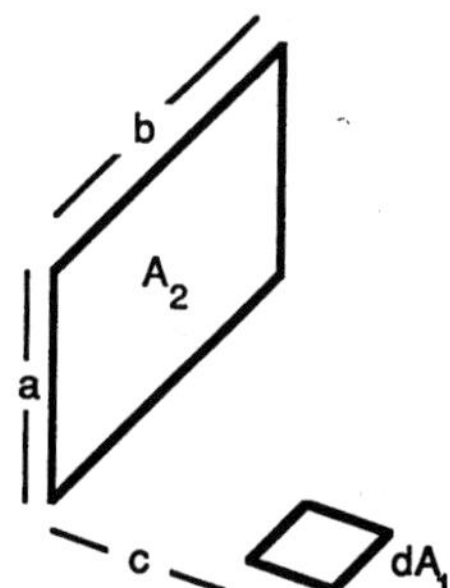

$$F_{1-2}(2\pi) = \tan^{-1}(1/Y) - ZY\tan^{-1}Z$$

V. **Configuration 5, strip parallel to rectangular plane,**
    $X = b/c$; $Y = a/c$; $W^2 = 1+X^2$; $Z^2 = 1+Y^2$.

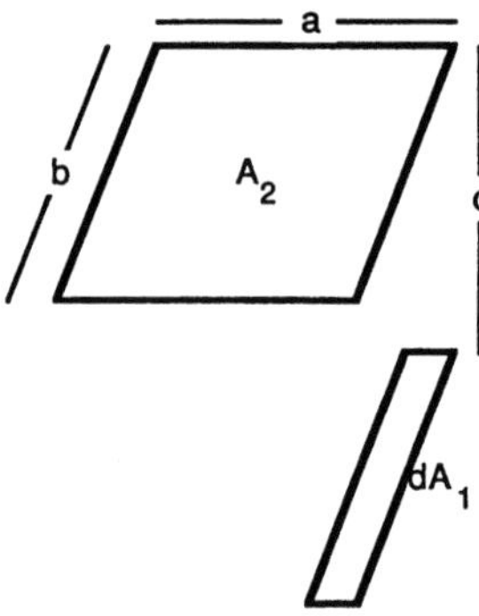

$$F_{1-2}(\pi X) = W\tan^{-1}(Y/W) - \tan^{-1}Y + (XY/Z)\tan^{-1}(X/Z)$$

VI. Configuration 6, strip perpendicular to rectangular plane,
$X = a/b$; $Y = c/b$; $Z^2 = X^2 + Y^2$.

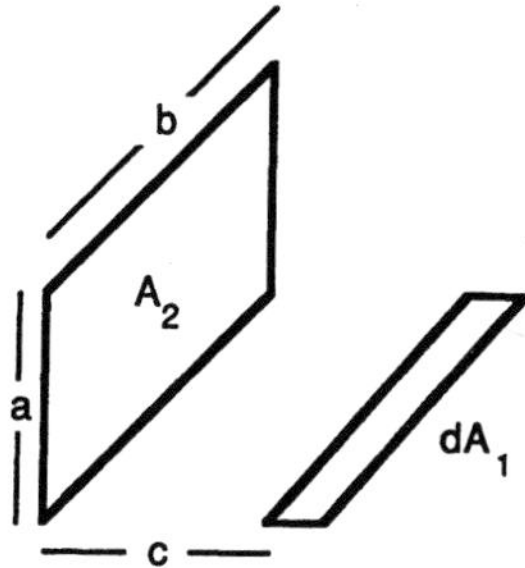

$$F_{1-2}(\pi) = \tan^{-1}(1/Y) + (Y/2)\ln\left[\frac{Y^2(Z^2+1)}{Z^2(Y^2+1)}\right] - (Y/Z)\tan^{-1}(1/Z)$$

VII. Configuration 7, small sphere and a rectangular plane,
$X = b/c$ and $Y = a/c$

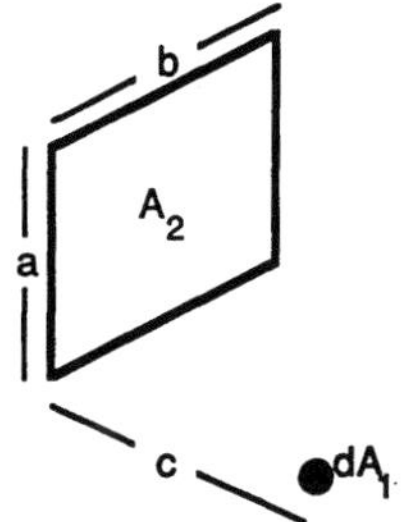

$$F_{1-2}(4\pi) = \tan^{-1}\left[\frac{x(y - \cos\emptyset)}{\sqrt{(1+x^2+y^2-2y\cos\emptyset)}}\right] + \tan^{-1}\left[(x\cos\emptyset)/\sqrt{(1+x^2)}\right]$$

and for $\emptyset = 90°$,

$$F_{1-2}(4\pi) = \tan^{-1}\left[xy/\sqrt{(1+x^2+y^2)}\right]$$

$$\lim_{x\to\infty} F_{1-2} = (1/4\pi)\tan^{-1}(y)$$

$$\lim_{\substack{x\to\infty \\ y\to\infty}} F_{1-2} = 1/8$$

Note: angle factors for more complex configurations can be built
using the above. For example, the angle factor from a small
animal in a barn to the ceiling of the barn can be built using
four applications of configuration 3 above. The ceiling is
divided into four quadrants, intersecting above the location of
the animal. The angle factor from the animal to each quadrant of
the ceiling is calculated, and the four angle factors added to
obtain the angle factor from the animal to the entire ceiling.
The animal need not be centered below the middle of the ceiling
for this technique to work.

Also note that reciprocity applies in calculating F_{2-1} for each
configuration above.

For additional angle factors see, for example: Sparrow, E.M. and
R.D. Cess, 1978, Radiation heat transfer, Hemisphere Publishing
Corp., Washington, 366 pp.

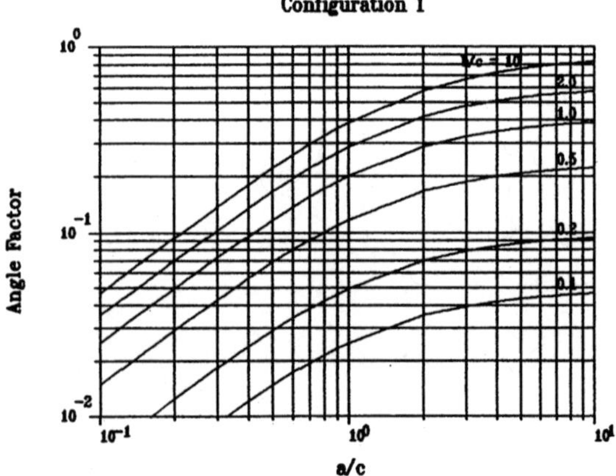

Configuration 1

Angle factors for configuration 1, for b/c of 10, 2, 1, 0.5, 0.2 and 0.1 (top to bottom).

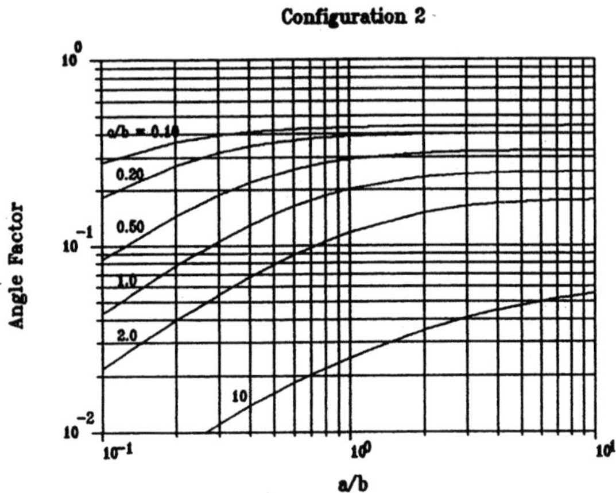

Configuration 2

Angle factors for configuration 2, for c/b of 0.1, 0.2, 0.5, 1, 2, and 10 (top to bottom).

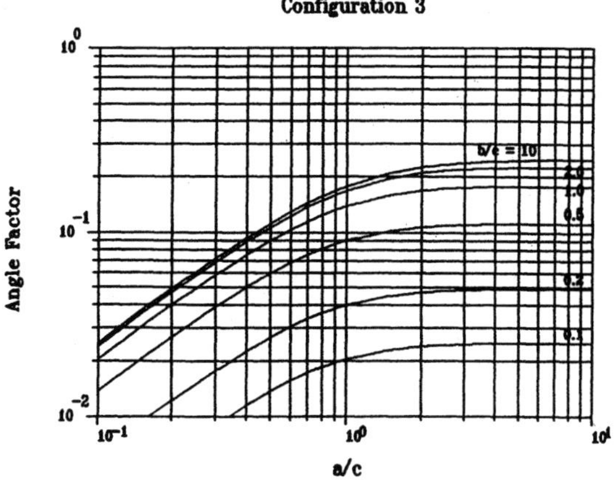

Angle factors for configuration 3, for b/c of 10, 2, 1, 0.5, 0.2 and 0.1 (top to bottom).

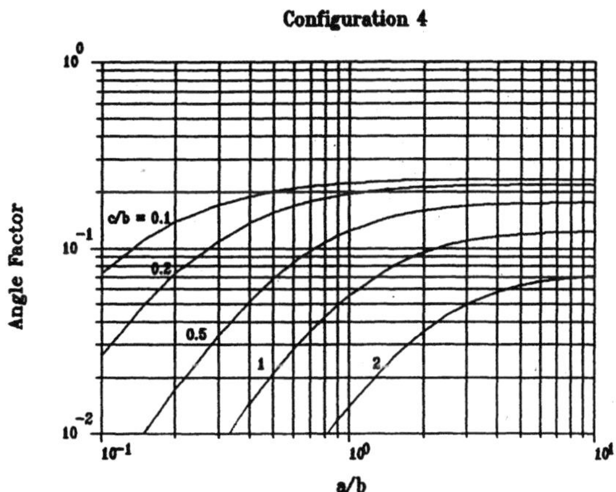

Angle factors for configuration 4, for c/b of 0.1, 0.2, 0.5, 1 and 2 (top to bottom). Factors for c/b of 10 are less than 0.01.

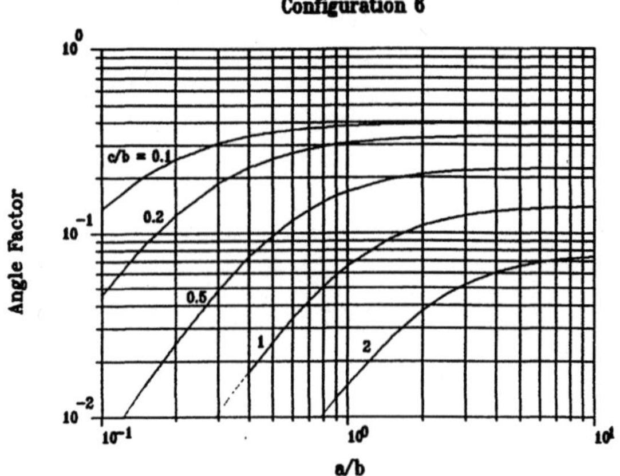

Angle factors for configuration 5, for a/c of 10, 2, 1, 0.5, 0.2 and 0.1 (top to bottom).

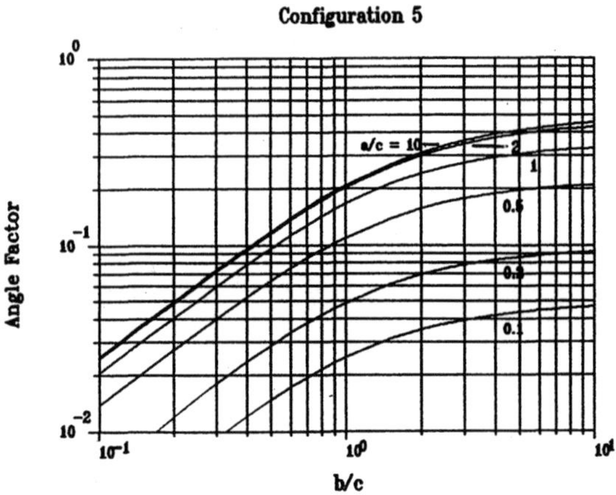

Angle factors for configuration 6, for c/b of 0.1, 0.2, 0.5, 1 and 2 (top to bottom). Factors for c/b of 10 are less than 0.01.

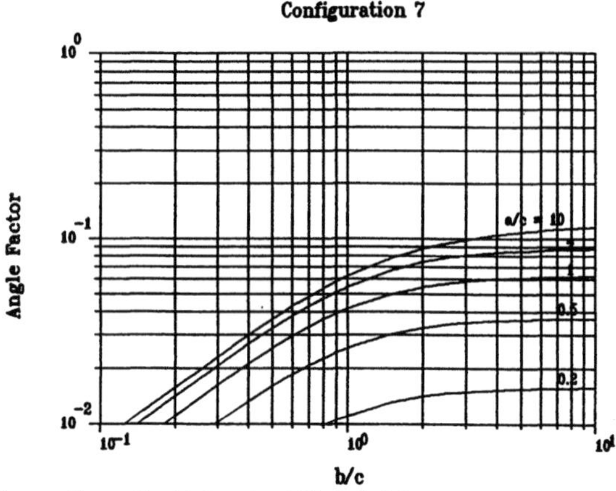

Angle factors for configuration 7, for a/c of 10, 2, 1, 0.5 and 0.2 (top to bottom). Factors for a/c of 0.1 are less than 0.01.

396

SURFACE CONDUCTANCES AND RESISTANCES FOR AIR
(h IN W/M^{2}K; R IN M^{2}K/W)

A. Standard data for still air, as inside a building:

Orientation of surface	Heat Flow Direction	Surface Emittance					
		0.90		0.20		0.05	
		h	R	h	R	h	R
vertical	horizontal	8.29	0.12	4.20	0.24	3.35	0.30
horizontal	upward	9.26	0.11	5.17	0.19	4.32	0.23
	downward	6.13	0.16	2.10	0.48	1.25	0.80
45 degree slope	upward	9.09	0.11	5.00	0.20	4.15	0.24
	downward	7.50	0.13	3.41	0.29	2.56	0.39

B. Moving air, as outside a building, surface in any orientation:

Wind Velocity, m/s	h	R	
6.7	34.08	0.030	(for winter)
3.4	22.72	0.044	(for summer)

NOTES: 1. Based on data in the ASHRAE Handbook of Fundamentals.

2. Assumes air temperature and mean radiant temperature of the surroundings are identical.

3. Values are based on a surface-air temperature difference of 5.5 K, and a surface temperature of 21 C.

4. Does not apply to ventilated attics above ceilings in summer conditions.

5. Values are based on smooth surfaces. For effect of surface roughness, see the ASHRAE Handbook of Fundamentals.

6. A plane air space does not have both surface resistances and air space resistance. For air space resistance, see Appendix 3-6.

7. For more precise values of surface coefficients between still air and glass, use

$$h_i = 1.77(T_g - T_a)^{0.25} + \epsilon\sigma(T_g^4 - T_a^4)/(T_g - T_a),$$

where T_g and T_a are the glass and air temperatures, K, respectively, ϵ is the surface emittance of the glass, and h_i is the surface coefficient, W/m^{2}K.

THERMAL RESISTANCES OF PLANE AIR SPACES
Taken from the 1985 ASHRAE Handbook of Fundamentals, American Society of Heating, Refrigerating and Air Conditioning Engineers, Atlanta GA
Used by permission

Table 2A Thermal Resistances of Plane[a] Air Spaces[b,c]
$m^2 \cdot °C/W$

Position of Air Space	Direction of Heat Flow	Mean Temp.[d] (°C)	Temp Diff.[d] (°C)	12.7-mm Air Space[c] Value of $E^{d,e}$					19.1-mm Air Space[c] Value of $E^{d,e}$				
				0.03	0.05	0.2	0.5	0.82	0.03	0.05	0.2	0.5	0.82
Horiz.	Up	32.2	5.6	0.37	0.36	0.27	0.17	0.13	0.41	0.39	0.28	0.18	0.13
		10.0	16.7	0.29	0.28	0.23	0.17	0.13	0.30	0.29	0.24	0.17	0.14
		10.0	5.6	0.37	0.36	0.28	0.20	0.15	0.40	0.39	0.30	0.20	0.15
		-17.8	11.1	0.30	0.30	0.26	0.20	0.16	0.32	0.32	0.27	0.20	0.16
		-17.8	5.6	0.37	0.36	0.30	0.22	0.18	0.39	0.38	0.31	0.23	0.18
		-45.6	11.1	0.30	0.29	0.26	0.22	0.18	0.31	0.31	0.27	0.22	0.19
		-45.6	5.6	0.36	0.35	0.31	0.25	0.20	0.38	0.37	0.32	0.26	0.21
45° Slope	Up	32.2	5.6	0.43	0.41	0.29	0.19	0.13	0.52	0.49	0.33	0.20	0.14
		10.0	16.7	0.36	0.35	0.27	0.19	0.15	0.35	0.34	0.27	0.19	0.14
		10.0	5.6	0.45	0.43	0.32	0.21	0.16	0.51	0.48	0.35	0.23	0.17
		-17.8	11.1	0.39	0.38	0.31	0.23	0.18	0.37	0.36	0.30	0.23	0.18
		-17.8	5.6	0.46	0.45	0.36	0.25	0.19	0.48	0.46	0.37	0.26	0.20
		-45.5	11.1	0.37	0.36	0.31	0.25	0.21	0.36	0.35	0.31	0.25	0.20
		-45.6	5.6	0.46	0.45	0.38	0.29	0.23	0.45	0.43	0.37	0.25	0.23
Vertical	Horiz. →	32.2	5.6	0.43	0.41	0.29	0.19	0.14	0.62	0.57	0.37	0.21	0.15
		10.0	16.7	0.45	0.43	0.32	0.22	0.16	0.51	0.49	0.35	0.23	0.17
		10.0	5.6	0.47	0.45	0.33	0.22	0.16	0.65	0.61	0.41	0.25	0.18
		-17.8	11.1	0.50	0.48	0.38	0.26	0.20	0.55	0.53	0.41	0.28	0.21
		-17.8	5.6	0.52	0.50	0.39	0.27	0.20	0.66	0.63	0.46	0.30	0.22
		-45.6	11.1	0.51	0.50	0.41	0.31	0.24	0.51	0.50	0.42	0.31	0.24
		-45.6	5.6	0.56	0.55	0.45	0.33	0.26	0.65	0.63	0.51	0.36	0.27
45° Slope	Down	32.2	5.6	0.44	0.41	0.29	0.19	0.14	0.62	0.58	0.37	0.21	0.15
		10.0	16.7	0.46	0.44	0.33	0.22	0.16	0.60	0.57	0.39	0.24	0.17
		10.0	5.6	0.47	0.45	0.33	0.22	0.16	0.67	0.63	0.42	0.26	0.18
		-17.8	11.1	0.51	0.49	0.39	0.27	0.20	0.66	0.63	0.46	0.30	0.22
		-17.8	5.6	0.52	0.50	0.39	0.27	0.20	0.73	0.69	0.49	0.32	0.23
		-45.6	11.1	0.56	0.54	0.44	0.33	0.25	0.67	0.64	0.51	0.36	0.28
		-45.6	5.6	0.57	0.56	0.45	0.33	0.26	0.77	0.74	0.57	0.39	0.29
Horiz.	Down	32.2	5.6	0.44	0.41	0.29	0.19	0.14	0.62	0.58	0.37	0.21	0.15
		10.0	16.7	0.47	0.45	0.33	0.22	0.16	0.66	0.62	0.42	0.25	0.18
		10.0	5.6	0.47	0.45	0.33	0.22	0.16	0.68	0.63	0.42	0.26	0.18
		-17.8	11.1	0.52	0.50	0.39	0.27	0.20	0.74	0.70	0.50	0.32	0.23
		-17.8	5.6	0.52	0.50	0.39	0.27	0.20	0.75	0.71	0.51	0.32	0.23
		-45.6	11.1	0.57	0.55	0.45	0.33	0.26	0.81	0.78	0.59	0.40	0.30
		-45.6	5.6	0.58	0.56	0.46	0.33	0.26	0.83	0.79	0.60	0.40	0.30

Position of Air Space	Direction of Heat Flow	Mean Temp.[d] (°C)	Temp Diff.[a] (°C)	38.1 mm Air Space[c] Value of $E^{d,e}$					88.9 mm Air Space[c] Value of $E^{d,e}$				
				0.03	0.05	0.2	0.5	0.82	0.03	0.05	0.2	0.5	0.82
Horiz	Up	32.2	5.6	0.45	0.42	0.30	0.19	0.14	0.50	0.47	0.32	0.20	0.14
		10.0	16.7	0.33	0.32	0.26	0.18	0.14	0.27	0.35	0.28	0.19	0.15
		10.0	5.6	0.44	0.42	0.32	0.21	0.16	0.49	0.47	0.34	0.23	0.16
		-17.8	11.1	0.35	0.34	0.29	0.22	0.17	0.40	0.38	0.32	0.23	0.18
		-17.8	5.6	0.43	0.41	0.33	0.24	0.19	0.48	0.46	0.36	0.26	0.20
		-45.6	11.1	0.34	0.34	0.30	0.24	0.20	0.39	0.38	0.33	0.26	0.21
		-45.6	5.6	0.42	0.41	0.35	0.27	0.22	0.47	0.45	0.38	0.29	0.23
45° Slope	Up	32.2	5.6	0.51	0.48	0.33	0.20	0.14	0.56	0.52	0.35	0.21	0.14
		10.0	16.7	0.38	0.36	0.28	0.20	0.15	0.40	0.38	0.29	0.20	0.15
		10.0	5.6	0.51	0.48	0.35	0.23	0.17	0.55	0.52	0.37	0.24	0.17
		-17.8	11.1	0.40	0.39	0.32	0.24	0.18	0.43	0.41	0.33	0.24	0.19
		-17.8	5.6	0.49	0.47	0.37	0.26	0.20	0.52	0.51	0.39	0.27	0.20
		-45.6	11.1	0.39	0.38	0.33	0.26	0.21	0.41	0.40	0.35	0.27	0.22
		-45.6	5.6	0.48	0.46	0.39	0.30	0.24	0.51	0.49	0.41	0.31	0.24
Vertical	Horiz. →	32.2	5.6	0.70	0.64	0.40	0.22	0.15	0.65	0.60	0.38	0.22	0.15
		10.0	16.7	0.45	0.43	0.32	0.22	0.16	0.47	0.45	0.33	0.22	0.16
		10.0	5.6	0.67	0.62	0.42	0.26	0.18	0.64	0.60	0.41	0.25	0.18
		-17.8	11.1	0.49	0.47	0.37	0.26	0.20	0.51	0.49	0.38	0.27	0.20
		-17.8	5.6	0.62	0.59	0.44	0.29	0.22	0.61	0.59	0.44	0.29	0.22
		-45.6	11.1	0.46	0.45	0.38	0.29	0.23	0.50	0.48	0.40	0.30	0.24
		-45.6	5.6	0.58	0.56	0.46	0.34	0.26	0.60	0.58	0.47	0.34	0.26
45° Slope	Down	32.2	5.6	0.89	0.80	0.45	0.24	0.16	0.85	0.76	0.44	0.24	0.16
		10.0	16.7	0.63	0.59	0.41	0.25	0.18	0.62	0.58	0.40	0.25	0.18
		10.0	5.6	0.90	0.82	0.50	0.28	0.19	0.83	0.77	0.48	0.28	0.19
		-17.8	11.1	0.68	0.64	0.47	0.31	0.22	0.67	0.64	0.47	0.31	0.22
		-17.8	5.6	0.87	0.81	0.56	0.34	0.24	0.81	0.76	0.53	0.33	0.24
		-45.6	11.1	0.54	0.62	0.49	0.35	0.27	0.66	0.64	0.51	0.36	0.28
		-45.6	5.6	0.82	0.79	0.60	0.40	0.30	0.79	0.76	0.58	0.40	0.30
Horiz.	Down	32.2	5.6	1.07	0.94	0.49	0.25	0.17	1.77	1.44	0.60	0.28	0.18
		10.0	16.7	1.10	0.99	0.56	0.30	0.20	1.69	1.44	0.68	0.33	0.21
		10.0	5.6	1.16	1.04	0.58	0.30	0.20	1.96	1.63	0.72	0.34	0.22
		-17.8	11.1	1.24	1.13	0.69	0.39	0.26	1.92	1.68	0.86	0.43	0.29
		-17.8	5.6	1.29	1.17	0.70	0.39	0.27	2.11	1.82	0.89	0.44	0.29
		-45.6	11.1	1.36	1.27	0.84	0.50	0.35	2.05	1.85	1.06	0.57	0.38
		-45.6	5.6	1.42	1.32	0.86	0.51	0.35	2.28	2.03	1.12	0.59	0.39

SKETCHES OF COMMON WALL CONSTRUCTION TYPES
Taken in part from the 1985 ASHRAE Handbook of Fundamentals,
American Society of Heating, Refrigerating and Air Conditioning Engineers, Atlanta GA
Used by permission

<u>standard wood-framed outside wall</u>

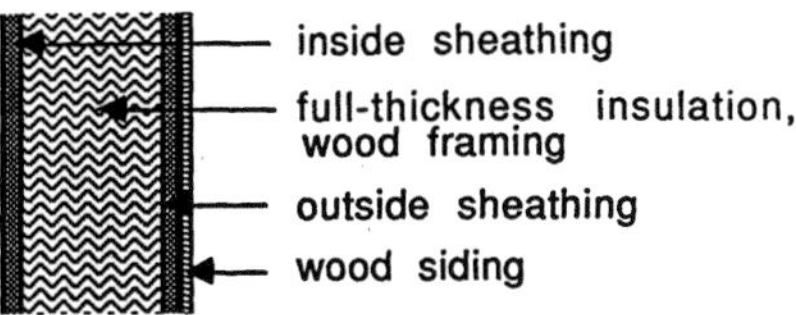

<u>standard wood-framed partition wall</u>

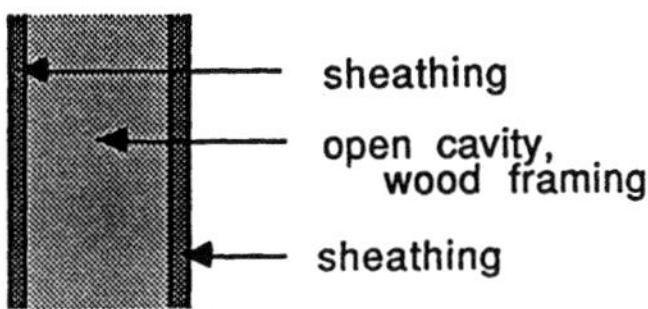

<u>insulated concrete block wall</u>

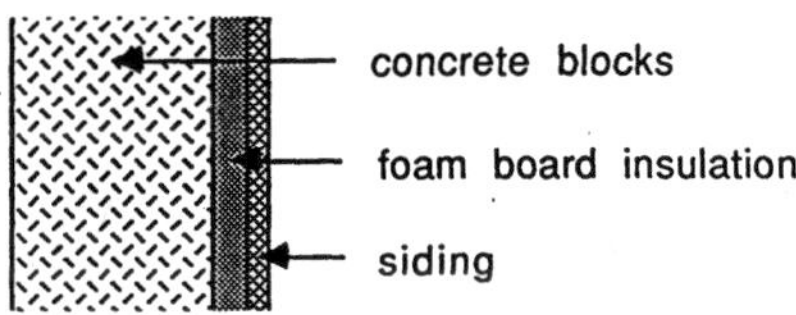

<u>metal-sided wall with wood framing</u>

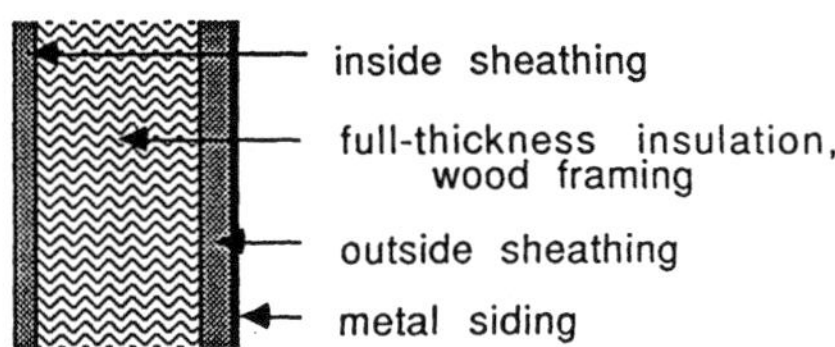

APPENDIX 4-2
OVERALL THERMAL RESISTANCES OF GLAZINGS

A. Windows (vertical), no indoor shades, no storm sash

Type	Thermal Resistance, m²K/W	
	Winter	Summer
Single glass, clear,		
surface emittance of 0.84	0.16	0.17
0.60	0.17	0.18
0.40	0.19	0.20
0.20	0.22	0.23
Insulating glass, double,		
5 mm air space, 3 mm glass	0.29	0.27
6 mm air space, 3 mm glass	0.30	0.29
13 mm air space, 6 mm glass	0.36	0.31
Insulating glass, triple,		
6 mm air space, 3 mm glass	0.45	0.40

B. Windows (vertical), no indoor shades, glass outdoor storm sash
with 25 mm air space

Type	Thermal Resistance, m²K/W	
	Winter	Summer
Single glass, clear,		
surface emittance of 0.84	0.43	0.36
0.60	0.37	0.29
0.40	0.40	0.29
0.20	0.43	0.36
Insulating glass, double,		
5 mm air space, 3 mm glass	0.48	0.43
6 mm air space, 3 mm glass	0.50	0.45
13 mm air space, 6 mm glass	0.56	0.45
Insulating glass, triple,		
6 mm air space, 3 mm glass	0.67	0.56

C. Adjustment factors for windows (multiply R-values in parts A
and B by these factors)

	Wood Frame	Metal Frame
Single glass	1.05 - 1.18	0.91 - 1.00
Double glass	1.00 - 1.11	0.77 - 0.83
Triple glass	1.00 - 1.05	0.67 - 0.77
Storm sash applied over single glass	1.00 - 1.11	0.71 - 0.83
Storm sash applied over double or triple glass	1.00 - 1.05	0.67 - 0.77

D. Greenhouse Glazings

Material (includes surface resistance)	Unit Area Thermal Resistance, m²K/W
Glass, single	0.16
Glass, double, 6 mm air space	0.25
Polyethylene film, single	0.14
Polyethylene film, double, 100 mm space	0.25
Fiberglass reinforced panel	0.15
Double panel, acrylic or polycarbonate	0.35

Notes:

a. Winter conditions are -18 C outdoor air, 21 C indoor air, and
   24 km/hr (6.67 m/s) wind speed.
b. Summer conditions are 32 C outdoor air, 24 C indoor air, solar
   radiation of 782 W/m², and 12 km/hr (3.33 m/s) wind speed.
c. Manufacturer's data should be consulted for more accurate data
   for specific products, especially for thermally improved
   windows.
d. Data obtained from the ASHRAE Handbook of Fundamentals.
   Consult that reference for more detail and other glazing
   configurations.

OVERALL THERMAL RESISTANCES OF DOORS

A. Wood doors, no storm doors

| | Thermal Resistance, m²K/W | |
Type	Winter	Summer
Flush doors		
hollow core, 35 mm thick	0.37	0.39
hollow core, 44 mm thick	0.38	0.40
with single glazing	0.28	0.33
solid core, 35 mm thick	0.45	0.46
solid core, 44 mm thick	0.53	0.55
with single glazing	0.38	0.40
with double glazing	0.48	0.49
solid core, 57 mm thick	0.65	0.68
with single glazing	0.43	0.44
with double glazing	0.53	0.55
Panel doors		
11 mm panels, 35 mm thick door	0.31	0.33
11 mm panels, 44 mm thick door	0.33	0.34
with single glazing	0.26	0.28
with double glazing	0.35	0.37
29 mm panels, 44 mm thick door	0.45	0.46
with single glazing	0.29	0.30
with double glazing	0.40	0.42

B. Wood doors, with storm doors, winter conditions only

| | Thermal Resistance, m²K/W | |
Type	wood storm door	metal storm door
Flush doors		
hollow core, 35 mm thick	0.59	0.55
hollow core, 44 mm thick	0.61	0.55
with single glazing	0.53	0.49
solid core, 35 mm thick	0.68	0.63
solid core, 44 mm thick		
with single glazing	0.61	0.55
with double glazing	0.70	0.65
solid core, 57 mm thick	0.88	0.84
with single glazing	0.65	0.61
with double glazing	0.76	0.70
Panel doors		
11 mm panels, 35 mm thick door	0.53	0.48
11 mm panels, 44 mm thick door	0.55	0.49
with single glazing	0.49	0.43
with double glazing	0.57	0.52
29 mm panels, 44 mm thick door	0.68	0.63
with single glazing	0.52	0.46
with double glazing	0.63	0.57

C. Steel doors

	Thermal Resistance, m²K/W	
Type	Winter	Summer
Solid urethane foam core, no thermal break, 44 mm thick, no storm door	0.44	0.45
Solid urethane foam core, with thermal break, 44 mm thick, no storm door	0.93	0.98
wood storm door	1.10	
metal storm door	1.03	

Notes:

a. Data is based on nominal door size of 1.12 m by 2.03 m.
b. Winter conditions are −17.8 C outdoor air temperature, 21.1 C
   indoor air temperature, and 24 km/hr (6.67 m/s) wind speed.
c. Summer conditions are 31.7 C outdoor air temperature, 23.9 C
   indoor air temperature, and 12 km/hr (3.33 m/s) wind speed.
d. Wood storm door values are for 50% glass area; metal storm door
   values are for any glass area.
e. Flush door values are for 17% glass area; double glazing with
   6.35 mm air space.
f. Panel door values are for 55% panel area; glazed panel door
   values are for 33% glass area and 22% panel area; double
   glazing with 6.35 mm air space.
g. Data obtained from the ASHRAE Handbook of Fundamentals.

APPENDIX 5-1
ANIMAL HEAT AND MOISTURE PRODUCTION

Moisture production (MP), latent heat production (LHP), sensible heat production (SHP) and total heat production (THP) of livestock. Adapted from data reference D270.4 of the American Society of Agricultural Engineers, and other sources. Heat of vaporization of water taken as 2410 kJ/kg (at animal body temperature). All data are taken from housing studies.

NOTE: Although these data are presented as design numbers; such data are frequently updated. Actual designs of environmental control systems should incorporate the latest research information. Such data would be available, for example, in reports in the Transactions of the American Society of Agricultural Engineers, Canadian Agricultural Engineering, and the Journal of Agricultural Engineering Research.

Animal	Air Temperature	MP mg/kg s	LHP W/kg	SHP W/kg	THP W/kg
Dairy cow,	-1 C	0.21	0.5	1.9	2.4
500 kg	10	0.28	0.7	1.5	2.2
	15	0.36	0.9	1.2	2.1
	21	0.36	0.9	1.1	2.0
	27	0.50	1.3	0.6	1.9
Brown Swiss and					
Holstein, 16 wk	10	0.56	1.4	2.3	3.7
	27	0.83	2.0	1.5	3.5
32 wk	10	0.33	0.9	1.5	2.4
	27	0.61	1.5	1.1	2.6
48 wk	10	0.33	0.7	1.5	2.2
	27	0.53	1.2	1.0	2.2
Jersey, 16 wk	10	0.67	1.6	2.5	4.1
	27	1.06	2.5	1.4	3.9
32 wk	10	0.42	1.0	1.8	2.8
	27	0.69	1.7	1.0	2.7
48 wk	10	0.36	0.9	1.6	2.5
	27	0.64	1.2	1.0	2.2
Calf, Ayreshire					
male, 39 kg	3	0.19	0.4	2.5	2.9
	23	0.19	0.4	2.0	2.4
40 kg	3	0.19	0.4	2.4	2.8
	23	0.19	0.4	1.9	2.3
45 kg	3	0.19	0.4	2.6	3.0
	23	0.19	0.4	2.0	2.4
Laying hen,	8	0.72	1.7	5.2	6.9
leghorn,	12	0.82	2.0	4.6	6.6
1.8 kg (a)	18	0.97	2.3	4.5	6.8
	28	1.19	2.9	3.7	6.6
Broilers, 0.1 kg	29	6.11	15.0	4.5	19.5
0.4 kg	24	3.47	8.5	6.5	15.0
0.7 kg	16	2.92	7.0	6.0	13.0
	27	2.92	7.0	3.0	10.0
1.0 kg	16	2.22	5.5	5.0	10.5
	27	2.64	6.5	3.0	9.5
1.5 kg	16	2.08	5.0	4.5	9.5
	27	2.50	6.0	3.0	9.0
2.0 kg	16	1.81	4.5	4.0	8.5
Growing/finishing	5	0.69	1.7	4.2	5.9
pigs, solid	10	0.75	1.4	4.0	5.4
floor, 20 kg	15	0.86	2.0	3.0	5.0
	20	1.03	2.5	2.3	4.8
	25	1.31	3.1	1.6	4.8
	30	1.75	4.2	0.6	4.8

40 kg	5	0.42	1.0	3.0	4.0
	10	0.44	1.1	2.5	3.6
	15	0.53	1.3	2.0	3.3
	20	0.61	1.5	1.6	3.1
	25	0.78	1.8	1.2	3.0
	30	1.00	2.4	0.6	3.0
60 kg	5	0.33	0.8	2.5	3.3
	10	0.36	0.9	2.0	2.9
	15	0.39	0.9	1.7	2.6
	20	0.47	1.1	1.3	2.4
	25	0.56	1.3	1.0	2.3
	30	0.75	1.8	0.5	2.3
80 kg	5	0.31	0.7	2.2	2.9
	10	0.31	0.7	1.8	2.5
	15	0.33	0.8	1.5	2.3
	20	0.39	0.9	1.2	2.1
	25	0.47	1.15	0.85	2.0
	30	0.61	1.4	0.5	1.9
100 kg	5	0.26	0.6	2.0	2.6
	10	0.28	0.7	1.6	2.3
	15	0.31	0.7	1.3	2.0
	20	0.33	0.8	1.1	1.9
	25	0.39	1.0	0.8	1.8
	30	0.50	1.2	0.5	1.7
Gilts, sows and boars, solid floor, 140 kg	5	0.22	0.5	1.8	2.3
	10	0.22	0.5	1.5	2.0
	15	0.23	0.6	1.2	1.8
	20	0.26	0.6	1.0	1.6
	25	0.31	0.7	0.8	1.5
	30	0.36	0.9	0.5	1.4
180 kg	5	0.18	0.4	1.7	2.1
	10	0.18	0.4	1.4	1.8
	15	0.18	0.4	1.2	1.6
	20	0.19	0.4	1.0	1.4
	25	0.22	0.5	0.8	1.3
	30	0.27	0.7	0.6	1.3

a. Moisture production estimated from data in ASAE data D270.4, then scaled according to:

Timmons, M.B. and R.S. Gates. 1985. Risk analysis methodology applied to environmental control options for animal housing: Part I, poultry layers. Paper No. 85-4507, American Society of Agricultural Engineers, St. Joseph MI.

Moisture calculated as equal to respired moisture, plus 13% of defecated moisture, with the sum increased by 15% because of the higher egg production of today's laying hens (the original data was published in 1966). The sensible heat production is also scaled by 15% for the same reason.

Station	Winter Air Temp. 99%	Winter Air Temp. 97.5%	Summer Air Temp. 1%	Summer Air Temp. 2.5%	MCWB 1%	MCWB 2.5%	DWB 1%	DWB 2.5%
Alabama								
Auburn	-8	-6	36	24	25	24	26	26
Huntsville	-12	-9	35	34	24	23	26	25
Mobile	-4	-2	35	34	25	25	27	26
Alaska								
Anchorage	-31	-28	20	19	15	14	16	15
Fairbanks	-46	-44	28	26	17	16	18	17
Juneau	-20	-17	23	21	16	14	16	15
Arizona								
Flagstaff	-19	-16	29	28	13	13	17	16
Phoenix	-1	1	43	42	22	22	24	24
Tucson	-2	0	40	39	19	19	22	22
Arkansas								
Fayetteville	-14	-11	36	34	24	23	25	24
Hot Springs	-8	-5	38	36	25	25	27	26
Little Rock	-9	-7	37	36	26	25	27	26
California								
Bakersfield	-1	0	40	38	21	21	23	22
Riverside	-2	0	38	37	20	19	22	22
Ukiah	-3	-2	37	35	21	20	21	20
Colorado								
Boulder	-17	-13	34	33	15	15	18	17
Denver	-21	-17	34	33	15	15	18	17
Fort Collins	-23	-20	34	33	15	15	18	17
Connecticut								
Bridgeport	-14	-13	30	29	23	22	24	23
New Haven	-16	-14	31	29	24	23	24	24
Waterbury	-20	-17	31	29	22	21	24	23
Delaware								
Dover	-12	-9	33	32	24	24	26	25
Wilmington	-12	-10	33	32	23	23	25	24
Florida								
Gainesville	-2	-1	35	34	25	25	27	26
Orlando	2	3	34	34	24	24	26	26
Tampa	2	4	33	33	25	25	26	26
Georgia								
Atlanta	-8	-6	34	33	23	23	25	24
Brunswick	-2	0	33	32	26	26	27	26
Columbus	-6	-4	35	34	24	24	26	26
Hawaii								
Hilo	16	17	29	28	23	22	24	23
Honolulu	17	17	31	30	23	23	24	24
Idaho								
Boise	-16	-12	36	34	18	18	20	19
Moscow	-22	-18	32	31	17	17	18	18
Twin Falls	-19	-17	37	35	16	16	18	17
Illinois								
Chicago	-21	-18	34	33	23	23	25	24
Peoria	-22	-20	33	32	24	23	26	24
Springfield	-19	-17	34	33	24	23	26	25
Indiana								
Evansville	-16	-13	35	34	24	24	26	26
Fort Wayne	-20	-17	33	32	23	22	25	24
Indianapolis	-19	-17	33	32	23	23	26	24
Iowa								
Ames	-24	-21	34	32	24	23	26	24
Des Moines	-23	-21	34	33	24	23	26	25
Ottumwa	-22	-20	34	33	24	23	26	25
Kansas								
Chanute	-16	-14	38	36	23	23	26	25
Hutchinson	-16	-13	39	37	22	22	25	24
Liberal	-17	-14	37	36	20	20	23	22
Kentucky								
Ashland	-15	-12	34	33	24	23	26	25
Lexington	-16	-13	34	33	23	23	25	24
Madisonville	-15	-12	36	34	24	24	26	26

Louisana							
Bogalusa	-4	-2	35	34	25	25	27	27
Minden	-7	-4	37	36	25	24	26	26
Shreveport	-7	-4	37	36	25	24	26	26
Maine							
Augusta	-22	-19	31	29	23	21	23	22
Caribou	-28	-25	29	27	21	19	22	21
Portland	-21	-18	31	29	22	22	23	22
Maryland							
Cumberland	-14	-12	33	32	24	23	25	24
Frederick	-13	-11	34	33	24	24	26	25
Salisbury	-11	-9	34	33	24	24	26	25
Massachusetts							
Fall River	-15	-13	31	29	22	22	23	23
Gloucester	-17	-15	32	30	23	22	24	23
Springfield	-21	-18	32	31	22	22	24	23
Michigan							
Alpena	-24	-21	32	29	21	21	23	22
Detroit	-16	-14	33	31	23	22	24	23
Lansing	-19	-17	32	31	23	22	24	23
Minnesota							
Duluth	-29	-27	29	28	21	20	22	21
Rochester	-27	-24	32	31	23	22	25	24
St. Cloud	-26	-24	33	31	23	22	24	23
Mississippi							
Biloxi	-2	-1	34	33	26	26	28	27
Jackson	-6	-4	36	35	24	24	26	26
Tupelo	-10	-7	36	34	25	25	27	26
Missouri							
Joplin	-14	-12	38	36	23	23	26	25
Rolla	-16	-13	34	33	25	24	26	25
Springfield	16	-13	36	34	23	23	26	25
Montana							
Bozeman	-29	-26	32	31	16	16	17	17
Cut Bank	-32	-29	31	29	16	16	18	17
Lewiston	-30	-27	32	31	17	16	18	17
Nebraska							
Fremont	-21	-19	37	35	24	23	26	25
Lincoln	-21	-19	37	35	24	23	26	25
Scottsbluff	-22	-19	35	33	18	18	21	20
Nevada							
Carson City	-16	-13	34	33	16	15	17	16
Ely	-23	-20	31	29	13	13	16	15
Las Vegas	-4	-2	42	41	19	18	22	21
New Hampshire							
Berlin	-26	-23	31	29	22	21	23	22
Concord	-22	-19	32	31	22	21	23	23
Portsmouth	-19	-17	32	29	23	22	24	23
New Jersey							
Paterson	-14	-12	34	33	23	23	25	24
Trenton	-12	-10	33	31	24	23	26	24
Vineland	-13	-12	33	32	24	23	26	24
New Mexico							
Albuquerque	-11	-9	36	34	16	16	19	18
Gallup	-18	-15	32	32	15	14	18	17
Roswell	-11	-8	38	37	19	19	22	21
New York							
Albany	-21	-18	33	31	23	22	24	23
Ithaca	-21	-18	31	29	22	22	23	23
Riverhead	-14	-12	30	28	22	22	24	23
North Carolina							
Raleigh	-9	-7	34	33	24	24	26	25
Rocky Mount	-8	-6	34	33	24	24	26	26
Wilmington	-5	-3	34	33	26	26	27	27
North Dakota							
Bismarck	-31	-28	35	33	20	20	23	22
Grand Forks	-32	-30	33	31	21	21	23	22
Minot	-31	-29	33	32	20	19	22	21
Ohio							
Cincinnati	-17	-14	33	32	23	22	25	24
Cleveland	-18	-15	33	32	23	23	25	24
Columbus	-18	-15	33	32	23	22	25	24
Oklahoma							
McAlester	-10	-7	37	36	23	23	25	24
Stillwater	-13	-11	38	36	23	23	25	24
Tulsa	-13	-11	38	37	23	24	26	26

Oregon								
Corvallis	-8	-6	33	32	19	19	21	19
Medford	-7	-5	37	34	20	19	21	20
Salem	-8	-5	33	31	20	19	21	20
Pennsylvania								
Erie	-16	-13	31	29	23	22	24	23
Harrisburg	-14	-12	34	33	24	23	25	24
Philadelphia	-12	-10	34	32	24	23	25	24
Rhode Island								
Providence	-15	-13	32	30	23	22	24	23
South Carolina								
Charleston	-4	-2	34	33	26	26	27	27
Columbia	-7	-4	36	35	24	24	26	26
Spartanburg	-8	-6	34	33	23	23	25	24
South Dakota								
Brookings	-27	-25	35	33	23	22	25	24
Peirre	-26	-23	37	35	22	22	24	23
Rapid City	-24	-22	35	33	19	18	22	21
Tennessee								
Athens	-11	-8	35	33	23	23	25	24
Knoxville	-11	-7	34	33	23	23	26	26
Memphis	-11	-8	37	35	25	24	27	26
Texas								
Amarillo	-14	-12	37	35	19	19	22	21
Brownsville	2	4	34	34	25	25	27	26
Plainview	-13	-11	37	36	20	20	22	22
Utah								
Logan	-19	-17	34	33	17	16	18	18
Provo	-17	-14	37	36	17	17	19	18
Richfield	-19	-15	34	33	16	16	19	18
Vermont								
Barre	-27	-24	29	27	22	21	23	22
Burlington	-24	-22	31	29	22	21	23	22
Rutland	-25	-22	31	29	22	21	23	22
Virginia								
Norfolk	-7	-6	34	33	25	24	26	26
Richmond	-10	-8	35	33	24	24	26	26
Roanoke	-11	-9	34	33	22	22	24	23
Washington								
Bellingham	-12	-9	27	25	19	18	20	18
Spokane	-21	-17	34	32	18	17	18	18
Yakima	-19	-15	36	34	18	18	20	19
West Virginia								
Clarksburg	-14	-12	33	32	23	23	24	24
Huntington	-15	-12	34	33	24	23	26	25
Wheeling	-17	-15	32	30	22	22	23	23
Wisconsin								
Ashland	-29	-27	29	28	21	20	22	21
Beloit	-22	-19	33	32	24	24	26	25
Madison	-24	-22	33	31	23	23	25	24
Wyoming								
Cheyenne	-23	-18	32	30	14	14	17	17
Rawlins	-24	-20	30	28	14	14	17	16
Sheridan	-26	-22	34	33	17	17	19	18
Alberta								
Calgary	-33	-31	29	27	17	16	18	17
Edmonton	-34	-32	29	28	19	18	20	19
Medicine Hat	-34	-31	34	32	19	18	21	20
British Columbia								
Dawson Creek	-38	-36	28	26	18	17	19	18
Trail	-21	-18	33	32	19	18	20	19
Vancouver	-9	-7	26	25	19	18	20	19
Manitoba								
Churchill	-41	-39	27	25	19	18	19	18
Dauphin	-35	-33	31	29	22	21	23	22
Winnipeg	-34	-33	32	30	23	22	24	23
New Brunswick								
Edmundston	-29	-27	31	28	21	20	23	22
Fredericton	-27	-24	32	29	22	21	23	22
Saint John	-24	-22	27	25	19	18	21	20
Newfoundland								
Gander	-21	-18	28	26	19	18	21	19
Goose Bay	-33	-31	29	27	19	18	20	19

	a	b						
Northwest Territories								
Inuvik	-49	-47	26	25	17	16	18	17
Nova Scotia								
Halifax	-17	-15	26	24	19	18	21	19
Sydney	-18	-16	28	27	21	20	22	21
Yarmouth	-15	-13	23	22	18	18	20	19
Ontario								
Hamilton	-19	-17	31	30	23	22	24	23
Sarnia	-18	-16	31	30	23	22	24	23
Timmins	-36	-34	31	29	21	20	22	21
Prince Edward Island								
Summerside	-22	-20	27	26	21	20	22	21
Quebec								
Hull	-28	-26	32	31	22	22	24	23
Montreal	-27	-23	31	29	23	22	24	23
Quebec	-28	-26	31	29	22	21	23	22
Saskatchewan								
Regina	-36	-34	33	31	21	20	22	21
Saskatoon	-37	-35	32	30	20	19	21	20
Yorkton	-37	-34	31	29	21	20	22	21
Yukon Territory								
Whitehorse	-43	-42	27	25	15	14	16	15

a. Mean Coincident Wet Bulb Temperature
b. Design Wet Bulb Temperature

Data abstracted from the ASHRAE Handbook of Fundamentals

WEATHER BIN DATA FOR SELECTED STATIONS

Data listed are hours per year during which air temperature is expected to be in the range indicated. Data are for stations with 24 hourly observations per day for at least 5 years.

Station		Temperature Ranges, C						

Huntsville, Alabama

	$-\infty$	-34.4	-28.9	-23.3	-17.8	-12.2	-6.7	-1.1
	-34.4	-28.9	-23.3	-17.8	-12.2	-6.7	-1.1	4.4
hrs:	0	0	0	2	16	79	350	871

	4.4	10.0	15.6	21.1	26.7	32.2	37.8
	10.0	15.6	21.1	26.7	32.2	37.8	$+\infty$
hrs:	1182	1291	1645	1975	1076	273	0

Anchorage, Alaska

	$-\infty$	-34.4	-28.9	-23.3	-17.8	-12.2	-6.7	-1.1
	-34.4	-28.9	-23.3	-17.8	-12.2	-6.7	-1.1	4.4
hrs:	1	9	68	271	554	921	1361	1528

	4.4	10.0	15.6	21.1	26.7	32.2	37.8
	10.0	15.6	21.1	26.7	32.2	37.8	$+\infty$
hrs:	1431	1932	629	53	2	0	0

Tucson, Arizona

	$-\infty$	-34.4	-28.9	-23.3	-17.8	-12.2	-6.7	-1.1
	-34.4	-28.9	-23.3	-17.8	-12.2	-6.7	-1.1	4.4
hrs:	0	0	0	0	0	1	25	296

	4.4	10.0	15.6	21.1	26.7	32.2	37.8
	10.0	15.6	21.1	26.7	32.2	37.8	$+\infty$
hrs:	1042	1589	1584	1830	1452	829	112

Little Rock, Arkansas

	$-\infty$	-34.4	-28.9	-23.3	-17.8	-12.2	-6.7	-1.1
	-34.4	-28.9	-23.3	-17.8	-12.2	-6.7	-1.1	4.4
hrs:	0	0	0	0	7	60	334	859

	4.4	10.0	15.6	21.1	26.7	32.2	37.8
	10.0	15.6	21.1	26.7	32.2	37.8	$+\infty$
hrs:	1217	1311	1574	1939	1143	311	5

Merced, California

	$-\infty$	-34.4	-28.9	-23.3	-17.8	-12.2	-6.7	-1.1
	-34.4	-28.9	-23.3	-17.8	-12.2	-6.7	-1.1	4.4
hrs:	0	0	0	0	0	0	33	556

	4.4	10.0	15.6	21.1	26.7	32.2	37.8
	10.0	15.6	21.1	26.7	32.2	37.8	$+\infty$
hrs:	1585	2198	1769	1289	822	446	62

Denver, Colorado

	$-\infty$	-34.4	-28.9	-23.3	-17.8	-12.2	-6.7	-1.1
	-34.4	-28.9	-23.3	-17.8	-12.2	-6.7	-1.1	4.4
hrs:	0	1	8	35	137	380	948	1427

	4.4	10.0	15.6	21.1	26.7	32.2	37.8
	10.0	15.6	21.1	26.7	32.2	37.8	$+\infty$
hrs:	1481	1513	1411	876	465	78	0

Dover, Delaware

	$-\infty$	-34.4	-28.9	-23.3	-17.8	-12.2	-6.7	-1.1
	-34.4	-28.9	-23.3	-17.8	-12.2	-6.7	-1.1	4.4
hrs:	0	0	0	0	17	169	611	1387

	4.4	10.0	15.6	21.1	26.7	32.2	37.8
	10.0	15.6	21.1	26.7	32.2	37.8	$+\infty$
hrs:	1430	1369	1573	1547	591	65	1

Orlando, Florida

	$-\infty$	-34.4	-28.9	-23.3	-17.8	-12.2	-6.7	-1.1
	-34.4	-28.9	-23.3	-17.8	-12.2	-6.7	-1.1	4.4
hrs:	0	0	0	0	0	0	3	82

	4.4	10.0	15.6	21.1	26.7	32.2	37.8
	10.0	15.6	21.1	26.7	32.2	37.8	$+\infty$
hrs:	413	1011	1949	3435	1667	200	0

Atlanta, Georgia

	-∞	-34.4	-28.9	-23.3	-17.8	-12.2	-6.7	-1.1
	-34.4	-28.9	-23.3	-17.8	-12.2	-6.7	-1.1	4.4
hrs:	0	0	0	0	2	32	185	775

	4.4	10.0	15.6	21.1	26.7	32.2	37.8
	10.0	15.6	21.1	26.7	32.2	37.8	+∞
hrs:	1274	1483	1832	2041	979	156	1

Honolulu, Hawaii

	-∞	-34.4	-28.9	-23.3	-17.3	-12.2	-6.7	-1.1
	-34.4	-28.9	-23.3	-17.8	-12.2	-6.7	-1.1	4.4
hrs:	0	0	0	0	0	0	0	0

	4.4	10.0	15.6	21.1	26.7	32.2	37.8
	10.0	15.6	21.1	26.7	32.2	37.8	+∞
hrs:	0	20	1175	5614	1950	1	0

Pocatello, Idaho

	-∞	-34.4	-28.9	-23.3	-17.8	-12.2	-6.7	-1.1
	-34.4	-28.9	-23.3	-17.8	-12.2	-6.7	-1.1	4.4
hrs:	0	1	15	43	162	415	1057	1738

	4.4	10.0	15.6	21.1	26.7	32.2	37.8
	10.0	15.6	21.1	26.7	32.2	37.8	+∞
hrs:	1483	1338	1101	796	491	120	0

Springfield, Illinois

	-∞	-34.4	-28.9	-23.3	-17.8	-12.2	-6.7	-1.1
	-34.4	-28.9	-23.3	-17.8	-12.2	-6.7	-1.1	4.4
hrs:	0	0	2	35	144	314	797	1415

	4.4	10.0	15.6	21.1	26.7	32.2	37.8
	10.0	15.6	21.1	26.7	32.2	37.8	+∞
hrs:	1116	1121	1433	1469	765	147	2

South Bend, Indiana

	-∞	-34.4	-28.9	-23.3	-17.8	-12.2	-6.7	-1.1
	-34.4	-28.9	-23.3	-17.8	-12.2	-6.7	-1.1	4.4
hrs:	0	0	3	38	141	405	1025	1572

	4.4	10.0	15.6	21.1	26.7	32.2	37.8
	10.0	15.6	21.1	26.7	32.2	37.8	+∞
hrs:	1148	1199	1479	1211	483	56	0

Des Moines, Iowa

	-∞	-34.4	-28.9	-23.3	-17.8	-12.2	-6.7	-1.1
	-34.4	-28.9	-23.3	-17.8	-12.2	-6.7	-1.1	4.4
hrs:	0	0	16	114	273	506	929	1343

	4.4	10.0	15.6	21.1	26.7	32.2	37.8
	10.0	15.6	21.1	26.7	32.2	37.8	+∞
hrs:	993	1139	1475	1273	593	104	2

Wichita, Kansas

	-∞	-34.4	-28.9	-23.3	-17.8	-12.2	-6.7	-1.1
	-34.4	-28.9	-23.3	-17.8	-12.2	-6.7	-1.1	4.4
hrs:	0	0	0	8	69	266	662	1154

	4.4	10.0	15.6	21.1	26.7	32.2	37.8
	10.0	15.6	21.1	26.7	32.2	37.8	+∞
hrs:	1184	1219	1506	1528	857	276	27

Fort Knox, Kentucky

	-∞	-34.4	-28.9	-23.3	-17.8	-12.2	-6.7	-1.1
	-34.4	-28.9	-23.3	-17.8	-12.2	-6.7	-1.1	4.4
hrs:	0	0	1	13	65	213	649	1209

	4.4	10.0	15.6	21.1	26.7	32.2	37.8
	10.0	15.6	21.1	26.7	32.2	37.8	+∞
hrs:	1211	1279	1637	1621	779	83	0

New Orleans, Louisiana

	-∞	-34.4	-28.9	-23.3	-17.8	-12.2	-6.7	-1.1
	-34.4	-28.9	-23.3	-17.8	-12.2	-6.7	-1.1	4.4
hrs:	0	0	0	0	0	3	36	301

	4.4	10.0	15.6	21.1	26.7	32.2	37.8
	10.0	15.6	21.1	26.7	32.2	37.8	+∞
hrs:	808	1326	1703	2851	1583	149	0

Presque		−∞	-34.4	-28.9	-23.3	-17.8	-12.2	-6.7	-1.1
Isle,		−34.4	-28.9	-23.3	-17.8	-12.2	-6.7	-1.1	4.4
Maine	hrs:	0	3	65	257	530	829	1151	1488

		4.4	10.0	15.6	21.1	26.7	32.2	37.8
		10.0	15.6	21.1	26.7	32.2	37.8	+∞
	hrs:	1268	1413	1208	464	83	1	0

Laurel,		−∞	-34.4	-28.9	-23.3	-17.8	-12.2	-6.7	-1.1
Maryland		−34.4	-28.9	-23.3	-17.8	-12.2	-6.7	-1.1	4.4
	hrs:	0	0	0	0	3	80	468	1221

		4.4	10.0	15.6	21.1	26.7	32.2	37.8
		10.0	15.6	21.1	26.7	32.2	37.8	+∞
	hrs:	1443	1425	1572	1764	711	73	0

Spring-		−∞	-34.4	-28.9	-23.3	-17.8	-12.2	-6.7	-1.1
field,		−34.4	-28.9	-23.3	-17.8	-12.2	-6.7	-1.1	4.4
Mass.	hrs:	0	0	7	40	161	448	954	1598

		4.4	10.0	15.6	21.1	26.7	32.2	37.8
		10.0	15.6	21.1	26.7	32.2	37.8	+∞
	hrs:	1293	1389	1439	1027	370	34	0

Lansing,		−∞	-34.4	-28.9	-23.3	-17.8	-12.2	-6.7	-1.1
Michigan		−34.4	-28.9	-23.3	-17.8	-12.2	-6.7	-1.1	4.4
	hrs:	0	0	1	40	176	504	1139	1561

		4.4	10.0	15.6	21.1	26.7	32.2	37.8
		10.0	15.6	21.1	26.7	32.2	37.8	+∞
	hrs:	1153	1277	1466	1038	376	29	0

Duluth,		−∞	-34.4	-28.9	-23.3	-17.8	-12.2	-6.7	-1.1
Minnesota		−34.4	-28.9	-23.3	-17.8	-12.2	-6.7	-1.1	4.4
	hrs:	1	21	131	330	529	819	1223	1289

		4.4	10.0	15.6	21.1	26.7	32.2	37.8
		10.0	15.6	21.1	26.7	32.2	37.8	+∞
	hrs:	1193	1448	1118	525	130	3	0

Jackson,		−∞	-34.4	-28.9	-23.3	-17.8	-12.2	-6.7	-1.1
Mississippi		−34.4	-28.9	-23.3	-17.8	-12.2	-6.7	-1.1	4.4
	hrs:	0	0	0	0	4	11	140	561

		4.4	10.0	15.6	21.1	26.7	32.2	37.8
		10.0	15.6	21.1	26.7	32.2	37.8	+∞
	hrs:	1065	1321	1722	2279	1237	410	10

Columbia,		−∞	-34.4	-28.9	-23.3	-17.8	-12.2	-6.7	-1.1
Missouri		−34.4	-28.9	-23.3	-17.8	-12.2	-6.7	-1.1	4.4
	hrs:	0	0	0	25	102	292	702	1260

		4.4	10.0	15.6	21.1	26.7	32.2	37.8
		10.0	15.6	21.1	26.7	32.2	37.8	+∞
	hrs:	1155	1184	1515	1497	810	209	9

Billings,		−∞	-34.4	-28.9	-23.3	-17.8	-12.2	-6.7	-1.1
Montana		−34.4	-28.9	-23.3	-17.8	-12.2	-6.7	-1.1	4.4
	hrs:	0	4	52	134	237	395	835	1521

		4.4	10.0	15.6	21.1	26.7	32.2	37.8
		10.0	15.6	21.1	26.7	32.2	37.8	+∞
	hrs:	1546	1433	1255	811	424	110	3

Grand Island, Nebraska

−∞ / −34.4	−34.4 / −28.9	−28.9 / −23.3	−23.3 / −17.8	−17.8 / −12.2	−12.2 / −6.7	−6.7 / −1.1	−1.1 / 4.4
0	1	10	83	267	502	968	1299

4.4 / 10.0	10.0 / 15.6	15.6 / 21.1	21.1 / 26.7	26.7 / 32.2	32.2 / 37.8	37.8 / +∞
1092	1190	1380	1138	627	187	16

Ely, Nevada

−∞ / −34.4	−34.4 / −28.9	−28.9 / −23.3	−23.3 / −17.8	−17.8 / −12.2	−12.2 / −6.7	−6.7 / −1.1	−1.1 / 4.4
0	1	18	85	270	636	1175	1635

4.4 / 10.0	10.0 / 15.6	15.6 / 21.1	21.1 / 26.7	26.7 / 32.2	32.2 / 37.8	37.8 / +∞
1462	1300	989	724	451	14	0

Portsmouth, New Hampshire

−∞ / −34.4	−34.4 / −28.9	−28.9 / −23.3	−23.3 / −17.8	−17.8 / −12.2	−12.2 / −6.7	−6.7 / −1.1	−1.1 / 4.4
0	0	1	30	162	477	964	1607

4.4 / 10.0	10.0 / 15.6	15.6 / 21.1	21.1 / 26.7	26.7 / 32.2	32.2 / 37.8	37.8 / +∞
1384	1477	1494	870	274	20	0

Wrightstown, New Jersey

−∞ / −34.4	−34.4 / −28.9	−28.9 / −23.3	−23.3 / −17.8	−17.8 / −12.2	−12.2 / −6.7	−6.7 / −1.1	−1.1 / 4.4
0	0	0	0	35	238	687	1483

4.4 / 10.0	10.0 / 15.6	15.6 / 21.1	21.1 / 26.7	26.7 / 32.2	32.2 / 37.8	37.8 / +∞
1389	1379	1571	1371	556	51	0

Albuquerque, New Mexico

−∞ / −34.4	−34.4 / −28.9	−28.9 / −23.3	−23.3 / −17.8	−17.8 / −12.2	−12.2 / −6.7	−6.7 / −1.1	−1.1 / 4.4
0	0	0	1	11	99	535	1227

4.4 / 10.0	10.0 / 15.6	15.6 / 21.1	21.1 / 26.7	26.7 / 32.2	32.2 / 37.8	37.8 / +∞
1432	1347	1587	1390	864	265	2

Syracuse, New York

−∞ / −34.4	−34.4 / −28.9	−28.9 / −23.3	−23.3 / −17.8	−17.8 / −12.2	−12.2 / −6.7	−6.7 / −1.1	−1.1 / 4.4
0	0	1	31	171	492	954	1536

4.4 / 10.0	10.0 / 15.6	15.6 / 21.1	21.1 / 26.7	26.7 / 32.2	32.2 / 37.8	37.8 / +∞
1262	1376	1450	1070	386	31	0

Greensboro, North Carolina

−∞ / −34.4	−34.4 / −28.9	−28.9 / −23.3	−23.3 / −17.8	−17.8 / −12.2	−12.2 / −6.7	−6.7 / −1.1	−1.1 / 4.4
0	0	0	0	6	83	437	1126

4.4 / 10.0	10.0 / 15.6	15.6 / 21.1	21.1 / 26.7	26.7 / 32.2	32.2 / 37.8	37.8 / +∞
1303	1426	1809	1660	798	112	0

Bismarck, North Dakota

−∞ / −34.4	−34.4 / −28.9	−28.9 / −23.3	−23.3 / −17.8	−17.8 / −12.2	−12.2 / −6.7	−6.7 / −1.1	−1.1 / 4.4
3	34	147	352	550	686	990	1246

4.4 / 10.0	10.0 / 15.6	15.6 / 21.1	21.1 / 26.7	26.7 / 32.2	32.2 / 37.8	37.8 / +∞
1065	1213	1165	804	404	97	4

Dayton, Ohio

−∞ / −34.4	−34.4 / −28.9	−28.9 / −23.3	−23.3 / −17.8	−17.8 / −12.2	−12.2 / −6.7	−6.7 / −1.1	−1.1 / 4.4
0	0	1	24	111	336	816	1354

4.4 / 10.0	10.0 / 15.6	15.6 / 21.1	21.1 / 26.7	26.7 / 32.2	32.2 / 37.8	37.8 / +∞
1224	1250	1558	1394	637	55	0

Oklahoma City, Oklahoma

−∞ / −34.4	−34.4 / −28.9	−28.9 / −23.3	−23.3 / −17.8	−17.8 / −12.2	−12.2 / −6.7	−6.7 / −1.1	−1.1 / 4.4
0	0	0	1	25	134	457	984

4.4 / 10.0	10.0 / 15.6	15.6 / 21.1	21.1 / 26.7	26.7 / 32.2	32.2 / 37.8	37.8 / +∞
1201	1282	1549	1778	996	334	19

Eugene, Oregon

	$-\infty$	-34.4	-28.9	-23.3	-17.8	-12.2	-6.7	-1.1
	-34.4	-28.9	-23.3	-17.8	-12.2	-6.7	-1.1	4.4
hrs:	0	0	0	1	5	32	169	1168

	4.4	10.0	15.6	21.1	26.7	32.2	37.8
	10.0	15.6	21.1	26.7	32.2	37.8	$+\infty$
hrs:	2468	2527	1369	677	288	55	1

Harrisburg, Pennsylvania

	$-\infty$	-34.4	-28.9	-23.3	-17.8	-12.2	-6.7	-1.1
	-34.4	-28.9	-23.3	-17.8	-12.2	-6.7	-1.1	4.4
hrs:	0	0	0	1	33	208	672	1596

	4.4	10.0	15.6	21.1	26.7	32.2	37.8
	10.0	15.6	21.1	26.7	32.2	37.8	$+\infty$
hrs:	1378	1304	1492	1362	610	102	2

Warwick, Rhode Island

	$-\infty$	-34.4	-28.9	-23.3	-17.8	-12.2	-6.7	-1.1
	-34.4	-28.9	-23.3	-17.8	-12.2	-6.7	-1.1	4.4
hrs:	0	0	0	5	61	269	748	1526

	4.4	10.0	15.6	21.1	26.7	32.2	37.8
	10.0	15.6	21.1	26.7	32.2	37.8	$+\infty$
hrs:	1520	1547	1698	1099	270	17	0

Sumter, South Carolina

	$-\infty$	-34.4	-28.9	-23.3	-17.8	-12.2	-6.7	-1.1
	-34.4	-28.9	-23.3	-17.8	-12.2	-6.7	-1.1	4.4
hrs:	0	0	0	0	0	14	165	679

	4.4	10.0	15.6	21.1	26.7	32.2	37.8
	10.0	15.6	21.1	26.7	32.2	37.8	$+\infty$
hrs:	1134	1420	1747	2268	1112	219	2

Huron, South Dakota

	$-\infty$	-34.4	-28.9	-23.3	-17.8	-12.2	-6.7	-1.1
	-34.4	-28.9	-23.3	-17.8	-12.2	-6.7	-1.1	4.4
hrs:	0	12	93	262	467	691	1025	1205

	4.4	10.0	15.6	21.1	26.7	32.2	37.8
	10.0	15.6	21.1	26.7	32.2	37.8	$+\infty$
hrs:	992	1114	1256	980	515	139	9

Nashville, Tennessee

	$-\infty$	-34.4	-28.9	-23.3	-17.8	-12.2	-6.7	-1.1
	-34.4	-28.9	-23.3	-17.8	-12.2	-6.7	-1.1	4.4
hrs:	0	0	0	6	23	132	495	1043

	4.4	10.0	15.6	21.1	26.7	32.2	37.8
	10.0	15.6	21.1	26.7	32.2	37.8	$+\infty$
hrs:	1221	1341	1676	1689	935	196	3

Austin, Texas

	$-\infty$	-34.4	-28.9	-23.3	-17.8	-12.2	-6.7	-1.1
	-34.4	-28.9	-23.3	-17.8	-12.2	-6.7	-1.1	4.4
hrs:	0	0	0	0	0	6	77	430

	4.4	10.0	15.6	21.1	26.7	32.2	37.8
	10.0	15.6	21.1	26.7	32.2	37.8	$+\infty$
hrs:	910	1248	1614	2366	1449	640	20

Ogden, Utah

	$-\infty$	-34.4	-28.9	-23.3	-17.8	-12.2	-6.7	-1.1
	-34.4	-28.9	-23.3	-17.8	-12.2	-6.7	-1.1	4.4
hrs:	0	0	1	4	47	283	986	1665

	4.4	10.0	15.6	21.1	26.7	32.2	37.8
	10.0	15.6	21.1	26.7	32.2	37.8	$+\infty$
hrs:	1489	1274	1254	1011	599	147	0

Richmond, Virginia

	$-\infty$	-34.4	-28.9	-23.3	-17.8	-12.2	-6.7	-1.1
	-34.4	-28.9	-23.3	-17.8	-12.2	-6.7	-1.1	4.4
hrs:	0	0	0	0	2	94	447	1142

	4.4	10.0	15.6	21.1	26.7	32.2	37.8
	10.0	15.6	21.1	26.7	32.2	37.8	$+\infty$
hrs:	1358	1421	1680	1634	795	183	4

Moses Lake, Washington

	$-\infty$	-34.4	-28.9	-23.3	-17.8	-12.2	-6.7	-1.1
	-34.4	-28.9	-23.3	-17.8	-12.2	-6.7	-1.1	4.4
hrs:	0	0	2	15	63	201	749	1662

	4.4	10.0	15.6	21.1	26.7	32.2	37.8
	10.0	15.6	21.1	26.7	32.2	37.8	$+\infty$
hrs:	1583	1520	1351	934	497	173	10

Charleston,	$-\infty$	-34.4	-28.9	-23.3	-17.8	-12.2	-6.7	-1.1
West	-34.4	-28.9	-23.3	-17.8	-12.2	-6.7	-1.1	4.4
Virginia hrs:	0	0	0	3	36	183	576	1248

	4.4	10.0	15.6	21.1	26.7	32.2	37.8
	10.0	15.6	21.1	26.7	32.2	37.8	$+\infty$
hrs:	1259	1441	1739	1485	710	79	1

Madison,	$-\infty$	-34.4	-28.9	-23.3	-17.8	-12.2	-6.7	-1.1
Wisconsin	-34.4	-28.9	-23.3	-17.8	-12.2	-6.7	-1.1	4.4
hrs:	0	1	19	130	305	561	1121	1458

	4.4	10.0	15.6	21.1	26.7	32.2	37.8
	10.0	15.6	21.1	26.7	32.2	37.8	$+\infty$
hrs:	1033	1188	1416	1056	424	48	0

Casper,	$-\infty$	-34.4	-28.9	-23.3	-17.8	-12.2	-6.7	-1.1
Wyoming	-34.4	-28.9	-23.3	-17.8	-12.2	-6.7	-1.1	4.4
hrs:	0	3	23	84	195	522	1152	1585

	4.4	10.0	15.6	21.1	26.7	32.2	37.8
	10.0	15.6	21.1	26.7	32.2	37.8	$+\infty$
hrs:	1466	1266	1125	780	483	75	1

APPENDIX 7-1

HEATING DEGREE-DAY DATA (BASE 18.3 C) FOR SELECTED STATIONS

Station	Jul	Aug	Sep	Oct	Nov	Dec	Jan	Feb	Mar	Apr	May	Jun	Total
Huntsville Alabama	0	0	7	71	237	368	386	309	241	77	11	0	1,707
Anchorage Alaska	136	162	287	517	713	873	906	731	718	488	329	175	6,035
Tucson Arizona	0	0	0	14	128	226	262	191	134	42	3	0	1,000
Little Rock Arkansas	0	0	5	71	258	398	420	321	241	70	5	0	1,789
Fresno California	0	0	0	43	188	310	326	226	177	83	31	0	1,384
Denver Colorado	3	5	65	238	455	575	629	521	493	310	160	37	3,491
Wilmington Delaware	0	0	28	150	327	515	544	486	408	215	62	3	2,738
Orlando Florida	0	0	0	0	40	110	122	92	58	3	0	0	425
Atlanta Georgia	0	0	10	71	230	348	355	294	243	93	14	0	1,658
Honolulu Hawaii	0	0	0	0	0	0	0	0	0	0	0	0	0
Pocatello Idaho	0	0	96	274	500	648	736	588	503	308	177	78	3,908
Springfield Illinois	0	0	40	162	387	568	631	519	427	197	76	10	3,017
South Bend Indiana	0	3	62	207	432	625	678	594	518	292	133	33	3,577
Des Moines Iowa	0	5	55	202	465	684	777	646	537	272	117	22	3,782
Wichita Kansas	0	0	18	127	343	503	568	447	358	150	48	3	2,565
Lexington Kentucky	0	0	30	133	338	501	526	454	381	181	58	0	2,602
New Orleans Louisiana	0	0	0	11	107	179	202	143	107	22	0	0	771
Caribou Maine	43	64	187	379	580	853	939	817	727	477	260	102	5,428
Frederick Maryland	0	0	37	171	347	531	553	487	412	213	71	7	2,829
Pittsfield Mass.	14	33	122	291	462	684	744	664	591	367	181	58	4,211
Lansing Michigan	3	12	77	239	452	646	701	634	562	322	152	38	3,838
Duluth Minnesota	39	61	183	351	628	878	969	843	753	467	272	110	5,554
Jackson Mississippi	0	0	0	36	175	279	303	230	172	48	0	0	1,243
Columbia Missouri	0	0	30	139	362	537	598	486	398	180	67	7	2,804
Billings Montana	3	8	103	271	498	631	720	611	539	317	158	57	3,916
Grand Island Nebraska	0	3	60	212	463	651	730	605	504	257	117	25	3,627
Ely Nevada	16	24	130	329	522	658	727	597	543	373	253	125	4,297
Concord N. Hamp.	3	28	98	281	457	689	754	658	573	353	166	42	4,102
Newark N. Jersey	0	0	17	138	318	512	546	487	405	212	66	0	2,701
Albuquerque N. Mexico	0	0	7	127	357	482	517	391	331	160	45	0	2,417
Syracuse N. York	3	16	73	231	413	641	706	633	558	317	138	25	3,754
Greensboro N. Carolina	0	0	18	107	285	432	436	373	307	130	26	0	2,114
Bismark N. Dakota	19	16	123	321	602	813	949	801	668	358	183	65	4,918
Dayton Ohio	0	3	43	172	387	581	609	531	449	238	93	17	3,123

Location														Total
Oklahoma City Oklahoma	0	0	8	91	277	426	482	369	293	105	19	0	2,070
Eugene Oregon	19	19	72	203	325	399	446	348	327	237	155	75	2,625
Harrisburg Penn.	0	0	35	166	360	551	581	504	426	220	69	7	2,919
Providence R. Island	0	9	53	207	367	568	617	549	482	297	131	28	3,308
Columbia S. Carolina	0	0	0	47	192	321	317	261	198	45	0	0	1,381
Huron S. Dakota	5	7	92	282	563	796	904	753	625	333	160	48	4,568
Nashville Tennessee	0	0	17	88	275	407	432	358	284	105	22	0	1,988
Austin Texas	0	0	0	17	125	216	260	181	124	28	0	0	951
Salt Lake City Utah	0	0	45	233	472	601	651	506	424	255	129	47	3,363
Richmond Virginia	0	0	20	119	275	436	453	391	303	122	29	0	2,148
Yakima Washington	0	7	80	250	460	577	646	482	396	242	122	38	3,300
Charleston W. Virginia	0	0	35	141	328	481	489	428	360	167	53	5	2,487
Madison Wisconsin	14	22	97	263	517	739	818	708	618	343	172	57	4,368
Casper Wyoming	3	9	107	291	523	649	717	602	567	365	212	72	4,117
Calgary Alberta	61	103	223	399	617	772	875	766	704	443	265	162	5,390
Vancouver Br. Col.	45	48	122	253	365	437	479	402	376	278	172	87	3,064
Winnipeg Manitoba	21	39	179	379	695	976	1116	955	814	452	225	82	5,933
Fredericton N. Bruns.	43	38	130	329	508	773	856	766	651	418	226	78	4,816
Halifax N. Scotia	32	28	100	254	394	597	674	623	572	412	271	132	4,089
Ottawa Ontario	14	45	123	315	520	816	902	801	684	393	189	50	4,852
Montreal Quebec	5	24	92	289	490	773	870	767	653	380	176	38	4,557
Regina Sask.	43	52	200	412	713	951	1092	937	818	446	227	112	6,003

NOTE: Local microclimate effects can make a significant difference in heating degree-day data for a locality (perhaps as much as ten percent). Elevation, proximity to a body of water, local air drainage, and the heat island effect of cities are examples of factors which affect heating needs.

APPENDIX 7-2
PERMEANCE OF COMMON BUILDING MATERIALS TO WATER VAPOR TRANSMISSION

Material	thickness, mm	permeance, perms	resistance, reps
A. Construction materials			
concrete (1:2:4 mix)	1000	4.7	0.21
brick masonry	102	46	0.022
concrete block, cored,			
limestone aggregate	203	137	0.0073
gypsum wall board	9.5	2860	0.00035
structural insulating			
board, sheathing qual.	12.3	2360-5900	0.000424 to 0.000169
hardboard, tempered	3.2	290	0.0034
wood, sugar pine	1000	0.58-0.78	1.7 to 1.3
plywood, Douglas fir,			
exterior glue	6.4	40	0.025
B. Thermal Insulations			
mineral wool, unprotected	1000	245	0.0041
expanded polyurethane,			
R-11 exp., bd. stock	1000	0.58-2.3	1.72 to 0.43
exp. polystyrene, extruded	1000	1.7	0.57
exp. polystyrene, bead	1000	2.3-8.4	0.34 to 0.12
C. Foil, films and papers			
aluminum foil	0.025	0	
	0.009	2.9	0.345
polyethylene	0.051	9.1	0.11
	0.1	4.6	0.22
	0.15	3.4	0.29
	0.2	2.3	0.43
	0.25	1.7	0.59
roll roofing, saturated			
and coated (Dry Cup)		2.9	0.34
(Wet Cup)		14	0.071
blanket thermal insulation			
back up paper (Dry Cup)		23	0.043
(Wet Cup)		34-240	0.029 to 0.0042
D. Paints			
comercial latex paints			
vapor retarding	0.07	26	0.038
primer-sealer	0.031	360	0.0028
exterior acrylic	0.042	313	0.0032
semi-gloss vinyl acrylic	0.06	378	0.0026
exterior, white lead and oil			
on wood siding, 3 coats		17-57	0.06 - 0.02
exterior, white lead-zinc oxide			
and oil on wood, 3 coats		51	0.02

NOTE: These values are for design guidance only and should not be used for design specifications. Such specifications should be based on data from manufacturers of materials to be used.

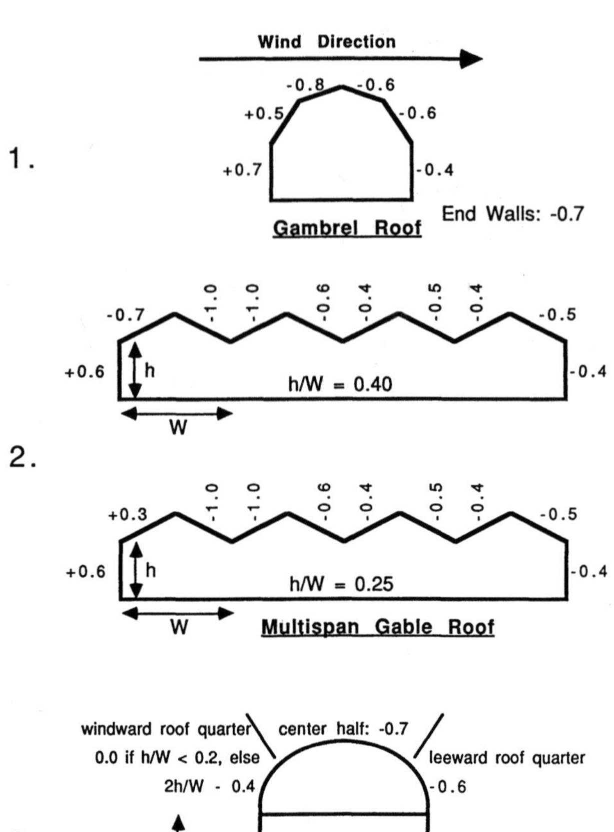

Wind Direction
-0.8
-0.6
+0.5
-0.6
+0.7
-0.4
1.
Gambrel Roof
End Walls: -0.7
-0.7
-1.0
-1.0
-0.6
-0.4
-0.5
-0.4
-0.5
+0.6
h
h/W = 0.40
-0.4
W
2.
+0.3
-1.0
-1.0
-0.6
-0.4
-0.5
-0.4
-0.5
+0.6
h
h/W = 0.25
-0.4
W
Multispan Gable Roof

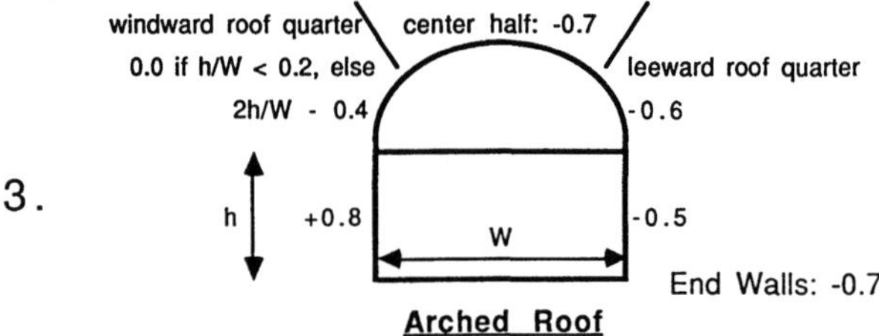

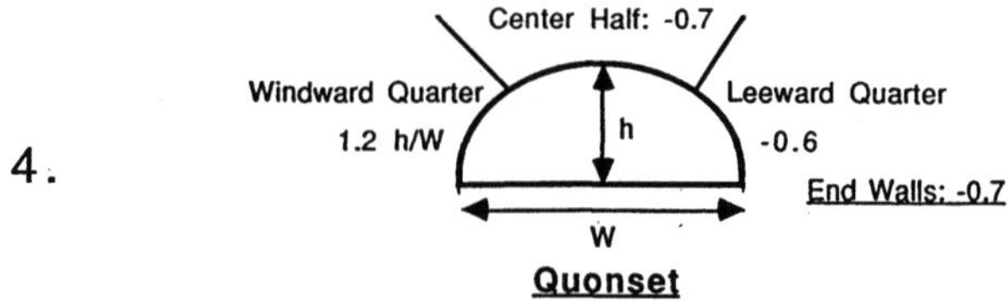

windward roof quarter
center half: -0.7
0.0 if h/W < 0.2, else
leeward roof quarter
2h/W - 0.4
-0.6
3.
h
+0.8
W
-0.5
End Walls: -0.7
Arched Roof
Center Half: -0.7
Windward Quarter
Leeward Quarter
1.2 h/W
h
-0.6
4.
W
End Walls: -0.7
Quonset

5.

-0.7 to +0.2
(10 to 30 degree roof slope)
-0.7
+0.7
-0.5
End Walls: -0.7
Gable Roof

-0.7
+0.7
-0.7
End Walls: -0.8
Monoslope Roof

6.

0.0
+0.7
-0.6
End Wall: -0.9

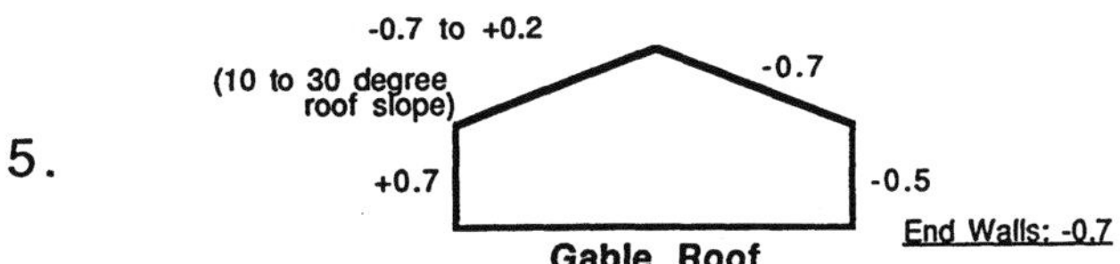

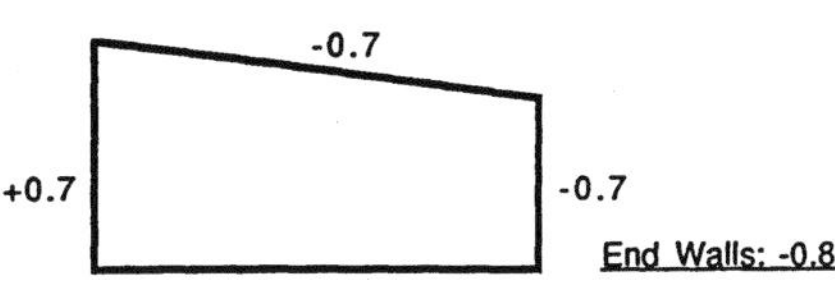

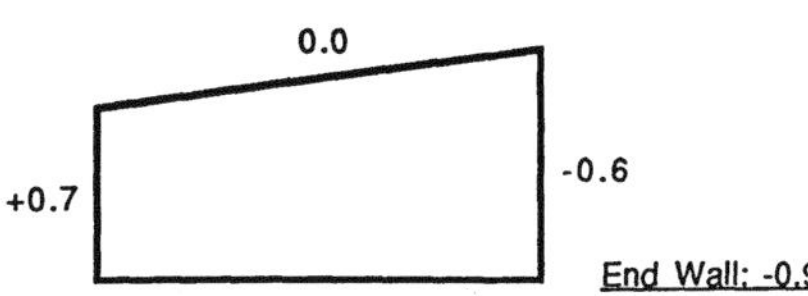

APPENDIX 11-2
COMBINED NATURAL VENTILATION DUE TO WIND AND
THERMAL BUOYANCY

Wind effects and thermal buoyancy each contribute to a static pressure difference between any point within a building, and without. It is thought the pressure differences add,

$$\Delta P_{total} = \Delta P_{wind} + \Delta P_{thermal}.$$

Wind and thermal buoyancy effects are described separately in Chapter 11,

$$\Delta P_{total} = \frac{1}{2}\rho_o v_w^2 (c_{pe} - c_{pi}) + g(\rho_o - \rho_i)(h_p - h).$$

If there is a flow path, such as ventilation opening n,

$$v_n = \sqrt{2\,\Delta P_{total}/\rho_n},$$

where v_n is air velocity through the opening at the elevation h, corresponding to ΔP_{total}, and v_w is the velocity of the wind. Air density, ρ_n, is that appropriate to the opening, equalling the density of indoor air for vents where flow is out of the building, and the density of outdoor air for vents where flow is in.

For vent n at elevation h_n,

$$v_n = \frac{\left| 2g(\Delta\rho/\rho_n)(h_p - h_n) + v_w^2(c_{pen} - c_{pi}) \right|^{3/2}}{2g(\Delta\rho/\rho_n)(h_p - h_n) + v_w^2(c_{pen} - c_{pi})}$$

Volumetric airflow rate through vent n, $\dot{V}_n$, is

$$(C_d)_n \int_{area} v_n\, dA$$

For continuity to be satisfied, the sum of all mass flows must be zero, or (for subscript m to indicate horizontal vents)

$$\sum_n \left[(C_d)_n \rho_n \int_{area} \frac{\left| 2g(\Delta\rho/\rho_n)(h_p - h_n) + v_w^2(c_{pen} - c_{pi}) \right|^{3/2}}{2g(\Delta\rho/\rho_n)(h_p - h_n) + v_w^2(c_{pen} - c_{pi})}\, dA_n \right] +$$

$$\sum_n \left[(C_d)_m \rho_m \frac{2g(\Delta\rho/\rho_m)(h_p - h_m) + v_w^2(c_{pem} - c_{pi})^{3/2}}{2g(\Delta\rho/\rho_m)(h_p - h_m) + v_w^2(c_{pem} - c_{pi})} W_m L_m = 0 \right.$$

Each horizontal vent (denoted by subscript m) is assumed to be rectangular with width W and length L. Ridge vents are typically horizontal openings. Each integral for airflow through vents which are not horizontal is of the form

$$\int_{area} \frac{\left| \phi_n(h_p - h_n) + a_n \right|^{3/2}}{[\phi_n(h_p - h_n) + \alpha_n]} \, dA_n,$$

where $\phi_n = 2g(\Delta\rho/\rho_n)$, and

$$\alpha_n = v_w^2(c_{pen} - c_{pi}).$$

If each non-horizontal ventilation opening is rectangular, $dA = Ldh$, where L is the length of the vent. This is the typical situation. If terms are grouped so

$$\Omega_n = h_p + \alpha_n/\phi_n,$$

after integration the continuity equation is

$$(-2/3)\Sigma(C_d)_n\rho_nL_n\phi_n^{1/2}\left|(\Omega_n - h_n)\right|^{5/2} (\Omega_n - h_n)^{-1}\Big|_{bottom}^{top} +$$

$$\Sigma(C_d)_m\rho_mL_mW_m\phi_m^{1/2}\left|\Omega_m - h_m\right|^{3/2} (\Omega_m - h_m)^{-1} = 0$$

The two unknowns are c_{pi}, the indoor wind pressure coefficient, and h_p, the elevation of the neutral plane.

A second relationship exists to determine c_{pi}, the continuity equation for wind ventilation by itself. Example 11-5 demonstrates the procedure to obtain c_{pi} independently.

When c_{pi} is known, h_p can be determined using an interative strategy as described in Section 11-4.

DYNAMIC PRESSURE LOSS COEFFICIENTS FOR SELECTED AIR DUCT FITTINGS
Taken from the 1985 AHRAE Handbook of Fundamentals, American Society of Heating, Refrigerating and Air Conditioning Engineers, Atlanta, GA.

Used by permission

FITTING LOSS COEFFICIENTS

NOTES

1. To calculate the total pressure loss coefficient for this fitting ($A_1 \neq A_o$) with a screen, use Eq. (B.1).

$$C_o = C_o' + \frac{C_s}{(A_1/A_o)^2}$$ (B.1)

where

C_o = combination fitting and screen resistance coefficient referenced to section *o*, dimensionless

C_o' = fitting resistance coefficient referenced to section *o*, dimensionless

C_s = screen resistance coefficient, dimensionless (see Fitting 7-8)

A_1 = area at section 1, mm^2

A_o = area at section *o*, mm^2

2. To calculate the total pressure loss coefficient for this fitting (constant cross-sectional area), use Eq. (B.2).

$$C_o = C_o' + C_s$$ (B.2)

3. For elbows with angles other than 90°, multiply by the following factor (K_θ).

θ	0	20	30	45	60	75	90	110	130	150	180
K_θ	0	0.31	0.45	0.60	0.78	0.90	1.00	1.13	1.20	1.28	1.40

4. To calculate *Reynolds number* (Re) use Eq. (B.3).

$$Re = \frac{\varrho D V_o}{\mu}$$ (B.3)

where

ϱ = density, kg/m^3

μ = viscosity, mPa•s

D = diameter, mm

V_o = velocity, m/s

For *standard air*, Re can be calculated by:

$$\text{Re} = 66.3\, D\, V_o \qquad (B.4)$$

5. For this fitting, determine the Reynolds number correction factor (K_{Re}) from the following table.

$Re \cdot 10^{-4}$	1	2	3	4	6	8	10	$\geqslant 14$
K_{Re}	1.40	1.26	1.19	1.14	1.09	1.06	1.04	1.0

6. The cross section to which the coefficients are referenced is indicated in the following tabulation.

Table No.	Type of Fitting	Cross Section to Which Coefficient Is Referenced
B-1	Entries	Downstream
B-2	Exits	Upstream
B-3	Elbows	Upstream or Downstream
B-4	Transitions (Diverging Flow)	Upstream
B-5	Transitions (Converging Flow)	Downstream
B-6	Junctions (Tees, Wyes)	Common
B-7	Obstructions	Upstream or Downstream

In the case of entries, exits, elbows, transitions and obstructions, the subscript o of A_o, Q_o and V_o shown on the sketches indicates the cross section to which the coefficients are referenced.

For converging and diverging junctions (tees, wyes) the subscript c of $C_{c,s}$ and $C_{c,b}$ signifies that the loss coefficient refers to the common section of the fitting. The subscripts s and b indicate that the loss coefficient is for the main and branch airflow paths.

The dynamic loss is based on the actual velocity in a rectangular duct, not the velocity in an equivalent round duct.

Table B-1   Local Loss Coefficients, ENTRIES

1-1   Duct Mounted in Wall (Hood, Non-Enclosing, Flanged and Unflanged)[1]

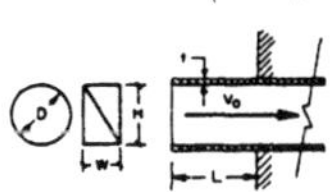

Rectangular: D = 2HW/(H + W)

	C_o					
	L/D					
t/D	0	0.002	0.01	0.05	0.2	0.5	$\geqslant 1.0$
≈ 0	0.50	0.57	0.68	0.80	0.92	1.0	1.0
0.02	0.50	0.51	0.52	0.55	0.66	0.72	0.72
$\geqslant 0.05$	0.50	0.50	0.50	0.50	0.50	0.50	0.50

Fitting coefficient (C_o) with screen at inlet.
Sharp Edge ($t/D \leqslant 0.05$):

$$C_o = 1 + C_s$$

Thick Edge ($t/D > 0.05$):

$$C_o = C_o' + C_s$$

where

C_o' = fitting coefficient (see Table above).
C_s = screen coefficient (see Fitting 7-8).

1-2   Smooth Converging Bellmouth without End Wall[1]

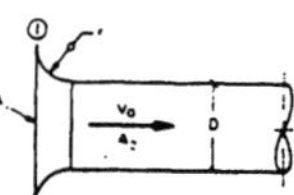

r/D	0	0.01	0.02	0.03	0.04	0.05
C_o	1.0	0.87	0.74	0.61	0.51	0.40
r/D	0.06	0.08	0.10	0.12	0.16	$\geqslant 0.20$
C_o	0.32	0.20	0.15	0.10	0.06	0.03

For loss coefficient with screen at section 1, see Note 1.

1-3   Smooth Converging Bellmouth with End Wall[1]

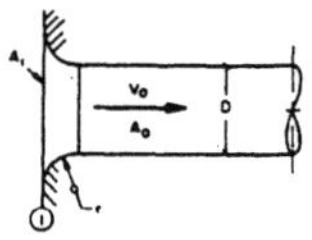

r/D	0	0.01	0.02	0.03	0.04	0.05
C_o	0.50	0r43	0.36	0.31	0.26	0.22
r/D	0.06	0.08	0.10	0.12	0.16	$\geqslant 0.20$
C_o	0.20	0.15	0.12	0.09	0.06	0.03

For loss coefficient with screen at section 1, see Note 1.

1-4   Conical Converging Bellmouth without End Wall, Round and Rectangular[1]

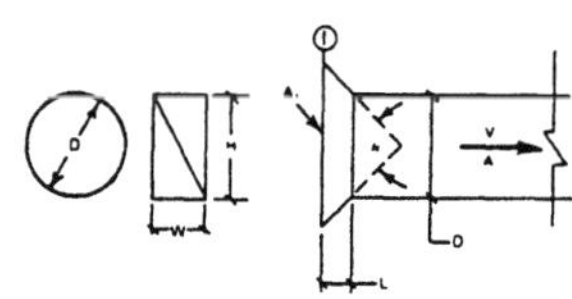

Rectangular: D = 2HW/(H + W)

$\dfrac{L}{D}$	C_o								
	θ, degrees								
	0	10	20	30	40	60	100	140	180
0.025	1.0	0.96	0.93	0.90	0.86	0.80	0.69	0.59	0.50
0.05	1.0	0.93	0.86	0.80	0.75	0.67	0.58	0.53	0.50
0.10	1.0	0.80	0.67	0.55	0.48	0.41	0.41	0.44	0.50
0.25	1.0	0.68	0.45	0.30	0.22	0.17	0.22	0.34	0.50
0.60	1.0	0.46	0.27	0.18	0.14	0.13	0.21	0.33	0.50
1.0	1.0	0.32	0.20	0.14	0.11	0.10	0.18	0.30	0.50

For loss coefficient with screen at section 1, see Note 1.

1-5 Conical Converging Bellmouth with End Wall, Round and Rectangular[1]

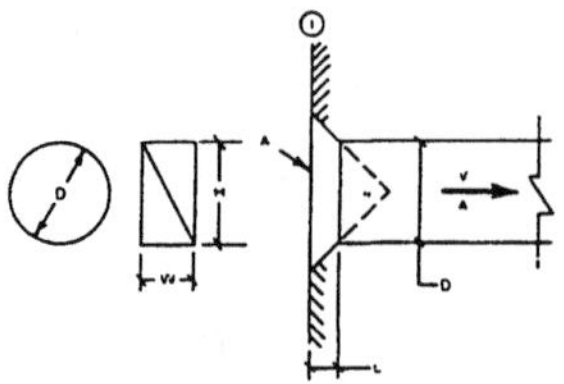

Rectangular: D = 2 HW/(H + W)

$\frac{L}{D}$	C_o								
	θ, degrees								
	0	10	20	30	40	60	100	140	180
0.025	0.50	0.47	0.45	0.43	0.41	0.40	0.42	0.45	0.50
0.05	0.50	0.45	0.41	0.36	0.33	0.30	0.35	0.42	0.50
0.075	0.50	0.42	0.35	0.30	0.26	0.23	0.30	0.40	0.50
0.10	0.50	0.39	0.32	0.25	0.22	0.18	0.27	0.38	0.50
0.15	0.50	0.37	0.27	0.20	0.16	0.15	0.25	0.37	0.50
0.60	0.50	0.27	0.18	0.13	0.11	0.12	0.23	0.36	0.50

For loss coefficient with screen at section 1, see Note 1.

1-6 Intake Hood[1]

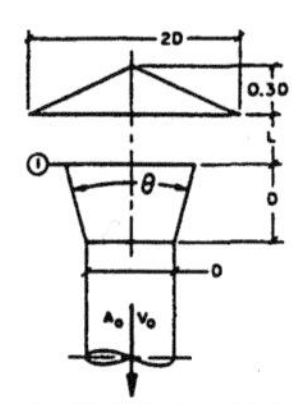

θ, deg	C_o								
	L/D								
	0.1	0.2	0.3	0.4	0.5	0.6	0.7	0.8	≥ 0.9
0	2.6	1.8	1.5	1.4	1.3	1.2	1.2	1.1	1.1
15	1.3	0.77	0.60	0.48	0.41	0.30	0.29	0.28	0.25

For hood loss coefficient with screen at section 1, see Note 1.

1-7 Hood, Tapered, Flanged or Unflanged[2]

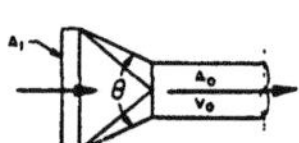

$A_1 \geqslant 2A_o$

θ is major angle for rectangular hoods

Hood Shape: Round									
θ	0°	20°	40°	60°	80°	100°	120°	140°	160°	180°
C_o	1.0	0.11	0.06	0.09	0.14	0.18	0.27	0.32	0.43	0.50

Hood Shape: Square or Rectangular									
θ	0°	20°	40°	60°	80°	100°	120°	140°	160°	180°
C_o	1.0	0.19	0.13	0.16	0.21	0.27	0.33	0.43	0.53	0.62

1-8 Air Inlet Openings on Side of Duct
a. First Opening[1]

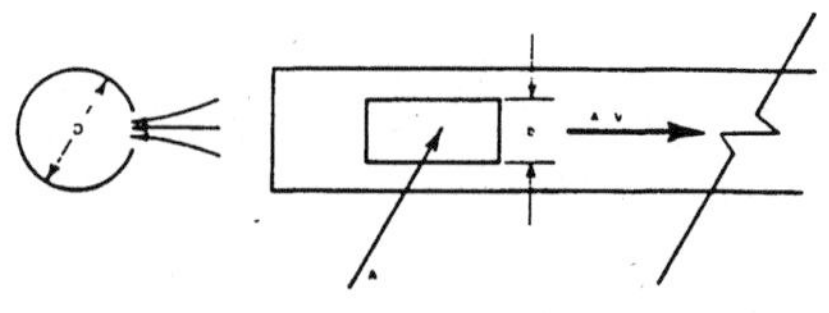

$\frac{b}{D_o}$	C_o							
	A_1/A_o							
	0.1	0.2	0.3	0.4	0.5	0.6	0.7	0.8
0.13	335	85	—	—	—	—	—	—
0.26	305	85	42	23	16	12	—	—
0.38	280	79	38	23	16	12	9.3	6.4
0.48	260	75	36	22	15	11	8.8	6.9
0.62	235	61	33	20	14	10	8.0	6.5
0.70	230	63	30	18	13	9.4	7.4	6.0

$\frac{b}{D_o}$	C_o							
	A_1/A_o							
	0.9	1.0	1.1	1.2	1.3	1.4	1.5	1.6
0.13	—	—	—	—	—	—	—	—
0.26	—	—	—	—	—	—	—	—
0.38	5.4	—	—	—	—	—	—	—
0.48	4.2	3.4	3.8	—	—	—	—	—
0.62	4.0	3.3	2.8	2.5	2.3	2.2	2.1	—
0.70	4.9	3.0	2.5	2.2	2.0	1.8	1.7	1.6

b. Succeeding Openings[3]

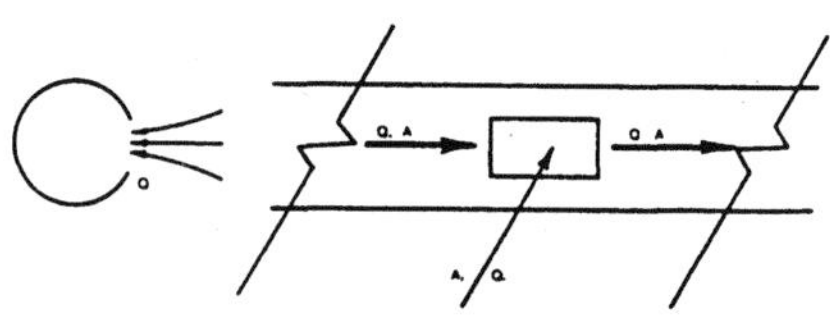

$\frac{A_b}{A_c}$	Branch (Opening), C_b				
	Q_b/Q_c				
	0.1	0.2	0.3	0.4	0.5
0.1	0.8	1.3	1.4	1.4	1.4
0.2	−1.4	0.9	1.3	1.4	1.4
0.4	−10	0.2	0.9	1.2	1.3
0.6	−21	−2.5	0.3	1.0	1.2

$\frac{A_b}{A_c}$	Main, C_s				
	Q_b/Q_c				
	0.1	0.2	0.3	0.4	0.5
0.1	0.1	−.1	−.8	−2.6	−6.6
0.2	0.1	0.2	0	−.6	−2.1
0.4	0.2	0.3	0.3	0.2	−.2
0.6	0.2	0.3	0.4	0.4	0.3

2-1 Exit, Abrupt, with Screen[1]

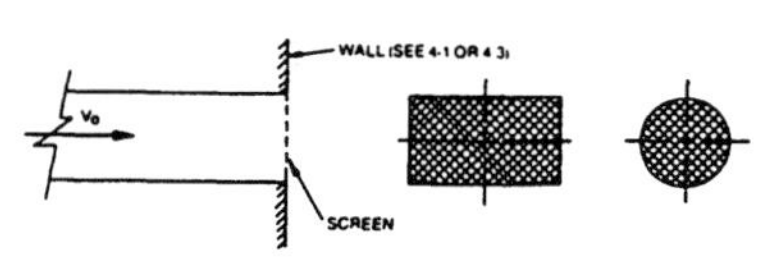

$C_o = 1 + C_s$, *where C_s is obtained from 7-8.*

If no screen is present, C_o is a function of velocity distribution and varies from 1 to 4.

2-2 Exit, Duct Flush with Wall, Flow along Wall[1]

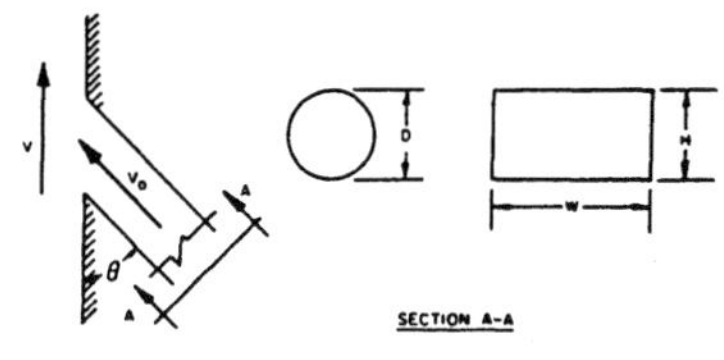

a. Round

	C_o				
		V/V_o			
θ, deg	0	0.5	1.0	1.5	2.0
30-45	1.0	1.0	1.1	1.3	1.6
60	1.0	0.90	1.1	1.4	1.6
90	1.0	0.80	0.95	1.4	1.7
120	1.0	0.80	0.95	1.3	1.7
150	1.0	0.82	0.83	1.0	1.3

b. Rectangular

		C_o				
			V/V_o			
Aspect Ratio (H/W)	θ, deg	0	0.5	1.0	1.5	2.0
0.1-0.2	30-90	1.0	0.95	1.2	1.5	1.8
	120	1.0	0.92	1.1	1.4	1.9
	150	1.0	0.75	0.95	1.4	1.8
0.5-2.0	30-45	1.0	1.0	1.1	1.3	1.6
	60	1.0	0.90	1.1	1.4	1.6
	90	1.0	0.80	0.95	1.4	1.7
	120	1.0	0.80	0.95	1.3	1.7
	150	1.0	0.82	0.83	1.0	1.3
5-10	45	1.0	0.92	0.93	1.1	1.3
	60	1.0	0.87	0.87	1.0	1.3
	90	1.0	0.82	0.80	0.97	1.2
	120	1.0	0.80	0.76	0.90	0.98

For loss coefficient with screen at outlet, see Note 2.

2-3 Exit, Conical, Round with Duct Upstream[1]

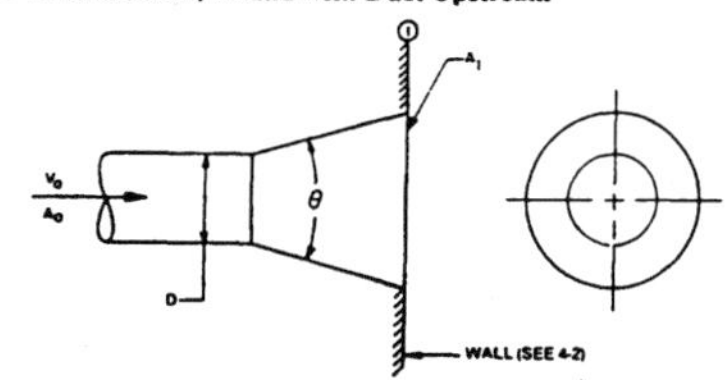

	C_o					
	θ, degrees					
A_1/A_o	8	10	14	20	30	45	≥60
2	0.36	0.33	0.37	0.51	0.90	1.0	1.0
4	0.24	0.21	0.28	0.40	0.70	0.99	1.0
6	0.20	0.19	0.26	0.37	0.67	0.99	1.0
10	0.18	0.16	0.24	0.36	0.68	0.99	1.0
16	0.16	0.16	0.20	0.36	0.66	0.99	1.0

For loss coefficient with screen at section 1, see Note 1

2-4 Exit, Plane Diffuser, Rectangular with Duct Upstream[1]

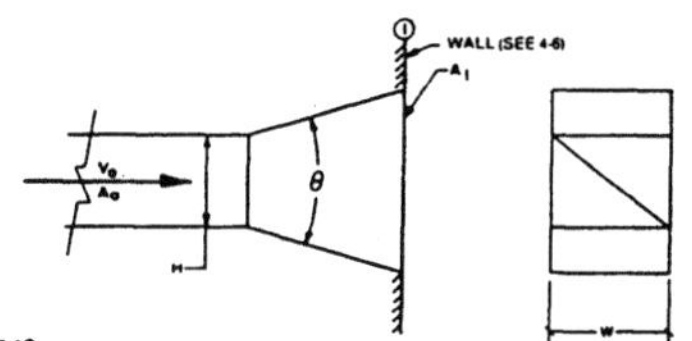

$0.5 \leqslant H/W \leqslant 2$

	C_o					
	θ, degrees					
A_1/A_o	8	10	14	20	30	34	≥60
2	0.50	0.51	0.56	0.63	0.80	0.96	1.0
4	0.34	0.38	0.48	0.63	0.76	0.91	1.0
6	0.32	0.34	0.41	0.56	0.70	0.84	0.96

For loss coefficient with screen at section 1, see Note 1.

2-5 Exit, Pyramidal Diffuser, with or without a Wall, Rectangular with Duct Upstream[1]

θ is larger of θ_1 and θ_2

	C_o					
	θ, degrees					
A_1/A_o	8	10	14	20	30	45	≥60
2	0.65	0.68	0.74	0.82	0.92	1.1	1.1
4	0.53	0.60	0.69	0.78	0.90	1.0	1.1
6	0.50	0.57	0.66	0.77	0.91	1.0	1.1
10	0.45	0.53	0.64	0.74	0.85	0.97	1.1

For loss coefficient with screen at section 1, see Note 1.

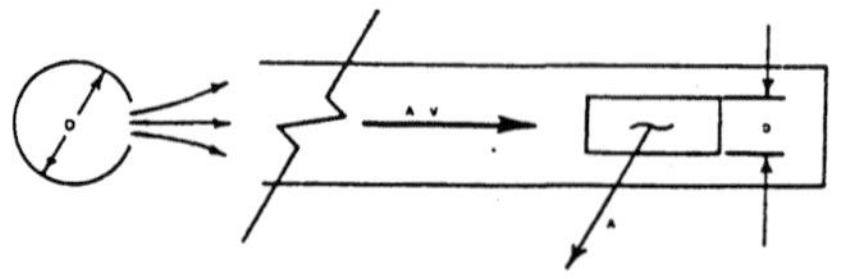

	C_o							
$\dfrac{b}{D_o}$	A_1/A_o							
	0.1	0.2	0.3	0.4	0.5	0.6	0.7	0.8
0.13	253	63	28	16	—	—	—	—
0.26	248	62	28	16	10	7.2	5.4	—
0.38	244	61	27	16	10	7.1	5.4	4.3
0.48	240	60	27	15	9.9	7.0	5.3	4.2
0.62	228	57	26	15	9.6	6.8	5.2	4.1
0.70	220	55	25	14	9.3	6.6	5.1	4.0

	C_o							
$\dfrac{b}{D_o}$	A_1/A_o							
	0.9	1.0	1.1	1.2	1.3	1.4	1.5	1.6
0.13	—	—	—	—	—	—	—	—
0.26	—	—	—	—	—	—	—	—
0.38	3.6	3.0	—	—	—	—	—	—
0.48	3.5	2.9	2.9	2.6	2.3	—	—	—
0.62	3.4	2.9	2.5	2.2	2.0	1.9	1.7	—
0.70	3.3	2.8	2.4	2.2	1.9	1.8	1.7	1.6

b. Succeeding Openings[3]

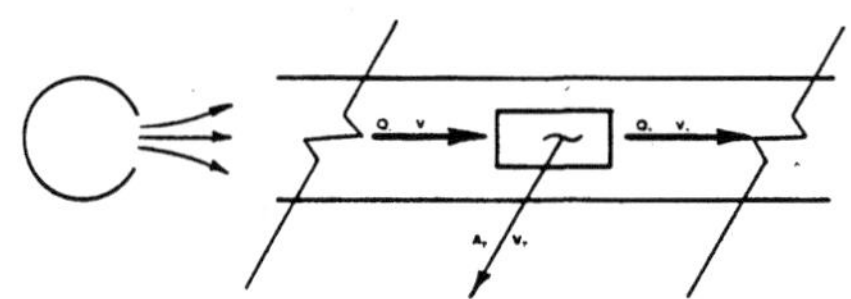

Branch (Opening), C_b									
V_b/V_c	0.4	0.6	0.8	1.0	1.2	1.4	1.6	1.8	2.0
C_b	1.8	1.7	1.7	1.8	1.9	2.1	2.3	2.6	3.0

Main, C_s					
V_s/V_c	0.4	0.5	0.6	0.8	1.0
C_c	0.06	0.01	−.03	−.06	−.03

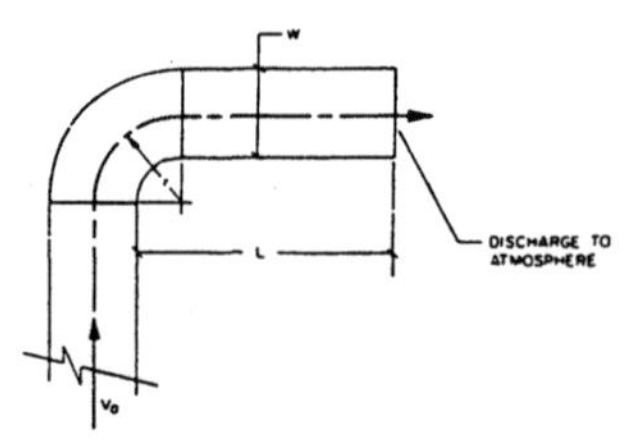

a. Rectangular

	C_o									
$\dfrac{r}{W}$	L/W									
	0	0.5	1.0	1.5	2.0	3.0	4.0	6.0	8.0	12.0
0	3.0	3.1	3.2	3.0	2.7	2.4	2.2	2.1	2.1	2.0
0.75	2.2	2.2	2.1	1.8	1.7	1.6	1.6	1.5	1.5	1.5
1.0	1.8	1.5	1.4	1.4	1.3	1.3	1.2	1.2	1.2	1.2
1.5	1.5	1.2	1.1	1.1	1.1	1.1	1.1	1.1	1.1	1.1
2.5	1.2	1.1	1.1	1.0	1.0	1.0	1.0	1.0	1.0	1.0

b. Round ($r/D = 1.0$)

| L/D | 0.9 | 1.3 |
| C_o | 1.5 | 1.4 |

For loss coefficient with screen at outlet, see Note 2.

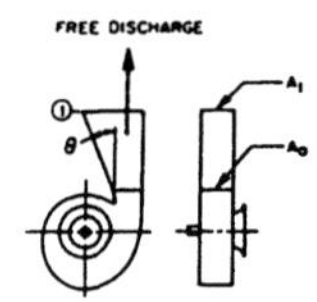

	C_o					
θ	A_1/A_o					
deg	1.5	2.0	2.5	3.0	3.5	4.0
10	0.51	0.34	0.25	0.21	0.18	0.17
15	0.54	0.36	0.27	0.24	0.22	0.20
20	0.55	0.38	0.31	0.27	0.25	0.24
25	0.59	0.43	0.37	0.35	0.33	0.33
30	0.63	0.50	0.46	0.44	0.43	0.42
35	0.65	0.56	0.53	0.52	0.51	0.50

For loss coefficient with screen at section 1, see Note 1.

2-9 Pyramidal Diffuser at Fan Outlet Without Ductwork[1]

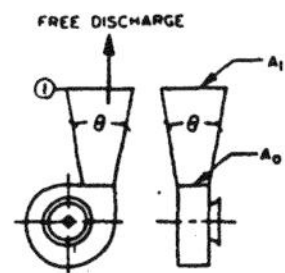

θ,			C_o			
			A_1/A_o			
deg	1.5	2.0	2.5	3.0	3.5	4.0
10	0.54	0.42	0.37	0.34	0.32	0.31
15	0.67	0.58	0.53	0.51	0.50	0.51
20	0.75	0.67	0.65	0.64	0.64	0.65
25	0.80	0.74	0.72	0.70	0.70	0.72
30	0.85	0.78	0.76	0.75	0.75	0.76

For loss coefficient with screen at section 1, see Note 1.

2-10 Exhaust Hood[1]

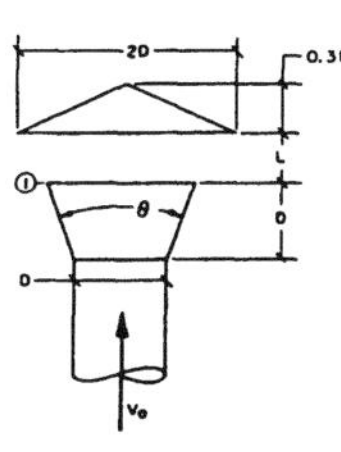

θ,				C_o						
				L/D						
deg	0.1	0.2	0.25	0.3	0.35	0.4	0.5	0.6	0.8	1.0
0	4.0	2.3	1.9	1.6	1.4	1.3	1.2	1.1	1.0	1.0
15	2.6	1.2	1.0	0.80	0.70	0.65	0.60	0.60	0.60	0.60

For hood loss coefficient with screen at section 1, see Note 1.

2-11 Discharge through Nozzle (Stackheads)[1]

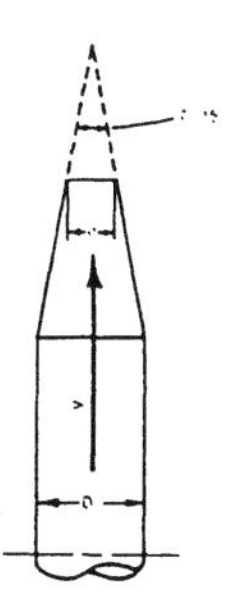

C_o	0.3	0.4	0.5	0.6	0.7	0.8	0.9	1.0
d/D_o	130	41	17	8.1	4.4	2.6	1.6	1.0

Table B-3　Local Loss Coefficients, ELBOWS

3-1 Elbow, Smooth Radius (Die Stamped), Round[1,4]

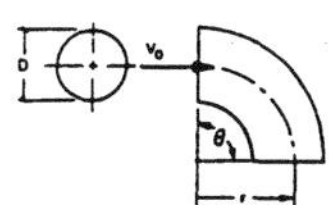

$$C_o = K_\theta C_o'$$

Coefficients for 90° Elbows:

r/D	0.5	0.75	1.0	1.5	2.0	2.5
C_o'	0.71	0.33	0.22	0.15	0.13	0.12

For angle correction factors K_θ, see Note 3.

3-2 Elbow, 90°; 3, 4 & 5-Pieces, Round[4]

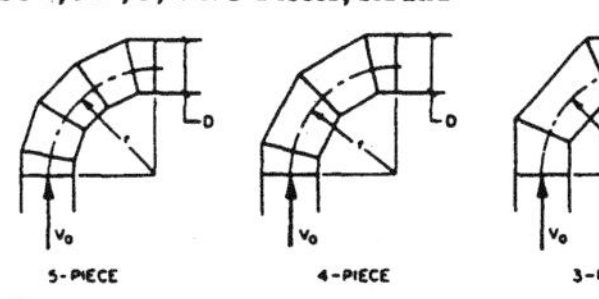

$$C_o = K_\theta C_o'$$

Coefficients for 90° Elbows:

No. of Pieces			C_o		
			r/D		
	0.5	0.75	1.0	1.5	2.0
5	—	0.46	0.33	0.24	0.19
4	—	0.50	0.37	0.27	0.24
3	0.90	0.54	0.42	0.34	0.33

For angle correction factors K_θ, see Note 3.

3-3 Elbow, Mitered, Round[1]

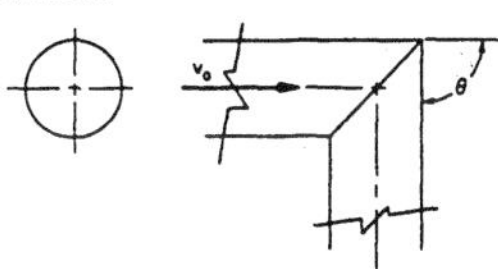

$$C_o = K_{Re} C_o'$$

θ, deg	20	30	45	60	75	90
C_o'	0.08	0.16	0.34	0.55	0.81	1.2

For Reynolds number correction factors K_{Re}, see Note 5.

3-4 Elbows, 30°, Z-Shaped, Round[1]

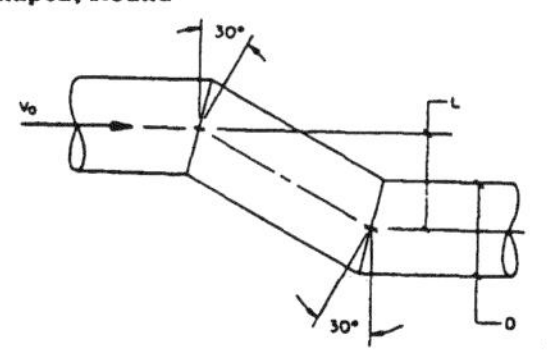

$$C_o = K_{Re} C_o'$$

L/D	0	0.5	1.0	1.5	2.0	2.5	3.0
C_o'	0	0.15	0.15	0.16	0.16	0.16	0.16

For Reynolds number correction factors K_{Re}, see Note 5.

3-5 Elbow, Smooth Radius without Vanes, Rectangular[1]

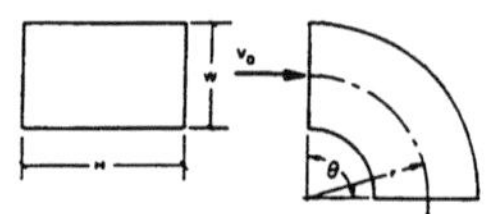

$$C_o = K_\theta K_{Re} C_o'$$

Coefficients for 90° Elbows:

| | C_o' | | | | | | | | | | |
| | H/W | | | | | | | | | | |
r/W	0.25	0.5	0.75	1.0	1.5	2.0	3.0	4.0	5.0	6.0	8.0
0.5	1.3	1.3	1.2	1.2	1.1	1.1	0.98	0.92	0.89	0.85	0.83
0.75	0.57	0.52	0.48	0.44	0.40	0.39	0.39	0.40	0.42	0.43	0.44
1.0	0.27	0.25	0.23	0.21	0.19	0.18	0.18	0.19	0.20	0.27	0.21
1.5	0.22	0.20	0.19	0.17	0.15	0.14	0.14	0.15	0.16	0.17	0.17
2.0	0.20	0.18	0.16	0.15	0.14	0.13	0.13	0.14	0.14	0.15	0.15

| | K_{Re} | | | | | | | | |
| | $Re \cdot 10^{-4}$ | | | | | | | | |
r/W	1	2	3	4	6	8	10	14	≥20
0.5	1.40	1.26	1.19	1.14	1.09	1.06	1.04	1.0	1.0
≥0.75	2.0	1.77	1.64	1.56	1.46	1.38	1.30	1.15	1.0

For elbows with angles other than 90°, see Note 3. To calculate Re, see Note 4.

3-6 Elbow, Mitered, Rectangular[1]

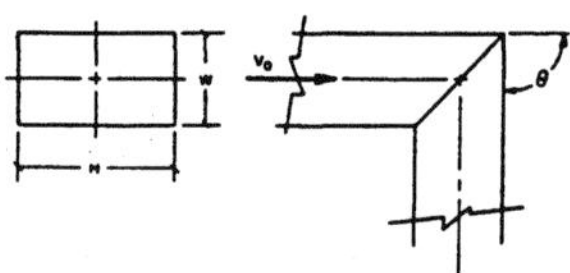

$$C_o = K_{Re} C_o'$$

| | H/W | | | | | | | | | | |
θ, deg	0.25	0.5	0.75	1.0	1.5	2.0	3.0	4.0	5.0	6.0	8.0
20	0.08	0.08	0.08	0.07	0.07	0.07	0.06	0.06	0.05	0.05	0.05
30	0.18	0.17	0.17	0.16	0.15	0.15	0.13	0.13	0.12	0.12	0.11
45	0.38	0.37	0.36	0.34	0.33	0.31	0.28	0.27	0.26	0.25	0.24
60	0.60	0.59	0.57	0.55	0.52	0.49	0.46	0.43	0.41	0.39	0.38
75	0.89	0.87	0.84	0.81	0.77	0.73	0.67	0.63	0.61	0.58	0.57
90	1.3	1.3	1.2	1.2	1.1	1.1	0.98	0.92	0.89	0.85	0.83

For Reynolds number correction factors K_{Re}, see Note 5.

3-7 Elbow, Smooth Radius with Splitter Vanes, Rectangular[4,5]

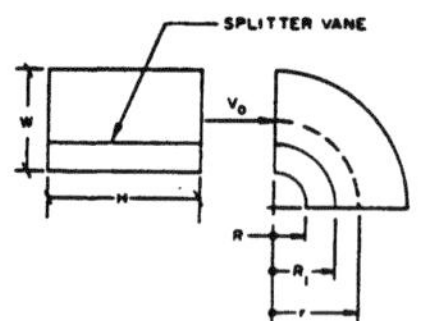

a. 1 Splitter Vane

$$C_o = K_\theta C_o'$$
$$R_1 = R/CR$$

where

R = throat radius
R_1 = splitter vane radius
CR = 'CURVE RATIO' (values from Table 3.7.a)
K_θ = angle factor (see Note 3 for values)

Table 3-7.a Coefficients for elbows with 1 splitter vane:

| | | | C_o' | | | | | | | | | | |
| | | | H/W | | | | | | | | | | |
R/W	r/W	CR	0.25	0.5	1.0	1.5	2.0	3.0	4.0	5.0	6.0	7.0	8.0
0.05	0.55	0.218	0.52	0.40	0.43	0.49	0.55	0.66	0.75	0.84	0.93	1.0	1.1
0.10	0.60	0.302	0.36	0.27	0.25	0.28	0.30	0.35	0.39	0.42	0.46	0.49	0.52
0.15	0.65	0.361	0.28	0.21	0.18	0.19	0.20	0.22	0.25	0.26	0.28	0.30	0.32
0.20	0.70	0.408	0.22	0.16	0.14	0.14	0.15	0.16	0.17	0.18	0.19	0.20	0.21
0.25	0.75	0.447	0.18	0.13	0.11	0.11	0.11	0.12	0.13	0.14	0.14	0.15	0.15
0.30	0.80	0.480	0.15	0.11	0.09	0.09	0.09	0.09	0.10	0.10	0.11	0.11	0.12
0.35	0.85	0.509	0.13	0.09	0.08	0.07	0.07	0.08	0.08	0.08	0.08	0.09	0.09
0.40	0.90	0.535	0.11	0.08	0.07	0.06	0.06	0.06	0.06	0.07	0.07	0.07	0.07
0.45	0.95	0.557	0.10	0.07	0.06	0.05	0.05	0.05	0.05	0.05	0.06	0.06	0.06
0.50	1.00	0.577	0.09	0.06	0.05	0.05	0.04	0.04	0.04	0.05	0.05	0.05	0.05

b. 2 Splitter Vanes

$$C_o = K_\theta C_o'$$
$$R_1 = R/CR$$
$$R_2 = R_1/CR = R/CR^2$$

where

R = throat radius
R_1 = splitter vane #1 radius
R_2 = splitter vane #2 radius
CR = 'CURVE RATIO' (values from Table 3.7.b)
K_θ = angle factor (see Note 3 for values)

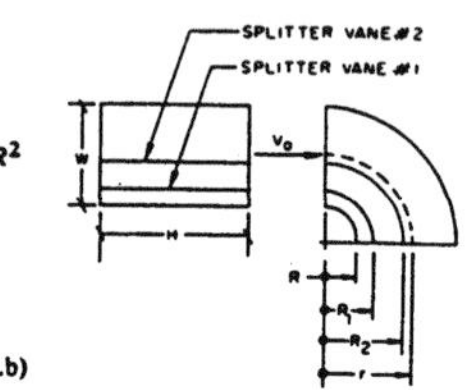

Table 3-7.b Coefficients for elbows with 2 splitter vanes:

| | | | C_o' | | | | | | | | | | |
| | | | H/W | | | | | | | | | | |
R/W	r/W	CR	0.25	0.5	1.0	1.5	2.0	3.0	4.0	5.0	6.0	7.0	8.0
0.05	0.55	0.362	0.26	0.20	0.22	0.25	0.28	0.33	0.37	0.41	0.45	0.48	0.51
0.10	0.60	0.450	0.17	0.13	0.11	0.12	0.13	0.15	0.16	0.17	0.19	0.20	0.21
0.15	0.65	0.507	0.12	0.09	0.08	0.08	0.08	0.09	0.10	0.10	0.11	0.11	0.11
0.20	0.70	0.550	0.09	0.07	0.06	0.05	0.06	0.06	0.06	0.06	0.07	0.07	0.07
0.25	0.75	0.585	0.08	0.05	0.04	0.04	0.04	0.04	0.05	0.05	0.05	0.05	0.05
0.30	0.80	0.613	0.06	0.04	0.03	0.03	0.03	0.03	0.03	0.03	0.04	0.04	0.04

c. 3 Splitter Vanes

$$C_o = K_\theta C_o'$$
$$R_1 = R/CR$$
$$R_2 = R_1/CR = R/CR^2$$
$$R_3 = R_2/CR = R/CR^3$$

where

R = throat radius
R_1 = splitter vane #1 radius
R_2 = splitter vane #2 radius
R_3 = splitter vane #3 radius
CR = 'CURVE RATIO' (values from Table 3.7.c)
K_θ = angle factor (see Note 3 for values)

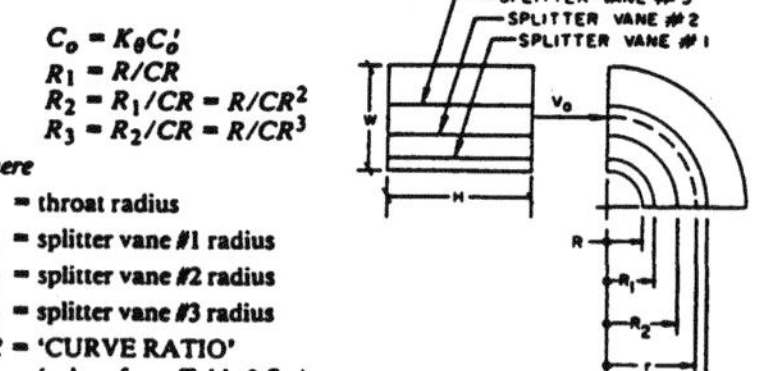

Table 3-7.c Coefficients for elbow with 3 splitter vanes:

| | | | C_o' | | | | | | | | | | |
| | | | H/W | | | | | | | | | | |
R/W	r/W	CR	0.25	0.5	1.0	1.5	2.0	3.0	4.0	5.0	6.0	7.0	8.0
0.05	0.55	0.467	0.11	0.10	0.12	0.13	0.14	0.16	0.18	0.19	0.21	0.22	0.23
0.10	0.60	0.549	0.07	0.05	0.06	0.06	0.06	0.07	0.07	0.08	0.08	0.08	0.09

3-8 Elbow, Mitered, with Single-Thickness Vanes, Rectangular[6]

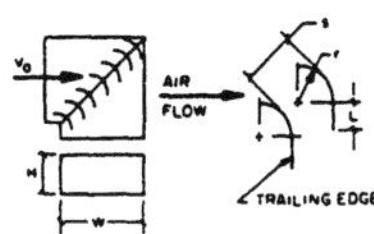

Design No.	Dimensions, mm			C_o
	r	s	L	
1[a]	50	40	20	0.12
2	115	60	0	0.15
3	115	80	40	0.18

[a]When extension of trailing edge is not provided for this vane, losses are approximately uncharged for single elbows, but increase considerably for elbows in series.

3-9 Elbow, Mitered, with Double-Thickness Vanes, Rectangular[6]

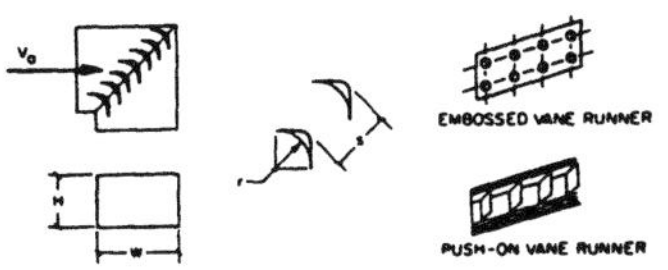

Design No.	Dimensions, mm		C_o — Velocity V_0, m/s				Remarks
	r	s	5	10	15	20	
1	50	40	0.27	0.22	0.19	0.17	Embossed Vane Runner
2	50	40	0.33	0.29	0.26	0.23	Push-On Vane Runner
3	50	55	0.38	0.31	0.27	0.24	Embossed Vane Runner
4	115	80	0.26	0.21	0.18	0.16	Embossed Vane Runner

3-10 Elbow, Converging or Diverging Flow, Rectangular[1]

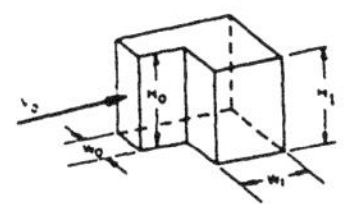

$$C_o = K_{Re} C_o'$$

H_o/W_o	C_o' — W_1/W_o					
	0.6	0.8	1.2	1.4	1.6	2.0
0.25	1.8	1.4	1.1	1.1	1.1	1.1
1.0	1.7	1.4	1.0	0.95	0.90	0.84
4.0	1.5	1.1	0.81	0.76	0.72	0.66
∞	1.5	1.0	0.69	0.63	0.60	0.55

For Reynolds number correction factors K_{Re}, See Note 5.

3-11 Elbows, 90°, Z-Shaped, Rectangular[1]

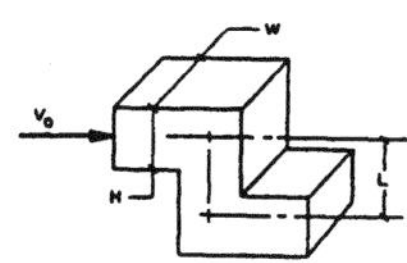

$$C_o = K K_{Re} C_o'$$

Coefficients for $W/H = 1.0$:

L/H	0	0.4	0.6	0.8	1.0	1.2	1.4	1.6	1.8	2.0
C_o'	0	0.62	0.90	1.6	2.6	3.6	4.0	4.2	4.2	4.2
L/H	2.4	2.8	3.2	4.0	5.0	6.0	7.0	9.0	10.0	∞
C_o'	3.7	3.3	3.2	3.1	2.9	2.8	2.7	2.6	2.5	2.3

For W/H values other than 1.0, apply the following factor:

W/H	0.25	0.50	0.75	1.0	1.5	2.0	3.0	4.0	6.0	8.0
K	1.10	1.07	1.04	1.0	0.95	0.90	0.83	0.78	0.72	0.70

For Reynolds number correction factors K_{Re}, see Note 5.

3-12 Combined 90° Elbows Lying in Different Planes, Rectangular[1]

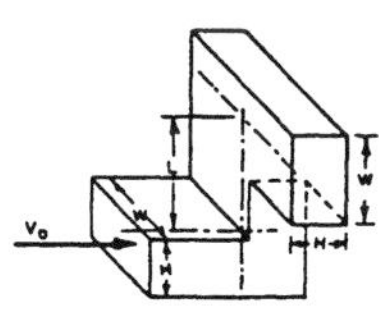

$$C_o = K K_{Re} C_o'$$

Coefficients for Square Ducts:

L/W	0	0.4	0.6	0.8	1.0	1.2	1.4	1.6	1.8	2.0
C_o'	1.2	2.4	2.9	3.3	3.4	3.4	3.4	3.3	3.2	3.1
L/W	2.4	2.8	3.2	4.0	5.0	6.0	7.0	9.0	10.0	∞
C_o'	3.2	3.2	3.2	3.0	2.9	2.8	2.7	2.5	2.4	2.3

Apply Following Factor for other than $H/W=1.0$:

H/W	0.25	0.50	0.75	1.0	1.5	2.0	3.0	4.0	6.0	8.0
K	1.10	1.07	1.04	1.0	0.95	0.90	0.83	0.78	0.72	0.70

For Reynolds number correction factors K_{Re}, see Note 5.

Table B-4   Local Loss Coefficients, TRANSITIONS (Diverging Flow)

4-1 Conical Diffuser, Round[1]

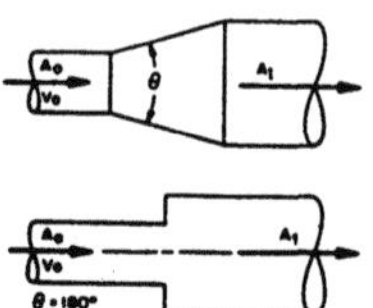

A_1/A_0	C_o									
	θ, degrees									
	8	12	16	20	30	45	60	90	120	180
2	0.11	0.11	0.14	0.19	0.32	0.33	0.33	0.32	0.31	0.30
4	0.15	0.17	0.23	0.30	0.46	0.61	0.68	0.64	0.63	0.62
6	0.17	0.20	0.27	0.33	0.48	0.66	0.77	0.74	0.73	0.72
10	0.19	0.23	0.29	0.38	0.59	0.76	0.80	0.83	0.84	0.83
⩾16	0.19	0.22	0.31	0.38	0.60	0.84	0.88	0.88	0.88	0.88

4-2   Conical Diffuser, Stepped, Round[1]

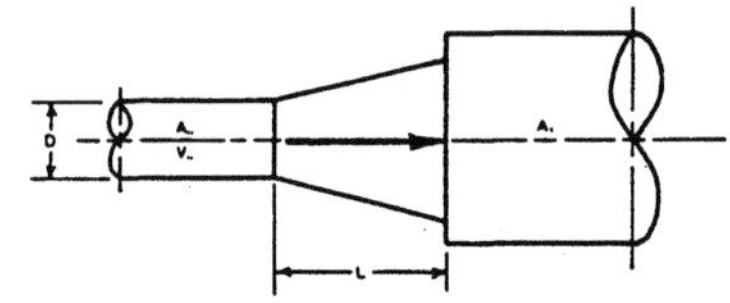

$\dfrac{A_1}{A_0}$	C_o									
	L/D_0									
	0.5	1.0	2.0	3.0	4.0	5.0	6.0	8.0	10	12	14
1.5	0.03	0.02	0.03	0.03	0.04	0.05	0.06	0.08	0.10	0.11	0.13
2.0	0.08	0.06	0.04	0.04	0.04	0.05	0.05	0.06	0.08	0.09	0.10
2.5	0.13	0.09	0.06	0.06	0.06	0.06	0.06	0.06	0.07	0.08	0.09
3.0	0.17	0.12	0.09	0.07	0.07	0.06	0.06	0.07	0.07	0.08	0.08
4.0	0.23	0.17	0.12	0.10	0.09	0.08	0.08	0.08	0.08	0.08	0.08
6.0	0.30	0.22	0.16	0.13	0.12	0.10	0.10	0.09	0.09	0.09	0.08
8.0	0.34	0.26	0.18	0.15	0.13	0.12	0.11	0.10	0.09	0.09	0.09
10	0.36	0.28	0.20	0.16	0.14	0.13	0.12	0.11	0.10	0.09	0.09
14	0.39	0.30	0.22	0.18	0.16	0.14	0.13	0.12	0.10	0.10	0.10
20	0.41	0.32	0.24	0.20	0.17	0.15	0.14	0.12	0.11	0.11	0.10

4-3 Pyramidal Diffuser, Rectangular[1]

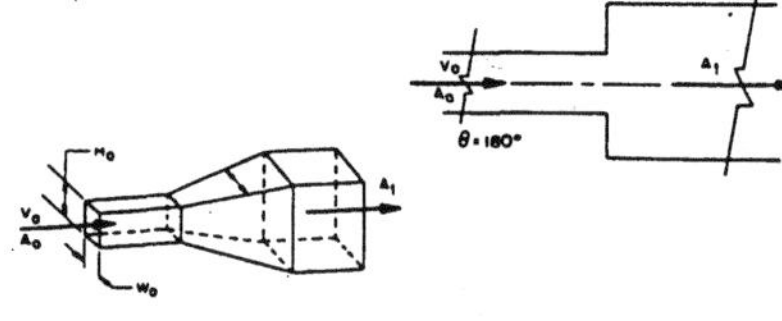

A_1/A_0	C_o								
	θ, degrees								
	8	10	14	20	30	45	60	90	180
2	0.14	0.15	0.20	0.25	0.30	0.33	0.33	0.33	0.30
4	0.20	0.25	0.34	0.45	0.52	0.58	0.62	0.64	0.64
6	0.21	0.30	0.42	0.53	0.63	0.72	0.78	0.79	0.79
⩾10	0.24	0.30	0.43	0.53	0.64	0.75	0.84	0.89	0.88

4-4   Pyramidal Diffuser, Stepped, Rectangular[1]

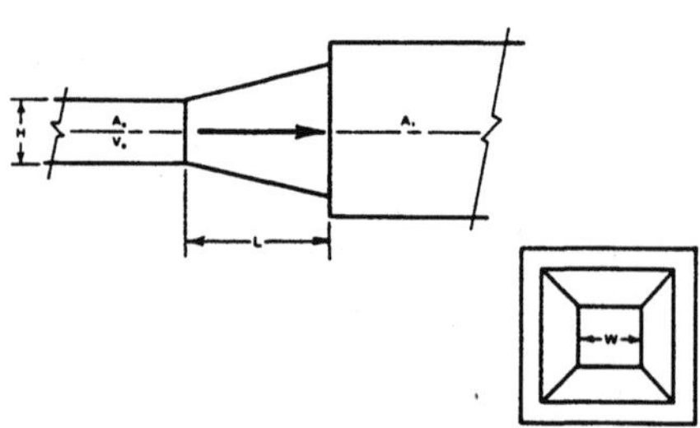

$$D = 2HW/(H + W)$$

$\dfrac{A_1}{A_0}$	C_o									
	L/D									
	0.5	1.0	2.0	3.0	4.0	5.0	6.0	8.0	10	12	14
1.5	0.04	0.03	0.03	0.04	0.05	0.05	0.06	0.08	0.10	0.11	0.13
2.0	0.11	0.08	0.06	0.06	0.06	0.06	0.07	0.07	0.08	0.09	0.10
2.5	0.16	0.13	0.09	0.08	0.08	0.07	0.08	0.07	0.08	0.08	0.09
3.0	0.21	0.17	0.12	0.10	0.09	0.09	0.09	0.09	0.09	0.09	0.09
4.0	0.27	0.22	0.17	0.14	0.12	0.11	0.11	0.11	0.11	0.10	0.10
6.0	0.36	0.28	0.21	0.18	0.16	0.15	0.14	0.13	0.12	0.12	0.11
8.0	0.41	0.32	0.24	0.21	0.18	0.17	0.16	0.14	0.13	0.12	0.12
10	0.44	0.35	0.26	0.22	0.20	0.18	0.17	0.15	0.14	0.13	0.13
14	0.47	0.37	0.28	0.24	0.21	0.20	0.18	0.16	0.15	0.14	0.14
20	0.49	0.40	0.30	0.26	0.23	0.21	0.19	0.17	0.16	0.15	0.14

4-5   Plane Diffuser, Rectangular[1]

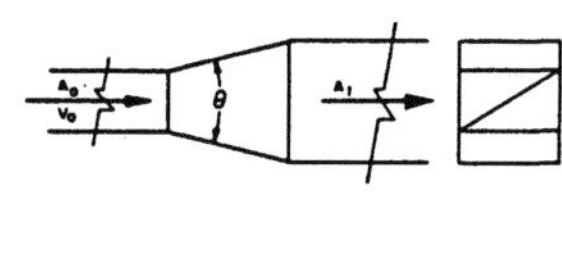

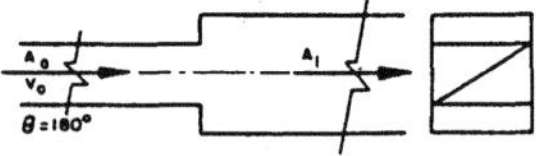

A_1/A_0	C_o								
	θ, degrees								
	8	10	14	20	30	45	60	90	180
2	0.14	0.14	0.13	0.15	0.24	0.35	0.37	0.38	0.35
4	0.19	0.17	0.18	0.25	0.42	0.60	0.68	0.70	0.66
6	0.22	0.21	0.21	0.30	0.48	0.65	0.76	0.83	0.80

4-6 Plane Diffuser, Stepped, Rectangular[1]

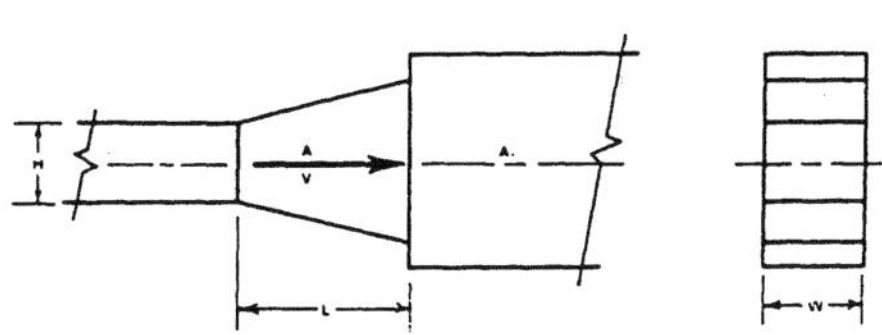

A_1/A_0	C_o										
	L/H										
	0.5	1.0	2.0	3.0	4.0	5.0	6.0	8.0	10	12	14
1.5	0.04	0.04	0.04	0.04	0.05	0.06	0.06	0.08	0.10	0.11	0.13
2.0	0.12	0.09	0.07	0.07	0.06	0.07	0.07	0.07	0.08	0.10	0.12
2.5	0.18	0.14	0.11	0.10	0.09	0.09	0.09	0.09	0.09	0.10	0.10
3.0	0.23	0.18	0.14	0.12	0.11	0.11	0.10	0.10	0.10	0.10	0.11
4.0	0.30	0.24	0.19	0.16	0.15	0.14	0.13	0.12	0.12	0.12	0.12
6.0	0.38	0.31	0.25	0.21	0.19	0.18	0.17	0.16	0.15	0.14	0.14
8.0	0.43	0.36	0.28	0.25	0.22	0.20	0.19	0.17	0.16	0.16	0.15
10	0.46	0.38	0.30	0.26	0.24	0.22	0.21	0.19	0.18	0.17	0.16
14	0.50	0.41	0.33	0.29	0.26	0.24	0.22	0.20	0.19	0.18	0.18
20	0.53	0.44	0.35	0.31	0.28	0.25	0.24	0.22	0.20	0.19	0.19

4-7 Transition Diffuser, Round to Rectangular or Rectangular to Round[1]

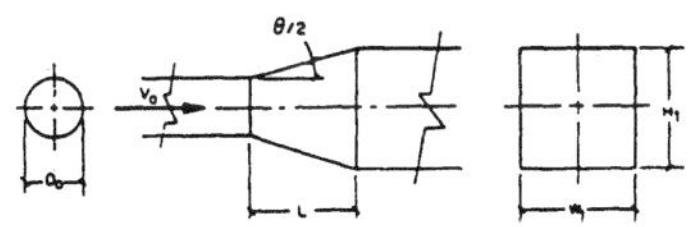

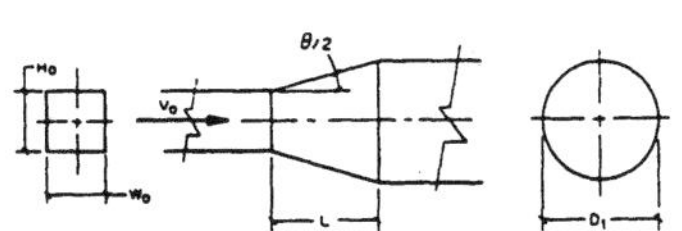

Use C_o from fitting 4-3 based on the following equivalent angles:

1. Transition from round to rectangular:

$$\tan(\theta/2) = (1.13\sqrt{H_1 W_1} - D_o)/2L$$

2. Transition from rectangular to round:

$$\tan(\theta/2) = (D_1 - 1.13\sqrt{H_o W_o})/2L$$

4-8 Plane Symmetric Diffuser at Fan Outlet with Ductwork[1]

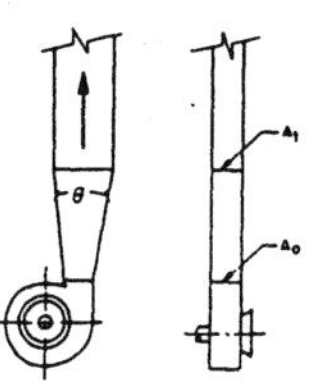

θ, deg	C_o					
	A_1/A_0					
	1.5	2.0	2.5	3.0	3.5	4.0
10	0.05	0.07	0.09	0.10	0.11	0.11
15	0.06	0.09	0.11	0.13	0.13	0.14
20	0.07	0.10	0.13	0.15	0.16	0.16
25	0.08	0.13	0.16	0.19	0.21	0.23
30	0.16	0.24	0.29	0.32	0.34	0.35
35	0.24	0.34	0.39	0.44	0.48	0.50

4-9 Plane Asymmetric Diffuser at Fan Outlet with Ductwork[1]

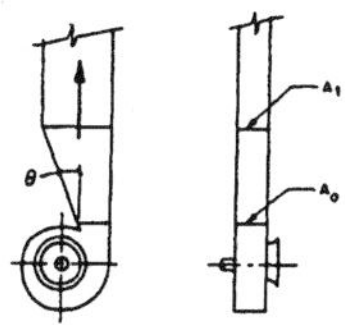

θ, deg	C_o					
	A_1/A_0					
	1.5	2.0	2.5	3.0	3.5	4.0
10	0.08	0.09	0.10	0.10	0.11	0.11
15	0.10	0.11	0.12	0.13	0.14	0.15
20	0.12	0.14	0.15	0.16	0.17	0.18
25	0.15	0.18	0.21	0.23	0.25	0.26
30	0.18	0.25	0.30	0.33	0.35	0.35
35	0.21	0.31	0.38	0.41	0.43	0.44

4-10 Plane Asymmetric Diffuser at Fan Outlet with Ductwork[1]

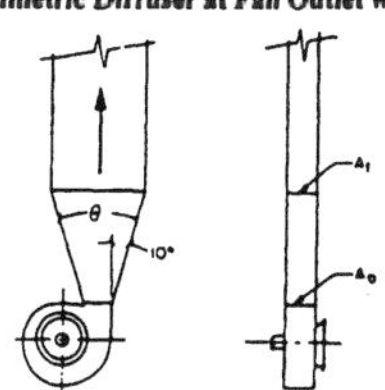

θ, deg	C_o					
	A_1/A_0					
	1.5	2.0	2.5	3.0	3.5	0.4
10°	0.05	0.08	0.11	0.13	0.13	0.14
15°	0.06	0.10	0.12	0.14	0.15	0.15
20°	0.07	0.11	0.14	0.15	0.16	0.16
25°	0.09	0.14	0.18	0.20	0.21	0.22
30°	0.13	0.18	0.23	0.26	0.28	0.29
35°	0.15	0.23	0.28	0.33	0.35	0.36

4-11 Plane Asymmetric Diffuser at Fan Outlet with Ductwork[1]

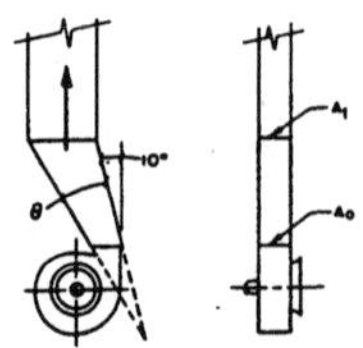

	C_o					
	A_1/A_o					
θ, deg	1.5	2.0	2.5	3.0	3.5	4.0
10°	0.11	0.13	0.14	0.14	0.14	0.14
15°	0.13	0.15	0.16	0.17	0.18	0.18
20°	0.19	0.22	0.24	0.26	0.28	0.30
25°	0.29	0.32	0.35	0.37	0.39	0.40
30°	0.36	0.42	0.46	0.49	0.51	0.51
35°	0.44	0.54	0.61	0.64	0.66	0.66

4-12 Pyramidal Diffuser at Fan Outlet with Ductwork[1]

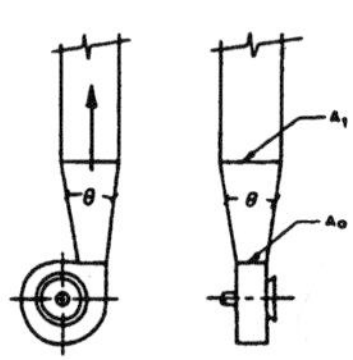

	C_o					
	A_1/A_o					
θ, deg	1.5	2.0	2.5	3.0	3.5	4.0
10	0.10	0.18	0.21	0.23	0.24	0.25
15	0.23	0.33	0.38	0.40	0.42	0.44
20	0.31	0.43	0.48	0.53	0.56	0.58
25	0.36	0.49	0.55	0.58	0.62	0.64
30	0.42	0.53	0.59	0.64	0.67	0.69

Table B-5   Local Loss Coefficients, TRANSITIONS (Converging Flow)

5-1 Contraction, Round & Rectangular[1]

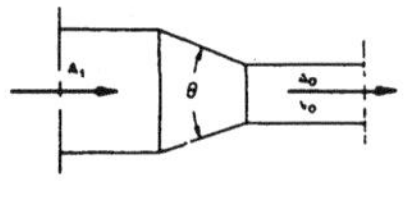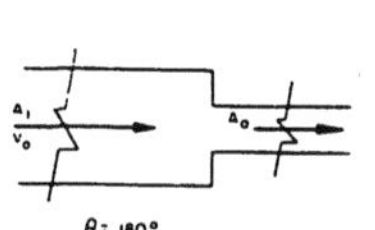

	C_o						
	θ, degrees						
A_1/A_o	10	15–40	50–60	90	120	150	180
2	0.05	0.05	0.06	0.12	0.18	0.24	0.26
4	0.05	0.04	0.07	0.17	0.27	0.35	0.41
6	0.05	0.04	0.07	0.18	0.28	0.36	0.42
10	0.05	0.05	0.08	0.19	0.29	0.37	0.43

5-2 Contraction, Conical, Round and Rectangular[1]

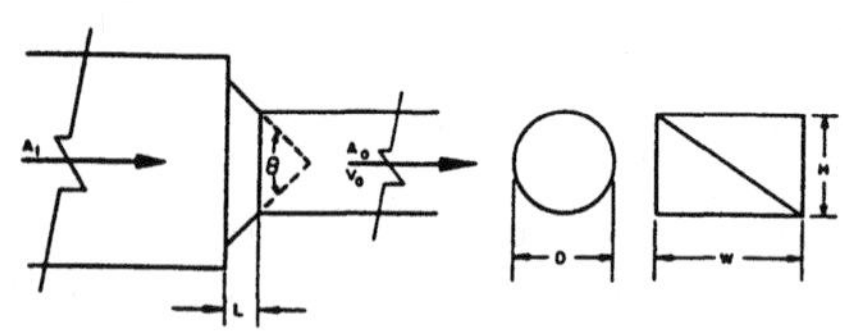

Rectangular: D = 2HW/(H + W)
θ = major angle
$C_o = K C_o'$

	C_o'								
	θ, degrees								
L/D	0	10	20	30	40	60	100	140	180
0.025	0.50	0.47	0.45	0.43	0.41	0.40	0.42	0.45	0.50
0.05	0.50	0.45	0.41	0.36	0.33	0.30	0.35	0.42	0.50
0.075	0.50	0.42	0.35	0.30	0.26	0.23	0.30	0.40	0.50
0.10	0.50	0.39	0.32	0.25	0.22	0.18	0.27	0.38	0.50
0.10	0.50	0.39	0.32	0.25	0.22	0.18	0.27	0.38	0.50
0.15	0.50	0.37	0.27	0.20	0.16	0.15	0.25	0.37	0.50
0.60	0.50	0.27	0.18	0.13	0.11	0.12	0.23	0.36	0.50

5-3 Contraction, Rectangular Slot to Round[1]

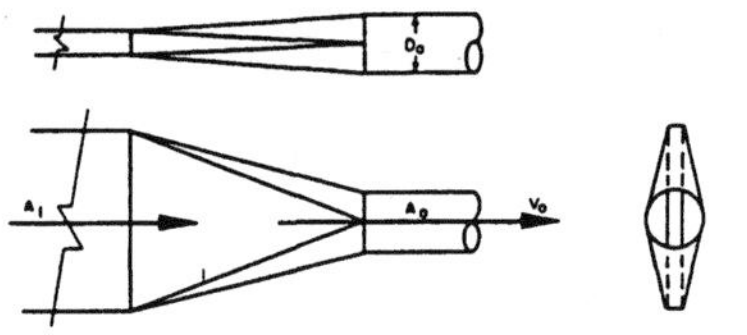

$A_o < A_1$

$Re \cdot 10^{-4}$	1	2	4	6	8	10	20	≥ 40
C_o	0.27	0.25	0.20	0.17	0.14	0.11	0.04	0

To calculate Re, See Note 4.

Table B-6   Local Loss Coefficients, JUNCTIONS (Tees, Wyes)

6-1 Converging Wye (30°), Round[1]

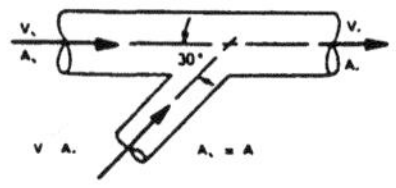

| Branch, $C_{c,b}$ | | | | | | |
| | | | A_b/A_c | | | |
Q_b/Q_c	0.1	0.2	0.3	0.4	0.6	0.8	1.0
0	-1.0	-1.0	-1.0	-1.0	-1.0	-1.0	-1.0
0.1	0.21	-.46	-.57	-.60	-.62	-.63	-.63
0.2	3.1	0.37	-.06	-.20	-.28	-.30	-.35
0.3	7.6	1.5	0.50	0.20	-.05	-.08	-.10
0.4	14	3.0	1.2	0.59	0.26	0.18	0.16
0.5	21	4.6	1.8	0.97	0.44	0.35	0.27
0.6	30	6.4	2.6	1.4	0.64	0.46	0.31
0.7	41	8.5	3.4	1.8	0.76	0.50	0.40
0.8	54	12	4.2	2.1	0.85	0.53	0.45
0.9	58	14	5.3	2.6	0.89	0.52	0.40
1.0	84	17	6.3	2.9	0.89	0.39	0.27

| Main, $C_{c,s}$ | | | | | | |
| Q_b | | | | A_b/A_c | | |
Q_c	0.1	0.2	0.3	0.4	0.6	0.8	1.0
0	0	0	0	0	0	0	0
0.1	0.02	0.11	0.13	0.15	0.16	0.17	0.17
0.2	-.33	0.01	0.13	0.19	0.24	0.27	0.29
0.3	-1.1	-.25	-.01	0.10	0.22	0.30	0.35
0.4	-2.2	-.75	-.30	-.05	0.17	0.26	0.36
0.5	-3.6	-1.4	-.70	-.35	0	0.21	0.32
0.6	-5.4	-2.4	-1.3	-.70	-.20	0.06	0.25
0.7	-7.6	-3.4	-2.0	-1.2	-.50	-.15	0.10
0.8	-10	-4.6	-2.7	-1.8	-.90	-.43	-.15
0.9	-13	-6.2	-3.7	-2.6	-1.4	-.80	-.45
1.0	-16	-7.7	-4.8	-3.4	-1.9	-1.2	-.75

6-2 Converging Wye (45°), Round[8]

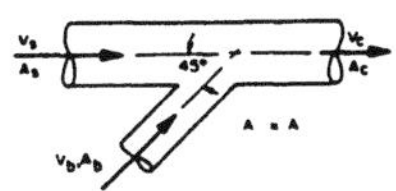

| Branch, $C_{c,b}$ | | | | | | |
| V_b | | | | A_b/A_c | | |
V_c	0.1	0.2	0.3	0.4	0.6	0.8	1.0
0.4	-.56	-.44	-.35	-.28	-.15	-.04	0.05
0.5	-.48	-.37	-.28	-.21	-.09	0.02	0.11
0.6	-.38	-.27	-.19	-.12	0	0.10	0.18
0.7	-.26	-.16	-.08	-.01	0.10	0.20	0.28
0.8	-.21	-.02	0.05	0.12	0.23	0.32	0.40
0.9	0.04	0.13	0.21	0.27	0.37	0.46	0.53
1.0	0.22	0.31	0.38	0.44	0.53	0.62	0.69
1.5	1.4	1.5	1.5	1.6	1.7	1.7	1.8
2.0	3.1	3.2	3.2	3.2	3.3	3.3	3.3
2.5	5.3	5.3	5.3	5.4	5.4	5.4	5.4
3.0	8.0	8.0	8.0	8.0	8.0	8.0	8.0

| Main, $C_{c,s}$ | | | | | | |
| V_s | | | | A_b/A_c | | |
V_c	0.1	0.2	0.3	0.4	0.6	0.8	1.0
0.1	-8.6	-4.1	-2.5	-1.7	-.97	-.58	-.34
0.2	-6.7	-3.1	-1.9	-1.3	-.67	-.36	-.18
0.3	-5.0	-2.2	-1.3	-.88	-.42	-.19	-.05
0.4	-3.5	-1.5	-.88	-.55	-.21	-.05	0.05
0.5	-2.3	-.95	-.51	-.28	-.06	0.06	0.13
0.6	-1.3	-.50	-.22	-.09	0.05	0.12	0.17
0.7	-.63	-.18	-.03	0.04	0.12	0.16	0.18
0.8	-.18	0.01	0.07	0.10	0.13	0.15	0.17
0.9	0.03	0.07	0.08	0.09	0.10	0.11	0.13
1.0	-.01	0	0	0.10	0.02	0.04	0.05

6-3 Converging Tee, Round[1]

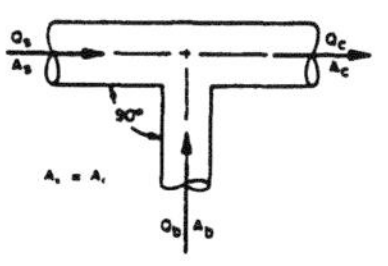

| Branch, $C_{c,b}$ | | | | | | |
| | | | A_b/A_c | | | |
Q_b/Q_c	0.1	0.2	0.3	0.4	0.6	0.8	1.0
0	-1.0	-1.0	-1.0	-.90	-.90	-.90	-.90
0.1	0.40	-.37	-.51	-.46	-.50	-.51	-.52
0.2	3.8	0.72	0.17	-.02	-.14	-.18	-.24
0.3	9.2	2.3	1.0	0.44	0.21	0.11	-.08
0.4	16	4.3	2.1	0.94	0.54	0.40	0.32
0.5	26	6.8	3.2	1.1	0.66	0.49	0.42
0.6	37	9.7	4.7	1.6	0.92	0.69	0.57
0.7	43	13	6.3	2.1	1.2	0.88	0.72
0.8	65	17	7.9	2.7	1.5	1.1	0.86
0.9	82	21	9.7	3.4	1.8	1.2	0.99
1.0	101	26	12	4.0	2.1	1.4	1.1

Main										
Q_b/Q_c	0	0.1	0.2	0.3	0.4	0.5	0.6	0.7	0.8	0.9	1.0
$C_{c,s}$	0	0.16	0.27	0.38	0.46	0.53	0.57	0.59	0.60	0.59	0.55

6-4 Converging Wye (30°), Conical, Round[9]

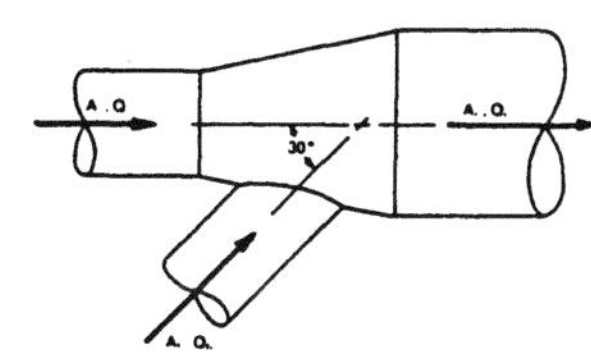

| Branch, $C_{c,b}$ | | | | | | | | | | | |
| $\dfrac{A_s}{A_c}$ | $\dfrac{A_b}{A_c}$ | | | | | Q_b/Q_s | | | | | |
		0.2	0.4	0.6	0.8	1.0	1.2	1.4	1.6	1.8	2.0
0.3	0.2	-2.4	-.11	1.8	3.4	4.8	6.0	7.1	8.0	8.9	9.7
	0.3	-2.8	-1.3	0.14	0.72	1.4	2.0	2.4	2.8	3.2	3.5
0.4	0.2	-1.4	0.61	2.3	3.8	5.2	6.3	7.3	8.3	9.1	9.8
	0.3	-1.8	-.54	0.42	1.2	1.8	2.3	2.7	3.1	3.4	3.7
	0.4	-1.9	-.89	-.17	0.36	0.76	1.1	1.3	1.5	1.7	1.9
0.5	0.2	-.82	0.97	2.6	4.0	5.3	6.4	7.4	8.3	9.1	9.9
	0.3	-1.2	-.15	0.71	1.4	2.0	2.5	2.9	3.3	3.6	3.9
	0.4	-1.4	-.54	0.06	0.50	0.85	1.1	1.3	1.5	1.7	1.8
	0.5	-1.4	-.66	-.15	0.21	0.48	0.68	0.84	0.97	1.1	1.2
0.6	0.2	-.52	1.2	2.7	4.1	5.3	6.4	7.4	8.3	9.1	9.9
	0.3	-.93	0.06	0.85	1.5	2.1	2.6	3.0	3.4	3.7	4.0
	0.4	-1.1	-.37	0.16	0.55	0.86	1.1	1.3	1.5	1.6	1.8
	0.5	-1.1	-.49	-.06	0.25	0.48	0.66	0.79	0.90	1.0	1.1
	0.6	-1.2	-.55	-.15	0.12	0.31	0.45	0.56	0.65	0.71	0.77
0.8	0.2	-.27	1.3	2.7	4.0	5.2	6.3	7.3	8.2	9.0	9.7
	0.3	-.67	0.18	0.90	1.5	2.0	2.5	2.9	3.3	3.6	4.0
	0.4	-.85	-.27	0.16	0.49	0.75	0.97	1.2	1.3	1.4	1.6
	0.5	-.90	-.40	-.07	0.18	0.36	0.50	0.61	0.70	0.78	0.84
	0.6	-.92	-.46	-.16	0.04	0.18	0.29	0.37	0.44	0.49	0.53
	0.7	-.93	-.49	-.21	-.03	0.10	0.19	0.25	0.30	0.34	0.37
	0.8	-.93	-.50	-.24	-.07	0.05	0.13	0.19	0.23	0.27	0.29
1.0	0.2	-.26	1.2	2.6	3.9	5.1	6.1	7.1	8.0	8.8	9.5
	0.3	-.65	0.12	0.79	1.4	1.9	2.4	2.8	3.1	3.5	3.8
	0.4	-.83	-.34	0.04	0.33	0.58	0.78	0.95	1.1	1.2	1.3
	0.5	-.89	-.48	-.20	0	0.15	0.27	0.37	0.45	0.51	0.57
	0.6	-.91	-.54	-.31	-.14	-.03	0.06	0.12	0.18	0.22	0.25
	0.8	-.93	-.59	-.38	-.25	-.16	-.10	-.06	-.03	-01	0.01
	1.0	-.93	-.60	-.40	-.28	-.20	-.14	-.11	-.08	-.07	-.06

Main, $C_{c,s}$

$\dfrac{A_s}{A_c}$	$\dfrac{A_b}{A_c}$	Q_b/Q_s									
		0.2	0.4	0.6	0.8	1.0	1.2	1.4	1.6	1.8	2.0
0.3	0.2	4.5	2.8	1.5	0.56	−.17	−.74	−1.2	−.16	−1.9	−2.1
	0.3	4.6	3.1	2.0	1.2	0.57	0.08	−.30	−.62	−.89	−1.1
0.4	0.2	1.6	0.85	0.16	−.43	−.92	−1.3	−1.7	−1.9	−2.2	−2.4
	0.3	1.7	1.1	0.58	0.13	−.24	−.56	−.82	−1.1	−1.3	−1.4
	0.4	1.8	1.3	0.80	0.42	0.11	−.15	−.37	−.55	−.72	−.86
0.5	0.2	0.67	0.18	−.33	−.79	−1.2	−1.5	−1.8	−2.1	−2.3	−2.5
	0.3	0.75	0.42	0.07	−.25	−.54	−.80	−1.0	−1.2	−1.4	−1.5
	0.4	0.80	0.55	0.28	0.03	−.20	−.40	−.57	−.73	−.86	−.98
	0.5	0.82	0.62	0.41	0.20	0.02	−.15	−.29	−.42	−.53	−.63
0.6	0.2	0.26	−.11	−.54	−.95	−1.3	−1.6	−1.9	−2.1	−2.4	−2.5
	0.3	0.34	0.13	−.14	−.42	−.67	−.90	−1.1	−1.3	−1.4	−1.6
	0.4	0.39	0.25	0.06	−.14	−.33	−.51	−.66	−.80	−.93	−1.0
	0.5	0.41	0.32	0.18	0.03	−1.2	−.26	−.38	−.50	−.60	−.69
	0.6	0.43	0.37	0.26	0.14	0.02	−.09	−.19	−.29	−.37	−.45
0.8	0.2	−.01	−.30	−.67	−1.1	−1.4	−1.7	−2.0	−2.2	−2.4	−2.6
	0.3	0.07	−.07	−.29	−.58	−.76	−.97	−1.2	−1.3	−1.5	−1.6
	0.4	0.11	0.05	−.09	−.26	−.42	−.58	−.72	−.85	−.97	−1.1
	0.5	0.14	0.12	0.03	−.09	−.21	−.34	−.45	−.55	−.64	−.73
	0.6	0.15	0.17	0.11	0.02	−.07	−.17	−.26	−.34	−.42	−.49
	0.7	0.17	0.21	0.17	0.11	0.03	−.05	−.12	−.19	−.26	−.32
	0.8	0.17	0.23	0.22	0.17	0.11	0.05	−.02	−.07	−.13	−.18
1.0	0.2	−.05	−.33	−.70	−1.1	−1.4	−1.7	−2.0	−2.2	−2.4	−2.6
	0.3	0.03	−.10	−.31	−.55	−.78	−.98	−1.2	−1.3	−1.5	−1.6
	0.4	0.07	0.02	−.12	−.28	−.44	−.59	−.73	−.86	−.98	−1.1
	0.5	0.09	0.09	0.01	−.11	−.23	−.35	−.46	−.56	−.65	−.74
	0.6	0.11	0.14	0.09	0	−.09	−.18	−.27	−.35	−.43	−.50
	0.8	0.13	0.20	0.19	0.15	0.09	0.03	−.03	−.08	−.14	−.19
	1.0	0.14	0.24	0.25	0.24	0.20	0.16	0.12	0.08	0.04	0

6-5 Converging Wye (45°), Conical, Round[9]

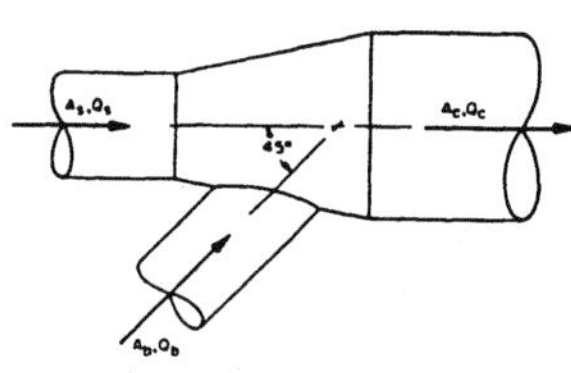

Main, $C_{c,s}$

$\dfrac{A_s}{A_c}$	$\dfrac{A_b}{A_c}$	Q_b/Q_s									
		0.2	0.4	0.6	0.8	1.0	1.2	1.4	1.6	1.8	2.0
0.3	0.2	5.3	−.01	2.0	1.1	0.34	−.20	−.61	−.93	−1.2	−1.4
	0.3	5.4	3.7	2.5	1.6	1.0	0.53	0.16	−.14	−.38	−.58
0.4	0.2	1.9	1.1	0.46	−.07	−.49	−.83	−1.1	−1.3	−1.5	−1.7
	0.3	2.0	1.4	0.81	0.42	0.08	−.20	−.43	−.62	−.78	−.92
	0.4	2.0	1.5	1.0	0.68	0.39	0.16	−.04	−.21	−.35	−.47
0.5	0.2	0.77	0.34	−.09	−.48	−.81	−1.1	1.3	−1.5	−1.7	−1.8
	0.3	0.85	0.56	0.25	−.03	−.27	−.48	−.67	−.82	−.96	−1.1
	0.4	0.88	0.66	0.43	0.21	0.02	−.15	−.30	−.42	−.54	−.64
	0.5	0.91	0.73	0.54	0.36	0.21	0.06	−.06	−.17	−.26	−.35
0.6	0.2	0.30	0	−.34	−.67	−.96	−1.2	−1.4	−1.6	−1.8	−1.9
	0.3	0.37	0.21	−.02	−.24	−.44	−.63	−.79	−.93	−1.1	−1.2
	0.4	0.40	0.31	0.16	−0.1	−.16	−.30	−.43	−.54	−.64	−.73
	0.5	0.43	0.37	0.26	0.14	0.02	−.09	−.20	−.29	−.37	−.45
	0.6	0.44	0.41	0.33	0.24	0.14	0.05	−.03	−.11	−.18	−.25
0.8	0.2	−.06	−.27	−.57	−.86	−1.1	−1.4	−1.6	−1.7	−1.9	−2.0
	0.3	0	−.08	−.25	−.43	−.62	−.78	−.93	−1.1	−1.2	−1.3
	0.4	0.04	0.02	−.08	−.21	−.34	−.46	−.57	−.67	−.77	−.85
	0.5	0.06	0.08	0.02	−.06	−.16	−.25	−.34	−.42	−.50	−.57
	0.6	0.07	0.12	0.09	0.03	−.04	−.11	−.18	−.25	−.31	−.37
	0.7	0.08	0.15	0.14	0.10	0.05	−.01	−.07	−.12	−.17	−.22
	0.8	0.09	0.17	0.18	0.16	0.11	0.07	0.02	−.02	−.07	−.11
1.0	0.2	−.19	−.39	−.67	−.96	−1.2	−1.5	−1.6	−1.8	−2.0	−2.1
	0.3	−.12	−.19	−.35	−.54	−.71	−.87	−1.0	−1.2	−1.3	−1.4
	0.4	−.09	−.10	−.19	−.31	−.43	−.55	−.66	−.77	−.86	−.94
	0.5	−.07	−.04	−.09	−.17	−.26	−.35	−.44	−.52	−.59	−.66
	0.6	−.06	0	−.02	−.07	−.14	−.21	−.28	−.34	−.40	−.46
	0.8	−.04	0.06	0.07	0.05	0.02	−.03	−.07	−.12	−.16	−.20
	1.0	−.03	0.09	0.13	0.13	0.11	0.08	0.06	0.03	−.01	−.03

6-6 Converging Wye, Rectangular[1]

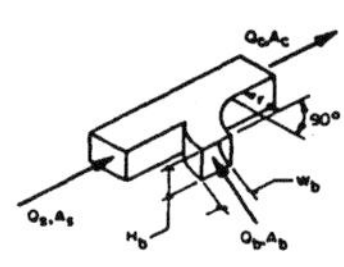

Branch, $C_{c,b}$

$\dfrac{A_s}{A_c}$	$\dfrac{A_b}{A_c}$	Q_b/Q_s									
		0.2	0.4	0.6	0.8	1.0	1.2	1.4	1.6	1.8	2.0
0.3	0.2	−2.4	−.01	2.0	3.8	5.3	6.6	7.8	8.9	9.8	11
	0.3	−2.8	−1.2	0.12	1.1	1.9	2.6	3.2	3.7	4.2	4.6
0.4	0.2	−1.2	0.93	2.8	4.5	5.9	7.2	8.4	9.5	10	11
	0.3	−1.6	−.27	0.81	1.7	2.4	3.0	3.6	4.1	4.5	4.9
	0.4	−1.8	−.72	0.07	0.66	1.1	1.5	1.8	2.1	2.3	2.5
0.5	0.2	−.46	1.5	3.3	4.9	6.4	7.7	8.8	9.9	11	12
	0.3	−.94	0.25	1 2	2.0	2.7	3.3	3.8	4.2	4.7	5.0
	0.4	−1.1	−.24	0.42	0.92	1.3	1.6	1.9	2.1	2.3	2.5
	0.5	−1.2	−.38	0.18	0.58	0.88	1.1	1.3	1.5	1.6	1.7
0.6	0.2	−.55	1.3	3.1	4.7	6.1	7.4	8.6	9.6	11	12
	0.3	−1.1	0	0.88	1.6	2.3	2.8	3.3	3.7	4.1	4.5
	0.4	−1.2	−.48	0.10	0.54	0.89	1.2	1.4	1.6	1.8	2.0
	0.5	−1.3	−.62	−.14	0.21	0.47	0.68	0.85	0.99	1.1	1.2
	0.6	−1.3	−.69	−.26	0.04	0.26	0.42	0.57	0.66	0.75	0.82
0.8	0.2	0.06	1.8	3.5	5.1	6.5	7.8	8.9	10	11	12
	0.3	−.52	0.35	1.1	1.7	2.3	2.8	3.2	3.6	3.9	4.2
	0.4	−.67	−.05	0.43	0.80	1.1	1.4	1.6	1.8	1.9	2.1
	0.6	−.75	−.27	0.05	0.28	0.45	0.58	0.68	0.76	0.83	0.88
	0.7	−.77	−.31	−.02	0.18	0.32	0.43	0.50	0.56	0.61	0.65
	0.8	−.78	−.34	−.07	0.12	0.24	0.33	0.39	0.44	0.47	0.50
1.0	0.2	0.40	2.1	3.7	5.2	6.6	7.8	9.0	11	11	12
	0.3	−.21	0.54	1.2	1.8	2.3	2.7	3.1	3.7	3.7	4.0
	0.4	−.33	0.21	0.62	0.96	1.2	1.5	1.7	2.0	2.0	2.1
	0.5	−.38	0.05	0.37	0.60	0.79	0.93	1.1	1.2	1.2	1.3
	0.6	−.41	−.02	0.23	0.42	0.55	0.66	0.73	0.80	0.85	0.89
	0.8	−.44	−.10	0.11	0.24	0.33	0.39	0.43	0.46	0.47	0.48
	1.0	−.46	−.14	0.05	0.16	0.23	0.27	0.29	0.30	0.30	0.29

$r/w_b = 1$

Branch, $C_{c,b}$

A_b/A_s	A_b/A_c	Q_b/Q_c								
		0.1	0.2	0.3	0.4	0.5	0.6	0.7	0.8	0.9
0.25	0.25	−.50	0	0.50	1.2	2.2	3.7	5.8	8.4	11
0.33	0.25	−1.2	−.40	0.40	1.6	3.0	4.8	6.8	8.9	11
0.5	0.5	−.50	−.20	0	0.25	0.45	0.70	1.0	1.5	2.0
0.67	0.5	−1.0	−.60	−.20	0.10	0.30	0.60	1.0	1.5	2.0
1.0	0.5	−2.2	−1.5	−.95	−.50	0	0.40	0.80	1.3	1.9
1.0	1.0	−.60	−.30	−.10	−.04	0.13	0.21	0.29	0.36	0.42
1.33	1.0	−1.2	−.80	−.40	−.20	0	0.16	0.24	0.32	0.38
2.0	1.0	−2.1	−1.4	−.90	−.50	−.20	0	0.20	0.25	0.30

Main, $C_{c,s}$

A_s/A_c	A_b/A_c	Q_b/Q_c								
		0.1	0.2	0.3	0.4	0.5	0.6	0.7	0.8	0.9
0.75	0.25	0.30	0.30	0.20	−.10	−.45	−.92	−1.5	−2.0	−2.6
1.0	0.5	0.17	0.16	0.10	0	−0.08	−.18	−.27	−.37	−.46
0.75	0.5	0.27	0.35	0.32	0.25	0.12	−.03	−.23	−.42	−.58
0.5	0.5	1.2	1.1	0.90	0.65	0.35	0	−.40	−.80	−1.3
1.0	1.0	0.18	0.24	0.27	0.26	0.23	0.18	0.10	0	−.12
0.75	1.0	0.75	0.36	0.38	0.35	0.27	0.18	0.05	−.08	−.22
0.5	1.0	0.80	0.87	0.80	0.68	0.55	0.40	0.25	0.08	−.10

6-7 Converging Tee, Round Tap to Rectangular Main[7]

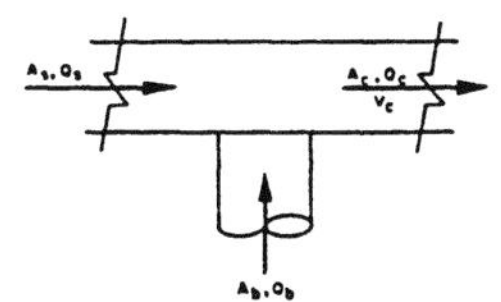

A_b/A_s	A_s/A_c	A_b/A_c
0.5	1.0	0.5

Branch, $C_{c,b}$										
V_c					Q_b/Q_c					
m/s	0.1	0.2	0.3	0.4	0.5	0.6	0.7	0.8	0.9	1.0
<6	−.63	−.55	0.13	0.23	0.78	1.30	1.93	3.10	4.88	5.60
>6	−.49	−.21	0.23	0.60	1.27	2.06	2.75	3.70	4.93	5.95

For main coefficient ($C_{c,s}$), see Fitting 6-3.

6-8 Converging Tee, Rectangular, Main and Tap[7]

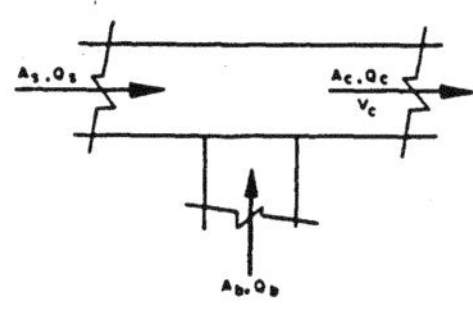

A_b/A_s	A_s/A_c	A_b/A_c
0.5	1.0	0.5

Branch, $C_{c,b}$										
V_c					Q_b/Q_c					
m/s	0.1	0.2	0.3	0.4	0.5	0.6	0.7	0.8	0.9	1.0
<6	−.75	−.53	−.03	0.33	1.03	1.10	2.15	2.93	4.18	4.78
>6	−.69	−.21	0.23	0.67	1.17	1.66	2.67	3.36	3.93	5.13

For main coefficient ($C_{c,s}$), see Fitting 6-3.

6-9 Converging Tee, 45° Entry, Rectangular, Main and Tap[7]

A_b/A_s	A_s/A_c	A_b/A_c
0.5	1.0	0.5

Branch, $C_{c,b}$										
V_c					Q_b/Q_c					
(m/s)	0.1	0.2	0.3	0.4	0.5	0.6	0.7	0.8	0.9	1.0
<6	−.83	−.68	−.30	0.28	0.55	1.03	1.50	1.93	2.50	3.03
>6	−.72	−.52	−.23	0.34	0.76	1.14	1.83	2.01	2.90	3.63

For main coefficient ($C_{c,s}$), see Fitting 6-3.

6-10 Diverging 90° Conical Tee, Round[10]

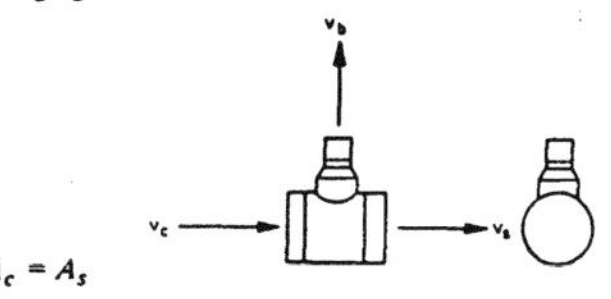

$A_c = A_s$

Branch											
V_b/V_c	0	0.2	0.4	0.6	0.8	1.0	1.2	1.4	1.6	1.8	2.0
$C_{c,b}$	1.0	0.85	0.74	0.62	0.52	0.42	0.36	0.32	0.32	0.37	0.52

For main loss coefficient ($C_{c,s}$), see Fitting 6-23.

6-11 Diverging 45° Conical Wye, Round[10]

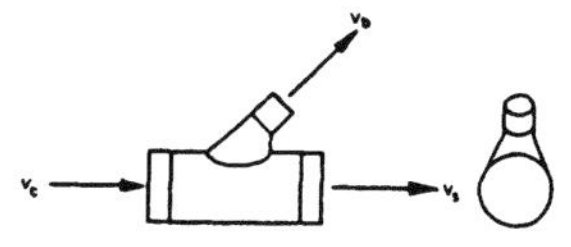

$A_c = A_s$

Branch											
V_b/V_c	0	0.2	0.4	0.6	0.8	1.0	1.2	1.4	1.6	1.8	2.0
$C_{c,b}$	1.0	0.84	0.61	0.41	0.27	0.17	0.12	0.12	0.14	0.18	0.27

For main loss coefficient ($C_{c,s}$), see Fitting 6-23.

6-12 Diverging 90° Tee, Round, Rolled 90° with 90° Elbow, Branch 90° to Main[10]

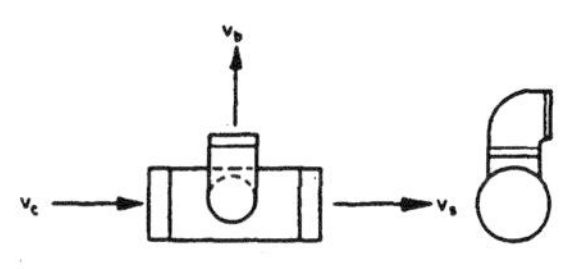

$A_c = A_s$

Branch											
V_b/V_c	0	0.2	0.4	0.6	0.8	1.0	1.2	1.4	1.6	1.8	2.0
$C_{c,b}$	1.0	1.03	1.08	1.18	1.33	1.56	1.86	2.2	2.6	3.0	3.4

For main loss coefficient ($C_{c,s}$), see Fitting 6-23.

6-13 Diverging 90° Tee, Round, Rolled 45° with 45° Elbow, Branch 90° to Main[10]

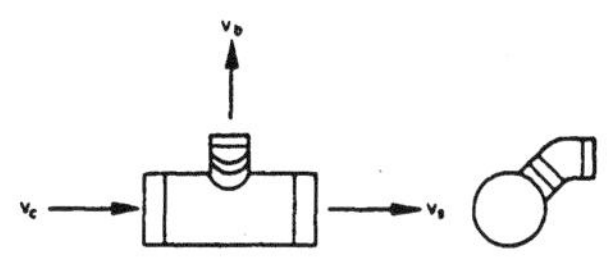

$A_c = A_s$

Branch											
V_b/V_c	0	0.2	0.4	0.6	0.8	1.0	1.2	1.4	1.6	1.8	2.0
$C_{c,b}$	1.0	1.32	1.51	1.60	1.65	1.74	1.87	2.0	2.2	2.5	2.7

For main loss coefficient ($C_{c,s}$), see Fitting 6-23.

6-14 Diverging 90° Conical Tee, Round, Rolled 45° with 45° Elbow, Branch 90° to Main[10]

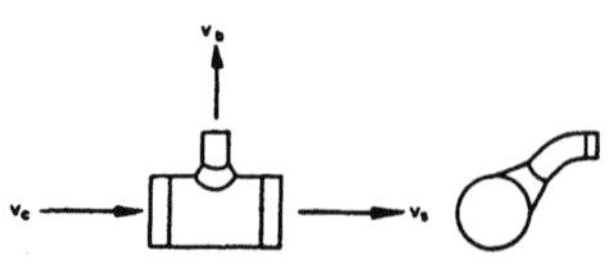

$A_c = A_s$

	Branch										
V_b/V_c	0	0.2	0.4	0.6	0.8	1.0	1.2	1.4	1.6	1.8	2.0
$C_{c,b}$	1.0	0.94	0.88	0.84	0.80	0.82	0.84	0.87	0.90	0.95	1.02

For main loss coefficient ($C_{c,s}$), see Fitting 6-23.

6-15 Diverging 45° Wye, Round, Rolled 45° with 60° Elbow, Branch 90° to Main[11]

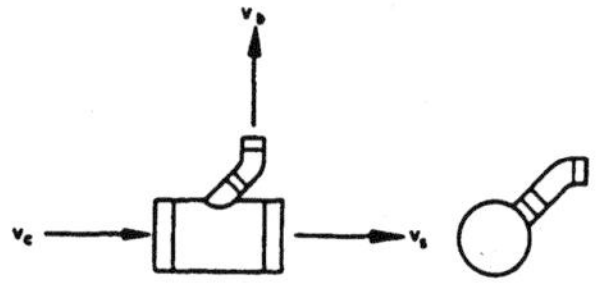

$A_c = A_s$

	Branch										
V_b/V_c	0	0.2	0.4	0.6	0.8	1.0	1.2	1.4	1.6	1.8	2.0
$C_{c,b}$	1.0	0.88	0.77	0.68	0.65	0.69	0.73	0.88	1.14	1.54	2.2

For main loss coefficient ($C_{c,s}$), see Fitting 6-23.

6-16 Diverging 45° Conical Wye, Round, Rolled 45° with 60° Elbow, Branch 90° to Main[10]

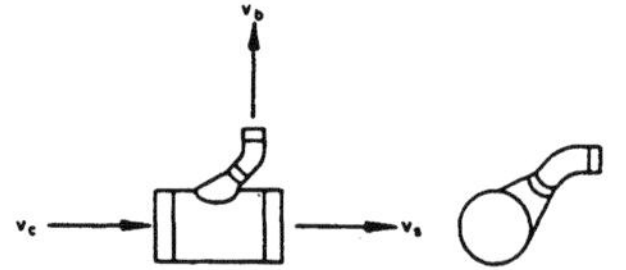

$A_c = A_s$

	Branch										
V_b/V_c	0	0.2	0.4	0.6	0.8	1.0	1.2	1.4	1.6	1.8	2.0
$C_{c,b}$	1.0	0.82	0.63	0.52	0.45	0.42	0.41	0.40	0.41	0.45	0.56

For main loss coefficient ($C_{c,s}$), see Fitting 6-23.

6-17 Diverging 45° Wye, Conical Main and Branch, with 45° Elbow, Branch 90° to Main[1]

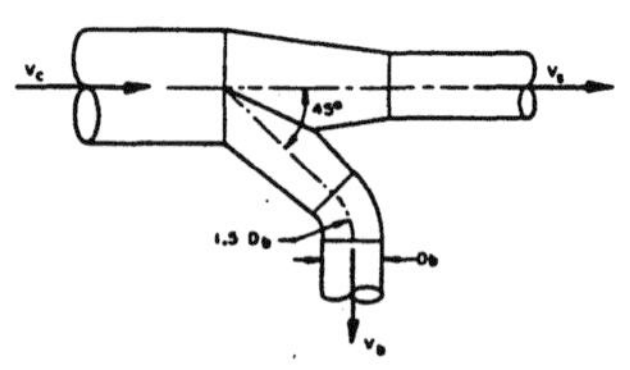

	Branch								
V_b/V_c	0.2	0.4	0.6	0.7	0.8	0.9	1.0	1.1	1.2
$C_{c,b}$	0.76	0.60	0.52	0.50	0.51	0.52	0.56	0.61	0.68
V_b/V_c	1.4	1.6	1.8	2.0	2.2	2.4	2.6	2.8	3.0
$C_{c,b}$	0.86	1.1	1.4	1.8	2.2	2.6	3.1	3.7	4.2

	Main									
V_s/V_c	0.2	0.4	0.6	0.8	1.0	1.2	1.4	1.6	1.8	2.0
$C_{c,s}$	0.14	0.06	0.05	0.09	0.18	0.30	0.46	0.64	0.84	1.0

6-18 Diverging 90° Tee, Round, Rolled 45° with 60° Elbow, Branch 45° to Main[10]

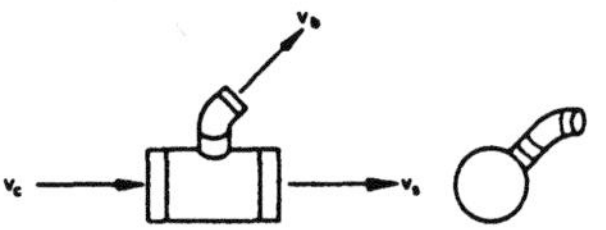

$A_c = A_s$

	Branch										
V_b/V_c	0	0.2	0.4	0.6	0.8	1.0	1.2	1.4	1.6	1.8	2.0
$C_{c,b}$	1.0	1.06	1.15	1.29	1.45	1.65	1.89	2.2	2.5	2.9	3.3

For main loss coefficient ($C_{c,s}$), see Fitting 6-23.

6-19 Diverging 90° Conical Tee, Round, Rolled 45° with 60° Elbow, Branch 45° to Main[10]

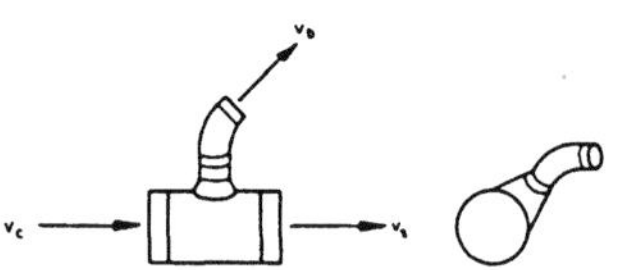

$A_c = A_s$

	Branch										
V_b/V_c	0	0.2	0.4	0.6	0.8	1.0	1.2	1.4	1.6	1.8	2.0
$C_{c,b}$	1.0	0.95	0.90	0.86	0.81	0.79	0.79	0.81	0.86	0.96	1.10

For main loss coefficient ($C_{c,s}$), see Fitting 6-23.

6-20 Diverging 45° Wye, Round, Rolled 45° with 30° Elbow, Branch 45° to Main[11]

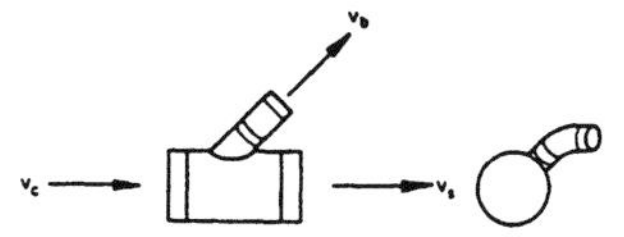

$$A_c = A_s$$

Branch											
V_b/V_c	0	0.2	0.4	0.6	0.8	1.0	1.2	1.4	1.6	1.8	2.0
$C_{c,b}$	1.0	0.84	0.72	0.62	0.54	0.50	0.56	0.71	0.92	1.22	1.66

For main loss coefficient ($C_{c,s}$), see Fitting 6-23.

6-21 Diverging 45° Conical Wye, Round, Rolled 45° with 30° Elbow, Branch 45° to Main[10]

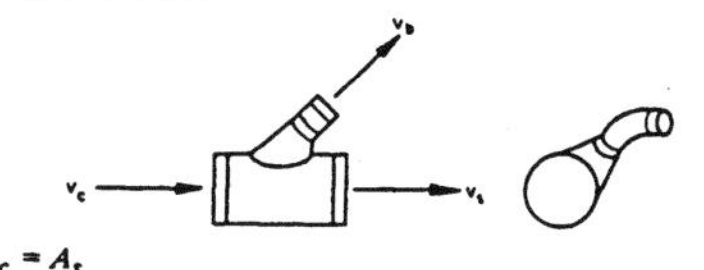

$$A_c = A_s$$

Branch											
V_b/V_c	0	0.2	0.4	0.6	0.8	1.0	1.2	1.4	1.6	1.8	2.0
$C_{c,b}$	1.0	0.93	0.71	0.55	0.44	0.42	0.42	0.44	0.47	0.54	0.62

For main loss coefficient ($C_{c,s}$), see Fitting 6-23.

6-22 Diverging Wye, Rectangular[1]

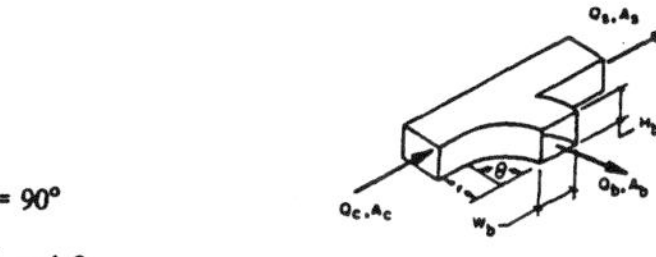

$$\theta = 90°$$

$$r/W_b = 1.0$$

Branch $C_{c,b}$											
$\dfrac{A_b}{A_s}$	$\dfrac{A_b}{A_c}$				Q_b/Q_c						
		0.1	0.2	0.3	0.4	0.5	0.6	0.7	0.8	0.9	
0.25	0.25	0.55	0.50	0.60	0.85	1.2	1.8	3.1	4.4	6.0	
0.33	0.25	0.35	0.35	0.50	0.80	1.3	2.0	2.8	3.8	5.0	
0.5	0.5	0.62	0.48	0.40	0.40	0.48	0.60	0.78	1.1	1.5	
0.67	0.5	0.52	0.40	0.32	0.30	0.34	0.44	0.62	0.92	1.4	
1.0	0.5	0.44	0.38	0.38	0.41	0.52	0.68	0.92	1.2	1.6	
1.0	1.0	0.67	0.55	0.46	0.37	0.32	0.29	0.29	0.30	0.37	
1.33	1.0	0.70	0.60	0.51	0.42	0.34	0.28	0.26	0.26	0.29	
2.0	1.0	0.60	0.52	0.43	0.33	0.24	0.17	0.15	0.17	0.21	

Main, $C_{c,s}$											
$\dfrac{A_b}{A_s}$	$\dfrac{A_b}{A_c}$				Q_b/Q_c						
		0.1	0.2	0.3	0.4	0.5	0.6	0.7	0.8	0.9	
0.25	0.25	−.01	−.03	−.01	0.05	0.13	0.21	0.29	0.38	0.46	
0.33	0.25	0.08	0	−.02	−.01	0.02	0.08	0.16	0.24	0.34	
0.5	0.5	−.03	−.06	−.05	0	0.06	0.12	0.19	0.27	0.35	
0.67	0.5	0.04	−.02	−.04	−.03	−.01	0.04	0.12	0.23	0.37	
1.0	0.5	0.72	0.48	0.28	0.13	0.05	0.04	0.09	0.18	0.30	
1.0	1.0	−.02	−.04	−.04	−.01	0.06	0.13	0.22	0.30	0.38	
1.33	1.0	0.10	0	0.01	−.03	−.01	0.03	0.10	0.20	0.30	
2.0	1.0	0.62	0.38	0.23	0.13	0.08	0.05	0.06	0.10	0.20	

6-23 Diverging Wye, Rectangular and Round[1]

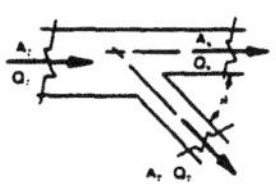

$A_c = A_s$; $H_b = H_c$, *where H is height of rectangular duct*

a. $\theta = 30°$

Branch, $C_{c,b}$									
				Q_b/Q_c					
A_b/A_c	0.1	0.2	0.3	0.4	0.5	0.6	0.7	0.8	0.9
0.8	0.75	0.55	0.40	0.28	0.21	0.16	0.15	0.16	0.19
0.7	0.72	0.51	0.36	0.25	0.18	0.15	0.16	0.20	0.26
0.6	0.69	0.46	0.31	0.21	0.17	0.16	0.20	0.28	0.39
0.5	0.65	0.41	0.26	0.19	0.18	0.22	0.32	0.47	0.67
0.4	0.59	0.33	0.21	0.20	0.27	0.40	0.62	0.92	1.3
0.3	0.55	0.28	0.24	0.38	0.76	1.3	2.0	—	—
0.2	0.40	0.26	0.58	1.3	2.5	—	—	—	—
0.1	0.28	1.5	—	—	—	—	—	—	—

b. $\theta = 45°$

Branch, $C_{c,b}$									
				Q_b/Q_c					
A_b/A_c	0.1	0.2	0.3	0.4	0.5	0.6	0.7	0.8	0.9
0.8	0.78	0.62	0.49	0.40	0.34	0.31	0.32	0.35	0.40
0.7	0.77	0.59	0.47	0.38	0.34	0.32	0.35	0.41	0.50
0.6	0.74	0.56	0.44	0.37	0.35	0.36	0.43	0.54	0.68
0.5	0.71	0.52	0.41	0.38	0.40	0.45	0.59	0.78	1.0
0.4	0.66	0.47	0.40	0.43	0.54	0.69	0.95	1.3	1.7
0.3	0.66	0.48	0.52	0.73	1.2	1.8	2.7	—	—
0.2	0.56	0.56	1.0	1.8	—	—	—	—	—
0.1	0.60	2.1	—	—	—	—	—	—	—

c. $\theta = 60°$

Branch, $C_{c,b}$									
				Q_b/Q_c					
A_b/A_c	0.1	0.2	0.3	0.4	0.5	0.6	0.7	0.8	0.9
0.8	0.83	0.71	0.62	0.56	0.52	0.50	0.53	0.60	0.68
0.7	0.82	0.69	0.61	0.56	0.54	0.54	0.60	0.70	0.82
0.6	0.81	0.68	0.60	0.58	0.58	0.61	0.72	0.87	1.1
0.5	0.79	0.66	0.61	0.62	0.68	0.76	0.94	1.2	1.5
0.4	0.76	0.65	0.65	0.74	0.89	1.1	1.4	1.8	2.3
0.3	0.80	0.75	0.89	1.2	1.8	2.6	3.5	—	—
0.2	0.77	0.96	1.6	2.5	—	—	—	—	—
0.1	1.0	2.9	—	—	—	—	—	—	—

d. $\theta = 90°$

Branch, $C_{c,b}$									
				Q_b/Q_c					
A_b/A_c	0.1	0.2	0.3	0.4	0.5	0.6	0.7	0.8	0.9
0.8	0.95	0.92	0.92	0.93	0.94	0.95	1.1	1.2	1.4
0.7	0.95	0.94	0.95	0.98	1.0	1.1	1.2	1.4	1.6
0.6	0.96	0.97	1.0	1.1	1.1	1.2	1.4	1.7	2.0
0.5	0.97	1.0	1.1	1.2	1.4	1.5	1.8	2.1	2.5
0.4	0.99	1.1	1.3	1.5	1.7	2.0	2.4	—	—
0.3	1.1	1.4	1.8	2.3	—	—	—	—	—
0.2	1.3	1.9	2.9	—	—	—	—	—	—
0.1	2.1	—	—	—	—	—	—	—	—

Main									
V_s/V_c	0	0.1	0.2	0.3	0.4	0.5	0.6	0.8	1.0
$C_{c,s}$	0.35	0.28	0.22	0.17	0.13	0.09	0.06	0.02	0

6-24 Diverging Wye, Rectangular[1]

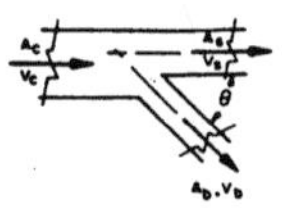

$$\theta = 15° \text{ to } 90°$$

$$A_c = A_s + A_b$$

Branch, $C_{c,b}$

θ, deg	V_b/V_c												
	0.1	0.2	0.3	0.4	0.5	0.6	0.8	1.0	1.2	1.4	1.6	1.8	2.0
15	0.81	0.65	0.51	0.38	0.28	0.20	0.11	0.06	0.14	0.30	0.51	0.76	1.0
30	0.84	0.69	0.56	0.44	0.34	0.26	0.19	0.15	0.15	0.30	0.51	0.76	1.0
45	0.87	0.74	0.63	0.54	0.45	0.38	0.29	0.24	0.23	0.30	0.51	0.76	1.0
60	0.90	0.82	0.79	0.66	0.59	0.53	0.43	0.36	0.33	0.39	0.51	0.76	1.0
90	1.0	1.0	1.0	1.0	1.0	1.0	1.0	1.0	1.0	1.0	1.0	1.0	1.0

Main, $C_{c,s}$

θ, deg	15—60	90				
$\dfrac{V_s}{V_c}$	A_s/A_c					
	0–1.0	0–0.4	0.5	0.6	0.7	≥0.8
0	1.0	1.0	1.0	1.0	1.0	1.0
0.1	0.81	0.81	0.81	0.81	0.81	0.81
0.2	0.64	0.64	0.64	0.64	0.64	0.64
0.3	0.50	0.50	0.52	0.52	0.50	0.50
0.4	0.36	0.36	0.40	0.38	0.37	0.36
0.5	0.25	0.25	0.30	0.28	0.27	0.25
0.6	0.16	0.16	0.23	0.20	0.18	0.16
0.8	0.04	0.04	0.17	0.10	0.07	0.04
1.0	0	0	0.20	0.10	0.05	0
1.2	0.07	0.07	0.36	0.21	0.14	0.07
1.4	0.39	0.39	0.79	0.59	0.39	—
1.6	0.90	0.90	1.4	1.2	—	—
1.8	1.8	1.8	2.4	—	—	—
2.0	3.2	3.2	4.0	—	—	—

6-25 Diverging Tee, Pyramidal Main and Take-off with 45° Leading Edge[12]

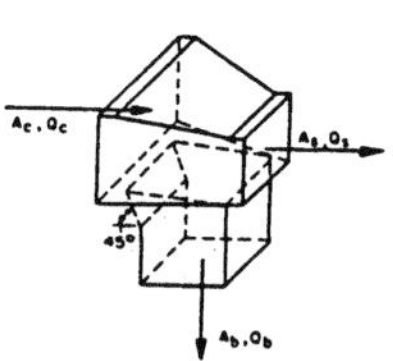

$$0.5 \leqslant A_b/A_c \leqslant 1.0$$
$$0.5 \leqslant A_s/A_c \leqslant 1.0$$

	Branch										
Q_b/Q_c	0	0.1	0.2	0.3	0.4	0.5	0.6	0.7	0.8	0.9	1.0
$C_{c,b}$	1.4	1.2	0.96	0.82	0.68	0.56	0.49	0.47	0.48	0.50	0.54

6-26 Diverging Tee, Rectangular Main to Round Tap[7]

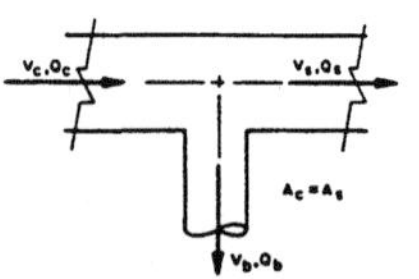

$$A_c = A_s$$

Branch, $C_{c,b}$

V_b/V_c	Q_b/Q_c								
	0.1	0.2	0.3	0.4	0.5	0.6	0.7	0.8	0.9
0.2	1.00								
0.4	1.01	1.07							
0.6	1.14	1.10	1.08						
0.8	1.18	1.31	1.12	1.13					
1.0	1.30	1.38	1.20	1.23	1.26				
1.2	1.46	1.58	1.45	1.31	1.39	1.48			
1.4	1.70	1.82	1.65	1.51	1.56	1.64	1.71		
1.6	1.93	2.06	2.00	1.85	1.70	1.76	1.80	1.88	
1.8	2.06	2.17	2.20	2.13	2.06	1.98	1.99	2.00	2.07

For main coefficient ($C_{c,s}$), see Fitting 6-23.

6-27 Diverging Tee, Rectangular Main to Round Tap[13]

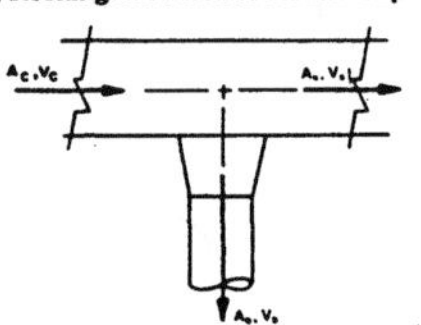

$$A_c = A_s$$

	Branch					
V_b/V_c	0.40	0.50	0.75	1.0	1.3	1.5
$C_{c,b}$	0.80	0.83	0.90	1.0	1.1	1.4

For main coefficient ($C_{c,s}$), see Fitting 6-23.

6-28 Diverging Tee, Rectangular Main and Tap[7]

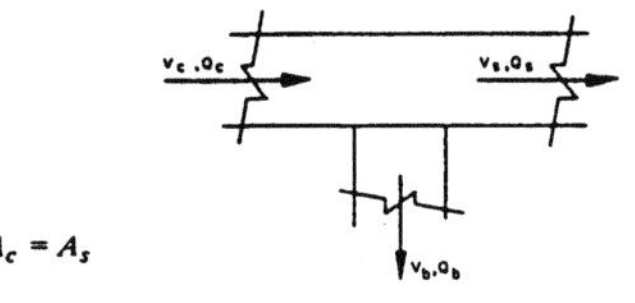

$$A_c = A_s$$

Branch, $C_{c,b}$

V_b/V_c	Q_b/Q_c								
	0.1	0.2	0.3	0.4	0.5	0.6	0.7	0.8	0.9
0.2	1.03								
0.4	1.04	1.01							
0.6	1.11	1.03	1.05						
0.8	1.16	1.21	1.17	1.12					
1.0	1.38	1.40	1.30	1.36	1.27				
1.2	1.52	1.61	1.68	1.91	1.47	1.66			
1.4	1.79	2.01	1.90	2.31	2.28	2.20	1.95		
1.6	2.07	2.28	2.13	2.71	2.99	2.81	2.09	2.20	
1.8	2.32	2.54	2.64	3.09	3.72	3.48	2.21	2.29	2.57

For main coefficient ($C_{c,s}$), see Fitting 6-23.

6-29 Diverging Tee, 45° Entry, Rectangular Main and Tap[7]

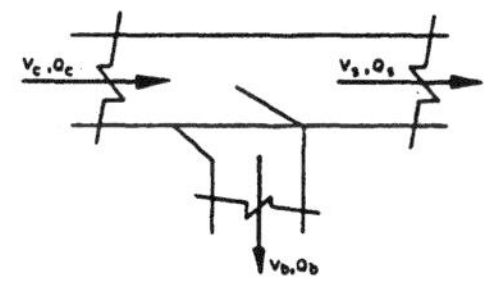

$A_c = A_s$

| | Branch, $C_{c,b}$ | | | | | | | | |
| | Q_b/Q_c | | | | | | | | |
V_b/V_c	0.1	0.2	0.3	0.4	0.5	0.6	0.7	0.8	0.9
0.2	0.91								
0.4	0.81	0.79							
0.6	0.77	0.72	0.70						
0.8	0.78	0.73	0.69	0.66					
1.0	0.78	0.98	0.85	0.79	0.74				
1.2	0.90	1.11	1.16	1.23	1.03	0.86			
1.4	1.19	1.22	1.26	1.29	1.54	1.25	0.92		
1.6	1.35	1.42	1.55	1.59	1.63	1.50	1.31	1.09	
1.8	1.44	1.50	1.75	1.74	1.72	2.24	1.63	1.40	1.17

For main coefficient ($C_{c,s}$), see Fitting 6-23.

6-30 Diverging Tee, 45° Entry, Rectangular Main and Tape, with Damper[7]

$A_c = A_s$

| | Branch, $C_{c,b}$ | | | | | | | | |
| | Q_b/Q_c | | | | | | | | |
V_b/V_c	0.1	0.2	0.3	0.4	0.5	0.6	0.7	0.8	0.9
0.2	0.61								
0.4	0.46	0.61							
0.6	0.43	0.50	0.54						
0.8	0.39	0.43	0.62	0.53					
1.0	0.34	0.57	0.77	0.73	0.68				
1.2	0.37	0.64	0.85	0.98	1.07	0.83			
1.4	0.57	0.71	1.04	1.16	1.54	1.36	1.18		
1.6	0.89	1.08	1.28	1.30	1.69	2.09	1.81	1.47	
1.8	1.33	1.34	2.04	1.78	1.90	2.40	2.77	2.23	1.92

For main coefficient ($C_{c,s}$), see Fitting 6-32.

6-31 Diverging Tee, Rectangular Main and Tap, with Damper[7]

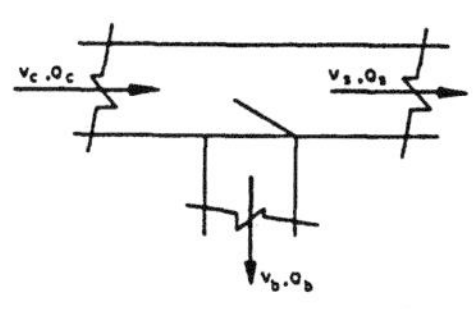

$A_c = A_s$

| | Branch, $C_{c,b}$ | | | | | | | | |
| | Q_b/Q_c | | | | | | | | |
V_b/V_c	0.1	0.2	0.3	0.4	0.5	0.6	0.7	0.8	0.9
0.2	0.58								
0.4	0.67	0.64							
0.6	0.78	0.76	0.75						
0.8	0.88	0.98	0.81	1.01					
1.0	1.12	1.05	1.08	1.18	1.29				
1.2	1.49	1.48	1.40	1.51	1.70	1.91			
1.4	2.10	2.21	2.25	2.29	2.32	2.48	2.53		
1.6	2.72	3.30	2.84	3.09	3.30	3.19	3.29	3.16	
1.8	3.42	4.58	3.65	3.92	4.20	4.15	4.14	4.10	4.05

For main coefficient ($C_{c,s}$), see Fitting 6-32.

6-32 Diverging Tee, Rectangular, with Extractor[7]

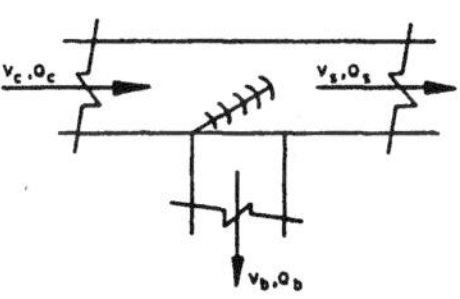

$A_c = A_s$

| | Branch, $C_{c,b}$ | | | | | | | | |
| | Q_b/Q_c | | | | | | | | |
V_b/V_c	0.1	0.2	0.3	0.4	0.5	0.6	0.7	0.8	0.9
0.2	0.60								
0.4	0.62	0.69							
0.6	0.74	0.80	0.82						
0.8	0.99	1.10	0.95	0.90					
1.0	1.48	1.12	1.41	1.24	1.21				
1.2	1.91	1.33	1.43	1.52	1.55	1.64			
1.4	2.47	1.67	1.70	2.04	1.86	1.98	2.47		
1.6	3.17	2.40	2.33	2.53	2.31	2.51	3.13	3.25	
1.8	3.85	3.37	2.89	3.23	3.09	3.03	3.30	3.74	4.11

	Main								
V_b/V_c	0.2	0.4	0.6	0.8	1.0	1.2	1.4	1.6	1.8
$C_{c,s}$	0.03	0.04	0.07	0.12	0.13	0.14	0.27	0.30	0.25

6-33 Symmetrical Wye, Dovetail, Rectangular[1]

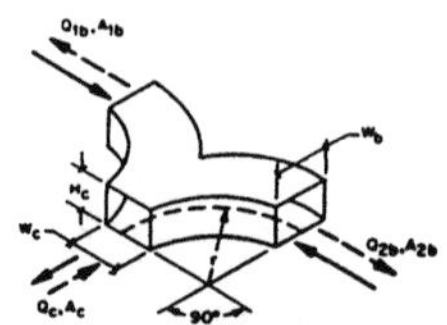

$$r/W_c = 1.5$$

$$Q_{1b}/Q_c = Q_{2b}/Q_c = 0.5$$

Converging		
A_{1b}/A_c or A_{2b}/A_c	0.50	1.0
$C_{c,1b}$ or $C_{c,2b}$	0.23	0.07
Diverging		
A_{1b}/A_c or A_{2b}/A_c	0.50	1.0
$C_{c,1b}$ or $C_{c,2b}$	0.30	0.25

6-34 Wye, Rectangular and Round[1]

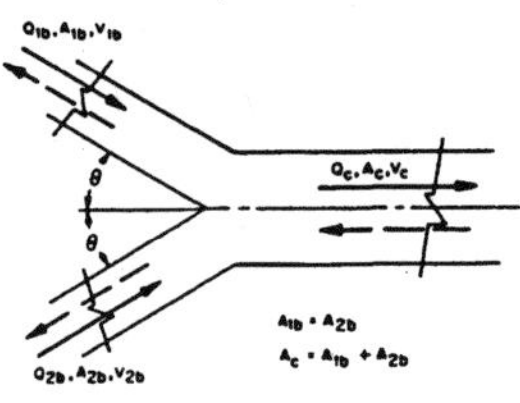

Converging

$C_{c,1b}$ or $C_{c,2b}$

θ, deg	Q_{1b}/Q_c or Q_{2b}/Q_c										
	0	0.1	0.2	0.3	0.4	0.5	0.6	0.7	0.8	0.9	1.0
15	-2.6	-1.9	-1.3	-.77	-.30	0.10	0.41	0.67	0.85	0.97	1.0
30	-2.1	-1.5	-1.0	-.53	-.10	0.20	0.69	0.91	1.1	1.4	1.6
45	-1.3	-.93	-.55	-.16	0.20	0.56	0.92	1.3	1.6	2.0	2.3

Diverging

$C_{c,1b}$ or $C_{c,2b}$

θ, deg	V_{1b}/V_c or V_{2b}/V_c												
	0.1	0.2	0.3	0.4	0.5	0.6	0.8	1.0	1.2	1.4	1.6	1.8	2.0
15	0.81	0.65	0.51	0.38	0.28	0.20	0.11	0.06	0.14	0.30	0.51	0.76	1.0
30	0.84	0.69	0.56	0.44	0.34	0.26	0.19	0.15	0.15	0.30	0.51	0.76	1.0
45	0.87	0.74	0.63	0.54	0.45	0.38	0.29	0.24	0.23	0.30	0.51	0.76	1.0
60	0.90	0.82	0.79	0.66	0.59	0.53	0.43	0.36	0.33	0.39	0.51	0.76	1.0
90	1.0	1.0	1.0	1.0	1.0	1.0	1.0	1.0	1.0	1.0	1.0	1.0	1.0

6-35 45° Double Wye, Rectangular and Round[1]

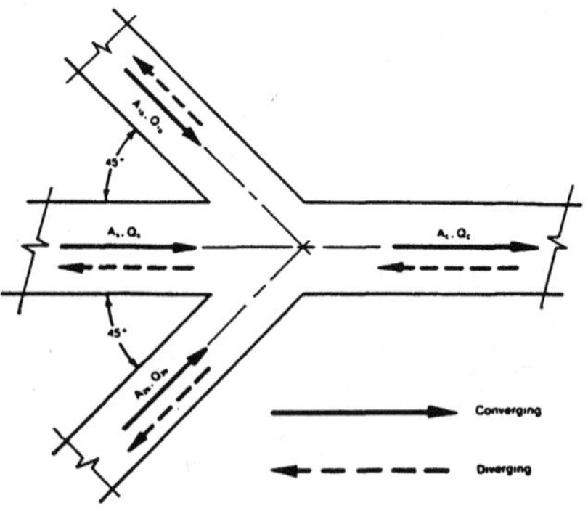

$$A_{1b} = A_{2b}$$
$$A_s = A_c$$

a. Converging Flow

Branch, $C_{c,b}$							
$\dfrac{Q_{2b}}{Q_{1b}}$	Q_{1b}/Q_c						
	0	0.1	0.2	0.3	0.4	0.5	0.6
$A_{1b}/A_c = 0.2$							
0.5	-1.0	-.36	0.59	1.8	3.2	4.9	6.8
1.0	-1.0	-.24	0.63	1.7	2.6	3.7	—
2.0	-1.0	-.19	0.21	0.04	—	—	—
$A_{1b}/A_c = 0.4$							
0.5	-1.0	-.48	-.02	0.58	0.92	1.3	16
1.0	-1.0	-.36	0.17	0.55	0.72	0.78	—
2.0	-1.0	-.18	0.16	-.06	—	—	—
$A_{1b}/A_c = 0.6$							
0.5	-1.0	-.50	-.07	0.31	0.60	0.82	0.92
1.0	-1.0	-.37	0.12	0.55	0.60	0.52	—
2.0	-1.0	-.18	0.26	0.16	—	—	—
$A_{1b}/A_c = 1.0$							
0.5	-1.0	-.51	-.09	0.25	0.50	0.65	0.64
1.0	-1.0	-.37	0.13	0.46	0.61	0.54	—
2.0	-1.0	-.15	0.38	0.42	—	—	—

Main, $C_{c,s}$											
$\dfrac{A_{2b}}{A_{1b}}$	Q_s/Q_c										
	0	0.1	0.2	0.3	0.4	0.5	0.6	0.7	0.8	0.9	1.0
$A_{1b}/A_c = 0.2$											
0.5 & 2.0	-2.9	-1.9	-1.3	-.80	-.56	-.23	-.01	0.16	0.22	0.15	0
1.0	-2.5	-1.9	-1.3	-.80	-.42	-.12	0.08	0.20	0.22	0.15	0
$A_{1b}/A_c = 0.4$											
0.5 & 2.0	-.98	-.61	-.30	-.05	0.14	0.26	0.33	0.34	0.28	0.17	0
1.0	-.77	-.44	-.16	0.05	0.21	0.31	0.36	0.35	0.29	0.17	0
$A_{1b}/A_c = 0.6$											
0.5 & 2.0	-.32	0.08	0.11	0.27	0.37	0.43	0.44	0.40	0.31	0.18	0
1.0	-.18	-.04	0.21	0.34	0.42	0.46	0.46	0.41	0.31	0.18	0
$A_{1b}/A_c = 1.0$											
0.5 & 2.0	0.11	0.36	0.46	0.53	0.57	0.56	0.52	0.44	0.33	0.18	0
1.0	0.29	0.42	0.51	0.57	0.58	0.58	0.54	0.45	0.33	0.18	0

b. Diverging Flow: Use Fitting 6-23.

6-36 90° Cross, Rectangular and Round[1]

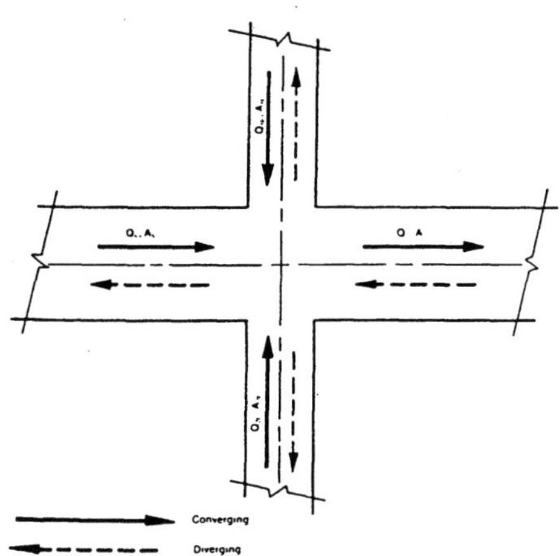

$A_{1b} = A_{2b}$

$A_s = A_c$

a. Converging Flow

			Branch, $C_{c,b}$				
$\frac{Q_{2b}}{Q_{1b}}$			Q_{1b}/Q_c or Q_{2b}/Q_c				
	0	0.1	0.2	0.3	0.4	0.5	0.6
			$A_{1b}/A_c = 0.2$				
0.5	−.85	−.10	1.1	2.7	4.8	7.3	10
1.0	−.85	−.05	1.4	3.1	5.1	7.4	—
2.0	−.85	−.31	1.8	3.4	—	—	—
			$A_{1b}/A_c = 0.4$				
0.5	−.85	−.29	0.34	1.0	1.8	2.6	3.4
1.0	−.85	−.14	0.60	1.3	2.1	2.7	—
2.0	−.85	0.12	1.0	1.7	—	—	—
			$A_{1b}/A_c = 0.6$				
0.5	−.85	−.32	0.20	0.72	1.2	1.7	2.1
1.0	−.85	−.18	0.46	1.0	1.5	1.9	—
2.0	−.85	0.09	0.88	1.4	—	—	—
			$A_{1b}/A_c = 0.8$				
0.5	−.85	−.33	0.13	0.61	1.0	1.4	1.7
1.0	−.85	−.18	0.41	0.91	1.3	1.5	—
2.0	−.85	0.08	0.83	1.3	—	—	—
			$A_{1b}/A_c = 1.0$				
0.5	−.85	−.34	0.13	0.56	0.93	1.3	1.5
1.0	−.85	−.19	0.39	0.86	1.2	1.4	—
2.0	−.85	0.07	0.81	1.2	—	—	—

			Main			
Q_s/Q_c	0	0.1	0.2	0.3	0.4	0.5
$C_{c,s}$	1.2	1.2	1.2	1.1	1.1	0.96
Q_s/Q_c	0.6		0.7	0.8	0.9	1.0
$C_{c,s}$	0.85		0.72	0.56	0.39	0.20

b. Diverging Flow: Use Fitting 6-23.

Table B-7 Local Loss Coefficients, OBSTRUCTIONS

7-1 Damper, Butterfly, Round[1]

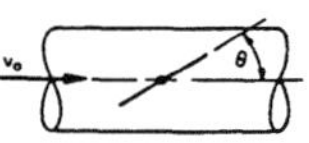

θ, deg	0	10	20	30	40	50	60
C_o	0.20	0.52	1.5	4.5	11	29	108

7-2 Damper, Butterfly, Rectangular[1]

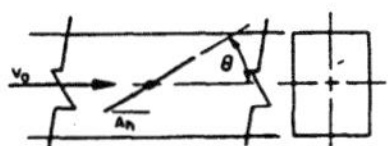

θ, deg	0	10	20	30	40	50	60
C_o	0.04	0.33	1.2	3.3	9.0	26	70

7-3 Damper, Gate, Round[1]

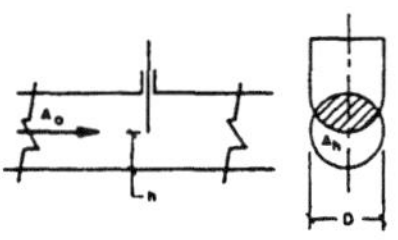

h/D	0.2	0.3	0.4	0.5	0.6	0.7	0.8	0.9
A_h/A_o	0.25	0.38	0.50	0.61	0.71	0.81	0.90	0.96
C_o	35	10	4.6	2.1	0.98	0.44	0.17	0.06

7-4 Damper, Gate, Rectangular[1]

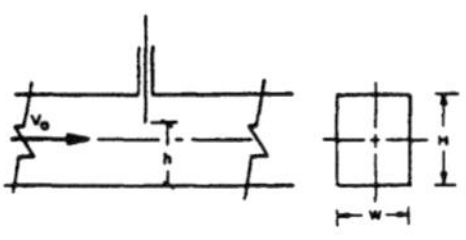

				h/H			
H/W	0.3	0.4	0.5	0.6	0.7	0.8	0.9
0.5	14	6.9	3.3	1.7	0.83	0.32	0.09
1.0	19	8.8	4.5	2.4	1.2	0.55	0.17
1.5	20	9.1	4.7	2.7	1.2	0.47	0.11
2.0	18	8.8	4.5	2.3	1.1	0.51	0.13

7-5 Damper, Butterfly, Airfoil Blade, Rectangular[1]

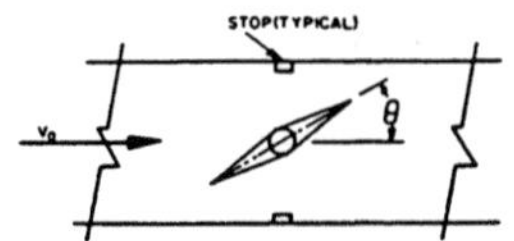

θ, deg	0	10	20	30	40	50	60
C_o	0.50	0.65	1.6	4.0	9.4	24	67

7-6 Damper, Rectangular, Parallel Blades[14]

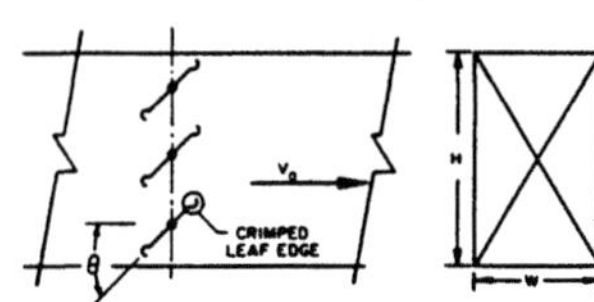

$$L/R = \frac{NW}{2\,(H + W)}$$

where

N = number of damper blades
W = duct dimension parallel to blade axis, mm
L = sum of damper blade lengths, mm
R = perimeter of duct, mm

$\frac{L}{R}$	C_o θ, degrees								
	80	70	60	50	40	30	20	10	0
0.3	116	32	14	9.0	5.0	2.3	1.4	0.79	0.52
0.4	152	38	16	9.0	5.0	2.4	1.5	0.85	0.52
0.5	188	45	18	9.0	5.0	2.4	1.5	0.92	0.52
0.6	245	45	21	9.0	5.4	2.4	1.5	0.92	0.52
0.8	284	55	22	9.0	5.4	2.5	1.5	0.92	0.52
1.0	361	65	24	10	5.4	2.6	1.6	1.0	0.52
1.5	576	102	28	10	5.4	2.7	1.6	1.0	0.52

7-7 Damper, Rectangular, Opposed Blades[14]

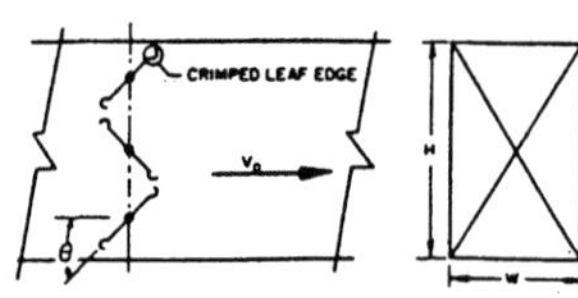

$$L/R = \frac{NW}{2\,(H + W)}$$

See Fitting 7-6 for definition of terms.

$\frac{L}{R}$	C_o θ, degrees								
	80	70	60	50	40	30	20	10	0
0.3	807	284	73	21	9.0	4.1	2.1	0.85	0.52
0.4	915	332	100	28	11	5.0	2.2	0.92	0.52
0.5	1045	377	122	33	13	5.4	2.3	1.0	0.52
0.6	1121	411	148	38	14	6.0	2.3	1.0	0.52
0.8	1299	495	188	54	18	6.6	2.4	1.1	0.52
1.0	1521	547	245	65	21	7.3	2.7	1.2	0.52
1.5	1654	677	361	107	28	9.0	3.2	1.4	0.52

7-8 Obstruction, Screen in Duct, Round, and Rectangular[1]

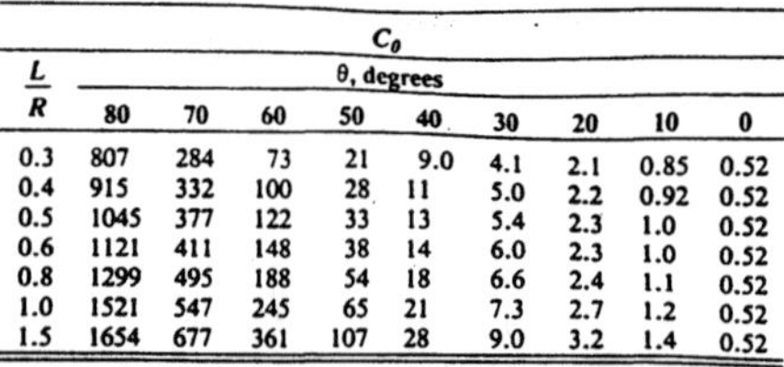

$$n = A_{or}/A_o$$

where

n = free area ratio of screen, dimensionless
A_{or} = total flow area of screen, mm^2
A_o = area of duct, mm^2

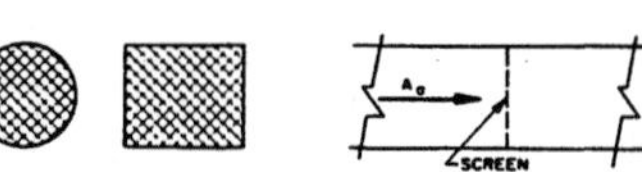

n	0.30	0.40	0.50	0.55	0.60	0.65	0.70	0.75	0.80	0.90	1.0
C_o	6.2	3.0	1.7	1.3	0.97	0.75	0.58	0.44	0.32	0.14	0

7-9 Obstruction, Perforated Plate, Thick, Round and Rectangular[1]

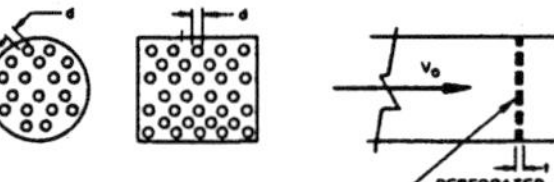

$t/d \geqslant 0.015$
$A_{or} = \pi\, d^2/4$
$n = \Sigma\, A_{or}/A_o$

where

A_o = area of duct, mm^2
A_{or} = orifice area, mm^2
d = diameter of perforated hole, mm
n = free area ratio of plate, dimensionless
t = plate thickness, mm

t/d	C_o n								
	0.20	0.25	0.30	0.40	0.50	0.60	.70	0.80	0.90
0.015	52	30	18	8.2	4.0	2.0	0.97	0.42	0.13
0.2	48	28	17	7.7	3.8	1.9	0.91	0.40	0.13
0.4	46	27	17	7.4	3.6	1.8	0.88	0.39	0.13
0.6	42	24	15	6.6	3.2	1.6	0.80	0.36	0.13

APPENDIX A
SELECTED UNITS CONVERSION FACTORS
Use the listed conversion factors to go from IP to SI, or vice-versa.

From	Times	To	From	Times	To
Btu, IT	1.055	kJ	0.9479	Btu, IT
Btu/ft^2	11.4	kJ/m^2	0.08772	Btu/ft^2
Btu/gal	0.2785	kJ/liter	3.591	Btu/gal
Btu/hr	0.2931	watt	3.412	Btu/hr
Btu/hrftF	1.729	W/mK	0.5783	Btu/hrftF
$Btu/hrft^2$	3.169	W/m^2	0.3155	$Btu/hrft^2$
$Btu/hrft^2F$	5.678	W/m^2K	0.1761	$Btu/hrft^2F$
Btu/lb	2.326	kJ/kg	0.4300	Btu/lb
Btu/lbF	4187	J/kgK	0.0002388	Btu/lbF
cfm/watt	0.4719	m^3/skW	2.119	cfm/watt
foot	0.3048	m	3.281	foot
ft/min	0.005080	m/s	196.9	ft/min
ft^3/min	0.0004719	m^3/s	2119	ft^3/min
$gr/hrft^2inHg/ft$	1.460	$ng/sm^2kPa/m$	0.6850	$gr/hrft^2inHg/ft$
$hrft^2F/Btu$	0.1761	m^2K/W	5.678	$hrft^2F/Btu$
inch water	248.8	Pa	0.004109	in. water
kWh	3.600	MJ	0.2778	kWh
langley	41.86	kJ/m^2	0.02389	langley
langley/s	41.86	kW/m^2	0.02389	langley/s
lb	4.448	N	0.2248	lb
lb/in^2	6895.	Pa	0.0001450	lb/in^2
lb/min	0.007560	kg/s	132.3	lb/min
lb/ft^3	16.02	kg/m^3	0.06243	lb/ft^3
mile	1.609	km	0.6214	mile
therm	105.5	MJ	0.009479	therm

Chapter 2

1. 17 mg
2. 13.4 C, 77% rh
3. 26 C
4. 0.011 kg/kg; 55.1 kJ/kg; 0.943 m^3/s; 13.6 C
7. 64 kW; 0.012 kg/s; 2.7 m^3/s
9. -5 C

Chapter 3

1. 24.5 C; 11.5 C; 9.2 W/m^2
2. 7.1 W/m^2K, or 1.3 W/mK
3. for example, at 0.1 m^3/s, h = 3 W/m^2K
4. 118 W/m^2 with no sun; 106 W/m^2 with sun
5. solve by setting up an energy balance on a differential element of air in the duct.
6. 92 C
8. 10.7 C
9. upper

Chapter 4

1. 8.7%
2. 2.2 m^2K/W
3. 1.16 m^2K/W
5. wall loses 65% of heat
8. 0.3 kg/s

Chapter 5

1. 150 kW latent heat; 230 kW sensible heat; 380 kW total heat

Chapter 6

1. -12 C is estimated barn temperature at -30 C outdoors; supplemental heat need is approximately 150 kW
4. for outside rh of 80%, inside air temperature of 12 C, ventilation rate is 5.3 m^3/s

Chapter 8

1. for 30 mm width, x = 3.6 m; jet width = 0.7 m and pressure difference = 11 Pa
2. 0.012 m^3/s, 32 degress, 4.7 m, 56 Pa and 5.9 m/s
8. per cow: infiltration = $0.0028\Delta P^{0.67}$
9. 13 Pa

Chapter 9

1. 7 mm

Chapter 10

2. 13 C

Chapter 11

1. $v_1 = v_{wind} [(c_{pe1} - c_{pe2})/(1+(A_1/A_2)^2)]^{1/2}$
7. 7 m^3/s; 8 m^3/s
8. 15 Pa
9. -0.26

Chapter 12

1. approximately 37 Pa
3. 0.5 m^3/s
7. branch: 200 mm

Index

A

Absolute roughness, 352, 353
Absorptance, radiant energy, 80, 152
Acceptable Weather Space, 304
Adiabatic mixing, 27, 38
Adiabatic saturation, 18
Air
    composition, atmospheric, 7
    cooling, 31
    dehumidification, 24, 39
    density, 14
    distribution, 261
    dry, 7
    ducts, see Air duct design
    gas constant, 7
    infiltration, 245, 254, 263
    inlets, 242, 249
    jets, 230
    mixed flow, 281
    mixing, 261, 265
        adiabatic, 27, 38
        dispersed plug flow, 280
        factor, 282
        idealized, 279
        incomplete, 276
        perfect, 280
        plug flow, 280
        zones in series, 278
    molecular weight, 7
    outlets, 255, 256
    pressure, see Atmospheric pressure
    properties, 64
    psychrometrics, see Psychrometrics
    recirculation, 270, 277
    saturation, water vapor, 8, 18
    space, thermal resistance, 99,
        Appendix 3-6
Air distribution, 261
Air duct design
    constant velocity method, 368
    critical path method, 368
    equal friction method, 368
    perforated polyethylene ducts, 231,
        236, 379
    static regain method, 368
    velocity reduction method, 368

Air inlets, 242, 249
Air Movement and Control Association
    (AMCA), 367
Airflow
    calculating rates of, 173 et seq.
    patterns, 262, 265, 268, 270, 277
    potential flow, 256
    recirculation, 270, 277
    short circuiting, 264, 276
    slotted inlets, 244
    stability 265, 268, 270
    visualization, 278
AMCA, see Air Movement and Control
    Association
Animal growth chambers, see Growth
    chambers
Animal laboratories, 5
Angle factors, 80 et seq., Appendix 3-4
Archimedes number, 265
Atmospheric pressure
    standard, 7
    as function of elevation, 7, 359

B

Baffles, see slotted inlets
BALANCE, 95
Balances, heat and mass,
    143, 146, 158, 167
Basement heat loss, 120
Bernoulli's Equation, 226, 228, 347
Bin data, weather, Appendix 6-2
Bin method, for temperature, 197
Black body radiation, 78
Blower door, 247
Boyle's Law, 7

C

Calorimetry, 148
Carbon dioxide
    ambient, 7
    balance, 166, 167
    effects, 167
    levels during ventilation, 166, 167
    production, 167

Cartesian coordinates, conduction in, 52
Ceilings, heat flow through, 115
Charles' Law, 7
Chimney effect, see Natural ventilation, stack effect
Climate space, 304
Climatic design temperature, 174
Coanda effect, 230
Coefficient of contraction, see Coefficient of discharge
Coefficient of convective heat transfer, 63
Coefficient of discharge, 227 et seq., 244, 322, 380
  sharp-edged orifice, 230
Coefficient of velocity, see Coefficient of discharge
Cold housing, see Natural ventilation
Colebrook Equation, 352
Comfort zone, 147
Computer programs, descriptions
  BALANCE, 95
  DUTYFACT, 298
  FRICTION, 358
  GRAPHID, v
  PLUS, 46
  POLYNOM, 188
  PSYFUNC, 41
  PSYPROC, 41
  RVALUE, 123
  SYSCHAR1, 253
  SYSCHAR2, 273
  VENTGRPH, 189
  WEATHER, 309
  XCHANGER, 137
Condensation
  on wall surfaces, 211
  within walls, 213, 216
Conditioning lines, 305
Conductance
  thermal, 56, 157
  moisture, 214
Conduction heat transfer, 51, 56
  basement heat loss, 120
  cartesian coordinates, 52
  cylindrical coordinates, 52, 57, 69, 74
  in common materials, 52
  perimeter heat loss, 158
Conductivity
  thermal, 51
  moisture, 214
Continuity, mass flow, see Flow continuity

Contraction coefficient, see Coefficient of discharge
Control
  computerized, 251
  derivative, 289
  hysteresis, 291
  integral, 289
  on/off, 288
  proportional, 288
  temperature, 287
Control volume, 143
Controlled Environment Agriculture (CEA), see Greenhouses
Convective heat transfer
  forced, 50, 71
  forced, coefficient c, 72
  natural, 50, 63, 66, 67, 68
Conversion factors, Appendix A
Cooling
  sensible, 22
  evaporative, 31, 34
Cylindrical coordinates, heat conduction in, 52, 54, 57, 69

D

Dalton's Law, 7
Darcy-Weisbach Equation, 352
Degree day, for heating, see Heating Degree Days
Degree of saturation, 11, 14
Dehumidification, 24, 39
Density of air, 14
Design temperature, 174
Dew point hygrometers, 21
Dew point temperature, 15
Diffusion, moisture, 214
Discharge Coefficient, see Coefficient of discharge
Doors, heat flow through, 117, Appendix 4-3
Dry bulb temperature, 8
Dry cup method, 214
Duty factors, fans, 290
DUTYFACT, 298
Dynamic losses, 347, 359, 372
  reference sections, 360

E

Effective emittance of cavity, 99
Effectiveness, heat exchangers, 132

Electric analogs to heat flow, 58
Electromagnetic spectrum, 51, 77
Emission, radiant energy, 78
Emissivity, radiant energy, see Emittance
Emittance, radiant energy, 78, Appendix
   3-3
Energy balance, 144, 158, 176
Energy conservation, 99, 111
Enthalpy, moist air, 16, 18
Entrainment, jets 233
Environment
   factors, 1-5
Environment control Effectiveness Index
   (EEI), 303
Environmental chambers, see Growth
   chambers
Evaporative cooling, 31, 34, 158
Exhaust ventilation, 225
Extinction coefficient, glass, 152

F

Fans
   control, 192, 251, 290, 299
   efficiency, VER, 294
   placement, 264
   selection of, 251
   staging, 191, 251
Fan Laws, 378
Fan operating cost, 294, 377
Fan power, 377
Fenestrations, 115, 117
Fick's Law, 214
Fittings, air duct
   loss coefficients, 360
Floors, heat flow through, 118, 158
Flow continuity, 226, 332, 340
Fluid mechanics, 226
Forced convection, 71 et seq.
Fourier Equation, 52
Fourier Law of heat conduction, 56
Free stall housing, see Natural ventilation
Friction chart, airflow, 355
FRICTION, 358
Friction factor, 352
   corrections, 356
Friction losses, 347, 352, 370
Fuels
   cost of seasonal heating, 207
   heating value, 207

G

Gas constant
   air, 7
   water vapor, 7
Glazing, heat flow through, 115
Glazing, thermal resistance, Appendix 4-2
Grashof Number, 64
Gray body radiation, 78
Ground temperature, 122
Greenhouses, 3
   control, 312
   evapotranspiration, 158, 200
   night curtain, 99, 111
   ventilation, 200, 312
Growth chambers, 4, 25

H

Heat
   balances, 69, 88, 89, 143, 158
   conduction equation, 51
   exchangers, 128 et seq.
   exchangers, counter flow, 129, 133
   exchangers, parallel flow, 128, 133
   gain, 147, 150, 151, 157
   latent, 158, 163
   loss, 157
   loss factor (HLF) for a building, 207
   production, 147, 148, 150,
      Appendix 5-1
   production as a function of body
      weight, 149
   sensible, 22, 158
   solar heat gain, 151
   supplemental, 157, 196
   total, 167
   vaporization, of water, 17
   value of fuel (HVF), 207
Heat of vaporization, water, 17
Heat transfer
   air space, 99
   basements, 120
   building heat loss, 157
   ceilings, 115
   conductive, 49, 51 et seq., 107 et seq.
   convective, 50, 63 et seq.
   doors, 117
   evaporation, 158
   floors on grade, 118
   glazings, 115
   mixed mode, 88 et seq.
   modes, 49

radiative, 51, 77 et seq.
steady state, 49, 53
steady periodic, 49
transient, 49, 53
walls, 107
Heating Degree Days, 206
data, Appendix 7-1
Heating loads, 207
Heating, sensible, 22
Heating value of fuels, 207
Homeostasis, 147
Homeothermy, 147
Humidistat, 22, 186
Humidity
Controlling with ventilation, 146, 164
Measuring, 21
Ratio, 9, 10
Relative, 8
Hydraulic diameter, 71, 357

I

Index of refraction, 153
Infiltration, air, 245, 254, 263
Infrared, see Long wave radiation
Inlets, air 242, 249
Inlet Jet Momentum Number, see Jet
Momentum Number
Insulation, 107, 205
conductances, conductivity values,
Appendix 3-1, Appendix 3-2
optimal thickness, 205
temperature/vapor pressure
gradients, 217
Irradiation, 86
Isothermal air jets, see Jets, air
Iterative solutions, 90, 94, 135, 136, 138,
328, 333, 340, 352

J

Jets, air
entrainment, 233
free, 230
heat transfer, 241
nonisothermal, 238
reattachment, 267
stability, 266, 270
temperature decay, wall jets, 237
temperature profile, wall jets 237
throw, 232
velocity decay, free round jets, 235

velocity profile, free round jets, 235
velocity decay, wall jets, 231
velocity profile, wall jets, 231
wall, 230
zones, 231
Jet momentum number, 271

L

Laminar air flow, 65
Laplace Equation, 53
Laplacian, 52
Log Mean Temperature Difference
(LMTD), 131
Long wave (thermal) radiation,
51, 77 et seq.
Loss coefficients, pressure, 360

M

Mass balance, 146, 163
Maximum ventilation, calculating, 176
Mean Radiant Temperature (MRT),
99, 113
Mechanical (forced) ventilation, 1,
Chapters 6-9
Minimum ventilation, calculating, 181
Mixing factor, 282
Models, air mixing, see Air mixing
Moist air, definition, 8
Moisture
balance, 146, 163, 165
control, 166
production, 163, Appendix 5-1
removal by ventilation, 164
Mole fraction, 9
Molecular weight, air and water, 7
Momentum transfer, 261
Moody Diagram, 353
Motor efficiency, 377
MRT, see Mean Radiant Temperature

N

Natural (free) convective heat transfer, 63
Natural ventilation
applications, 2
combined, 342, Appendix 11-2
neutral pressure plane, 326, 329,
Appendix 11-2
rules of thumb, 320
stack effect, 324

thermal buoyancy, empirical
    approach, 322
wind effect, 263, 334
wind effect, empirical approach, 337
wind effect, pressure coefficient
    method, 338
Negative pressure ventilation, 225
Net radiation, 85, 126
Neutral pressure plane, 329
Neutral pressure ventilation, 225
Nonisothermal wall jets, see Jets, air
NTU method, heat exchangers, 132
Nusselt number, 63

O

Optimum insulation thickness, 205
Outdoor design temperatures, 174, 211,
    Appendix 6-1
Outlets, air, 255, 264

P

Parallel thermal circuits, 60
Perfect gas law, 7, 9
Perimeter heat loss factor, 119, 158
Permeability, 214
Permeance, 214
    data, Appendix 7-2
Perms, unit of permeance, 214, Appendix
    7-2
Physiological effects, 147, 148
Plant growth chambers, see Growth
    chambers
PLUS, 46
Poisson Equation, 52
Polyethylene tubes, perforated,
    231, 236, 379
Poly-tubes, see Polyethylene tubes
POLYNOM, 150, 188
Polynomial regression, 150, 187
Positive pressure ventilation, 225
Power, requirements for fans, 377
Prandtl number, 64
Pressure
    atmospheric, 7
    coefficients, 263, 334
    loss coefficients, 360, Appendix 12-1
    losses, dynamic, see Dynamic losses
    losses, friction, see Friction losses
    standard, 7
    static, 229, 348 et seq.

total, 229, 347 et seq.
    velocity, 229, 348 et seq.
    velocity, computational form, 358
Product storage, 4
Production space, 305
Properties, dry air, 64
Psychrometer, 21
PSYFUNC, 41
PSYPROC, 41
Psychrometric chart, 10, 38
    Mollier form, 11
Psychrometric properties
    degree of saturation, 11
    density, 14
    dew point temperature, 15
    dry bulb temperature, 8
    enthalpy, 16
    example calculations, 19
    humidity ratio, 9
    measurements of, 21
    relative humidity, 8
    specific volume, 14
    water vapor saturation partial
        pressure, 8
    wet bulb depression, 32
    wet bulb temperature, 17
Psychrometrics, 7, et seq.
    adiabatic mixing, 27
    analysis of processes, 22, 34
    computerized procedures, 26, 30, 33,
        39, 41
    cooling with dehumidification, 24
    evaporative cooling, 31
    Integrated processes, 34
    sensible heating and cooling, 22

Q

Quadrature in natural ventilation, 342

R

R values for heat transmission,
    56, 108, 116
Radiation, thermal
    absorptance, 80, 152
    black body, 78
    emittance, 78, 80
    gray body, 78
    heat exchange, 85
    long wave, 77
    peak intensity, 77

polarization, 153
reflectance, 79, 153, 155
short wave, see Solar radiation
sol-air temperature, 126
solar, see Solar radiation
transmittance, 79, 155
wavelengths, 51, 77
Radiosity, 86
Recirculation (in air mixing), 270, 277
Reflectance, 79, 153, 155
Reflectivity, see Reflectance
Refraction, index of, 153
Relative humidity, 8
Relative roughness, 352
Reps, a unit of moisture flow
    resistance, 214
Resistance, thermal
    air spaces, 99
    heat flow, 56, 57, 58, 74, 108, 109,
        114, 116
    water vapor flow, 214
Reynold's number, 71
    computational form, 358
Ridge vents, 320, 321
RVALUE, 123

S

Saturation, degree of,
    see Degree of saturation
Sensible
    energy balances, 144, 158, 176
    heat production, 147, 148
    heating, 22
    cooling, 22
Sensors, 21, 186
Series thermal circuits, 58, 114
Shape factors, see Angle factors
Short circuiting, 276
Short wave radiation, see Solar radiation
Sky temperature, 89
Slotted inlets, Chapters 8, 9
    air flow from, 244
    control, 271
Snell's Law, 153
Snow in naturally ventilated barns, 321
Sol-air temperature, 126
Solar radiation, 151
Specific heat
    dry air, 16, 134, 157
    water, 135
    water vapor, 16

Specific volume, 14, 15
Spectrum, radiation, 51, 77
Staging of ventilation, 191, 290
Stagnation, wind, 334, 336
Stack effect, see Natural ventilation,
    stack effect
Standard atmosphere, 7
Static pressure, 229, 348
Static pressure recovery, 351
Steady-state conditions, 107, 225
Stefan-Boltzmann Constant, 78
Stefan-Boltzmann Equation, 78
Stokes' Equation, 155
Straight-line Law, 40
Supplemental heating, 157, 196
Surface coefficients
    combined, convective and radiative,
      97, 98, 112, 126
    convective, 63, 71
Surface resistance, 69
SYSCHAR1, 253
SYSCHAR2, 273
System characteristics, 247, 273, 351, 363
System effect factors, 366

T

Temperature
    control with ventilation, 159, 161
    dew point, 15
    dry bulb, 8
    fields, 53
    gradients, 52, 53
    ground, 122
    log mean difference, 131
    outdoor design, 174, 197, 211, 290,
      297, Appendix 6-1
    probabilities, 297
    sky, 89
    sol-air, 126
    surface, 109
    wet bulb, 17
Temperature profiles, wall jets, 237
Thermal conductance, 56
Thermal conduction, 49
Thermal conductivity, 52
Thermal convection, 50
Thermal diffusivity, 52
Thermal radiation, see Radiation, thermal
Thermal radiation exchange with
    gases, 100

Thermal resistance, 56
    of plane air spaces, 99
Thermal stress, 147
Thermodynamics, 49
Total pressure, 229, 347
Transmittance, radiant energy, 79, 155

U

U-values for heat transmission, 56
Unit area thermal conductance, 56
Unit area thermal resistance, 56

V

Vapor barriers, 220
Vapor transmission (water), 214, 216
Velocity coefficient, see Coefficient
    of discharge
Velocity pressure, 229, 358
    computational form, 358
Velocity profile
    free jets, 230
    wall jets, 237
VENTGRPH, 189
Ventilation
    average, 293
    cost of, 294, 377
    effectiveness, see Environment control
        Effectiveness Index
    efficiency, see Ventilation
        Efficiency Ratio
    exhaust, 225
    graphs, 182 et seq.
    greenhouse, 200, 312
    heat exchange, 157
    maximum rates, 173, 176
    minimum rates, 173, 181
    mechanical (forced), 1, Chapters 6-9
    moisture removal, 164
    natural, 2, 139
    negative pressure, 225
    neutral pressure, 225
    positive pressure, 225
    staging, 191, 290
    suitability index, see Environment
        control Effectiveness Index
    system characteristics, 247
    temperature set points, 193
Ventilation Efficiency Ratio, 294
Ventilation graphs, 182 et seq.
VER, see Ventilation Efficiency Ratio

Visualization, air flow, 278

W

Wall jets
    convective coefficients, 241
    entrainment, 234
    reattachment, 267
    stability, 266, 270
    temperature profiles, 237
    throw, 232
    velocity profiles, 231
Walls
    condensation on surfaces, 211
    condensation within, 213
    heat flow through, 107
    sections, Appendix 4-1
    temperature gradients, 53
    vapor pressure gradients, 214
Water, heat of vaporization, 17
Water, molecular weight, 7
Water vapor pressure
    saturated, 8, 10, 216
    unsaturated, 8
Wavelength spectrum, 51, 77
WEATHER, 309
WEATHER.DAT data file, 309
Weather data, for design, 174, 197, 211,
    290, 297, Appendix 6-1
Weather bin data, Appendix 6-2
Weather space, see Acceptable
    Weather Space
Wet bulb depression, 32
Wet bulb temperature, 17
Wet bulb thermometer, 21
Wet cup method, 214
Wien's law, 77
Wind
    induced natural ventilation, see Natural
        ventilation, wind effect
    pressure coefficients, 263, 337,
        Appendix 11-1
    pressure, induced by, 335
    speed, 334
    speed, corrected for height, 337
Windows, heat flow through, 115,
    Appendix 4-2

X

XCHANGER, 137

Z

Zones, jet flow, 231
Zoo animals, 5

INDEX OF EXAMPLES

2-1.   Humidity ratio calculation, 10
2-2.   Degree of saturation calculation, 14
2-3.   Specific volume calculation, 15
2-4.   Dew point temperature calculation, 16
2-5.   Enthalpy content calculation, 17
2-6.   Wet bulb temperature calculation, 19
2-7.   Sensible cooling, calculation of, 22
2-8.   Sensible cooling, psychrometric chart, 23
2-9.   Sensible and latent heat removal, use of psychrometric chart, 25
2-10. Sensible and latent hear removal, calculation of, 26
2-11. Adiabatic mixing of two air streams, psychrometric chart 29
2-12. Adiabatic mixing of two air streams, calculation of, 30
2-13. Evaporative cooling, psychrometric chart, 32
2-14. Evaporative cooling, calculation of, 33
2-15. Evaporative cooling and adiabatic mixing, 34
2-16. Computer program development for psychrometric calculations, 42
3-1.   Temperature field in homogeneous solid, cartesian coordinates, 53
3-2.   Temperature field of homogeneous solid, cylindrical coordinates, 54
3-3.   Calculating temperature field and heat flow, conduction, 56
3-4.   Conduction heat flow, cylindrical coordinates, 57
3-5.   Heat conduction, series thermal circuit, 59
3-6.   Heat conduction through a wall, 61

3-7.   Convective heat transfer (natural) from heating pipe, 66
3-8.   Convective heat transfer (natural) from duct, 67
3-9.   Heat flow from insulated pipe, 68
3-10. Convective heat transfer (forced) inside air duct, 72
3-11. Heat loss from air duct, 73
3-12. Radiant flux calculation, 79
3-13. Angle factor calculation for radiant energy exchange, 82
3-14. Angle factor calculation for radiant energy exchange, 82
3-15. Radiant energy exchange, 86
3-16. Mixed mode heat transfer, 89
3-17. Mixed mode heat transfer, 92
3-18. Use of Program BALANCE, 96
3-19. Thermal resistance of a wall cavity, 100
4-1.   R-value of a wall, 109
4-2.   R-Value of a wall, 109
4-3.   R-value of a greenhouse night curtain, 111
4-4.   Wall thermal properties design, 113
4-5.   R-value calculation, glazing and wall, 116
4-6.   R-value calculation, door and wall, 117
4-7.   Basement heat loss, 122
4-8.   R-value calculation, glazing and wall, 124
4-9.   Sol-air temperature calculation, 127
4-10. Heat exchanger unit area thermal conductance, 129
4-11. Log Mean Temperature Difference (LMTD) calculation, 131
4-12. Heat exchanger Ntu calculation, 134

4-13. Heat exchanger Ntu calculation, 136

4-14. Use of Program XCHANGER, 138

5-1. Dairy cow sensible heat production, 149

5-2. Polynomial regression, dairy cow sensible heat production, 150

5-3. Sensible heat gain from lights, 151

5-4. Solar absorptance in glass, 153

5-5. Solar reflectance from glass, 154

5-6. Absorptance, reflectance and transmittance through glass, 155

5-7. Ventilation rates for temperature control, 160

5-8. Temperature control in dairy barn, 161

5-9. Temperature control in dairy barn, 161

5-10. Moisture production from swine, 164

5-11. Moisture production from dairy cows, 164

5-12. Ventilation rate calculation for poultry house, 165

5-13. Ventilation for carbon dioxide control in poultry house, 167

5-14. Calculating air exchange in greenhouse, 169

6-1. Calculating maximum ventilation rate, dairy barn, 177

6-2. Calculating maximum ventilation rate, dairy barn, 180

6-3. Calculating minimum ventilation rate, dairy barn, 184

6-4. Use of Program POLYNOM, 188

6-5. Use of Program VENTGRPH, 190

6-6. Fan staging, 195

6-7. Ventilation system design, dairy barn, 197

6-8. Ventilation for greenhouse, 200

7-1. Use of Heating Degree Days to calculate heating cost, 208

7-2. Optimum insulation thickness, 208

7-3. Wall R-values to prevent surface condensation of water, 212

7-4. Calculating water vapor diffusion through a wall, 215

7-5. Condensation of water vapor in a wall, 217

7-6. Designing for a vapor barrier, 221

8-1. Use of mass continuit equation, 227

8-2. Bernoulli equation for air flow through opening, 229

8-3. Wall jet velocity calculation, 232

8-4. Wall jet velocity and entrainment calculation, 234

8-5. Free jet velocity calculation, 236

8-6. Wall jet temperature calculation, 239

8-7. Calculation of system characteristic graph, slotted inlets, 249

8-8. Use of Program SYSCHAR1, 254

9-1. Wind pressure calculation, 264

9-2. Wall jet thickness/detachment calculation, 268

9-3. Jet Momentum Number calculation, 269

9-4. Use of Program SYSCHAR2, 274

10-1. Fan operation cost, 294

10-2. Fan operation cost, 295

10-3. Use of Program DUTYFACT, 299

10-4. Use of Program DUTYFACT, effect of parameters, 302

10-5. Use of Program WEATHER, 310

11-1. Ventilation due to thermal buoyancy, empirical approach 322

11-2. Ventilation due to thermal buoyancy, stack effect approach 327

11-3. Wind speed correction for height, 337

11-4. Ventilation due to wind, empirical approach, 338

11-5. Ventilation due to wind, 341

12-1. Friction drop in air duct, 354

12-2. Friction drop in air duct, corrected, 356

12-3. Friction drop in rectangular air duct, 358

12-4. Use of Program FRICTION, 359

12-5. Air duct pressure drop calculation, 361

12-6. Air duct system characteristic graph, 364

12-7. Air duct design, 369

12-8. Fan laws application, 379